Insurance Restoration Contracting

– Startup to Success –

by Paul Bianchina

Craftsman Book Company
6058 Corte del Cedro / P.O. Box 6500 / Carlsbad, CA 92018

Acknowledgments

The author wishes to express his sincere gratitude to the many different people who helped take this book from concept to reality.

To the dedicated and hardworking team at Craftsman Book Company; especially to Genie Runyon, editor extraordinaire, for her thoughtful suggestions and superhuman powers of organization; and a special debt of gratitude to Laurence Jacobs, Editorial Manager, for all the things that simply make Laurence Laurence — you probably don't know it, but you're truly a joy to work with.

To Marcia Neu and Brandon Burton at Dri-Eaz®, for the wealth of photographs and written materials used both in the book and as background material. Dri-Eaz also deserves special recognition for all they've done to advance the field of insurance restoration into the recognized profession it is today.

Closer to home, a special thanks to Sandi McKinley for wading through the insurance policy chapter; to Kathy Ellis for all of her time spent reviewing and commenting on the strange world of mold; to Mark Ostrom, both for the great interview and for some of the most enjoyable years of my career; and to Diane Price, for being a rock of calm during some stormy times.

To the many companies and individuals that offered help, including:

> *Absorene; Airsled, Inc.; American Van Equipment; Bosch Tools; Curtis Dyna-Fog, Ltd.; Ebac Industrial Products, Inc.; Foremost Insurance Group; Institute of Inspection, Cleaning and Restoration Certification (IICRC); Omegasonics Corporation; On-Site Storage Solutions; ProRestore Products; pulsFOG; Ryobi Tools; Sonozaire; T.E.S. Drying System; U.S. Products; Wise Steps; Wm. W. Meyer & Sons, Inc.; Xactware Solutions Inc.; Zefon International; Zinsser; ZipWall, LLC.*

And finally, to Steve Manina. Simply put, you are and always will be the best partner — and the best friend — a person could ever ask for. As you've always said, it was quite a ride!

This book is dedicated to my wife, Rose, with love and appreciation for your infinite patience and unwavering support.

Library of Congress Cataloging-in-Publication Data

Bianchina, Paul.
Insurance restoration contracting : startup to success / by Paul Bianchina.
p. cm.
Includes index.
ISBN 978-1-57218-239-4
1. Construction industry--Management. 2. Buildings--Repair and reconstruction.
3. Insurance claims. 4. Small business--Management. I. Title.

HD9715.A2B496 2011
690.068--dc23

2011023146

Layout: Joan Hamilton

Contents

Introduction

Welcome to the world of insurance restoration. If you've reached for this book, it's obviously a topic that you're interested in, and you've come to the right place!

The field of insurance restoration, perhaps more than any other field of construction, offers some amazing opportunities: Steady work, even in a slow economy; a great paycheck; lots of variety; constant challenges; and the chance to create something for the future. And along the way, you'll have the opportunity to make a real difference in the lives of a lot of people. By any standards, that's not a bad career choice!

And, that's what *Insurance Restoration Contracting: Startup to Success* is all about. This book will start you off on the right foot along a path toward a very bright future in the thriving insurance restoration field. I've spent over 30 years in the field, so you'll benefit from my experience, including my mistakes. You'll enjoy the "Real Stories" included throughout, which give you an honest, practical, and often amusing look at the ins and outs of running a restoration company.

In chapter after chapter, packed with practical information, we'll cover it all:

- How to find work
- What jobs offer the best potential
- Which projects you ought to steer clear of
- What type of equipment you'll need, and when you'll need it

You'll learn about damage caused by fire, smoke, water and mold. You'll see how to work with insurance adjusters, and how to deal with distraught clients. You'll learn how to make money — and then how to collect it.

You'll learn some secrets about ozone deodorization, ultrasonic cleaning, and a little jet engine device that will fill a house with fog in minutes. You'll hear the story of a house that was almost squashed by snow, and another with an exploding roof, and yet another that was hit by a crane and knocked partway off its foundation.

But best of all, you'll read *how they were repaired!*

You're going to find out a lot about contents as well: How they're inventoried, tracked, wrapped, boxed, transported and stored. And along the way, we'll look at the very personal nature of those contents, and learn how to handle both owners and their belongings with empathy and compassion.

Insurance restoration is a rapidly changing field. In just a few short decades, it's gone from what was basically a sideline occupation to the well-respected construction specialty that it is today. With highly specialized equipment, sophisticated estimating programs, burgeoning national franchises, and insurance companies showing an increased reliance on Preferred Contractor Programs, you need *Insurance Restoration Contracting: Startup to Success* to help you start and thrive in this fascinating business.

Again, welcome to the world of insurance restoration. It's a world you'll be glad you entered . . .

1

Is Insurance Restoration the Business for You?

In some ways, insurance restoration is an odd thing — at least in other people's eyes. You're chatting with someone at a party, and they ask you what you do for a living. If you say you're in construction, they'll understand that. If you say you build new homes, they find that pretty interesting. If you say you remodel kitchens, or build room additions or decks, they can relate to that. If you're in one of the specialty trades, like electricians or plumbers or masons, they'll nod and probably ask you a question or two.

But insurance restoration? What in the world is *that*? You'll see a blank look at first. So you say something like, "You know, I fix fire and water damage, and mold; things like that." And their eyes kind of glaze over a bit and then they nod and wander off in search of someone more interesting to talk to.

And that's too bad. Because insurance restoration is actually one of the most interesting, exciting, and challenging trades in the construction industry. But apparently it's also one of the least known and understood, as well as one of the hardest to explain in a sentence or two. I guess if you *had* to put it into a single sentence, it would be something like this: *When damage occurs to a building, and that damage is covered by an insurance policy, the insurance restoration contractor is the person who steps in to do the work.*

Tip! *For simplicity and clarity, "he" has been used throughout this book when referring to adjusters. In decades past, insurance adjusting was definitely a male-dominated occupation. But all that's changed, and today you'll find you're just as likely to be working with a female adjuster as a male.*

For some contractors, insurance restoration work is just a sideline. They concentrate primarily on remodeling or new construction, and do insurance jobs as they happen to come up. The problem is that insurance restoration requires a big investment in equipment and training. You also need to establish and maintain working relationships with insurance companies. It can be tough to run both a remodeling company and an insurance restoration company at the same time. When you try to operate a business where you wear two hats, it often means you're not wearing either one of them very well.

This book is for the contractor looking to *specialize* in insurance restoration; someone who wants to start, operate, and prosper as a full-time insurance restoration contractor. And, as you'll see later in Chapter 25, that doesn't totally exclude other work. There are also a lot of good remodeling opportunities you can take advantage of along the way.

But before you make the decision to specialize in this field, let's take a realistic look at what you'll be getting into.

The Pros and Cons of Insurance Restoration Contracting

While some remodeling and new home contractors are perfectly happy working out of the back of their pickup, insurance restoration is different. It requires a definite commitment, an investment in equipment, and the time and effort to acquire some specific training. Since there's a learning curve with insurance restoration, many of today's restoration contractors move into this field deliberately, not accidentally. It helps if you take the time to understand the industry before you get involved.

Ten Reasons to Become an Insurance Restoration Contractor

1. ***It provides a steady flow of work:*** This is one of the biggest advantages to this area of construction. Damage, like the kitchen fire in Figure 1-1, is likely to occur in most everyone's home at one time or another. Once you're established in the field, you can count on jobs like these coming in pretty much continuously. Work doesn't slow down in the winter. The job flow is year-round, often picking up during cold weather due to increased fire risks created by the use of portable heaters, fireplaces and wood stoves, as well as damage to structures from ice, snow and frozen pipes.

2. ***It's recession-proof:*** If you're a remodeling contractor or a new-home builder, you know that when the economy slows down, the jobs start to dry up. People spend less money, and lender financing for new projects becomes harder to get. However, that's typically not the case with insurance restoration work. When damage occurs to a home, the insurance company has to pay for it, no matter what the economy's doing. In fact, during a slow economy, insurance work often increases. People stay home more, and that increases the opportunity for damage to occur in their homes.

3. ***It pays well:*** You'll find that you can charge a very fair rate for doing insurance restoration work. Insurance restoration estimates are very thorough. They'll probably be the most thorough that

Figure 1-1 This kitchen fire is an example of the insurance-related losses that happen daily in people's homes. These losses provide a steady flow of work for the restoration contractor.

you've ever prepared. While that may take some getting used to at first, it ensures that you rarely miss anything. But if you *do* miss something, the insurance company will typically allow you to bill for it later. Try that with your customer on a remodeling or new construction project.

4. ***Payment is virtually guaranteed:*** The insurance company issues a check to the client, so you rarely have to worry about getting paid. Once the estimate is accepted, the money is pretty much guaranteed.

5. ***You're helping people:*** This is one of the few fields in construction where you feel like you're coming to the rescue in what may be one of the worst days in a person's life. Figure 1-2 shows a major structural fire. You could have the opportunity to restore this family's home. You may not only be able to rebuild, but save their wedding pictures or an antique that's been handed down from father to son. Maybe you can step in and get their kitchen rebuilt before Christmas, so the family can celebrate in their own home. There's really no other area of construction where you have a chance to be this kind of hero.

6. ***Every job is different:*** Do you like variety? Insurance restoration will certainly give you that! Like the snow-load damage in Figure 1-3, every job presents its own unique challenges. Some jobs last for months; and some only take a day or two. Each homeowner, house, and loss is different. You'll definitely not get bored by the work.

Figure 1-2 A major structure fire represents a devastating loss for a family. The field of insurance restoration offers an opportunity to help people rebuild their homes and get back to their normal lives.

Figure 1-3 Snow loads put so much weight on this building, it caused this structural beam to shift dangerously out of alignment.

7. ***It's a fast-paced profession:*** Most contractors schedule jobs, like a kitchen remodel, several weeks down the road. But in our line of work, when a fire strikes or a pipe breaks, you're needed *now*. Events such as wildfires (see Figure 1-4) create some very fast-paced and interesting days.

8. ***It presents other opportunities:*** Unlike most areas of construction, insurance restoration offers lots of unique opportunities for income. For example, you can be hired as a consultant to an insurance company, or to a homeowner. Some insurance companies will hire you just to put a bid together. Or you might be called to be an expert witness in court. All this is on a per-hour basis, and usually pays pretty well.

9. ***There are franchises available:*** Unlike other types of construction, there are some very good franchise opportunities available in insurance restoration. Buying a franchise gives you a chance to establish

Figure 1-4 Natural disasters, such as the wildfire that roared through this community, leave little time for planning and scheduling your work flow. A major fire can result in dozens of new jobs, ranging from cleaning and deodorizing smoky interiors to completely rebuilding homes.

yourself quickly, and cuts down on the learning curve. I'll go into more detail about franchises in Chapter 18.

10. ***You're building a business you can sell:*** Construction is a relatively easy field to get into, so there are a lot of construction companies around. They come and go, but aren't necessarily sold. But insurance restoration is different. Because of the equipment you need for the specialized work and the insurance company contacts you make, it's an easier business package to sell when you're ready to retire.

Ten Reasons Not to Become an Insurance Restoration Contractor

While there are lots of good reasons to consider becoming an insurance restoration contractor, it's not a career for everyone. Let's take a look at some of the downsides you may not have thought about:

1. ***There's a lot of paperwork:*** Because you'll be dealing with insurance companies, things need to be done in a certain way. And unfortunately, each insurance company is different. If you're not a very organized person and you hate the thought of paperwork, this may not be the best field for you.
2. ***Some things will be out of your control:*** Here again, insurance companies have specific ways of doing things, and even specific pricing. You may have to adjust your thinking to be a little less independent

Figure 1-5 A large water loss, such as the one in this condo, requires many hours to put a bid together. A final computerized, unit-cost estimate for a job of this type will probably total 25 pages or more.

than before. Bids may have to be presented a little differently than you're used to. You may have to take photos of damaged areas in a certain way. In most cases, you'll also be limited in the amount of profit and overhead that you charge.

3. ***It requires very detailed estimates:*** Insurance companies require that all estimates be very detailed. For example, you may be used to doing an estimate that includes one price for all the drywall in the house, one price for all the trim, and one price for all the painting. You total those items up, add in your overhead and profit, and that's your bid for the homeowner. Insurance companies, on the other hand, require that the estimates be broken down room by room, with each cost listed separately. My estimate for repairs to the condo shown in Figure 1-5 was 25 pages long.

4. ***It takes time to become established:*** Insurance restoration pays well and provides year-round work. But, it does take a little bit of time to become established with the insurance companies. And you may be stuck with some of the less desirable jobs before you're trusted with some of the bigger stuff. Once you're established, however, you'll find a steady flow of work.

5. ***There's an investment in training and equipment:*** Today's insurance restoration contractor has to be knowledgeable about water, smoke, mold, and other issues. That requires an investment of both time and money in getting the proper training. You'll need to complete specific classes to get your certifications. And, in addition to your construction tools, you'll have a large investment in specialized equipment that you wouldn't need in other areas of construction. Figures 1-6 and 1-7 are examples of some of the special equipment used in restoration work.

6. ***It can be emotional work:*** Doing insurance restoration means entering your client's life at a difficult time. A flood may have just destroyed all their family photos or some heirloom furniture. A fire may have killed a beloved pet, or worse, injured a family member. While you have a real

(Photo courtesy of Dri-Eaz Products)

Figure 1-6 The EnviroBoss™ 1400 from Dri-Eaz, a leading manufacturer of restoration equipment. This machine can provide dehumidification, warm air, ventilation, and cooling as needed on restoration jobs. It's a good example of the new generation of equipment designed specifically for restoration work.

(Photo courtesy of Dri-Eaz Products)

Figure 1-7 A set of air movers on a cart, ready for transport to a water loss. Equipment of this type is essential for doing restoration work. And while it represents a sizable investment, it's also what makes this a very lucrative profession.

opportunity to help people and make a difference in their lives, working in this environment can be emotionally stressful for you as well. The home in Figure 1-8 was vandalized by an intruder. The family felt victimized and violated, as well as angry and frightened. That's a lot of emotion to take on with a job.

7. ***You have to react quickly:*** To build your reputation as an insurance restoration contractor, you have to be able to react quickly to emergency situations. That can make it tough to schedule your day and your crew — but it's a very important part of this field of construction. Some repairs, like removing the downed tree in Figure 1-9, can't wait for you to finish the remodel job your crew is on.

8. ***It's hard to advertise:*** One of the major problems with insurance restoration work is that people only need you in a time of emergency. Unlike contractors who do kitchen remodeling or room additions, it's difficult to advertise a service that most people hope they'll never need.

9. ***You may be competing against larger franchises:*** Buying a franchise might be a great way to get into the business. However, if a franchise already exists in your area, you'll have to compete against it; and big franchises have a lot of name recognition and national clout.

Figure 1-8 Dealing with certain losses, such as this vandalized home, can be tough on you as well as on your clients.

Figure 1-9 When a loss occurs, like the massive tree that came down across the roof of this home, you have to be prepared to respond quickly. If you can't get there soon, the homeowners will usually find someone else who can.

10. ***You need some cash flow:*** Insurance restoration pays well, and your money is pretty secure, but it can take time to get paid. So, you'll need either some cash reserves or a line of credit with your bank. This is especially important when you're first starting out.

An Overview of How Insurance Restoration Works

When a person buys a home, they typically buy a homeowner's insurance policy. This is not only good common sense, it's usually required by the lender. A homeowner's policy protects the owner of the house against a number of different things. It offers liability coverage if someone trips over a garden hose in the front yard, or payment for items that are stolen from the home. More important, the homeowner's policy also protects the homeowner against many types of damage that occur to the house. This damage, no matter the size or the type, is known in the industry simply as a *loss*. And that's where the insurance restoration contractor comes in.

Depending on the type of policy and its terms, such things as fire damage, water damage, storm damage, and other types of damage to the home are all covered by the policy. The exact details regarding coverage are listed in the policy; we'll talk more about that in Chapter 2. Also specified in the policy is the *deductible*. This is an amount that the homeowner must pay toward the loss, typically anywhere from $500 to $2,500 or more. Once again, we'll cover that in more detail in Chapter 2.

When a loss occurs, there are several different people that interact with the homeowner, and each other — agents from the insurance company, representatives from the company that will ultimately repair the damage, and often emergency service workers.

If it's not an emergency, the homeowner's first call is usually to their insurance agent. That may be someone they've been doing business with for a while, or an anonymous rep answering an 800 number in another state. In an emergency, such as a broken water pipe that's flooding the home, their call might be to the first plumber they find in the phone book — and then to the insurance company.

The insurance representative will initiate a claim for the homeowner. That sets a few different wheels in motion — one of which will be a request that the homeowner obtain an estimate for repairs. Ideally, this is where you, as an insurance restoration contractor, enter the picture.

Good restoration contractors build a network of contacts with insurance agents and insurance companies, establishing themselves on referral lists. Remember, this is a specialized business, and most homeowners won't know where to find you. They may not even know that companies such as yours exist. So referrals from insurance companies, as well as advertising in key areas, will help homeowners find you when they need you for that all-important estimate. We'll look into this phase of the business in more detail in the coming chapters.

The estimate is submitted to the insurance company, who then reviews it. For smaller jobs, the insurance company may simply review the estimate over the phone, either with the homeowner or the contractor who submitted the estimate.

On larger jobs, the insurance company will typically send out one of their Claims Representatives or Claims Adjusters. The adjuster will examine the damage, review the estimate, meet with the homeowner and the contractor, and determine if the estimate is fair.

Once the estimate is accepted, the homeowners will hire a contractor to have the repairs done. The homeowners can choose any contractor they want to perform the repairs. They may choose the contractor who wrote the estimate that was approved by the adjuster, or they may hire another contractor. They may also choose to do the repairs themselves, or in some situations, just collect the insurance money and not make repairs at all. This is their right in most states.

During the course of performing the repairs, additional damage may be discovered. If this happens, the contractor will write another estimate covering the additional work. The additional damage is known as a *supplemental loss*, and becomes part of the overall claim. The additional estimate is known as a *supplemental estimate*, or just a *supplemental*.

Tip! *While the insurance company may be providing the money to pay for damage repairs, never forget that your contract is with the homeowner, not the insurance company. I can't stress this enough, because it's a very important issue to understand — and it's an area where both contractors and homeowners often get confused. If everyone keeps this one detail in mind throughout the job, you'll avoid a lot of unnecessary problems.*

When the work is complete, the insurance company will pay the homeowner. The payment issued by the insurance company will be for the amount of the original estimate, plus the amount of any supplemental, minus the amount of the homeowner's deductible.

The homeowner will then pay the contractor. That payment will be done in accordance with the terms of the contract between the homeowner and the contractor. It's very important to understand that you're paid by the homeowner. You're *not* working for, or paid by, the insurance company.

A Typical Loss

Let's take a look at how a typical loss is handled financially. A homeowner has a kitchen fire. The contractor examines the damage, and writes an estimate for $19,000. The homeowner has an insurance policy in place, and the fire is a covered loss. The insurance policy has a $1,000 deductible.

The insurance adjuster examines the damage and the estimate, and approves everything. The homeowner signs a contract with the contractor, and the work starts. While the work is being done, some additional damage is discovered that was hidden in the walls. The estimate for this supplemental damage is $3,000. The contractor submits a supplemental estimate to the adjuster, who approves the additional work.

The contractor writes an addendum to the original contract with the homeowner, covering the additional $3,000.

This is how the finances break down:

Original estimate	$19,000
Supplemental damage estimate	+ 3,000
Total cost of repairs	$22,000
Less the deductible	– 1,000
Total paid by insurance company	$21,000

The insurance company is responsible for $21,000, and the homeowner is responsible for his portion, which is the $1,000 deductible

Who's Your Client?

Let's look at a couple more examples. Here's a typical job situation that any contractor will be familiar with: Imagine that you're a remodeling contractor and homeowners have hired you to remodel their kitchen. The total cost of the job is $30,000. They have $1,000 of their own money, and they're getting a bank loan for the other $29,000. You write up a contract in the amount of $30,000, regardless of where the money is coming from. Whatever the terms are in your contract for start date, payment schedules, etc., they're between you and the homeowners, *not* between you and the homeowners' bank. When a payment is due, the homeowners pay you, *not* the bank.

Now let's look at a fire restoration job. Once again, the cost of the job is $30,000. The homeowners have a $1,000 deductible, and the insurance company is paying the remaining $29,000. Again, your contract is for the entire $30,000, regardless of where the money's coming from. As in the previous example, whatever terms are in your contract for start date, payment schedules, etc., they're between you and the homeowners. When a payment is due, the *homeowners* pay you.

The Three Elements of a Loss

There are three separate elements that can be part of an insurance loss. Some losses have only one of these elements, some have two, and some have all three. These elements are:

1. ***Emergency response:*** An emergency response is required whenever there's sudden damage to a building that needs to be dealt with immediately. That response might be necessary in order to secure the building against weather or unauthorized entry; to stop water damage from spreading; to make electrical damage safe; or for any of a number of other reasons.

(Photo courtesy of Dri-Eaz Products)

Figure 1-10 Air movers in use to dry a building during an emergency response. It's the homeowners' responsibility to take whatever steps they can to prevent further damage from occurring. That process is known as *mitigating damage*. Emergency response is one of the three main elements of insurance restoration work.

Emergency responses are limited to just what needs to be done to secure and dry a building — to *mitigate damage*. This is called *loss mitigation*. Emergency responses are usually not estimated, but instead are billed on a time and materials basis. If an emergency response is required, it becomes part of the same overall loss, and only one deductible is paid by the homeowner. The emergency can cover damage both to the structure and to the contents. Figure 1-10 shows air movers being used to dry out a building to prevent further damage.

2. ***Structural damage:*** If there's structural damage to the building, this is estimated after the emergency work is completed and the building is stabilized. Structural work is rarely done on a time and materials basis; instead, it's almost always estimated. The estimates are very detailed, and are done using unit-cost methods.

   Structural estimates deal only with repair and cleaning of the structure itself, and don't include any of the contents. If structural work is required, it becomes part of the same overall loss as the emergency. Once again, only one deductible is paid by the homeowner. Figure 1-11 shows structural repairs in progress after fire damaged this building.

3. ***Content damage:*** The third element of the loss is damage to the contents. This includes the moving, packing, storing, cleaning, repair, disposal, deodorization, and replacement of content items involved in a covered loss.

Figure 1-11 Structural repairs, such as those for this fire-damaged roof, can be quite involved. They make up the second part of insurance restoration work.

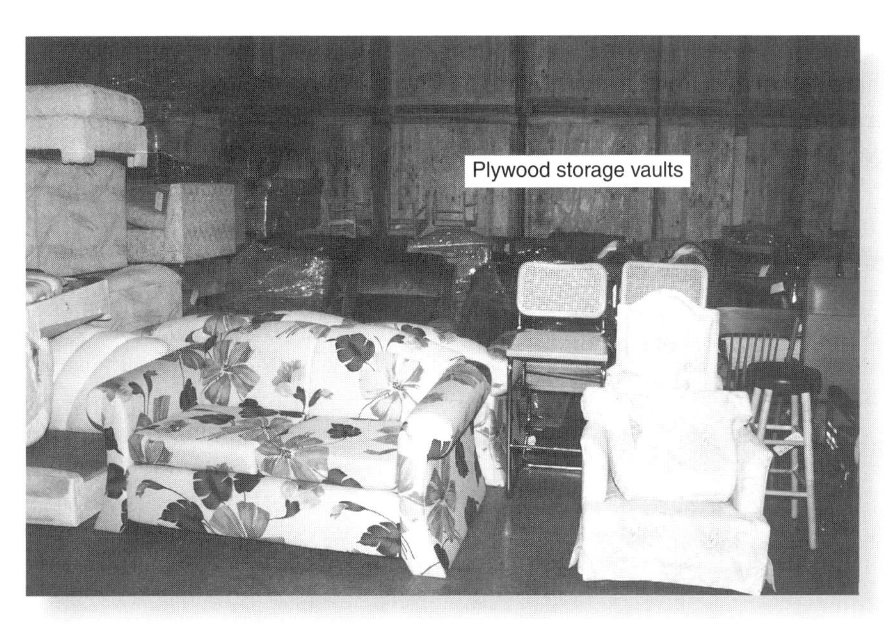

Figure 1-12 The cleaning and storage of contents represents the third main element of insurance restoration work. Here, cleaned contents are awaiting transport back to a home. You can see the plywood storage vaults for contents in the background.

Contents are typically handled on a time and materials basis, although some insurance companies have considered asking for estimates on content cleaning and moving. If content work is required, it becomes part of the same overall loss as the emergency and structural elements. Once again, only one deductible is paid by the homeowner. Figure 1-12 shows stored contents ready to be returned to refurbished homes.

In most cases, all of the elements of a loss are handled by the same restoration company. It's easier for both the insurance company and the homeowner. Being qualified to handle all three elements of insurance restoration work is very lucrative for the restoration contractor. That's a major reason why you should strive to establish yourself as someone who can provide as many services as possible to the clients and insurance companies you serve.

Understanding the Insurance Industry

The insurance industry is complicated. It would be impossible to fully explain all the ins and outs of how it works, including how it writes and interprets its policies. However, it's important for you, as a restoration contractor, to understand a little bit about the industry.

An insurance company is a business; its objective is to make a profit and provide a service at the same time. To increase or even just maintain that profit, insurance companies are constantly changing how they do business and what they cover. They try new things, often with mixed results.

A homeowner's insurance policy is a contract between the insurance company and the homeowner. Both have rights and obligations as part of that contract. It's important for the homeowner to know and understand what his policy covers and doesn't cover. Most people think that all homeowner's insurance policies are pretty much the same. However, they actually vary quite a bit, and in many cases, you get what you pay for.

Higher-priced policies issued by large national insurance companies will typically offer more complete coverage, with fewer exclusions. Most of these companies will have local claims adjusters, making review of the claim faster. That means the claim will get settled more quickly and easily, and often more fairly.

On the other hand, when a homeowner purchases a policy with a smaller premium amount, the insurance company simply can't offer the same level of coverage as they do with the more expensive policies. The lower cost means less coverage and probably fewer or no local service representatives. The company representatives may be located in another city or even another state. That means that the homeowner can't deal directly with a local person. And you, as the restoration contractor, end up dealing with "claims specialists" who work from photos of the damage without ever visiting the home in person. That makes your job that much harder.

A Look at Insurance Restoration Jobs

One of the best things about insurance restoration is the variety of work that you do. Every day brings a new challenge, and a different job. Let's look at the types of work by major category that an insurance restoration contractor is typically called upon to perform. This is just a quick overview, and we'll be looking at all these work categories in a lot more detail in the coming chapters.

A large percentage of your jobs will come from water, fire, and smoke damage. Other areas, such as mold remediation and trauma scene cleanup, are fields that you probably won't be dealing with right away. However, they offer some excellent expansion opportunities for your business at a later date.

Some of this may intimidate you at first, but don't worry about it. That's normal. You were no doubt intimidated the first time you faced a large room addition, a complete home remodel, or even a complicated deck project. Like any other field of construction, you'll start small, learn the techniques, and gradually acquire the skills, tools and equipment. While you're learning, you'll build your confidence — and your business.

Figure 1-13 A broken pipe on a second floor caused a massive water leak in this home. The water eventually made its way into the exterior walls and behind the siding, resulting in this rather dramatic ice sculpture in the front yard.

Water Damage

Water damage makes up the majority of the jobs you'll be working on. For most restoration companies, water losses probably account for about 75 to 80 percent of their workload.

Water damage can come from a number of different sources. It might be a broken washing machine hose (incidentally, that's one of the most common sources of water damage), or a broken ice maker line. Other sources include loose or damaged fittings on sink stops or water heater flex lines, and overflowing sinks or washing machines. If you live in a cold climate, you'll also be seeing a lot of damage from frozen pipes (see Figure 1-13).

What you typically won't be dealing with is actual flooding, at least not for the insurance companies. Flooding from an overflowing river or similar natural disaster is considered *ground water*, and isn't normally a covered loss unless the homeowner has special flood insurance. We'll look at those coverage issues in more detail in the next chapter.

One of the nice things about water restoration, besides the steady work and nice paychecks, is that there are very comprehensive training classes available in that field. You can learn all the proper procedures, how to use the various types of equipment, and even how to estimate the loss and bill the client.

Fire Damage

Fire damage is another big part of restoration work. Fires come in all shapes and sizes, from a minor home kitchen fire to a large commercial fire loss or major structural loss involving several units in an apartment complex.

Fire repair involves several different phases. First, there's the evaluation of the work to be done. That involves deciding which components are a total loss and which components are salvageable. In this area you may be working hand-in-hand with structural engineers to determine cost-effective repair methods that'll be acceptable to both the insurance company and the building department. You'll often be working with cause-and-origin inspectors hired by the insurance company to determine how the fire began (the cause), and where it started (the origin). See Figure 1-14. Occasionally, you may be working with law enforcement investigators dealing with arson. Next, there's the demolition work, which is one of the dirtier aspects of the trade. When everything has been cleared away and cleaned up, then you can begin the approved repair work.

Figure 1-14 This aging wall heater was the cause of the fire in this home. As a restoration contractor, you'll be dealing with fires of all sizes and types.

Smoke Damage

To reverse the old saying, "Where there's fire, there's smoke." Smoke damage and smoke odor is another aspect of restoration work. These can both be very challenging. Some smoke damage is as obvious as wiping soot off a wall. Odors, such as the permeating odor of burnt meat, are both difficult to find and difficult to treat.

As with water damage, dealing with smoke damage and smoke odor is something you can learn about in classes. There are excellent classes put on all over the country where you can learn about the different types of smoke and how to both clean and deodorize it. We'll look at training classes in more detail later in the chapter. You'll also find more information about training in Appendix B.

Storm Damage

Storms damage homes in a lot of different ways. Wind can blow shingles off a roof or topple a tree and crush the roof. Snow and ice can overload structural components (look again at Figure 1-3). Hail can dent and damage all sorts of things; and lightning can create all kinds of havoc.

The type, frequency, and severity of storm damage varies from region to region, and from season to season. You may go for long periods with no storm-related calls, then be hit with dozens of them at once. It's another aspect of insurance restoration that keeps you on your toes!

Vehicle Damage

Vehicle damage is an interesting part of insurance restoration. It can be caused by a drunk driver who barrels across a lawn and into someone's house; or it can be caused by a homeowner accidentally hitting the gas instead of the brake as he pulls into his own garage. Vehicles do various amounts of structural damage when they hit a building, as well as damage to fences, landscaping, and outbuildings.

Vandalism

Depending on where you live, vandalism may range from a minor problem to one that's much more serious. Vandalism can take many forms. There's graffiti, or

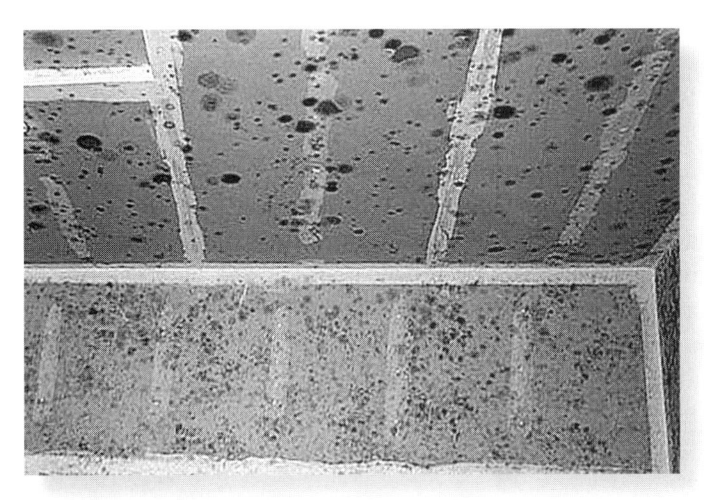

Figure 1-15 Mold is a tough and often dangerous problem in water-damaged homes. Many insurance restoration contractors also offer mold remediation services.

property damage of various degrees, caused by kids. There's damage caused by people who break into vacant homes and decide to live there for awhile, or burglars who come in, steal, destroy and leave. Some damage to homes is caused by angry tenants who've been evicted; and sadly, vandalism related to foreclosures is a growing problem as well.

Content Damage

Most of the insurance restoration work you'll be involved with occurs in occupied homes. So in addition to the damage to the structure, you'll also encounter a lot of damage to contents.

Content restoration ranges from simply wiping off furniture, to the cleaning and complex restoration of antiques and very expensive area rugs. Furniture can be damaged by fire and heat. Clothes and bedding may be contaminated with smoke. An antique might suffer water damage from sitting on a wet carpet, or furniture might be scratched or damaged by falling debris. You may even find yourself dealing with dozens of boxes of important business records from a company that was flooded, or hundreds of smoke-stained photographs.

The cleaning and restoration of contents is something you'll learn about in specialized training classes. In the beginning, while you're getting your business up and running, you may prefer to subcontract the restoration of contents to other companies.

In most circumstances, the moving and storage of contents are both considered part of the covered loss. Again, this is something you might want to leave to others initially, but it's an area that represents excellent expansion opportunities.

Mold Remediation

Mold remediation, like that shown in Figure 1-15, is a complex issue that's been in the news a lot in recent years. It's a highly specialized area of insurance restoration, requiring specific types of equipment and training. It can be very lucrative, but it carries a lot of liability as well. This is another area of restoration that you should seriously consider subcontracting initially, and then perhaps move into later after you're well established.

Trauma Scene Cleanup

Trauma scene cleanup, like mold remediation, is a highly specialized line of work. Trauma scenes include murders, suicides, accidents, and deaths from natural causes.

REAL STORIES:

The Best Piece of Advice We Ever Got!

When just starting out in the business, our company was called to respond to an emergency at a condo with a lot of water damage. We immediately began tearing out carpet, drywall, and any other wet materials.

Several days later, we had lunch with the insurance adjuster who was assigned to the job. He advised us that much of what we tore out could have been saved. An experienced restoration contractor would have known that.

Lucky for us, he was a very nice guy who wanted to help us out and see a new company survive. "Remember guys," he told us while discussing our overzealous demolition work, "I can hire *any* contractor to come in and tear things out. What I'm looking for is a contractor who can come in and *save* things."

That turned out to be the single best piece of advice we ever received. In fact, it became the guiding philosophy for our company from that point on!

Due to the nature of the work, and the danger of handling some of the materials at a trauma scene, this area isn't for everyone. However, if you're interested, there are several places where you can get the necessary training. Companies qualified to do this type of work are relatively scarce, so trauma-scene cleanup is one of the highest paid fields of insurance restoration.

There's a Reason It's Called "Restoration"

There's a very good reason why this particular field of construction is known as insurance *restoration* and not insurance *repair*.

Have you ever been hired to do remodeling on a classic older home? Let's say you're working in the living room and it has beautiful, original mahogany flooring. You remove a small cabinet and find that the hardwood flooring doesn't extend under the cabinet. How are you going to handle this situation? You're not going to tell the homeowner that you need to rip out and replace all of the expensive hardwood flooring because it needs a small patch. And you certainly don't want to fill in the missing section with lesser-quality prefinished wood flooring from the home center! Instead, you'd look for a cost-effective method to fill in the missing flooring and preserve the original appearance of the home. In this case, you'd have to find or mill some mahogany to the size of the original, stain it to match, and then refinish the entire floor.

Insurance restoration works much the same way — except you do this on a daily basis. If that same house had a water loss, and a small part of the mahogany hardwood flooring was damaged, an estimate would be written to restore the house (which includes that flooring) back to *pre-loss condition*. But the insurance company wouldn't authorize a completely new hardwood floor in order to repair a small area. And the homeowner wouldn't want a patch that doesn't match. So you, as the restoration contractor, would seek out the best solution to please both parties. As in the previous example, you'd find some matching mahogany, stain it, and then refinish the entire floor.

Balancing the needs of all parties — the insurance company, the homeowner, and you — is at the heart of insurance restoration. You can't just rip everything out and start over. Your objective is to save or restore items by matching materials and finding creative compromises and solutions. And most important, you keep the concept of "repair the home to a pre-loss condition" in your mind. This is the challenging part of our field of construction. And for many, it's one of the most rewarding.

The Three-Legged Stool

Sometimes, the best way to view and understand something is through a simple analogy. If you can compare a complicated subject to something simple, it often makes a lot more sense. For your insurance restoration company, an analogy that works well is the three-legged stool.

First of all, visualize a stool made of the finest materials. Those materials are honesty, integrity, a commitment to excellent customer service, and a group of conscientious and well-trained employees.

For a company (your stool) to thrive, it needs to be profitable enough in the good times to be stable and survive when times are a little tougher. To keep your stool both stable and profitable, you need to make sure it has three stout and sturdy legs under it. Those three legs are: the Emergency Division, the Structural Division, and the Cleaning/Contents Division.

Those three divisions, separate but interlinked, create a solid three-legged stool that'll prosper in good times and stay stable in tough times. It's a structure that allows your company to provide diverse services and excellent client and adjuster care.

The Three Divisions

The three divisions each have separate areas of responsibility within the company. However, in many circumstances, they also overlap, so they need to interact well with each other.

For a small restoration company with only a couple of employees, you may not initially need formal divisions. All of the employees will be trained to do the same work, usually under the direct supervision of the owner. But as the company grows, creating separate divisions is a good idea. That way, each division can specialize in its own set of duties and responsibilities, working under a supervisor who reports to either the General Manager or the owner.

The Emergency Division

As you'll see in Chapters 16 and 17, emergency calls are a huge part of restoration work. The Emergency Division handles all of these calls. They're typically the first ones to see a new job. They're responsible for the initial drying of wet structures, and for securing burnt or storm-damaged buildings. They maintain the emergency vans, and are responsible for tracking and maintaining all of the drying equipment and other specialized tools and equipment. The Emergency Division is usually the one that handles mold remediation projects as well.

The Emergency Division typically consists of a supervisor and one or more emergency technicians. While work is being done on an emergency job, crew members from other divisions will often help out. When that happens, they're under the charge of the Emergency Division supervisor.

The Structural Division

Once the emergency portion of the work is done, and the estimator has estimated and sold the job, the Structural Division takes over. This division handles demolition, plans and building permits, and all phases of the rebuilding. They're responsible for all the lumber, hardware, and other supplies stocked in the warehouse. They typically inventory and maintain all of the construction tools and construction vehicles.

Employees within the Structural Division include the supervisor, carpenters and helpers, as well as specialty trades, such as painters and drywall installers. This division may also include one or more estimators; or estimating may be done by the company owner or by a person who's not attached to any of the three divisions.

The Cleaning/Contents Division

Most restoration jobs also require cleaning, both of the structure and the contents. This is where the Cleaning/Contents Division comes in. They take care of smoke cleaning and deodorization. They handle carpet cleaning, and the final cleaning of the jobsite at the end of the project. They take care of the packing and removal of contents from the jobsite. They're responsible for inventorying contents, as well as its storage and redelivery. They also handle the cleaning, drying, and restoration of contents. The Cleaning/Contents Division maintains and is responsible for the carpet cleaning van, other cleaning vans, and other cleaning equipment, as well as their division supplies.

Employees within the Cleaning/Contents Division include the supervisor, carpet cleaning technicians, and cleaners. Larger restoration companies will also have contents crews, who deal primarily with the restoration, inventorying, storage, and transporting of contents. Also, in some larger companies, the Cleaning/Contents Division may have its own estimator.

We'll be looking at what each division does in more detail in the coming chapters.

Specialized Training

Today, the equipment and products used in the removal of such things as moisture, mold, and smoke odors from the home are very effective. New and ever more sophisticated products are being introduced on a regular basis. You need specialized training to correctly use all this technology. You'll want to start taking some specialized training classes as soon as possible.

There are a number of very good courses available. They may seem a little expensive at first, but the knowledge you'll gain is absolutely essential. And, take it from me, the classes will pay for themselves with the first costly mistake you avoid!

For example, in a typical multi-day water restoration class, you'll learn about both liquid water and water vapor. You'll learn some of the basic science behind the movement of moisture. You'll see how moisture reacts in different environments, and how it reacts with different materials. You'll learn about moisture testing, and all of the different restorative drying equipment that's available.

In some classes, such as the ones offered by Dri-Eaz (for a fee), the instructors actually flood a test house and then offer hands-on training in the different drying techniques. You not only learn about how to dry different materials, but you'll also get a much better idea of which pieces of equipment you need in order to get your business started, and which ones you can consider adding at a later date.

But classes offer a lot more than just how-to training. Well-taught classes also offer information on the business of running a company. You'll learn about pricing and marketing your services, and pick up some very useful forms and other written materials you can use in your business. They'll also tell you about some of the liability issues surrounding insurance restoration work.

Finally, you'll have the opportunity to interact with owners, managers, and staff from other restoration companies. People come from all over the country to attend these classes. A typical training course will have several new restoration contractors in it, as well as several of the more experienced folks looking for a refresher course or continuing education credits.

These classes give you a chance to swap stories and information. You'll often get some very good suggestions on different ways of doing things — that aren't in the class curriculum. As with any industry, the problems and challenges faced in insurance restoration are shared by pretty much everyone in the business. It's always great to hear how someone else solved the very same problem that you've faced. And, you can have a chance to share your solutions.

At the end of almost every training class, you'll have the opportunity to take an *Institute of Inspection, Cleaning and Restoration Certification* (IICRC)-approved exam. Passing these exams and receiving IICRC certification is an important and well-recognized achievement in the industry. In addition to authenticating your knowledge in the field, IICRC certification is an important endorsement and marketing tool for your company, especially to insurance adjusters.

The Institute of Inspection, Cleaning and Restoration Certification

The IICRC was formed in 1972 as a non-profit independent certification body. They help set and maintain standards for people working in the restoration industry. Their goals are to establish technical proficiency standards, monitor testing and certification, and promote the ethical behavior of restoration contractors and others in the industry.

The IICRC is actually owned by 16 different trade associations from all around the world, with no single controlling interest. This arrangement has helped them remain neutral and independent. They maintain high ethical standards, and have become a very well-respected group.

The IICRC doesn't have any training facilities or teachers of their own. Instead, they approve and authorize schools and instructors to teach IICRC-certified material and conduct testing. As of this writing there are approximately 140 IICRC-certified instructors, teaching a wide variety of classes all over this country, and in the international community as well.

Tip! *Once you, an employee, or a member of your staff becomes certified, be sure and use that achievement as a marketing opportunity. Do a press release right away (see Chapter 21). Also, send an announcement letter to insurance agents, adjusters, property managers, and others that you do business with.* Never miss a marketing opportunity!

Because the IICRC is so well recognized within the industry, it's important that you and your company become certified members of their organization. The IICRC currently offers over 20 different certifications in everything from water damage to various types of rug and fabric cleaning. This allows you to choose, from among a wide selection of training and certifications, the courses that will most benefit your company.

Upon completion of any certified class, you can pay a small fee and take the certification exam. The exams are given at the conclusion of the course, or you can take it later if you prefer. Taking it at the end of the class is ideal, since the material is still fresh in your mind and you don't have to make special arrangements for testing at another time.

When you successfully pass the exam — a score of 75 percent or better is required — you'll become individually certified in that particular aspect of restoration. Once you or one of your employees is certified, you can then apply to have your company certified as well.

Individual and company certification is not only good for you personally, it helps maintain high standards throughout the industry. And, it demonstrates to insurance adjusters and clients that you are knowledgeable, trained, and committed to your industry.

For more information on IICRC certification courses and their course descriptions, see Appendix B at the end of the book.

2

Understanding the Basics of Insurance Policies

If you're going to be doing insurance restoration work, you need a basic understanding of insurance policies and how they work. Insurance policies are very complex, but don't let your eyes glaze over just yet. Sit back, pour yourself a cup of coffee, and let's look at just the basics.

Perhaps your most important need in understanding insurance policies is to know which losses will or won't be covered by a policy. That's something you must be certain about before you undertake the work.

Imagine, for example, that you're called out on a water loss, and there are also some mold issues. You go ahead and repair the house and deal with the mold. But later, you find out that the repairs were covered, but the mold wasn't. The insurance company won't pay for your time, and the homeowner tells you he can't afford it. Now what?

These types of incidents happen, so the more you know about how policies work, the better prepared you'll be to prevent them from happening to you.

Another reason you need to understand insurance policies is, unfortunately, most homeowners don't have a clue about exactly what's covered by their policy. They've never looked at it; and they've long forgotten what their agent told them when he sold it to them. If you can help them understand what you can do for them in regard to their loss, you'll be providing them with a helpful service. That'll enhance your standing in their eyes, and they'll see you as the type of competent, qualified restoration contractor that they'd like to work with.

Having said that, I now need to clarify just how much help you can provide them, with a disclaimer and a warning about interpreting someone else's insurance policy.

A Disclaimer and a Warning

First, the disclaimer: When you finish reading this book, you won't be an insurance expert. We're only going to be looking at a basic overview of insurance policies in this chapter. The intention is to give you a better understanding of how most homeowners, adjusters, and even the courts look at insurance coverage issues. *This is not a complete course on insurance policies, and it's not intended to be.*

> ***Tip!*** *Offer your clients the best information you can about their policy — but* always *tell them to consult with their insurance agent, their insurance adjuster, or their attorney for specific information.*

As a restoration contractor, you'll often be asked policy questions. Do your best to answer your client's questions and help them out wherever you can. That's all part of good customer service. But here's the warning: Don't ever attempt to decipher a person's actual policy for them. And *never* tell them anything that could be misconstrued as legal advice.

Understanding the Terminology

As in every profession, the insurance industry has its own terminology. There are certain terms you'll come across that are key to understanding insurance policies and the people you'll be working with. Let's look at some of the more important terms you'll be hearing.

Agents and Adjusters

Your clients may sometimes get confused about the difference between an *agent* and an *adjuster*. They are two completely different occupations within the insurance industry, so it's good to understand the responsibilities of each one.

An *agent* is the person who sells the homeowner his insurance policy. Agents may represent one single company, or they may represent several different ones. They may also represent several different types of insurance, such as homeowner's policies, renter's policies, auto, health, life, and others.

Typically, insurance agents are very well versed in the policies issued by the company or companies that they represent. It's the agent's job to understand what the policyholder needs in order to be adequately covered, and then to provide the appropriate policies.

An insurance claims *adjuster*, also known as a claims representative, only enters the picture when a loss occurs. It's his or her job to interpret the policy. Adjusters make the final decisions on coverage issues, and determine how much the insurance company will pay toward repairing the loss. We'll look at insurance adjusters and what they do in a lot more detail in Chapter 22.

Named Perils and Covered Losses

Within the typical homeowner's insurance policy are sections on *named perils*. A *peril* is something that could conceivably happen to a structure, such as a fire, a broken pipe or damage from a wind storm. The insurance company lists or names these perils in the policy, hence the term *named perils*. They're essentially saying, "If it's on the list and it happens to your house, it's covered. If it's not on the list and it happens to your house, it's probably not covered."

While the term *named perils* is often seen in the policies, you probably won't hear it used very often by insurance agents and adjusters. Instead, they tend to refer to *covered losses*. As we saw in Chapter 1, a loss is something that happens to a structure — a fire, storm damage, etc. *Covered* simply means that the insurance company will pay for the damages, subject to all the specific terms of the homeowner's policy.

So let's clarify that terminology a bit more:

- **Named peril:** A peril that's listed in the policy as something that *could* theoretically happen to a structure; and if that peril is listed and does happen, the insurance company should pay for it.
- **Covered loss:** When that named peril actually *does* happen, and the insurance company makes the determination that they actually *will* pay for it, then the *named peril* becomes a *covered loss*.

Sudden and Accidental vs. Maintenance Issues

Another phrase you'll hear quite often is *sudden and accidental*. This is a very important concept to understand, because it affects a lot of what you'll be doing as a restoration contractor.

In most instances, for a loss to be covered it has to be considered to have happened suddenly. In other words, it isn't something that's been going on for a long time. It also has to have occurred accidentally, which obviously means that you didn't do it on purpose.

Sudden and accidental is especially important with losses involving water or losses related to the exterior of the home, such as shingle damage. If a water pipe freezes and bursts, that's obviously both sudden and accidental, and would typically be a covered loss. See Figure 2-1. On the other hand, if a pipe has been dripping for a long time and causes damage, as in Figure 2-2, that would usually be considered a maintenance issue. Even if the homeowner wasn't aware of the dripping pipe, it isn't something that would typically be covered.

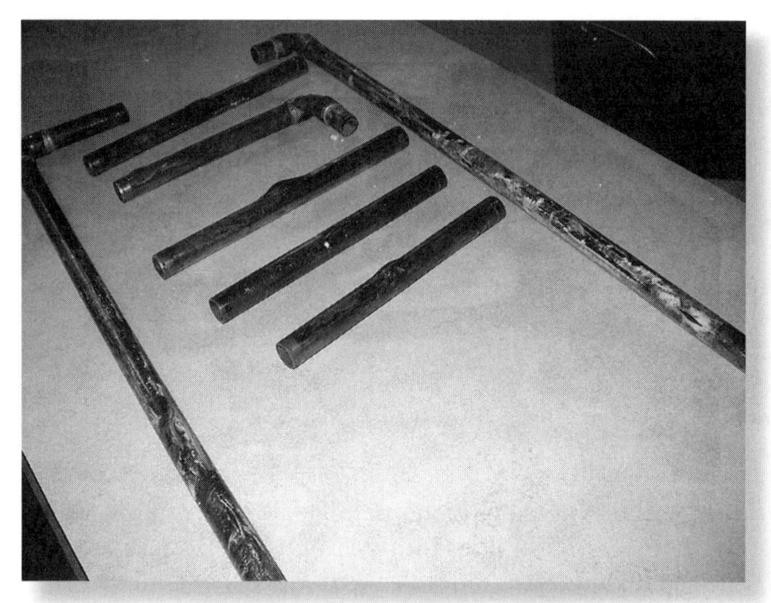

Figure 2-1 A frozen pipe is a clear example of a *sudden and accidental* loss. The outward puckering of the metal is characteristic of a freeze break in a copper pipe. These broken copper pipes all came from the same home.

Figure 2-2 The old and deteriorating galvanized plumbing in this bathroom is a clear example of a maintenance issue. This leak had been going on for quite a while, and was *not* sudden and accidental. The insurance company denied this claim.

Let's look at an example of damage to roofing shingles. If the wind blows the shingles off, it's *sudden and accidental,* and that would typically be covered. But if the shingles start leaking because they're old and well beyond their established useful life, that would be considered a maintenance problem and the damage probably wouldn't be covered.

The Basics of Named Perils

The typical homeowner's insurance policy will list several named perils. Policies are written in different ways with different formats and terminology, so what follows is only a general listing. And, as we mentioned earlier, there are bare-bones policies and top-of-the-line policies. In most cases, the more the policyholder pays for the coverage, the more coverage they receive. The following are the named perils most commonly found in homeowner's insurance policies, but these aren't all the possible named perils — and not every policy will cover all of these.

Water

Water damage can be a tough one sometimes, so you really need to work closely with the adjuster to clarify any coverage issues. As mentioned earlier, water damage is usually covered if it's considered sudden and accidental. This covers water coming from a broken pipe; overflowing from a sink, toilet, or other fixture; from an appliance, such as an icemaker or a washing machine (Figure 2-3); from

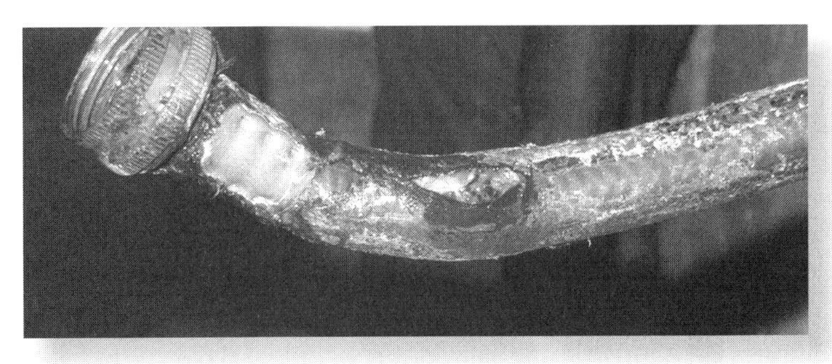

Figure 2-3 Splits in rubber washing machine hoses, such as this one, are a leading cause of residential water losses each year.

Figure 2-4 The water in this toilet tank froze after the furnace failed and the home was without heat. As the water froze, it expanded, causing the tank to crack.

Figure 2-5 The weight of the snow on this deck overhang caused it to collapse, tearing off the rear fascia of the home.

a broken fire sprinkler line, hot water heating system, or air conditioner; as well as breaks caused by frozen water in lines, fixtures, or appliances, such as the frozen water in the toilet tank shown in Figure 2-4.

As we've seen, water damage includes more than water in its liquid state. It typically includes sudden and accidental damage caused by water in its vapor form, meaning steam, as well as water in its solid form, ice. Losses caused by a broken steam line or malfunctioning steam heating system are usually covered. Most policies will also cover damage from the weight or the action of ice, snow, sleet, and freezing rain (see Figure 2-5).

What isn't covered is *ground water* damage, unless the building owner has a specific flood insurance policy. This includes flooding, either from a river overflowing its banks, or from ongoing rains. It also includes damage caused by snow and ice *after it melts*. That's an important distinction. If snow or ice builds up on a roof and causes damage, that's covered. But if snow or ice builds up around a house and then causes foundation damage as it melts, that's usually *not* covered.

Fire

Fire peril includes any type of accidental fire caused by the homeowner, like the one in Figure 2-6, or by someone else. It also covers damage caused by natural disasters, such as wildfires (Figure 2-7). Arson is covered as well — unless the fire is started by the property owner or someone he hires.

Figure 2-6 This fire was caused when a homeowner discarded his cigarette and it rolled underneath the wooden deck and caught it on fire. The fire did a massive amount of structural damage to the home. The insurance company determined that the cause was accidental, and covered all of the damage.

Smoke

Smoke damage is most often the result of a covered fire loss. This usually includes smoke from sources outside the home, such as a wildfire or a fire in an adjacent structure, even if it wasn't the fire itself that damaged the policyholder's structure.

Also covered is what's known as a *puffback*. This is the release of smoke and soot into a structure caused by a malfunctioning furnace or boiler. Smoke that comes from a fireplace, chimney, or flue may or may not be covered, depending on the cause and the specific terms of the policy. Not covered within this category is smoke from cigarettes, cigars and pipes.

Figure 2-7 Fortunately, only this home's deck was damaged by the wildfire that swept through this area. Damage such as this is typically covered.

Wind Storms

Wind storms cover anything related to wind, such as shingles blown off a roof, damage to fences, patio covers, outbuildings, and other structures, like the barn shown in Figure 2-8. It also covers damage caused by a falling tree limb or even an entire tree that's been blown over by the wind.

There are some interesting distinctions made with regard to wind damage, and damage caused by wind-driven rain or snow that you need to know about. Let's look at two examples of what could happen when a wind storm is followed by a rain storm.

In our first example, the wind causes some shingles to blow off a roof, or maybe a tree limb is blown down and creates a hole in the roof. In either case, the initial damage to the roof was the result of the wind storm. Then it starts to rain and water gets inside the building. The rain does additional damage to the interior. In this example, the damage to the roof is covered, and the water damage to the interior would typically be covered as well.

Figure 2-8 Whole sections of the roof of this hay barn blew off in a sudden wind storm.

In our second example, we'll say that the homeowner left some windows or a skylight open. That same wind storm comes through, but there's no damage to the roof. A rain storm follows the wind, and the wind kicks up again, causing the rain to blow in through the open window or skylight, damaging the interior of the home. In this scenario, the water damage would probably *not* be covered.

Hail

Hail damage is a covered loss, and it can be a tough one to deal with. Hail causes a lot of widely scattered damage, from broken window screens to severely dented roofing

REAL STORIES:
I Guess I'll Be Late for Work!

Toward the end of a particularly bad snow storm, we received a call from a woman who had just heard a loud crash outside her house. She went out to investigate, and found that the weight of all the snow on her carport had collapsed it — right onto her SUV!

When we arrived, we found that the vehicle was now the only thing holding up the roof of the carport. If the roof were to collapse completely, it would not only do a lot of additional damage to the SUV, but it would also damage the contents stored in the carport.

Our first task was to shovel away all the snow around the carport so that we had room to work. Then, working carefully around the perimeter of the unstable roof structure, we built temporary walls to stabilize it and keep it from dropping any further. Next, we slowly and carefully jacked up the temporary walls, adding more lumber as we went. Finally, when the roof was a couple of inches above the SUV, we had a towing company come in and tow the vehicle out of danger.

We built additional temporary supports so we could clear out the contents of the carport. Then we reversed the process, slowly lowering the supports and the roof until it was completely down, and could be safely cut up and hauled away.

The homeowner stayed very calm through the whole ordeal, and even snapped a few pictures as the work progressed. "One thing's for sure," she laughed. "I've got a great excuse for being late for work!"

REAL STORIES:

Now That's an Unusual Fire!

Fires can be caused by all sorts of things. And typically, unless they're intentionally set, the insurance company will cover the damage.

That was proven by one of the most unusual fires we ever encountered. A homeowner called to say that one side of his house had caught fire and burned. The fire damaged the siding, trim, a window, some electrical wiring, and even the wall insulation. But the question was — what caused the fire? There was no sign of arson, no lightning or wildfires in the area, and no negligence on the part of the homeowner. In fact, he hadn't even been home at the time.

The insurance company sent out a cause-and-origin inspector to see what he could find. After quite a bit of research, the inspector came up with the following explanation: The house was located in a wooded area heavily populated by squirrels. There were some overhead power lines running very near the house.

A squirrel, running along one of the power lines, encountered a transformer and was electrocuted; he caught fire, and fell to the ground next to the house. The squirrel's burning fur ignited some dry brush next to the house, which in turn ignited the side of the structure.

The insurance company covered all of the repairs. And the homeowner ended up with a great story about the day a flaming squirrel attacked his home!

and siding. It's the adjuster's job to decide the extent of the coverage, which we'll look at in more detail in Chapter 10.

As with wind damage, if the hail causes the structure to be opened up in such a way that interior damage occurs, then the interior damage will typically be covered as well. This might include water that comes in through a broken window, or even window coverings that are damaged when the window breaks.

Lightning

This is a storm-related area of coverage that's typically covered. Lightning can do all sorts of interesting things to a house, from structural damage to damage to electrical wiring and appliances. Typically, if there's any kind of a link between the damage and the massive electrical discharge that accompanies a lightning storm, it'll be covered. That generally includes content items such as computers and other home office equipment; televisions and home entertainment systems; and other types of home electronics.

Vehicle and Aircraft

Damage that occurs to a structure from the impact of a vehicle or an aircraft is typically covered. This also covers damage from parts of the vehicle or aircraft that come loose and impact the home.

Vehicle damage to the home and landscaping are both considered part of the structural coverage. Any damage to the home's contents caused by the impact is

Figure 2-9 This photo shows massive graffiti damage in an unoccupied home. Although most of the damage was just done with paint, it was so extensive that many of the materials in the home had to be removed and replaced.

also covered under the same claim. Typically, the homeowner's insurance policy will cover the cost of the damage, and then they'll seek to recover those costs from the company that insures the driver of the car.

Here's another one of those interesting coverage twists: While damage that's caused by a vehicle is covered, it may not be covered if the vehicle is owned or driven by the owner of the damaged building. In other words, if you pull into your garage one day and accidentally hit the gas pedal instead of the brake, plowing into the back wall of the garage, that damage may not be covered by your homeowner's insurance. But there's some good news — the damage probably would be covered by your auto insurance policy.

Vandalism

Vandalism coverage includes breaking and entering, as well as graffiti. See Figure 2-9. It also covers willful damage to public or private property or contents, and damage done by people who enter or occupy a home without authorization (as in Figure 2-10). It's important to note that most policies don't cover vandalism if the home has been left unoccupied for more than 60 days.

Figure 2-10 In addition to breaking holes in the drywall, the vandals in this home took the time to create a little "art" display on one of the walls.

Riots, Civil Commotions, Explosions, and Volcanic Eruptions

In the unlikely event that a riot or a volcano damages a home, it's probably covered. However, if the damage is caused by an earthquake, that's probably not covered unless you have specific earthquake coverage.

Content Coverage

Contents are all of the items in a building that aren't physically attached to the structure. They can range from major items like furniture, electronics and clothing, to the smallest of items such as books and toiletries.

Coverage for contents is typically included as part of the basic homeowner's policy. The value of the contents coverage is typically set as a percentage of the value of the structural policy. For example, if the insurance company sets the value of the contents coverage at 10 percent of the value of the structural coverage, and the house is insured for $300,000, then the contents are automatically insured for $30,000.

If the homeowners feel that's not enough coverage, they can increase the contents coverage in the policy for an additional fee. They can also insure unusually expensive items, such antiques, paintings, jewelry, stamp collections, etc., separately. These separate policies are known as *riders* or *endorsements* to the original policy.

We'll go into contents coverage in a lot more detail in Chapters 13 and 14.

Outbuildings

Outbuildings are separate, detached buildings located on the same property as the primary building that's covered by the insurance policy. Examples of outbuildings include garages, barns, shops, sheds and greenhouses. See Figure 2-11.

As with contents, outbuildings are typically covered as a percentage of the value of the primary covered building. Outbuilding coverage is usually an aggregate amount — which means, it's the total for all the buildings combined.

Let's say you have a house valued at $400,000, and the outbuilding coverage established by the insurance company is 10 percent of the value of the house. Your outbuilding coverage is $40,000. That's the aggregate — or total — value of the outbuilding coverage, no matter how many outbuildings you have.

If your only outbuilding is a detached garage, and it burns down, you have $40,000 in coverage. But if you have a detached garage, a barn and two shops, and a wildfire burned all four buildings to the ground, you still only have an aggregate value of $40,000 outbuilding coverage for all of them.

As with contents coverage, you can purchase additional coverage for outbuildings if you need it. In the example above, you would obviously want to do that.

Figure 2-11 Most policies automatically extend coverage to outbuildings, such as this garden shed that was in the path of a falling tree.

Figure 2-12 Despite repeated warnings, people continually dispose of hot fireplace ashes by putting them in paper bags or cardboard boxes and setting them outside. This fire, which started on the deck right outside a sliding glass door, started in just that way.

Figure 2-13 The hot fireplace ashes caught the deck on fire, then the wood siding. The fire spread up the wall and into the attic, where it did extensive damage to the roof structure.

Is Stupidity Covered?

Thankfully, stupidity and ignorance are covered. If they weren't, most insurance restoration contractors would be out of work.

Once a person has a homeowner's insurance policy in place, the named perils we discussed are covered in virtually any circumstance that meets the sudden and accidental criteria. That includes pretty much everything except deliberate actions on the part of the homeowner.

The truth is, many of the losses you'll see are caused by carelessness on the part of homeowners. They may leave candles burning in the bedroom and forget that the window's open. When a breeze blows the curtains over the candles and they catch fire, that's covered.

They may leave a frying pan on the stove when they go answer the phone, and while chatting away the hot oil catches fire. That's typically covered as well.

They may cut down a tree in the backyard and have it fall right onto their fence. Or let their bathtub or sink overflow, or do any of a thousand other "How stupid was that!" things.

The damage to the home in Figures 2-12 and 2-13 was caused by putting hot ashes from a fireplace in a paper bag and setting the bag out on the deck. As long as a person doesn't cause the damage intentionally, it's almost certainly covered.

Deductibles

As explained earlier, every homeowner's insurance policy will have a deductible associated with it. This is an amount of money that the policyholder will have to pay out of his own pocket.

Since deductibles can cause all kinds of problems for you, this is an area that you need to understand.

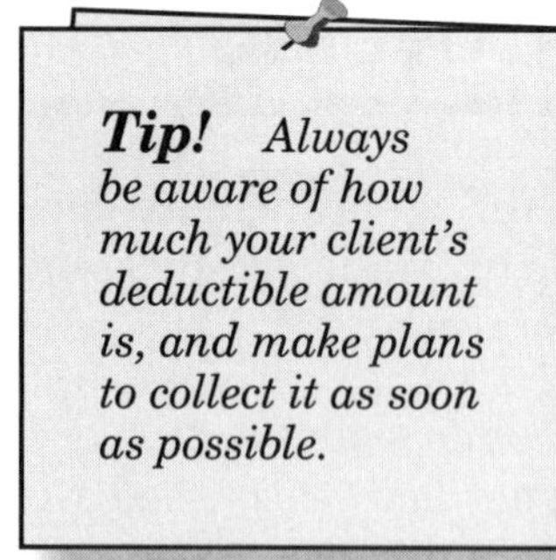

A deductible is a means for the insurance companies to limit the number of smaller claims that policyholders file. For example, if a homeowner had a small water leak in the bathroom and didn't have a deductible, he might decide to file a claim for the loss, figuring that he would get some new flooring at the insurance company's expense.

On the other hand, if he has a $500 deductible and the new flooring only costs $400, he won't file a claim because he has nothing to gain. Even if the flooring is $600, most homeowners will still figure that it's not worth filing a claim for only a $100 reimbursement.

For many years, $250 was the most common deductible amount. But in recent years, insurance companies have had large losses from hurricanes and other natural disasters, and diminished returns on their investments in the stock market. As a result of these and other factors, most companies have raised their deductible amounts.

Today, $500 and $1,000 deductibles are much more common than $250. Some policies have even higher deductibles. In fact, a few now have deductibles that are based on a percentage of the insured value of the home.

This works fine for the homeowner, as the higher deductible amount is offset by a lower policy premium. And, because many homeowners assume that nothing bad will ever happen to their home, they often choose to have a higher deductible to save money on the cost of their premiums. The problem for them — and, ultimately, for you — is that if they *do* sustain a loss to their home, they may not have the necessary cash readily available to pay their deductible. We'll look at that in more detail in Chapter 25.

You Can't Profit From a Loss

A basic concept of insurance coverage is that a policyholder can't profit from a loss. In other words, he can't expect the insurance company to pay him more than his loss was actually worth at the time that the loss occurred.

Depreciation

One way of preventing a policyholder from profiting from a loss is through what is known as *depreciation*. Depreciation is based on the concept that most things have a useful lifespan, and are therefore worth less (or have less useful life remaining) after they've been used, than when they're brand new. That's perfectly logical, but it can be a little confusing when it comes to insurance restoration.

In the world of auto insurance, depreciation is fairly easy to understand. If a person has a brand new car, valued at $20,000, and he has an accident resulting in a total loss to the car, the person would receive a payment from his auto insurance company for $20,000.

But assume he owned that $20,000 car for five years before he totaled it in an accident. What's its value then? The insurance company would refer to one of the used-car value guides to determine the car's value. They would look at the same year, make, and model of car, with the same features and the same amount of mileage, and in comparable condition.

If a comparable car's value is $11,000, that's what the policyholder would receive. It wouldn't be fair to expect to receive the full $20,000 that he paid for the car when it was brand new, because at the time of the accident the car was no longer worth that amount.

The difference between what the car was worth when it was new and what it's worth at the time of the loss is the car's depreciation. In our example, the car had depreciated $9,000 from its original value of $20,000 to its current value of $11,000.

The same concept of depreciation applies to homeowner's policies, but it's often more difficult to understand. Let's say a person has brand new carpeting in his home, and the carpeting gets damaged due to a covered loss, such as a fire. As with the new car example, since the carpet's brand new, the policyholder would receive the full value of that carpet.

But, if the carpet is 10 years old, it would have to be depreciated from its original value, due to age and wear and tear. As in the car example, the homeowner would receive the *current* (depreciated) value of the carpet, not the *new* value.

Where the confusion comes in, is that there's no simple used-carpet value guide to refer to, like there is with used cars. Instead, the insurance company has to make an educated guess of the carpet's current value, based on the expected life of that particular type and brand of carpeting.

If that carpet has an average life expectancy of 20 years, and it was 10 years old at the time of the loss, it has depreciated by half its life, or 50 percent. If the same carpet has a value of $25 a square yard new, then the insurance company would settle on a *depreciated* value of $12.50 a square yard — 50 percent of the $25 value — and they would pay the policyholder accordingly.

Carpeting, and other floor covering, is a very common area where you'll see depreciation applied, but it can actually be applied to any components in the house that can wear out over time. For example, you might see depreciation applied to roofing shingles, water heaters, kitchen appliances, and other items that will wear out over time. Where you typically won't see it applied is with the actual structure of the house, such as the framing or foundation. Another area where deductibles usually don't become an issue is with items that have virtually no wear, such as toilets, windows, or ceramic tile.

Cash Value Policies

A *cash value* homeowner's policy, also known as a *fair market value* policy, is one that pays a policyholder at the current (depreciated) value of the home or the contents. In other words, it takes depreciation into account for just about everything that's damaged, just as we described in the examples of the car and the carpet.

The advantage of a cash value policy is that it's less expensive to purchase. Homeowners with tight budgets often opt for this type of reduced coverage. You'll also sometimes find the cash value policy coupled with a high deductible amount. That makes the coverage even more affordable, but it can lead to problems when it comes to repairing damage to the home.

The disadvantage of this type of coverage is obvious: Building materials and contents are replaced at the current value — today's cost minus depreciation. That makes it very difficult to repair a loss, and the bigger the loss is, the tougher it becomes. In the case of a substantial loss, such as a large structure fire, repairing the home to a pre-loss condition becomes almost impossible, unless the homeowner has a substantial cash reserve to draw upon.

Full Replacement Policies

Many homeowner's insurance policies in place today are *full replacement* policies. They cost more, but if a loss occurs under this type of policy, all of the damaged is covered at *full replacement value*, and at today's market cost. So, in the example of the carpeting, there would be no depreciation. The carpet would be replaced at today's value, regardless of age or condition. That works well for both you and the homeowner.

Because not all policies or companies handle depreciation in the same way, you'll want to discuss the issue of depreciation with the adjuster up front, to be sure you're clear on how it affects the specific job you're working on.

Building Code Upgrades

You'll occasionally come across code upgrade situations with older homes. The home may have been built in accordance with the building codes that were in force at the time of construction, but it no longer meets the code requirements that are in force today. See Figure 2-14. That can be an expensive upgrade that may or may not be covered by the policy. We'll look at code upgrade situations in more detail in Chapter 22.

Luckily, a growing number of homeowner's insurance policies now offer code upgrade coverage. This coverage pays for the cost of any building code upgrades required to repair a building to today's standards. If you're bidding work on an older home, be sure and check with the adjuster regarding the issue of building code upgrades.

Figure 2-14 Older homes present a number of code upgrade issues. This home, built in the 1920s, suffered a water loss, and extensive repairs were required. There was also mold damage due to the water loss. The home had issues with undersized framing, a rock foundation, inadequate insulation, and other problems. Situations such as these require careful handling, and close cooperation with the adjuster, the homeowners, and the building department.

Mold Coverage

The issue of mold in homes has received a lot of media attention in recent years, often resulting in confusion and misinformation on the part of homeowners. While it's been a source of concern for the homeowners, it's been a nightmare for insurance companies. They've had to deal with a number of very expensive lawsuits and repair claims involving mold infestations.

The end result is that most insurance companies now exclude mold coverage from their policies. Those exclusions can present a real problem for both you and the building owners that you work with. This is an area where you need to be especially cautious, because you can easily open yourself and your company up to some very expensive litigation.

Right now, at this point in the book, just be aware that mold is probably not a covered loss. We'll take a much closer look at mold issues, and what to do about them, in Chapter 5.

Outside Living Expenses

All homeowner's insurance policies contain some type of provision for outside living expenses. This means that if the loss is bad enough that the home can't be occupied, the insurance company will pay to have the occupants live somewhere else temporarily.

Typically, a home is considered unlivable if it doesn't have heat, cooking facilities, or sanitation facilities. However, in some cases, it may still have all three of those things, but the bedrooms or other living spaces are so affected by the loss that the house is still considered uninhabitable.

The way insurance companies handle outside living expenses can vary greatly from job to job. If the house can't be occupied for a relatively short period of time — less than about a week, for example — the occupants may be put up at a local hotel. In addition, each occupant of the home would be entitled to be reimbursed for eating out three meals per day.

REAL STORIES:
You Went Where?

We once worked on a home that was damaged by a small kitchen fire. We discussed the repairs with the owners, decided on the materials, paint colors, and other details, and went to work. The owners checked in periodically by phone, but we never saw them on the job until the very end.

Not having our clients come by and check on the job regularly is pretty unusual. But everything went smoothly, they were delighted with the finished product, and we got paid without incident. As a result, we didn't think any more about it.

Several weeks later, we were on another job with the same adjuster who had handled the kitchen fire. I commented on how well the fire repairs had gone, and how unusual it had been to not have the homeowners looking in more often. That's when I learned the reason for their absence.

"They were in Vegas," he laughed. "They booked a flight, and got themselves a nice upscale hotel room and had a great time. Then they tried to bill the insurance company for everything. They said that since they were out of the house and entitled to outside living expenses, they might as well have them in Las Vegas! Never mentioned it to me beforehand — just took off, and presented me with all the bills when they got back!"

Needless to say, the insurance company settled with them for their outside living expenses — but for a much more reasonable amount.

If the repairs will take longer, then the insurance company may opt to pay for a hotel room or rooms with a kitchen. The understanding here is that the policyholders will buy their own groceries and cook their own meals, since they would be doing that at home anyway.

For long-term repairs lasting several months, the most common solution is for the insurance company to help the homeowners rent an apartment or a house. In that case, the insurance company will also typically pay the cost of moving any contents from the old home to the temporary new one, and then back again. If most of their contents are ruined as a result of the loss, the homeowners may buy new ones while they're in their temporary home. In that case, the insurance company will usually pay the cost of having the new contents moved into the home after the repairs are complete.

Temporary living arrangements can take a variety of other forms as well. You'll sometimes encounter homeowners who don't want to leave their property. They may have security issues, they may have animals to care for, or there may be other considerations that make them want to stay close to home. If that's the case, insurance companies are usually pretty flexible about how they handle the outside living arrangements.

For example, if the policyholder has a recreational vehicle, they may choose to set it up near the house and live there. The insurance company may pay to rent a trailer and have it put on the property temporarily, as opposed to renting them a home elsewhere. Insurance companies have even been known to pay for the cost of minor renovations to outbuildings in order to create temporary living space during the course of repairs.

Arrangements for temporary living accommodations are between the homeowner and the insurance company. If it's a satisfactory arrangement for the homeowner, and it makes financial sense for the insurance company, just about anything goes. For your part, you'll want to be ready and willing to help with relocation, temporary renovations, temporary water and power for a trailer or motor home, or whatever else might be required of you.

REAL STORIES:

Love That Room Service!

When I was a teenager, we had a fire in our home that affected about half the house. It was the damage to our contents, especially irreplaceable things like photo albums, that was upsetting, to my parents in particular. When I decided to go into insurance restoration work, I think that experience really helped me to empathize with our clients.

But I have to admit, the experience wasn't all bad. The repairs took about a month, and the insurance company put us up in a nice hotel a few miles away from where we lived. My parents had a room, and I had another one all to myself. It was like being on vacation! My girlfriend and my buddies would come over, and we'd lounge by the pool for hours. My family had breakfast in the hotel restaurant every morning, and dinner at either the hotel or at another restaurant in the area in the evening. I never had to clean up or even make my bed!

However, by the time the repairs were done, we were all a little tired of hotel living. It had been fun, but we were ready to be back in our own home. There was one thing I really missed about hotel living — being able to pick up the phone and order a steak sandwich from room service while watching a game on my very own TV! For a teenage boy, that was the ultimate luxury!

Job Delays

One thing you need to be aware of is that outside living expenses can be quite expensive for the insurance company — even if their clients don't take off for Las Vegas. The companies are obligated to pay, but they obviously don't want to spend more money than necessary.

You'll often be asked to estimate a timeframe for repairs, so that the insurance company can allocate outside living expenses accordingly. Do the best you can, but try not to underestimate. If you say three weeks and it takes four, it can become an issue for both the insurance company and the homeowner. On the other hand, if you say four weeks and it takes you three, you're a hero and you've further enhanced the reputation of your company.

Losses on Other than a Primary Residence

While most of the losses you'll be working on will be in the policyholder's primary residence, that's not always going to be the case. You'll also find yourself dealing with losses to rental properties and vacation homes. Both of these present some unique challenges that you need to be aware of.

If a home is being used as a rental, the insurance policy that was taken out by the property owner will usually cover only the structure itself. It typically *won't* cover anything belonging to the tenant. Neither will it cover the tenant's outside living expenses. This is a very important point.

If there's a fire in a rental home that damages both the structure and the tenant's contents, the contents need to be removed from the home in order to do the repairs. But, sometimes the tenants don't have a renter's policy that covers the cost of moving the salvageable contents, or the cost of disposing of the items that are ruined. You need to know who's responsible before you do anything.

Never step in and start dealing with the tenant's contents on your own! For one thing, you don't have any type of signed agreement with the tenants, so chances are you won't get paid for the labor of moving their stuff. More importantly, if they claim that something is missing or damaged and you're at fault, you'll definitely have some liability issues.

In some cases, the property owner will have to step in and pay for the cost of removing the tenant's belongings. In other cases, if debris from the fire is mingled with the tenant's ruined contents, the insurance company may just allow you to remove and dispose of everything at the same time. But here again, be sure that someone — the building owner, the insurance company, or the tenant — has given you *written* authorization to dispose of the tenant's belongings.

Here's another side of that situation that'll come up: You're asked to deal with a loss that leaves the rental home unlivable, but the tenants don't have a renter's policy that'll cover their outside living expenses. You need to make repairs. You can't do the work while they're there, but they can't afford to leave. In this event, you need to inform the insurance company and the building owner of the problem, and then step away. This is something that they need to work out among themselves. Once it's settled, and the tenants are out, you can come back in and do your work.

As a restoration contractor, you're also going to run into losses that occur in condominiums and commercial buildings. Each of these loss situations will present their own specific policy challenges. We'll look at both in more detail in Chapter 22.

What if it Isn't Covered?

With all this talk about insurance coverage, you might be wondering what happens when you encounter a loss, or part of a loss, that *isn't* covered?

At that point, it's just like any other repair job that you would be asked to work on. You are, after all, a contractor. You can choose to put together an estimate for the portion of the repair that isn't covered by insurance, or you can work out a time and materials arrangement with the homeowner. (Be aware that some jurisdictions have restrictions on T & M contracts with homeowners.)

Either way, you'll need a contract or some other type of payment agreement that complies with the laws of your state. The bottom line is, insurance coverage or no insurance coverage, you'll need to work out all of the job specifications and all of the payment details with the homeowner before the repair work gets underway.

In the next chapter, we'll start looking at water damage and the equipment you'll need to deal with that part of restoration work.

3

Understanding Water & Water Restoration Equipment

Drying buildings and repairing water damage will almost certainly become the biggest part of your insurance restoration business. Water-related losses happen far more frequently than fires, storm damage, or any other type of loss. Water does a surprising amount of damage, and that damage can be much more extensive than you might think.

For a restoration contractor, water restoration jobs are both good and bad. On the good side, water restoration is very lucrative. It's also cleaner and generally easier than fire repair. The average water job takes less time than the average fire job; and callbacks are less as well. Perhaps most importantly, you have the opportunity to do the initial drying of the structure. Structural drying has become quite specialized in recent years, so there are fewer people doing it. That translates to more work for you, and a higher rate of return.

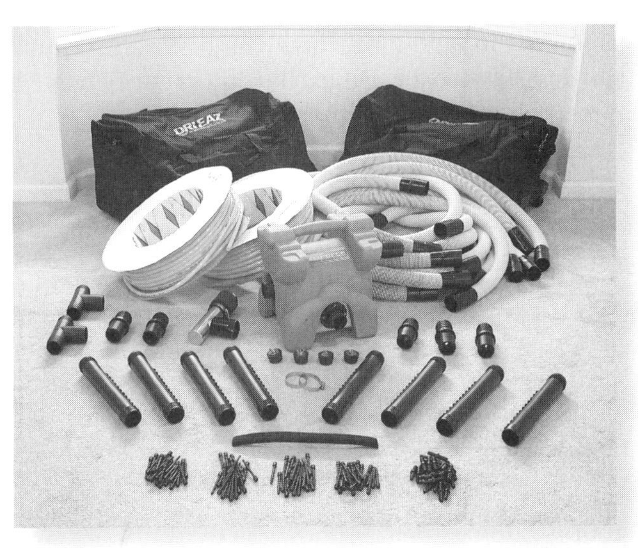

(Photo courtesy of Dri-Eaz Products)

Figure 3-1 This sophisticated DriForce® wall drying system is an example of the specialized restoration equipment now on the market. This equipment is specifically designed for restorative drying, and represents an investment in both money and training for the restoration contractor.

The downside of water restoration work is the need for specialized drying equipment, like the equipment shown in Figure 3-1. This equipment represents the largest investment you'll be making in your new company. In addition to the financial commitment, water restoration equipment of this type requires some specialized training to use it correctly and effectively. It also needs ongoing maintenance, and a safe and secure storage area.

But that's really not too much of a downside. As you'll learn later in the chapter, the equipment pays for itself fairly quickly — and it helps set you apart as a true professional in this field. So, thankfully, the good *far* outweighs the bad!

What Is Restorative Drying?

The process of getting a house dry enough to safely repair is known in the industry as *restorative drying*. You may be under the impression that drying a building is little more than mopping up the water and turning up the heat until everything dries. But it's far more involved than that. The field of restorative drying has evolved over the last couple of decades. Doing it right is now a combination of both art and science.

In the comprehensive book *Guide to Restorative Drying, Revised Edition*, from the Dri-Eaz Education Series, you'll find the following definition:

> *"Restorative drying is a process used to recover from many forms of water intrusion by assessment, documentation, controlled drying and selective replacement."*

That's restorative drying in a nutshell: You need to first assess and document what happened; then get the structure dry; and last, repair those areas that need it.

We'll go over the assessment, documentation, and repair in Chapter 4. In this chapter, we'll take a look at some of the different types of drying equipment on the market. We'll examine how the equipment operates, and why you use different pieces of equipment in different circumstances. We'll also look at how the different equipment works together to get the house dry. Understanding the equipment, how it works and what it does, will make the next chapter a lot easier.

IICRC Certification Classes

First, let's talk about training classes. As we discussed earlier, it's very important for you to have specific training in some of the different areas of restoration. For water restoration, you'll initially want to take the IICRC's three-day class: *Water Damage Restoration Technician (WRT)*. This will give you a solid foundation in restorative drying, from the science of moisture movement to how to set up and operate the drying equipment.

Here's the IICRC's official course description:

> *"WRT: (three-day course) The Water Damage Restoration Technician course is designed to teach restoration personnel that perform remediation work to give them a better concept of water damage, its effects, and techniques for drying of structures. This course will give residential and*

> *commercial maintenance personnel the background to understand the procedures necessary to deal with water losses, sewer backflows, and contamination such as mold."*

Another very good IICRC water damage remediation course to consider is Applied Structural Drying (ASD):

> *"ASD: (three-day course) The Applied Structural Drying course covers the effective, efficient and timely drying of water-damaged structures and contents, using comprehensive classroom and hands-on training, in order to facilitate appropriate decision-making within a restorative drying environment. Students will experience live hands-on use of instruments, extraction systems, drying equipment, and chemistry use in an actual flooded building situation."*

For a complete and comprehensive set of classes on water restoration, you can also take a combined WRT/ASD course. This combo class is five days long, and allows you take the IICRC exams for both certifications at the same time.

Why Use Specialized Equipment?

The drying equipment that you need to successfully start and run a restoration company will cost you several thousand dollars. A small restoration company can easily have tens of thousands of dollars invested in equipment. You have to decide how much to invest in right away, and how quickly you build up your inventory after that.

Initially, a very basic investment in water testing gear and drying equipment will set you back some $12,000 to $15,000. A more-complete setup, which will allow you to handle several jobs at once, as well as losses such as hardwood floors, would be in the $24,000 to $27,000 range. A medium-sized company will typically have a $50,000 to $75,000 investment in drying and other specialized equipment, and larger companies may carry an equipment inventory in the six-figure range.

Controlled Drying

Why do you need all this equipment? The answer is in the Dri-Eaz restorative drying definition: It's needed for *controlled drying*. Pay close attention to those words, because they form the heart of what proper drying is all about.

Water covers about 70 percent of the Earth, and it's essential to our health and the health of our surroundings. Water is also one of the primary ways that pathogens are transmitted. Pathogen is a word that you'll hear a lot in restoration, so here's a simple definition:

> *A pathogen is any disease-producing agent, especially a virus, bacterium, or other microorganism, that can cause disease in humans, animals and plants.*

So, while water is essential to our environment, sometimes what's in it can prove unhealthy or even deadly. That's especially true in closed environments, like our homes. When a home gets wet, it's essential that it gets dried quickly and thoroughly, before any detrimental health effects can occur.

However, you can't dry everything down to zero percent moisture, because most of the materials in a home should contain some moisture at all times. If you managed to take *all* the moisture out, you'd do substantial damage to wood, paper, fabrics, and just about everything in the home.

And too-rapidly lowering the moisture content of many materials, from framing lumber to wood furniture to the pictures on the walls, can cause warping, splitting, and other damage. This damage is called *secondary damage*, since it occurs after the initial moisture damage. Secondary damage can cause almost as many problems as the initial water damage.

That's why *controlled* drying is so important. You remove excess moisture in a way that you can monitor and have control over, so that you don't do any additional damage to the home and its contents. And the only way to do that effectively is to use the proper drying equipment.

Your Equipment Investment

To do restoration work, you need to know the temperature and humidity levels inside the house. You also need to be able to accurately locate moisture, and determine how much is there. And then, you need to be able to successfully remove it. That takes the right equipment.

In spite of its expense, the equipment is a huge part of what makes insurance restoration profitable. You're essentially renting your equipment to the insurance company every time you set it up on a job. So before you panic too much about your initial outlay of money, remember that the equipment will quickly pay for itself.

Sample Payback Schedule

Let's look at how long it will take you to recoup your investment in a piece of equipment — an air mover, for example. There's a wide range of prices for air movers, but the average cost for an upper-end model is $300 to $350. Typical rates for having the air mover on a job range from $25 to $35 per day, and an air mover will be on an average water loss job three to four days. Let's see how payback breaks down for a professional-grade air mover:

Average initial air mover investment	$350
Average daily rental rate	$ 30
Number of days needed to recover initial investment	11.66
Average days an air mover is needed per water loss	3
Number of jobs needed to recover initial investment	3.88

You can find the current daily rental rate for a piece of equipment in most unit-cost estimating programs. By knowing the daily rate, you can quickly figure how long it will take to pay for your investment in any given piece of equipment. You can see that it would only take about four jobs to recoup your investment in the air mover in our example.

Protecting Your Investment

Once you have your equipment, you need to protect your investment. Your employees should all understand exactly how to operate and maintain every piece of company equipment. They also have to be taught how to properly move and store the equipment. Don't tolerate any abuse or lackadaisical treatment of this gear. It's far too expensive and important to your company's success.

Here are recommendations for the treatment of your water restoration equipment:

1. ***Training:*** Make sure that any employee who'll be handling equipment is properly trained. That includes training on how to set it up, how to operate it, and how to shut it down correctly. It also includes how to properly lift it, move it, transport it, and store it. You should conduct regular training classes for your entire staff. Also, make sure that new employees are trained as soon as they're hired.

2. ***Maintenance:*** Designate one or more people for equipment maintenance. That includes inspecting its condition and making minor repairs. Malfunctioning equipment should be taken out of service immediately. If necessary, have it repaired by a qualified technician. All equipment should be cleaned as soon as it comes back from a job.

   If your equipment has been in a mold remediation or contaminated water environment, it *must* be cleaned before it goes out to another job. Cleaning and sanitizing your equipment is a separate charge that's billed as part of the mold or contaminated water loss.

3. ***Accountability:*** Make it very clear to all your employees what will happen if the equipment is damaged through abuse or negligence. You might want to consider a two-warning system. The first time they damage the equipment, they'll receive a written warning. The second time it happens they'll receive a stronger warning, and be held liable for repair costs (if your state allows that). The third time will result in termination. If any damage is determined to have been caused intentionally, that employee is subject to immediate termination. Be sure that the terms of your warning and termination policies are listed in your employee handbook.

4. ***Insurance:*** It's not a very comfortable feeling to walk off and leave thousands of dollars of equipment in someone's home. However, it's something you're going to have to do on a daily basis. So you want to check with your insurance agent to be sure that all of your equipment is covered.

   You should have complete coverage for your equipment while it's stored at your shop, while it's in transit, and while it's in use on the job. The coverage should include theft, vandalism, fire and water damage, and damage that may occur in an automobile accident while in transit. It should also include liability coverage in case your equipment accidentally injures a person or property while it's in use. If your regular contractor's policy doesn't cover all this, then talk with your agent about purchasing a rider to your policy that extends the coverage.

The Basic Drying Process

When you take the water restoration classes, you'll be introduced to the science of psychrometrics. Psychrometrics deals with the properties of temperature and moisture in the air, both of which are important to understanding how moisture moves and how you can get rid of it.

The topic of psychrometrics can get quite involved, and is something that's best left for your classroom time. But a basic overview of the subject will help you understand the overall drying process better, as well as how the various pieces of your equipment work together.

Humidity, Air Flow, and Temperature

When you deal with a wet structure, you're dealing with three important elements:

- **Humidity**
- **Air flow**
- **Temperature**

In the restoration world, these are commonly known by an easy-to-remember acronym: HAT.

Humidity

Humidity is the amount of moisture that's in the air. The *relative humidity* is a percentage of humidity based on what the air is *capable* of holding. For example, when you hear that the relative humidity is 35 percent, that means that the air is holding 35 percent of the moisture that it's capable of holding. If you live in an area with 35 percent relative humidity, you may feel the effects of dry skin, or you may see the wood in antique furniture shrinking and even splitting due to the dryness of the air.

On the other hand, if the relative humidity is 80 percent, then the air contains 80 percent of the moisture that it's capable of holding. If you're living in an area with 80 percent humidity, you no doubt feel the effects of all that moisture. The air seems heavy, and your wood furniture may swell from the moisture-laden air. You can measure the humidity in the air with a digital hygrometer, like the one in Figure 3-2.

Figure 3-2 The display on this digital hygrometer is showing slightly in excess of 100 percent relative humidity. That means that the air in the severely flooded house where we took this reading is completely saturated, and not capable of holding any more moisture.

Temperature

When the temperature increases, the molecules in the air expand. As they expand, they become capable of absorbing more moisture vapor. Generally, warm air holds more moisture than cooler air. That's also why you commonly see humidity levels lower in the winter than they are in the summer.

Air Flow

When air moves over a surface, it carries moisture away with it. This is known as evaporation. Try this simple test: Put a little bit of water on the back of your hand. Now blow across the water. The movement of the air will evaporate the moisture off the back of your hand. You can actually feel the evaporative effect, since the movement of the air is also drawing off body heat as it evaporates the water.

The Three States of Water

As you know, water exists in three different states. You're probably most familiar with the liquid state, plain water. But water can also be a gas, water vapor, and in this form it's what we commonly think of as *moisture in the air.* Steam is another example of water in a gaseous state. And ice is water in a frozen or solid state.

Here's an easy way to understand all three states of water, and how they interact. First, visualize being in a kitchen on a very cold winter day. Imagine that the kitchen has a single-pane, aluminum-frame window. There's a pot of water boiling on the stove.

That pot of boiling water is releasing steam into the air. That steam is water vapor. The warm water vapor is naturally drawn toward the cold window. When the water vapor meets the cold window, it condenses against the glass. As it condenses, it turns into water droplets. The water droplets run down the glass and encounter the aluminum frame, which is even colder than the glass. There, the water freezes into its solid state, ice.

Five Basic Steps in Drying a Structure

There are five basic steps in drying a structure after a water-related loss:

1. ***Extraction of liquid water:*** First, remove the excess water. That's done in two different ways: One way is to remove the saturated materials, such as tearing out wet carpet padding, insulation, and other highly porous materials. The other way is to use extraction equipment to physically suck up as much liquid as possible.

2. ***Evaporation:*** Here's where air flow comes into play. Using high-velocity air movers, you introduce a high-speed flow of air over the wet surfaces. The moving air evaporates the moisture, turning it to a water vapor which is then released into the air.

3. ***Removal of water vapor:*** Now that the liquid water has become a vapor, you can remove it from the air using one or more dehumidifiers designed specifically for this purpose.

4. ***Temperature control:*** Remember that warm air holds more moisture than cool air. So warming up the house will aid in the air's ability to absorb the water vapor. That makes it easier for the dehumidifiers to do their job. On the other hand, raising the temperature too much or too quickly can lead to secondary damage. So, controlling the indoor temperature is an important part of the drying process. In most situations, the temperature is maintained at a minimum level of about 70 degrees, up to a maximum of around 85 degrees.

5. ***Ventilation control:*** If the relative humidity outside the house is low enough and the air temperature is warm enough, you can aid the drying process through ventilation control. This is done by opening windows and turning on the existing ventilation fans inside the house. Existing ventilation fans include bath fans and kitchen range hoods. The ventilation fans pull moist air out of the house while drawing dryer air in, which can be beneficial in the right circumstances. This drying process, known as an *open drying play*, requires careful monitoring of the conditions outside the house.

Tip! *When you attend training classes and other events, build relationships with other restoration contractors that are outside your area. Not only are they a great source of information and support, but some of them may be in the market for new equipment at the same time you are. You can go in together to purchase equipment and take advantage of the quantity discount. That way, you can both save more money than either of you would be able to if you bought on your own.*

Equipment Prices and Purchasing Options

In the following sections, you'll see an overview of the various types of equipment currently on the market. For reference purposes, an average price range for each type of equipment is also included. These are the full retail prices as of the time of this writing.

To save money, you may be able to find used or reconditioned equipment, or equipment being sold at closeout if it's being discontinued or replaced by a newer model. Some manufacturers and distributors also have periodic sales on equipment that's cosmetically damaged. It may be dented, dinged, or scratched, but in otherwise perfect working order.

Another money-saving strategy is to buy in quantity, if you can. You may be able to negotiate a discount if you're buying several pieces of equipment at one time. Always

> ***Tip!*** *Don't be afraid to work something out with your competition. There'll be times when your competitors' business will be slow, and they may have equipment available to rent to you. On the flip side, there may be times when they get busy and want to rent some things from you. If you have a local competitor that you're on friendly terms with, see if you can't work out a mutually-beneficial arrangement.*

work closely with your sales representatives. Don't be afraid to ask what sales, discounts, or reduced-rate deals they may be able to offer you.

Some of the more expensive equipment may be available on a lease option as well. If you decide to lease, be sure you understand the exact terms of the lease, and what your rights and obligations are. Also, see if any portion of your lease payments can be applied against the price if you decide to exercise your option and purchase the equipment later.

And finally, don't overlook the possibility of renting some of your equipment, especially when you're first starting out. The drawback is that it eliminates the profit you could have made by owning your equipment and renting it to the insurance company. However, if it means the difference between being able to respond to a job or having to turn it down, renting makes sense. Renting allows you to get the equipment you need when you need it, so you can take on more, or larger, jobs than you would have been able to do without it.

Renting is also a good idea in times of large natural disasters. You may have severe weather come through your area, leaving you with dozens of wet houses to dry out. That can quickly stretch your equipment thin. Rather than investing in additional equipment that you may not use again for awhile, rent it.

You can rent specialized drying equipment from large rental yards, from some suppliers, and sometimes directly from the manufacturers. In large cities you may have ready access to what you need, while in smaller areas the equipment may be more difficult to find.

It's a good idea to be ready for those times when you need extra equipment. Check with local rental yards as well as with your suppliers, so you know what's available and how to get hold of it. Don't wait until you need something to track it down. By then it's usually too late!

Moisture Detection Equipment

In the next chapter we'll discuss some of the obvious ways you can determine where water has gone after an intrusion. It's very difficult to know exactly how extensive the water movement is, how much water there is, and what hidden cavities it may have moved into. You need good moisture detection equipment to track water movement — and that's what we'll look at now.

Today's moisture detection equipment is quite sophisticated, but fortunately it's also easy to use and very reliable. There are several different pieces of equipment available; what you invest in depends on your budget and your specific needs.

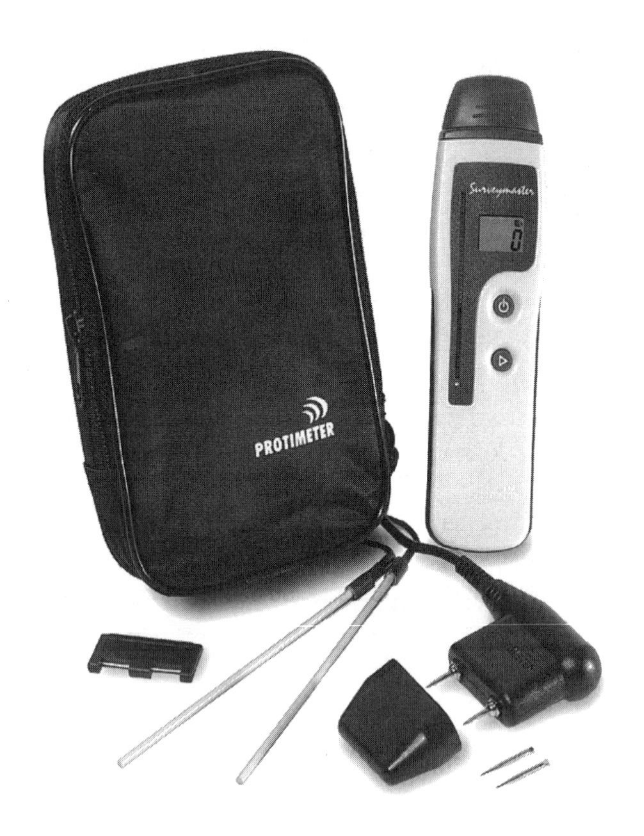

(Photo courtesy of Dri-Eaz Products)

Figure 3-3 This type of penetrating moisture meter has two small pins under the protective cap on the end of the meter. There are also two small pins on a flexible cable, two long probes on a separate cable, and a carrying case.

This equipment is both sensitive and expensive. Treat it carefully, and *always* keep it in its case when not in use. If your employees have access to it, be sure they understand how to use and store it; and make them aware there are consequences if it's mishandled!

Penetrating Moisture Meters ($450 to $600)

A penetrating moisture meter, Figure 3-3, has two small, sharp pins on one end that can be pushed into the material you want to test. The meter works on the principle of electrical conductance. If moisture is present in the material, it will complete an electrical circuit between the two pins. The amount of moisture that's detected is then displayed.

To use the penetrating meter, simply turn it on and press the pins into the surface you wish to test. The relative amount of moisture present appears on the meter's scale. Some types also have interchangeable long and short probes. You can use the long probes to test inside wall cavities and other areas that are too deep to reach with the short probes. This is especially useful for testing wall insulation or inside a floor cavity to see if there's moisture present.

Some types of meters use an indicator needle and a scale to measure the moisture. Newer models have a digital readout screen to show moisture levels. Some penetrating meters also use a sliding, color-coded scale for quick reference, going from green (dry) through red (excessive moisture detected). There may also be an audible alarm that sounds when moisture is detected.

Because penetrating moisture meters directly read the presence of moisture across their pins, they're considered very accurate. Their downside is that they leave two tiny holes from the probe pins in the material each time it's tested.

Some of the newer models of penetrating meters offer both penetrating and non-penetrating capabilities in a single meter. We'll look at those shortly.

Hammer-Probe Meter ($200 to $250)

Hammer-probe meters are similar to penetrating meters, utilizing two probes that indicate the presence of moisture when they complete a circuit. The difference is that

(Photo courtesy of Dri-Eaz Products)

Figure 3-4 This flooring probe, for carpet and other types of flooring, has a pistol grip that's easy to hold. The long handle allows you to quickly check large areas of flooring. It has a light and an audible alarm to indicate moisture. The pins are retractable for storage.

the pins are larger and longer, and there's a sliding weight that's used to hammer the pins into the surface being tested.

To use the hammer probe, position the pins against the material to be tested. Using the sliding weight, hammer the probes into the material until they reach the desired depth. Then plug the cable from the probe into your moisture meter and read the results.

The advantage to this type of meter is that you can drive the pins deep into the material, which lets you test the depth of the moisture penetration. This is especially helpful when checking saturation levels of subfloors, framing lumber, and other thick materials. The obvious disadvantage is that it leaves two holes that are even larger than the ones left by the penetrating meter.

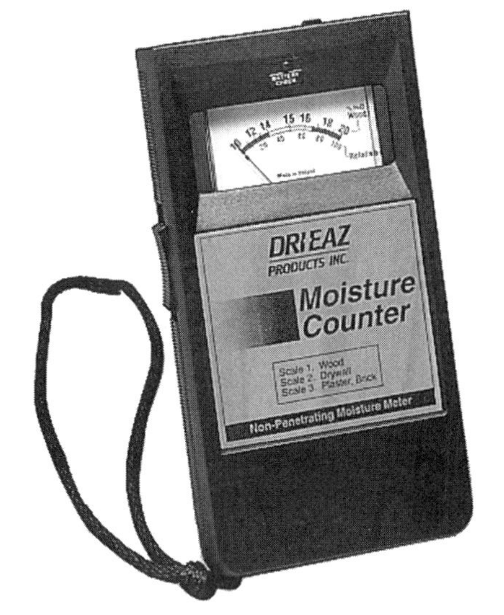

(Photo courtesy of Dri-Eaz Products)

Figure 3-5 This non-penetrating moisture meter uses the electrical circuit between two rubber pads on the back to indicate the presence of moisture. The moisture content is measured on the scale on the front of the meter. A switch on the top adjusts the meter for different densities of material, such as wood, drywall or concrete.

Flooring Probe Meter ($150 to $200)

The flooring probe meter is also similar to a penetrating meter, but longer. It has two probe pins at the end of a long handle (see Figure 3-4), and can provide quick evaluations of large areas of carpet and other flooring.

To use the flooring probe, simply extend the retractable pins and press them against the floor. If moisture is found, you'll receive a warning tone, an alarm light, or both.

You can use a flooring probe while walking across the floor, saving you from the bending and crawling you'd have to do with a conventional penetrating meter. While not as accurate as a regular penetrating meter, they're great for rapidly checking the migration of the water across large expanses of floor area.

Non-Penetrating Meter ($350 to $400)

Also called noninvasive or pinless meters, non-penetrating meters measure moisture levels without creating holes. Some types of non-penetrating meters have pads on the back that act similar to the pins on penetrating meters. See Figure 3-5. When moisture is present, an electrical circuit is completed and the meter indicates the relative amount of moisture detected. Newer types of non-penetrating meters, like the one in Figure 3-6, utilize radio frequencies to find the moisture.

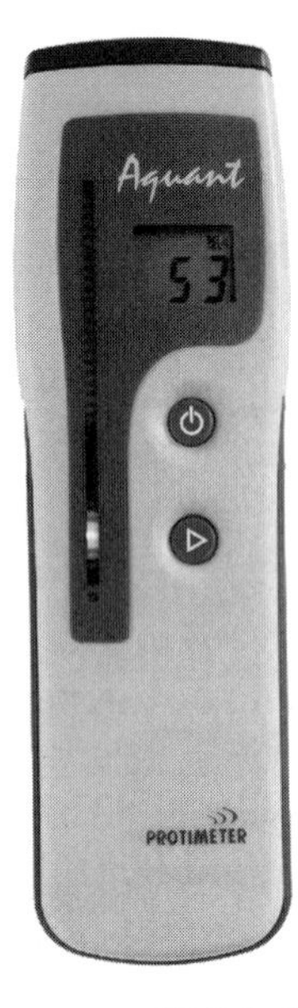

(Photo courtesy of Dri-Eaz Products)

Figure 3-6 This type of non-penetrating moisture meter uses radio frequencies to identify moisture and displays the result on a digital readout.

Using a non-penetrating meter is very simple. Just turn the meter on, put it in contact with the surface you want to meter, and read the results. Depending on the type of meter, it may have a needle that moves to indicate the relative amount of moisture detected, or it may have different-colored lights. Newer types also have a digital readout display. Most types also have an audible alarm in addition to the visual meter.

The advantage to this type of meter is obvious — no holes to worry about. It's also faster to use, especially for initial evaluations of large areas, such as floors. On the down side, they're not as accurate as penetrating meters. Some of the older types can also be fooled by surface condensation or by some types of conductive surfaces, such as metallic wallcoverings. Also, be careful when reading around the edges of openings, as the meter will sometimes detect the metal drywall corners just below the surface.

Thermo-Hygrometer ($50 to $500)

A thermo-hygrometer, like the one in Figure 3-7, reads both temperature and relative humidity. It allows you to quickly determine the interior and exterior temperatures, as well as the interior and exterior humidity levels. These measurements are critical to determining your drying strategy, as well as for monitoring the effectiveness and progress of your drying efforts. They're also essential when trying to determine whether or not you can use outside air in the drying process.

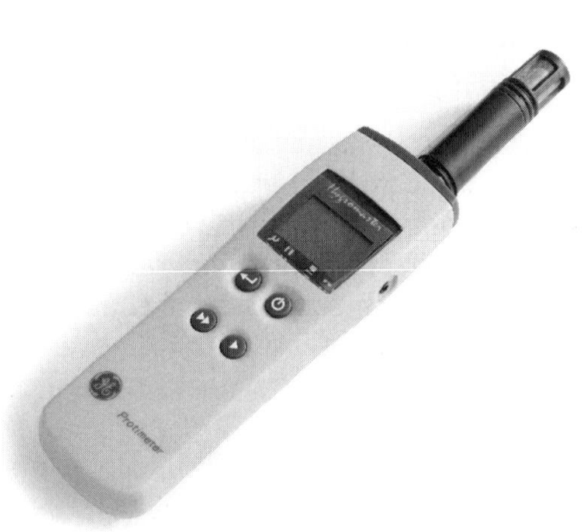

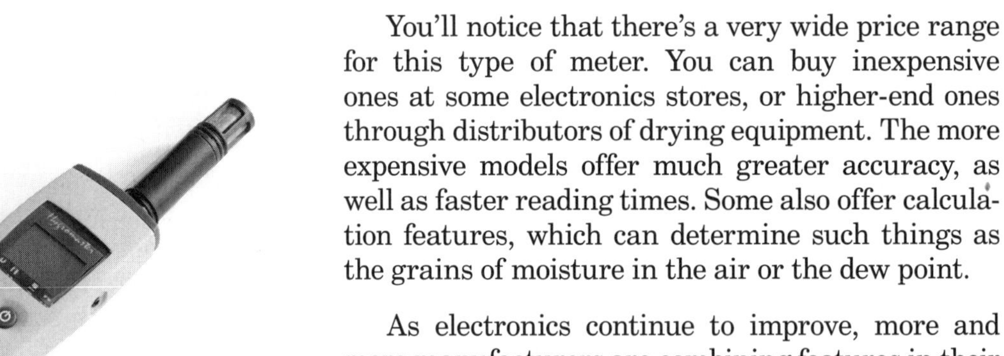

(Photo courtesy of Dri-Eaz Products)

Figure 3-7 A digital thermo-hygrometer can be used for readings both inside and outside the building. The sensor probe at the top of the instrument reads the temperature of the air and the amount of moisture it contains.

You'll notice that there's a very wide price range for this type of meter. You can buy inexpensive ones at some electronics stores, or higher-end ones through distributors of drying equipment. The more expensive models offer much greater accuracy, as well as faster reading times. Some also offer calculation features, which can determine such things as the grains of moisture in the air or the dew point.

As electronics continue to improve, more and more manufacturers are combining features in their meters. It's now possible to buy one meter that combines thermo-hygrometer functions with penetrating and non-penetrating moisture detection.

Combo Kits ($1,100 to $1,900)

You can save yourself some money and add some convenience by buying meters in a combo kit, like the one shown in Figure 3-8. Different kits offer different combinations. You may find a penetrating

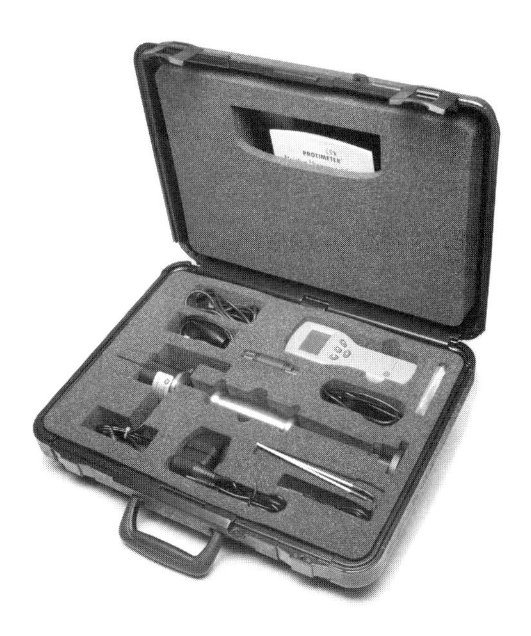

(Photo courtesy of Dri-Eaz Products)

Figure 3-8 A combination kit of restoration instruments. This particular kit contains a meter that does both penetrating and non-penetrating readings (top right) and a hammer probe (center). There are also long probes for checking inside wall cavities (lower right), and cables and other accessories. One advantage to a kit like this is the fitted, padded carrying case.

meter combined with a hammer probe, or perhaps a moisture meter combined with a thermo-hygrometer. Combo kits offer a good value, and usually include a fitted case that keeps everything together and well protected. Select a kit that matches both your budget and your needs.

Water Extraction Equipment ($2,100 to $2,900)

One of the very first steps in drying a home is to remove as much of the standing water as possible. To do that you'll need a water extractor or water extraction vacuum. You may hear them referred to by various trade names, such as *Flood Pumper* or *Flood King*.

The typical water extractor consists of a metal or, more commonly, a polyethylene tank, with wheels and a handle, like the *Flood Pumper* shown in Figure 3-9. Inside the tank is a powerful vacuum suction motor, which is used to draw water into the machine.

You'll find that the vacuum motors are commonly rated in *inches of water lift*. Imagine putting a tube into a pool of water, then turning on the vacuum and measuring how high it can pull the water up the tube. That's water lift. There's a lot of science and even a little controversy behind those calculations, but basically the higher the water lift number, the more powerful the extractor. That makes it pretty easy to compare one extractor to another. Commercial water extractors used for restoration projects should have a suction capacity of 100 to 160 inches of water lift. To put that into perspective, the average high-end wet and dry shop vacuum has a water-lift rating between 50 and 70 inches, or about half that of a commercial water extractor.

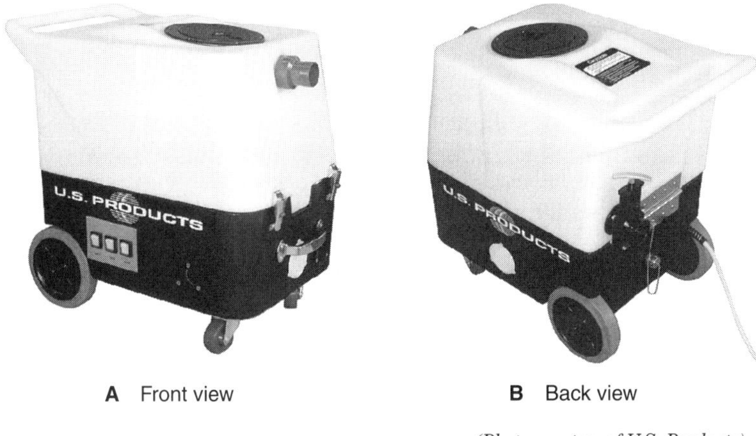

A Front view **B** Back view

(Photo courtesy of U.S. Products)

Figure 3-9 The Flood Pumper, a commercial-grade water extractor made specifically for the restoration industry. This unit features dual pumps, a lightweight polyethylene tank, and easy maneuverability. The tank discharge valve and the connection for the pump-out hose (bottom, under cap) are located in the rear (B).

Inside the water extractor tank is a discharge pump. Discharge pumps usually average around $^1/_3$ horsepower, with a flow rate of about 30 to 40 gallons per minute (GPM). The tank also contains a float mechanism and a filter.

To set up the extractor, attach the flexible, crush-proof hose to the inlet connection point on the outside of the tank. Most intake hoses are 1½ inches in diameter and 25 feet long. A steel or stainless steel wand attaches to the other end of the hose. The wand is curved and about 4 to 5 feet long, making it easy to use from a standing position. The end of the wand has a 12- to 14-inch-wide head. Two different heads are commonly available, one for use on carpeting, and one for use on hard floors.

Tip! *Be careful where you put the end of the discharge hose. The water is pumped out of the hose with quite a bit of force, and can uproot landscaping, damage a lawn, or even cut a furrow in a gravel driveway. Also, be sure you locate it so the water coming out can't make its way back into the house.*

Next, attach the discharge hose to the discharge connection point on the tank. This is usually a crush-proof hose, 1½ inches in diameter. Some manufacturers use a collapsible flat hose instead, similar to a lighter-duty fire hose. Others utilize a ¾-inch garden hose. For convenience, most discharge hoses are 50 feet, either in one piece or by combining two 25-foot hoses. Route the discharge hose to a safe location outside the house.

To use the extractor, turn on the vacuum pump and run the head of the wand over the floor and suck up the water. The water is drawn into the tank. Depending on the model of the extractor, the tank will have a holding capacity of 10 to 20 gallons. When

REAL STORIES:

Watch That Connection!

Our crew had responded to a water loss call one afternoon, at a local home with a nicely-finished basement. The basement's three rooms were being used as a home office setup, and the owner had just discovered that a pipe in the ceiling over one of the offices had frozen and burst. Water was everywhere!

Our first task was to move furniture, and then to get the standing water sucked up and out of there as fast as possible so we could start the drying process. For that task, we brought our water extraction vacuum downstairs.

We had a new employee operating the extractor that day. He set the machine up in the room closest to the stairs, and ran the discharge hose up to a good location outside. Then he took the vacuum hose over to the wettest room, turned everything on, and started vacuuming water out of the carpets.

Everything just the way he'd been taught. Well, almost everything.

Suddenly, he heard one of the other crew members yelling at him. He ran into the other room, just in time to see someone diving for the OFF switch on the extractor. When our conscientious new employee set everything up, he forgot to make sure that the discharge hose was securely connected to the extractor! It wasn't. So when the pump kicked in to pump the water out of the tank, the pressure blew the loose hose right off the machine. All of the water in the tank came gushing out, reflooding the room he'd just finished sucking the water out of!

Our red-faced employee set about extracting all that water for a second time. But not before double-, then triple-checking, the hose connection!

you draw in more than the tank's holding capacity, the internal float mechanism activates the discharge pump, pumping the water out of the tank.

As you can imagine, having both a suction motor and a discharge pump in the same machine draws a lot of power — usually in excess of 20 amps. To make it easier to use the extractor without blowing a circuit, most manufacturers set it up with two separate cords and two separate switches. The cords are long enough to run them to two different locations and plug into two different circuits.

Tip! *20 gallons of water weighs over 165 pounds. Keep that in mind if you decide not to set up and use the discharge hose. You'll need to plan in advance how to get a full extractor out of the house to empty it.*

There are several manufacturers that make very good water extraction equipment for use in restoration. They all work pretty much the same, and are relatively close to one another in price. So, what you want to consider is the lift capacity of the vacuum motor, and the size of the holding tanks. You also want to know what accessories they come with, how easy they are to set up and use, and how easy they are to transport. These machines are about 80 to 90 pounds empty, and you want your crews to be able to get them onto and off of your truck and into the wet house easily.

Weighted Extraction Heads ($300 to $3,500)

One accessory available for water extractors is a weighted extraction head. The added weight of the extractor head forces more water out of highly absorbent materials, such as carpet and carpet pad. It's like stepping on a wet sponge — the weight of your foot forces water out of the sponge.

At the lower end of the price range are simple weighted heads used on the end of the wand. They typically have wheels on them, and allow you to use your own body weight for extra extraction.

At the higher end of the pricing scale is a water extractor that you can stand on and ride, shown in Figure 3-10. The added weight of the person on the machine forces more water out of the carpet and pad, making extraction faster and much less tiring. A pickup head is built into the machine, which attaches to the hose leading to the extraction equipment. These machines are especially useful for large areas where you want to extract the water without having to remove the carpet and pad.

(Photo courtesy of Dri-Eaz Products)

Figure 3-10 The advantage to an extractor that you can stand on and ride is that it's faster and less fatiguing for the operator. The extractor is controlled by the joystick located between the handles, and folds for storage.

Hardwood Floor Extractors ($550)

At least one company makes a very nice system for extracting water trapped inside hardwood

(Photo courtesy of Dri-Eaz Products)

Figure 3-11 The Dri-Eaz Rescue Mat® system has mats that seal down against a hardwood floor. The mats are connected by hoses to the extractor (left center), which draws water out of the flooring.

flooring. The Dri-Eaz Rescue Mat® system, shown in Figure 3-11, uses clear PVC mats with hose connectors on the top that fit standard size extraction hoses. The mats range in size from 23 by 30 inches to 30 by 46 inches. Multiple mats can be connected together to cover larger areas.

To use this system, place the mats on the wet hardwood floor. A gasket around the edge of each mat forms a seal against the floor. Attach the hose from a water extractor to each mat. Using the extractor, the water is drawn up through the flooring into the mat, and then out to the extractor. When as much water as possible has been drawn out, remove the hose from the extractor and attach it to a high-speed air mover with a hose adaptor. The air stream from the air mover creates a negative pressure, which will draw even more moisture from the flooring.

Figure 3-12 A truck-mounted carpet cleaning machine can also be used for water extraction. A connection to the engine of the van drives the carpet cleaner.

Extraction with Truck-Mounted Carpet Cleaning Equipment ($25,000 and up)

Some restoration contractors also have truck-mounted carpet cleaning equipment (Figure 3-12). Those that do can use the extraction side of the machine to extract water. The advantage is that you don't have to buy an additional machine just for water extrac-

(Photo courtesy of Dri-Eaz Products)

Figure 3-13 A commercial-grade air mover. The motor and squirrel cage fan are located behind the protective screen. The air exits through the wide snout, close to the floor.

tion. On the downside, it means having to take your very expensive carpet cleaning truck off to a jobsite whenever you need to extract water. Also, in a dirty environment it can mean drawing dirt and debris up into your carpet cleaner, which can damage the machine and the filters.

Air Movers ($150 to $550)

Once you've removed the standing water, the next step in the drying process is to evaporate the remaining moisture. This is done with an air mover, like the one in Figure 3-13.

As the name implies, an air mover moves air. An electric motor powers a set of fan blades that create a high-speed stream of air. There are different sizes and designs of air movers, with motors ranging from about $^{1}/_{3}$ to 1 horsepower. Depending on the design and the motor size, an air mover can generate an air stream of approximately 2,000 to 3,500 cubic feet per minute (CFM). Some types have only one speed, but most have a two- or three-speed selector switch that allows you to regulate the amount of air generated to match specific drying situations.

The air movers commonly used for restoration work have a horizontally-mounted squirrel cage fan that pushes air through a wide snout located low on the front of the unit. This wide-snout design, sometimes called a carpet dryer, creates a broad, concentrated stream of air, right at floor level. That makes these units ideal for floor-drying applications. The low snout also fits under carpeting, which is commonly needed in drying situations.

These air movers usually have four stout rubber feet on the bottom to keep them stable, as well as a built-in carrying handle. Some types also have additional feet on the back as well. The extra feet allow you to tip the unit into different positions to direct the angle of the air stream. This is very useful for directing air at walls, stairs, and other areas up off the floor.

(Photo courtesy of Dri-Eaz Products)

Figure 3-14 An axial fan. These high-speed, in-line fans draw air in through one end and push it out through the other end, usually through a hose.

Another type of air mover is the axial or in-line fan, shown in Figure 3-14. This type of fan utilizes blades instead of a squirrel cage, and directs the air through a round opening that's perpendicular to the floor. Axial fans produce a less concentrated air stream,

(Photo courtesy of Dri-Eaz Products)

Figure 3-15 An air mover that's been fitted with a multi-hose splitter, also called a snout adapter. This converts the snout of the air mover so that it can accept, one, two, or three 4-inch diameter flexible hoses. Note the cap on the floor, to the right of the front of the air mover. It can be used to block off one of the 4-inch hose ports if desired. There's a similar cap on the other side.

but operate at speeds that can reach more than 5,200 CFM, so the air is dispersed over a wider area. They're typically used for ventilating large spaces, either for drying or for the removal of smoke or other odors. Some types have a collar on them so that you can attach a 10- or 12-inch diameter hose that allows you to direct the air where you need it.

There are also axial fans designed so that they can be placed with their fan blades parallel to the floor. This allows a highly concentrated stream of air to be blown directly down at the floor, which works well for drying very wet areas. Another type of axial fan has wheels that allow you to move it across hard floors, making it easier to dry large areas.

Air Mover Accessories ($50 to $450)

(Photo courtesy of Dri-Eaz Products)

Figure 3-16 This type of air-mover snout adapter allows the use of a single 10-inch diameter hose. This is particularly useful for pushing air into crawlspaces without having to subject the air mover to a dirty environment that could potentially harm the motor.

Air movers are among the most important and versatile pieces of equipment you'll own as a restoration contractor. To add to their versatility, there are different accessories available for the wide-snout air movers. Some of these accessories are fairly universal, and will fit most types of air movers. Others are proprietary, and fit only the air movers made by a specific manufacturer.

Most accessories are designed to better concentrate or direct the air flow. One particularly useful accessory is the multi-hose splitter, shown in Figure 3-15. This hard plastic adapter fits into the snout of the air mover, converting the single wide snout into three smaller air outlets. You can clamp a 4-inch diameter hose to each of the outlets, and direct the air from these smaller hoses where you need it. This arrangement is ideal for getting airflow into cabinets and cabinet toe kicks, behind toilets and other obstacles, into ceiling and wall cavities, and into any other confined or enclosed space that needs drying.

A variation of the multi-hose adapter is the large diameter, single-hose adapter shown in Figure 3-16. This 10-inch diameter hose attaches over the outside of the air mover's snout with a flexible band. You can then direct the 15-foot hose into confined areas, such as crawlspaces, attics, or even into stair cavities.

(Photo courtesy of Dri-Eaz Products)

Figure 3-17 A 48-inch wide boot fitted into the end of an air mover. This boot is being used to push air into a wall cavity through holes that have been drilled in the drywall.

(Photo courtesy of Dri-Eaz Products)

Figure 3-18 A carpet clamp attached to the front of an air mover. These clamps are essential accessories for holding carpet in place while drying. Some types of air movers come with them already attached. If they don't, you can add them as an option.

Another useful accessory is the wide boot, shown in Figure 3-17. It's 48-inches wide, hard plastic, and fits into the snout. The wide boot has two important uses: You can use it to direct the air mover's air flow over a wide span, for drying areas such as floors. And, it's also ideal for drying walls. By simply removing the baseboards and drilling a series of holes low on the wall, you can push the boot against the wall and force air up into the wall cavities.

The carpet clamp (Figure 3-18) attaches to the front of the air mover, either with screws or by clamping it to the handle. It's designed to hold carpet in place while drying. Place the air mover with the snout under the carpet, then slide the carpet clamp bar up. Drape the edge of the carpet over the snout of the air mover, then lower the bar down and lock it in place. The clamp keeps the carpet from flapping around and getting damaged, and allows a layer of air to circulate under the carpet.

In addition to renting air movers to the insurance company, you can price out and rent the attachments, such as wide boots and hose splitters, separately. So, when you purchase accessories, you not only improve the speed and effectiveness of the drying, you're also able generate extra income.

(Photo courtesy of Ebac Industrial Products)

Figure 3-19 A small refrigerant dehumidifier with a metal case and folding handle. This is a good general-purpose dehumidifier for a number of smaller applications.

Dehumidifiers

Now that you've released the moisture vapor into the air, you need to trap it and get it out of the house. That's where the dehumidifiers come in, and they're an extremely important part of your drying equipment arsenal. See Figure 3-19.

There are two basic types of dehumidifiers — refrigerant and desiccant. Refrigerant dehumidifiers are more common and less expensive, and are used quite a bit more often than desiccants. However, there are some situations where a desiccant dehumidifier is the only effective option.

Refrigerant Dehumidifiers

The operation of a refrigerant dehumidifier is similar to that of an air conditioner or a refrigerator. There are four main components: a compressor, a condenser, an evaporator, and a fan. Commercial-grade dehumidifiers made for use in the restoration industry have a fifth component, a pump-out system.

The compressor, condenser, and evaporator are connected together with tubing in a closed loop. Inside the tubing there's a refrigerant. A refrigerant is a material that has the capability of being converted easily back and forth between a liquid and a gas.

In a refrigerant dehumidifier, the refrigerant enters the compressor as a cool, low-pressure gas. The compressor then compresses the gas, which packs the molecules tightly together. This in turn raises both the pressure and the temperature of the refrigerant. The refrigerant then leaves the compressor as a hot, high pressure gas.

The gas then moves into the condenser, also known as a *hot coil*. The condenser has a series of tubes with fins, very much like those on a car radiator. It's located near the fan, which draws a stream of air across the tubes. All of these finned tubes, together with the air flow from the fan, allow heat from the gas to dissipate. As the heat dissipates, the refrigerant changes from a gas to a liquid. This liquid is now much cooler, but still under high pressure.

In the final step of the process, the cool, high-pressure liquid enters the evaporator coil, also known as a *cold coil*. The evaporator causes the pressure in the liquid to drop, and the liquid then evaporates back into a cool gas. This cool, low-pressure gas moves back into the compressor and the cycle is repeated.

How does all the movement of the refrigerant take moisture out of the air? Visualize a glass of ice water on a hot summer day. The water inside the glass is colder than the surrounding air outside the glass. When the warm air outside the glass comes into contact with the cold surface of the glass, water vapor is drawn out of the air. That vapor condenses on the cold glass, and turns from water vapor — a gas — into water — a liquid. The liquid shows up as water droplets, which eventually run down the side of the glass and puddle on the table below.

The exact same principle is at work with a dehumidifier. The unit's internal fan is drawing moist room air into the dehumidifier. The moist air passes over the evaporator coils. The coils, as you saw above, have been cooled by the action of the refrigerant passing through them. As with the cold glass on the warm day, the moisture vapor that's in the air condenses on the cold evaporator coils. From there it drips off into a collection tray, just like the water on the glass dripping off onto the table.

On the other side of the dehumidifier, the air that's been drawn into the unit by the fan flows back out into the room. Along the way, it passes across the condenser coils. That air movement, as just described, pulls heat off the coils and out of the refrigerant. So the cool, wet air that was drawn in on one side of the dehumidifier comes out the other side as warm, dry air, which helps with the overall drying process in the room.

There are some variations in how the water is collected and how the coils are heated and cooled. But the underlying principle remains the same for all refrigerant dehumidifiers.

Inexpensive dehumidifiers for home use simply have a collection bucket for the moisture that the machine gathers. Periodically, the bucket has to be taken out of the machine and dumped. There are two obvious disadvantages to this type of a dehumidifier in a restoration environment. For one thing, you would have to keep coming back to the house every few hours to empty the bucket. And, since the bucket is open on top, it'll continually release a small amount of water vapor back into the air. That means that you'll be collecting and disposing some of the same water over and over again.

Commercial-grade restoration dehumidifiers have a discharge pump. Some types of discharge pumps run continuously at a low speed, while others are activated by a timer or a float. The pump pushes the collected water through a length of small-diameter tubing, allowing you to dispose of it easily. You can direct the tubing to the outside of the house, or, more commonly, into a sink drain.

Shopping for Refrigerant Dehumidifiers ($1,300 to $3,600)

There are several different manufacturers of commercial-grade dehumidifiers. There's also a fairly wide spread in their cost, with some advantages and disadvantages at each end of the price spectrum. When you're ready to go shopping, here are some of the things to consider:

- **Type of Refrigerant:** Dehumidifiers are available with both conventional refrigerant and low-grain refrigerant (LGR). Conventional — also called standard — refrigerant has long been the industry standard. It has a humidity operating range that goes down to around 45 percent, which is adequate for many drying situations. If you need a lower humidity range than that, LGR dehumidifiers will operate down to 30 percent. So, an LGR dehumidifier will give you more flexibility. However, they're also more expensive than standard refrigerant units.
- **Water Removal Capacity:** Dehumidifiers are rated for their capacity to remove water, indicated in pints per day. Manufacturers provide the tested capacity, as rated by the Association of Home Appliance Manufacturers (AHAM), and they may also give their own maximum extraction capacities for the unit. Since the AHAM testing is done in a standardized environment of 80 degrees F and 60 percent relative humidity, theirs is the best number to use when comparing dehumidifiers.

  Obviously, higher-capacity dehumidifiers remove more water, so you can get the job dried faster. Removal rates range from around

65 pints per day to over 175 pints per day with some of the very high capacity machines. As you would expect, higher capacity machines also cost more.

- **Air Speed:** A dehumidifier needs to have adequate air flow over the coils in order to work effectively. Dehumidifiers are also rated in CFM, which is the capacity of the internal fan. Common CFM ratings for a dehumidifier range from around 160 to over 500. Once again, the higher the rating, the more expensive the machine will be.

- **General Construction:** You'll also want to consider the overall construction of the unit. Dehumidifiers are manufactured with steel or rotomolded polyethylene cases. The case material doesn't really affect the performance or even the price of the machine, so your choice of case material is a matter of preference. Some contractors feel that the metal cases are a little sturdier, especially on the larger units. On the other hand, rotomolded polyethylene cases, like the one in Figure 3-20, are lighter, and both dent- and scratch-resistant. Rotomolded cases are also more rounded, which lessens the chance of damaging buildings or contents during setup and use.

  Another consideration is the overall weight and portability of the unit. Weights range from around 80 pounds to over 175 pounds, which makes a big difference when you're loading and transporting the units. Also, you should look for large non-marring wheels, good balance, and a retractable or folding handle, if possible.

  Make sure the controls are conveniently located and easy to operate. It's also important that the dehumidifier have an hour meter for tracking usage as well as maintenance schedules. Look at the type of filter it uses — and how easy it is to change.

Desiccant Dehumidifiers

Desiccant dehumidifiers work quite a bit differently from refrigerant models. Instead of using refrigerant and coils to remove moisture from the air, desiccant dehumidifiers use a silica adsorbent material that attracts and holds moisture. This is the same type of material that you find in the small bags packed with electronics, clothes, shoes, and other items.

Silica is what is referred to as an *adsorbent* material, as opposed to an *absorbent* one. Absorbent materials physically change shape as they pick up and release moisture. Adsorbent materials don't, so they can be used very effectively inside a machine, such as a dehumidifier.

> ***Tip!*** *If you're thinking about investing in high-capacity dehumidifiers, be sure that the insurance companies you work with allow you to charge a higher rental rate for them. Most companies have determined the allowable rates for different sizes of dehumidifiers. If you use a dehumidifier that's too large for the job, you'll cost the insurance company additional money, billing them for the larger equipment. On the other hand, if your dehumidifiers are too small, they'll need to be on the job longer, which again costs the insurance company more money. They'll expect you to use, and to bill them for, the dehumidifier size that's most appropriate for the job.*

(Photo courtesy of Dri-Eaz Products)

Figure 3-20 A low-grain refrigerant (LGR) dehumidifier in a lightweight, rounded, rotomolded case. This high-capacity dehumidifier can be used on larger water losses, as well as on jobs where a lower relative humidity is required. Note the filter on the front, which is very easy to access for cleaning.

In a desiccant dehumidifier, a fan draws moist air into the unit. As the air enters, it's split into two different air streams. About 75 percent of the air passes directly over a rotating disc that's filled with silica gel. This is called the *process air*. The silica removes the moisture out of the air stream.

The other 25 percent of the air stream is directed to the lower part of the silica rotor. This is part of a process called *reactivation*, so this is called *reactivation air*. In the reactivation process, a heating element removes the stored moisture from the silica gel. This moisture is then exhausted to the outside of the house through a duct. The reactivation air is the part of the air stream needed to help cool down the rotor disk as it turns, extract the moisture from the silica, and move the wet air to the outside.

Dri-Eaz explains the process like this:

> *"Once the desiccant absorbs this humidity, how does the desiccant release its moisture? Think of putting a wet, fully saturated bath towel in your clothes dryer. The combination of heat and airflow dries the towel out quickly, and moisture is exhausted through ducting to the outside of your house. If you were to put your hand over the exhaust on your desiccant dehumidifier, it would feel exactly like the damp exhaust from your clothes dryer. Without airflow, the towel would simply get hot. Without heat, the towel would dry — but very slowly."*

The overall result of the process is that the moist air that's been drawn into the dehumidifier has been dried by the silica and warmed as part of the reactivation process. This warm, dry air then exits the dehumidifier and reenters the room, further aiding with the overall drying process.

Since a desiccant dehumidifier doesn't dry through a condensation process like a refrigerant dehumidifier does, it's unaffected by low air temperatures or low humidity. That's an advantage in certain situations. A desiccant dehumidifier will work very well when a building is cold, or when the humidity is low. In fact, since it relies on the silica adsorbent material to do the drying, the cooler and dryer the incoming air is, the better the desiccant dehumidifier will perform.

The air that comes out of a desiccant dehumidifier is dryer than the air processed through a refrigerant dehumidifier. That makes it ideal for drying books, papers, wood floors, and other objects with low moisture content. A desiccant dehumidifier also works better for drying an attic, crawlspace, or unheated outbuilding.

(Photo courtesy of Dri-Eaz Products)

Figure 3-21 A small, portable desiccant dehumidifier. This unit has a capacity of around 90 CFM. Note the four indentations on the top of the housing; this unit is designed to be stacked with additional units to increase drying capacity as needed.

Shopping for Desiccant Dehumidifiers ($3,100 to $11,000)

With a desiccant dehumidifier, you're primarily paying for air flow. The more air the dehumidifier can take in and effectively process, the faster it will dry a building. Smaller, more portable machines, like the one in Figure 3-21, may have an air flow of only around 100 CFM. On the other hand, high capacity desiccants will process 1,000 or even 1,200 CFM.

Beyond that, you'll want to look for a machine that's easy to set up and operate. Also, with larger machines, you'll want to see how portable they are and how they're transported. Check on how much hose comes with the machine you buy, and what other accessories might be included. Finally, make sure you find out who can service and repair the machine when the time comes. As you can imagine, these are sophisticated pieces of equipment, and service centers may be few and far between.

Thermal Energy System (TES™)

The Thermal Energy System, commonly known as *TES*, is a unique piece of equipment that adds heat into the drying equation. Temperature has always been part of the HAT acronym for understanding the principles of drying (remember Humidity, Airflow, and Temperature). But for the most part it's been limited to controlling the building's temperature so it stays within a specific range.

The TES equipment was developed to speed up the drying out of a building by accelerating evaporation — the phase change of liquid water into water vapor. Faster evaporation of the water will result in faster drying times. The idea behind the TES system is what the manufacturer calls *Directed Heat Drying*™, which allows heat to be directed to the wet areas without overheating the rest of the building.

Propane-Powered TES Units ($21,950 to $25,790)

The original TES system uses a propane-fired boiler to bring a heat exchange fluid to a high temperature. Using a high-volume pump, the heated fluid is moved through insulated hoses to a specially-designed *Thermal Exchanger* (TEX) box.

On top of the TEX box is a slot that accepts any standard air mover. Air generated by the air mover blows across the hot fluid inside the TEX box, and the heat is transferred into the room.

The TEX box looks something like a larger version of a conventional air mover, complete with a snout on the front and even a carpet clamp, so it's used in much the same way as an air mover. Heated air can be directed under carpet, or under tented areas for drying hardwood floors or substrates. Additional accessories also allow the heated air to be directed into wall cavities. For rapid heating of large areas, the TEX boxes can be used without any accessories.

As the moisture vapor is released into the air, thermostatically-controlled air movers vent the moisture to the outside using lay-flat ducts that run through windows. Dehumidifiers and evaporative air-handling systems can also be used to remove the water vapor.

The lower price is for a 200,000 BTU unit, and includes the boiler unit, 600 feet of hose, and four TEX boxes. The higher price is for a 250,000 BTU unit, with 700 feet of hose and five TEX boxes. (The air movers that provide the air flow aren't included in either price.)

Electric-Powered E-TES and E-TES SD Units ($2,195 to $2,395, and $2,295 to $2,495)

A smaller, more portable electric version of the TES drying system is also available. Known as E-TES, it works on the same principle. The difference is that E-TES uses one self-contained TEX box with an electric heating element inside. Once again, a separate air mover (not included) is inserted into a slot in the TEX box to provide air movement to distribute the heat.

The E-TES is available in 120-volt and 240-volt versions. The 120-volt version has two cords, and needs to be plugged into two different 120-volt circuits. The 240-volt model plugs directly into a dryer or stove circuit.

The E-TES SD (Smart Dry) has a built-in control panel that assists with monitoring and documentation. It includes inputs for moisture sensing probes, air and surface temperature sensors, data loggers, and other accessories.

Some Heated-Drying Considerations

The use of high heat for drying has been gaining acceptance among restoration contractors in recent years, but it's not without some controversy. Here are a couple of things to keep in mind:

- Limiting the heat to the wet areas is typically better than heating the entire structure. It limits potential secondary damage, keeps the structure more comfortable, and saves energy.
- Warm temperatures will stimulate microbial growth, but that growth is greatly inhibited once the temperatures get over about 100 degrees. When using heat in a wet environment where mold is present or suspected, rely on the advice of your hygienist for the correct temperature settings.
- Increased levels of heat, even well-directed heat, still has the potential to damage house plants and certain contents. When

purchasing any type of heated drying equipment, be sure you receive proper instruction on temperature ranges and potential damage.

- Heat can be a factor in whether or not occupants want to remain in the house during the drying process. Discuss this with both the occupants and the adjuster.
- Xactimate and some other estimating books and software have unit costs for some heating equipment, such as TES. However, it may be prudent to verify the pricing with your local adjusters beforehand.

Where Can You Buy the Equipment?

Thanks to the Internet, it's now much easier to locate equipment manufacturers and to compare what they have to offer. Some manufacturers will sell direct to contractors, others only sell through their approved distributors.

Even if you can buy directly from a manufacturer, it's still a very good idea to establish a relationship with a large supplier. Suppliers can help you out in a number of different ways. For one thing, they're familiar with more than one brand of equipment, and they can help you find what best suits your specific needs. Also, your supplier will be your source for chemicals, tools, and other items, not just a specific air mover or dehumidifier.

The right supplier also becomes a friend to your business. Since the success of his business depends on the success of his customers, a good supplier will help you out with technical questions, rush orders, and perhaps even financing.

Many IICRC training classes are sponsored by restoration equipment suppliers. That will often provide you with an opportunity to meet some of their staff, tour their warehouse, and see the equipment they carry. Be sure and leave yourself enough time before or after the class to sit down with someone from their company. Introduce yourself, discuss your needs, and perhaps see about opening an account.

How Much Equipment Do You Need?

When you're first starting out, the question becomes one of how much equipment you need versus how much equipment you can afford.

There's no hard and fast rule to what you should buy, or when you should buy it. That depends on the types of jobs you'll be doing. So when making your purchasing decisions, you might want to refer to the descriptions of the different types of water losses that we'll be covering next, in Chapter 4. The different classes of water losses include recommendations for how much equipment you'll need to handle each one. Other factors to consider when purchasing equipment include the amount of storage space you have, what your personal preferences are, and, of course, your financial situation. This is where your relationship with both your supplier and your banker come in.

Here's a suggestion for getting started:

- **Moisture Detection Equipment:** Invest in a high-quality penetrating moisture meter and a high-quality non-penetrating moisture meter right away. Don't skimp on these meters. They're crucial pieces of equipment, so get good ones! As soon as possible after that, buy a high-quality thermo-hygrometer that's made for restoration. A slide hammer probe and a carpet probe can wait until later.

- **Water Extraction Equipment:** You're going to need a water extractor at some point, perhaps even on your first water job. A medium-capacity one should be fine, and it'll serve your needs for quite some time; a smaller one is less expensive, but not by much, and a larger one may be more than you need on most jobs.

  You can get by with a carpet wand on almost all your jobs, so skip the hard-surface wand until you really need one. You can also skip the weighted heads until you're doing enough water jobs to justify one. Wait on the hardwood floor extractor until you have a large enough wood floor drying job to make the investment worthwhile.

- **Air Movers:** These are a must, and you're going to need several. The average water job requires anywhere from three to eight at a time, sometimes more. And that's per job — hopefully you'll have several jobs going at once. A good starting point would be five of the lower-end ones, and five of the medium-capacity ones. Then, plan on adding more as quickly as possible. Axial fans can wait until you have a definite need for them.

- **Air Mover Accessories:** These are almost as important as the air movers themselves. Start with two 48-inch boots and two multi-hose splitters. Add more of each whenever you can — you'll definitely use them.

  You'll definitely need carpet clamps. Air movers used to come equipped with them, but now most manufacturers have made them an option. Ideally, each of your air movers should have one, but initially plan on putting a clamp on at least every other one. The large hose adaptors can wait until you need them.

- **Dehumidifiers:** Here's the big investment. You need to have these, but they're pretty expensive when you're just getting started. But remember that you'll be renting them back, and they can pay for themselves fairly quickly. Initially, you might think about buying at least two medium-size refrigerant dehumidifiers. If you can afford it, also buy a third, smaller one, which will save you some money and accommodate your smaller jobs.

There's no need for a large-capacity unit right away. In fact, you may never need one unless you encounter some really large jobs. As far as a desiccant dehumidifier is concerned, that can wait until you have a job that warrants it.

- **TES and E-TES Equipment:** One or two of the affordable E-TES electric units make a nice addition to your arsenal of drying equipment. The full-size propane-fired TES unit represents a more sizeable investment, but if you have a good-size water job come up that warrants the expenditure, your rental income should recover the investment pretty quickly.

Remember that you're a restoration contractor now, and the investment you're making in this equipment is part of what sets you apart from the crowd. As a true professional in the field of restoration, these are the tools you'll be using to build a thriving business and a solid reputation. And even more than that, you'll be helping your friends and neighbors in your community.

It's a great investment all the way around!

4

Water Losses & Restorative Drying

The following statement is from the *Guide to Restorative Drying, Revised Edition*, from the Dri-Eaz Education Series. It gives you a very good overview of what's expected of you when you show up at a water loss.

> *"The goal of the restoration process is to transform an abnormally wet, potentially damaged structure into an environment that can be reoccupied, of equal or better appearance and cleanliness than before the intrusion occurred, and to do so in the most economical and efficient means possible."*

To better understand this, let's break the statement down. First, it says your goals with any wet building are:

- Dry the building;
- Make any necessary repairs;
- Return the building to the owners in a condition that's equal to, or better than, it was before the loss.

And second, you want to do the work in such a way that:

- It doesn't overly inconvenience the owners;
- It doesn't keep them out of the house any longer than necessary;
- Your crews and equipment aren't tied up any longer than necessary;
- You don't create unnecessary costs for the insurance company.

This sounds like a tall order, but it's really not. It just takes an understanding of the loss, and then a well-thought-out plan on how best to deal with it.

The Drying Process and the Repairs

A typical water loss breaks down into two parts. First, you'll inspect the home, come up with a drying plan, and then get the home dry enough to repair. Second, you'll evaluate the damage, write an estimate, and if you get the contract, make the necessary repairs.

On most jobs, these two parts are handled and billed separately. There are good reasons for doing it this way. In some cases, you may do the drying portion of a job, but not be hired to do the repairs. The homeowner may opt to do the repairs himself, or he may choose another contractor to do the work. There are also times when you'll be hired to do the repairs after another company has done the initial drying.

But more importantly, there may be a gap of several weeks, or even several months, between the time the drying process is completed and the time you complete the repairs. You don't want to wait until the end of the repair work to be paid for the drying work if it was done much earlier.

In this chapter, we'll cover the first of those two parts, the drying process.

Documentation

Part of the payment process is the need for good documentation. That's something that can't be overemphasized. Take good, clear notes and lots of pictures. You'll need both for billing, as well as for discussions with the adjuster. You'll also need them to protect yourself against possible damage claims that may come up later. This documentation applies to both the structure, which we'll cover in this chapter, and to the contents, which are covered in more detail in later chapters.

As you go through the structure, pay particular attention to potential problem areas. For example, say you're examining a water loss and there's some saturated drywall in the hallway outside the bathroom. The drywall is so wet that it's bowed in, and can be easily punctured just by pressing on it with your fingers. Also, the MDF baseboard is warped. However, the door casing in the same area is solid wood and is salvageable. You would start your documentation with a couple of photos of the overall hallway, as a general point of reference. Then take some pictures of the drywall itself. Show the bowing if you can, or how crumbly it is. Finally, take pictures of the baseboard and casing.

> ***Tip!*** *On a water loss, whether it's an emergency callout or not, you'll be billing for your time. That includes the time spent documenting the job for yourself and for the insurance company. So, in more ways than one, the time you spend taking notes and photos will never be time that's wasted.*

Then, make a couple of notes about the area. Here's an example:

> *"Drywall saturated and crumbling in hallway, to right of bathroom door. Wet from floor up to a height of 3 feet. Not salvageable; demo. Base is MDF, saturated and not salvageable. Demo; sample taken. Casing is salvageable; put in garage."*

Figure 4-1 These light fixtures are a good example of the type of parts you accumulate as you dismantle things on a water loss. If you lose or damage any of them, it can be expensive and time consuming trying to find replacements.

Keep Track of Those Parts and Pieces

During the drying process, you're constantly going to be dismantling things. Light fixtures, like the ones in Figure 4-1, will need to come down because of wet ceilings. Shelving will have to be removed from wet walls, along with all the little clips that hold it in place. Shower doors and tub surrounds will have to be taken apart because of wet drywall. Furniture and cabinets will have to be dismantled, along with all their hardware. The list is endless.

In the rush to do all that, especially during an emergency call, it's very likely that some small part will be lost. Someone will put a part on a window sill or on the floor, get distracted by something else, and it'll be forgotten or get swept up and discarded. And, as luck would have it, that part, whatever it is, will be irreplaceable.

You need to establish a system for collecting and storing all those little parts and pieces. Plastic bins with lids are a great solution. Buy them in a couple of different sizes, to fit different size items. Also buy several boxes of zip-lock plastic bags in various sizes, and a couple of boxes of black permanent markers. Warehouse stores such as Costco and Sam's Club have great deals on items like these.

REAL STORIES:
The Case of the $500 Hinge

Our crew was tearing out saturated drywall and insulation in the bathroom of a wet job, and had to remove the shower door and framework. They took it apart and stored all the pieces in the garage.

A couple of weeks later, with the repairs complete, the only thing left to do was reassemble the shower door. It was an easy task that turned into a monster. Why? Because one piece of the hardware was missing. It was a small part that made up the hinge assembly, and without it, the door wouldn't work. We talked to the subcontractor who normally did our shower doors to see if he could order a new part for us. Unfortunately, it was an older door and the parts were no longer available. We tried a local hardware store to see if they had something close that we could modify. No luck again.

In the end, we had to order a new $500 shower door and frame. In addition to the cost of the door, there was the cost of the time spent trying to track down the part; a delay in the completion of the job, resulting in a delay getting paid; and of course, there was the homeowner, who wasn't at all happy about the delays. All because we lost one tiny part that was worth about 50 cents.

Put a piece of tape on the outside of the bin, and write the client's name on it. Then put anything you dismantle into the bins. Small parts, such as hardware, shelf clips, hinges, etc., should go into a zip-lock bag inside the bin. On the outside of the bag, write the name of the client and what the parts are for: "Smith, ceiling light for green bedroom." Or, "Jones, master bathroom shower track."

The bins stack for easy storage in the garage, or whatever part of the house you're using for storage during the repairs. And when the job is done, just remove the tape with the client's name. Remove all the lids, and then nest the bins together and store them in your warehouse until the next job.

Ten Basic Steps in the Drying Process

When you're called out on a water loss, there are 10 basic steps that make up the drying process. Every job is different of course, but here are the basic guidelines.

Step 1 — Move Contents Out of Obvious Wet Areas

Before beginning any work on a water job, be certain that the source of the leak has been shut off. You don't want more water coming into the building as you're working to get it out. Also, make sure there's no danger from live electrical wires — have your electrician check the building if necessary.

Once the building is secure and you've verified that the leak has been stopped, the first step is pretty much common sense — move the contents out of wet areas. However, it's important that it's done quickly to help mitigate further damage. Besides the damage that water can do to furniture, wet furniture can also do damage to materials, such as carpet. Wood furniture, especially antiques, can transfer dyes and stains to carpeting and ruin otherwise salvageable carpeting (see Figure 4-2). Metal furniture — or even metal feet and metal glides on wood furniture — can leave rust stains on carpeting.

Move contents out of the water as quickly as possible. If items can't be moved immediately, raise them up on foam blocks. The blocks will lessen the possibility of damage, both to the items and to the flooring, until everything can be moved out and the flooring dried. Remember — you're a restoration contractor. Saving things is part of the job!

Tip! *You can charge for the time required to move, block and cover contents. You can also charge for the foam blocks and plastic sheeting.*

When the contents are moved out of the wet area, cover them with thin, clear plastic sheeting, also known as painter's plastic. It's available in rolls from any paint retailer. Covering the contents protects them from the dust and dirt kicked up by your air movers. This will save you a lot of cleaning headaches later, and may also prevent damage to the furniture. And, the extra care makes a positive impression on homeowners.

Figure 4-2 This living room carpet was saturated by a water loss. The water wicked up into the wooden table, causing dye to bleed out of the table and create the cross-shaped stain on the carpet. The stain was permanent, and the carpet had to be replaced.

Leave wet contents uncovered, at least for the time being. Covering wet furniture, especially upholstered pieces, can lead to mildew. Allow the air movers and dehumidifiers to draw out some of the moisture. As soon as they're dry, cover them. You may need to transport very wet furniture back to your shop for drying; we'll discuss this in more detail in Chapter 15.

If it's necessary to stack furniture due to limited space, be very careful how you do it! Keep some padded furniture-mover's blankets on your truck, and use them to protect anything you stack. If you arrive at the job and find furniture already moved or stacked up by someone else, document it with notes and photos.

Temperature and Humidity Readings

Another thing you need to do as soon as possible is set up your thermo-hygrometer. First take a reading outside the house. Pick a spot that isn't overly wet or exposed to direct sunlight or high winds. Give the instrument time to acclimate, and note the exterior temperature and humidity levels.

Once you have accurate outdoor readings, bring the instrument inside and set it up. Place it in a central section of the wet area of the house, but not directly by a heat vent or in the path of your air movers. Once again, give it time to acclimate and then note the interior temperature and humidity readings.

Figure 4-3 Using a wand and a water extractor to remove excess water from carpet and pad prior to drying.

Step 2 — Determine the Point of Origin of the Water

After the contents are moved, start assessing the damage to the house. You need to determine what caused the water leak, and where it first started.

Finding the point of origin will give you a logical starting place for your investigation. It's also where you generally find the most extensive damage. For example, if the icemaker line to the refrigerator broke, start in the kitchen and work out from there. If the washing machine overflowed, then start in the laundry room.

If the leak originated from a toilet, then you have some additional information to uncover. Water from the toilet's supply line is clean water, and you can proceed as you would with any other leak. However, if the water came from an overflowing toilet or some other type of sewage backup, then it has to be treated differently. We'll look at sewage spills in the next chapter.

Step 3 — Remove Standing Water

After deciding where the water leak started, the next step is a visual inspection of the damaged areas. You'll quickly find if there's standing water anywhere, including in saturated carpets. If there is, you need to start your water extraction immediately. If you're on the job by yourself, then you can stop your investigation at this point and begin the extraction process (Figure 4-3). If you have crew members with you, then have them start on the extraction while you proceed with the next steps in your evaluation process.

> ***Tip!*** *You can charge for the use of your extractor, either by the hour, by the number of gallons pumped, or by the number of square feet of area extracted. You can also bill for submersible pumps or trash pumps. This usually includes the rental charges and an hourly rate covering your time to set up and take down the pump and related hoses. Some adjusters will also allow you to charge for your time to pick up and return rental equipment, while others will insist that this time be covered as part of your overhead.*

Never use your extraction equipment in a crawlspace or an unfinished basement. Sucking up standing water in these areas will draw dirt and debris into the extractor and not only clog the filters, but ruin the vacuum and discharge motors as well. Instead, if the water appears relatively clean, you can use a submersible pump to remove most of it. For standing water that's dirty, rent a trash pump from a local rental yard. Get one

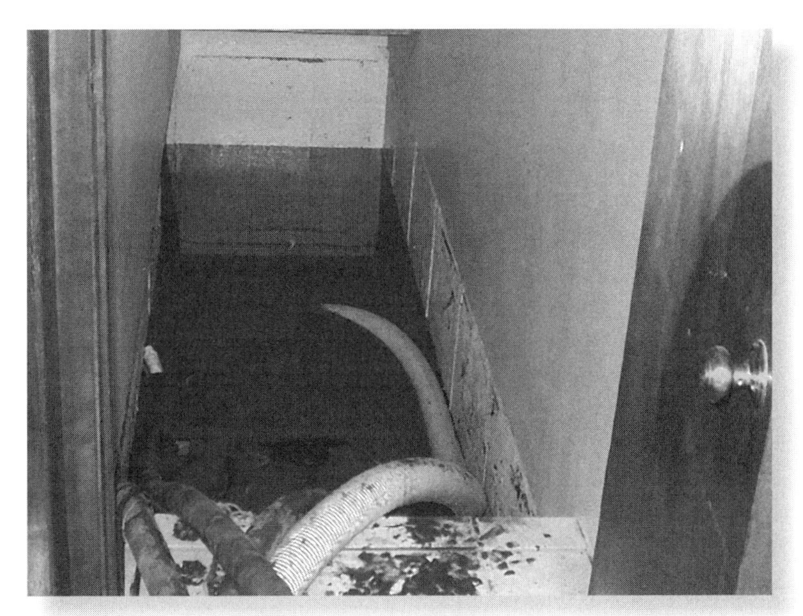

Figure 4-4 A flooded basement often contains a lot of floating debris. Instead of regular water extraction equipment, use a trash pump with a large-diameter hose and a screened pickup to remove the water. Trash pumps move large volumes of water quickly, making them well-suited for jobs like these.

with a large-diameter hose and a screened pickup. They're designed for handling dirty water, and are capable of moving large volumes quickly (see Figure 4-4).

Step 4 — Determine the Extent of the Wet Areas

Working out from the point of the initial leak, look for patterns indicating where the water traveled so you can find what damage it did along the way. You'll soon learn that few houses are actually level, and that water flows in some very unpredictable directions!

This process will help you decide what can be saved and what can't, as well as where to place your drying equipment. The following sections are an overview of common things to look for in a typical water loss. But remember, materials may react differently to water for a wide variety of reasons, so you need to evaluate each job carefully and methodically.

Figure 4-5 This wet carpet shows delamination of the backing, which is the mesh-looking material. The water has also caused the seam between two pieces of carpet to come apart, exposing the seam tape.

Carpeting

Water shows up as dark patches in carpet. If the carpet is wall-to-wall over a pad, the pad acts as a sponge below the carpeting, absorbing the water before it soaks the carpet. So the extent of the water may be greater than the pattern that you see on top.

The backings on many of today's carpets are installed with water-soluble adhesives. When the carpet gets wet, those backings can separate from the carpet, as shown in Figure 4-5. This process is known as *delamination*. You can't determine if carpet is delaminating by looking at the top — you have to see it from underneath. If the carpet butts up to a noncarpeted area, you may be able to inspect the underside of the carpet from there.

Figure 4-6 This is a water-damaged tack strip. The wood is saturated, and the nails have begun to discolor. It should be replaced, as it could potentially discolor the carpet when it's reinstalled.

Figure 4-7 Water from the saturated carpet (top of photo) has worked under the hardwood floor. The water damage to the hardwood shows up as dark lines along the edges and between the ends of the boards.

If it doesn't, you'll have to detach the carpet from the tack strip. Start at a corner, and use a carpet kicker to push the carpet toward the wall. That'll detach it from the tack strip, and allow you to pull it up without doing further damage. Small areas of delamination can usually be repaired. With large areas, it's better to replace the carpeting.

You also want to take a close look at the tack strip itself. (See Figure 4-6.) The small nails in the strip can transfer stains and rust marks to the back of the carpet fairly quickly. The longer the carpet is in contact with the wet strip, the greater the likelihood that stains will show up on the carpet. Pull up the wet tack strip as soon as possible.

If the carpet is a glue-down, laid directly over a subfloor or a concrete slab, not as much water will be absorbed, but it will travel farther over the carpet.

Hardwood Flooring

Water damage to hardwood flooring shows up in different ways, depending on the type of floor. Traditional hardwood flooring, with a good layer of finish on top, will handle a small amount of water quite well, almost like sheet vinyl. With too much water, however, the wood will begin to cup. Water will work up from below, showing up as dark areas between the seams, as in Figure 4-7. If it becomes soaked with enough water, the seams between the edges or the ends of the boards will begin lifting up.

Engineered-wood flooring, which has a thin layer of prefinished hardwood over a plywood base, usually absorbs water faster because there's no finish covering the seams. Laminate flooring absorbs water fairly easily, and will quickly start to warp.

Other Flooring

Look for standing water on vinyl flooring. Also, check for seams that are coming up. Individual vinyl floor tiles will have dark seams and the tiles will start coming loose. There are a lot more seams in a vinyl tile floor than there are in a floor covered with sheet vinyl. So be aware that the water may have caused additional damage to the subfloor or underlayment.

A ceramic tile floor, in addition to obvious standing water, will have dark grout joints that indicate where the water has soaked through to the subfloor or underlayment.

Drywall

Drywall is usually installed with a small gap between the floor and the bottom of the drywall sheet. This exposes the gypsum core to water that flows across a floor. Water damage to drywall shows up as dark patches spreading out in a fan shape, with the widest part of the fan closest to the water source. Prefinished drywall, like that used in many manufactured homes, also shows dark or discolored areas where it's been exposed to water. There may also be some lifting at the seams.

Other Wallcoverings

Wet wallpaper shows up as a dark patch, or you may notice that the seams are starting to come loose. Prefinished plywood sheet paneling will also show dark spots, depending on the type of finish on the paneling. With faux-finish paneling, which has a layer of imitation-wood vinyl on it, the vinyl may begin to delaminate from the underlying wood.

Real wood paneling will usually absorb more water than plywood paneling, because more of the wood is exposed to the moisture. This is especially true with individual boards that have been installed vertically, since the end grain of the wood is in contact with the water. Even if the wood has a surface finish over it, it'll quickly discolor and even warp. Open grain woods such as pine and cedar will show dark, jagged stains where the water has been drawn up into the wood.

Cabinets and Trim

You'll see water marks and discoloration in stained wood trim and cabinets. As with drywall, the wider part of the stain will be toward the source of the water. In Figure 4-8, you can see the fingers of moisture getting narrower as they move up the wood and away from the wet carpet. With painted wood, the water stains may be harder to see. It depends on the amount of water, how dirty the water is, and what color the wood is painted. Other obvious clues include warping, splitting wood, and rusted or discolored nail heads.

If the home's trim is made from MDF or another type of composite material, it'll absorb water very quickly. Once it does, it'll begin to swell. This happens fairly quickly, so it's easy to spot.

Figure 4-8 The water stains on the wood trim were caused by water wicking up from the carpet. Also notice the discoloration in the wood around the nail heads in the trim.

Figure 4-9 The two doors on this wet vanity cabinet are so badly swollen that they no longer close. The joints between the stiles and rail have separated, and the doors, if not the entire cabinet, will have to be replaced.

Tip! *Become a detective! Study the clues to find where the water traveled. Make your deductions based on what you can see, and what seems logical to you. The longer you do this, the more instinctive you'll be at finding hidden damage.*

If water flowed across the floor, cabinets will first show water stains on their toe kicks, since they're at floor level. Many cabinet toe kicks are made with veneered or painted MDF or other composites, so here again you'll want to look for signs of swelling.

Anything wood that's supposed to move freely will be a good indicator of moisture. Doors may begin to rub against their frames. Wood windows may bind and be hard to open. Cabinet doors may begin to twist or warp, and no longer lie flat against the face frame (see Figure 4-9).

In any wood items, such as bookcases, shelving, and furniture, you may also see swelling, lifting, or other signs of damage.

Metal

Rusted metal is an obvious sign of moisture. Some metal surfaces rust in a surprisingly short time, so they're often a better clue to moisture than you might think.

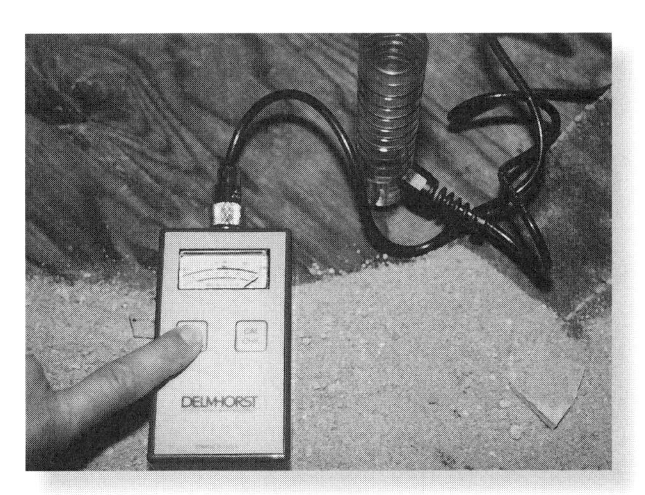

Figure 4-10 The indicator needle on this penetrating moisture meter is all the way to the right. That clearly indicates that the plywood subfloor being tested is saturated with moisture.

Masonry and Concrete

While masonry and concrete are much more resistant to moisture damage than other materials, they offer clues to the movement of the water in a house. Unsealed masonry and concrete will absorb a small amount of moisture, causing them to darken slightly. You can also look for effervescence on masonry and concrete surfaces. White rings or streaks are caused by moisture bringing salts to the surface, and can also be an indicator of water damage.

Step 5 — Search for Hidden Damage

You've no doubt spilled a glass of water. Remember how quickly the water moved, and how it disappeared into lots of different places? It's the same with a water loss, only on a much larger scale. For the next step in our process, you'll need a moisture meter to help figure out where all the water might have gone.

Start with a non-penetrating meter. Run the meter over the surfaces you think are wet, and, depending on the type of meter, read the scale, check the lights, or listen for the warning tone indicating moisture. This will help to verify and refine the visual clues from your initial evaluation.

Next, follow up with a penetrating meter to pinpoint the water more accurately (Figure 4-10). Pay particular attention to how the meter readings drop off as you approach the outer limits of where the water traveled. That'll give you a good idea of the degree of wetness and the extent of the damage.

As you read your meters, see if there are patterns that might lead you to believe that the water went into or under areas that you can't see. For example, can you trace the water across the kitchen floor toward the cabinets? Are you still getting fairly high readings at the point where the flooring meets the toe kick? If so, then there's a pretty good chance that the water went under the cabinets and appliances. That information will help you in the next steps, as you decide how and what to dry.

Tip! *The removal of wet padding is usually done on a square footage basis, so keep track of how much you tear out. The disposal of the wet pad is billed separately, as part of the debris removal from the job.*

Step 6 — Remove Wet Materials

Based on your initial inspection of the home, you should have a pretty good idea which materials can't be salvaged. Disposing of them is your next step.

Figure 4-11 The usual sequence for dealing with wet carpet and pad is to extract as much water as possible, then roll the carpet back, and remove and discard the wet pad. Dry the carpet and reinstall it over new pad.

Don't just start ripping things out — that can cause you all kinds of problems with the insurance company. Take several digital photos of the items and the areas where you intend to do tear out. Your photos should include overall shots of the area, plus close-ups of the specific damage that led to your decision to remove the materials. Let's look at some examples.

Carpet and Pad

You want to be very cautious about tearing out carpeting. In recent years, insurance companies have become increasingly skeptical of contractors ripping out and discarding carpeting before the adjuster has a chance to see it. So, when you encounter wet carpeting, attempt to dry it first, as outlined next. Then talk to the adjuster about whether or not it needs to come out.

Carpet pad, on the other hand, isn't an issue. Adjusters understand that the pad absorbs water like a sponge, and it's very difficult to dry. Since most padding is relatively inexpensive, it doesn't make economic sense to try and salvage it. Instead, roll back the carpet, cut the pad out just beyond the wet area (see Figure 4-11), and dispose of it. Be sure to keep samples of the damaged pad.

Drywall

Drywall that is crumbly or no longer feels structurally solid needs to come out. After you've seen a couple of wet houses, you'll begin to get a feel for which drywall can and which can't be saved.

There are two methods for removing drywall in a wet house. The one you use depends on your personal preference and, more importantly, the preference of the adjuster working with you.

- **Method 1 — Prepped for reinstallation:** Mark a chalk line on the drywall 4 feet up from the floor. Then, cut away the drywall below the line. Take the time to make plumb cuts down the center of a stud. This makes it easier and more economical for patching-in with a full 4-foot-wide piece of drywall. This method takes longer, but it leaves the remaining drywall prepped and ready for your patches and repairs.
- **Method 2 — Quick and dirty:** With this method, you simply break out whatever drywall has to come out. You can cut it with a saw, or simply hack it out with a hammer. This method takes much less time, and you get the wet drywall out of the house quickly — which is important during a water loss. The disadvantage is that you still have to make the cutbacks and do the prep work before you can do any drywall patching and repair work.

As a contractor, you probably feel that the first method is the correct way to do things. Oddly enough, many insurance companies disagree. They're paying you to mitigate the damage as fast as possible, and that includes removing the wet drywall quickly and then moving on.

Get some guidance on this subject from the adjusters you work with. See what their preferences are, and proceed accordingly.

Tip! *The removal of wet drywall is usually billed on a per-square-foot basis. So if you opt for demo method number one, it's going to take you longer and you won't earn as much on a unit-cost basis. Make sure you add some additional hours onto your drying bill to cover the time it took to do the cutbacks. On the other hand, if you demo using method number two, your time will probably be adequately covered in the unit-cost price. But when you do your estimate, you need to add in extra hours to cover the cost of the cutbacks and prep work for patching the drywall. You'll learn more about unit costs in Chapters 23 and 24.*

Insulation

Insulation is something you have to look at on a case-by-case basis. If it's saturated, it's not worth the time to try and salvage it. If it's not overly wet, however, you may be able to save it and reuse it.

Wall Insulation — First, remove the drywall to access the insulation. If the insulation is saturated, cut it above where it's wet, and discard it. If it's not overly wet, don't cut it, but simply pull it forward out of the cavity as best you can and leave it loose so that air can circulate around it.

Floor Insulation — Saturated floor insulation needs to come out as quickly as possible. Leaving it in place will hinder the drying process, and can also damage the framing. Once it's pulled out from between the floor joists, take it out of the crawlspace or basement right away. If you don't, your dehumidifiers will have to fight all the additional moisture. If the insulation is damp but not soaked, drop it down out of the cavities and let air circulate around it.

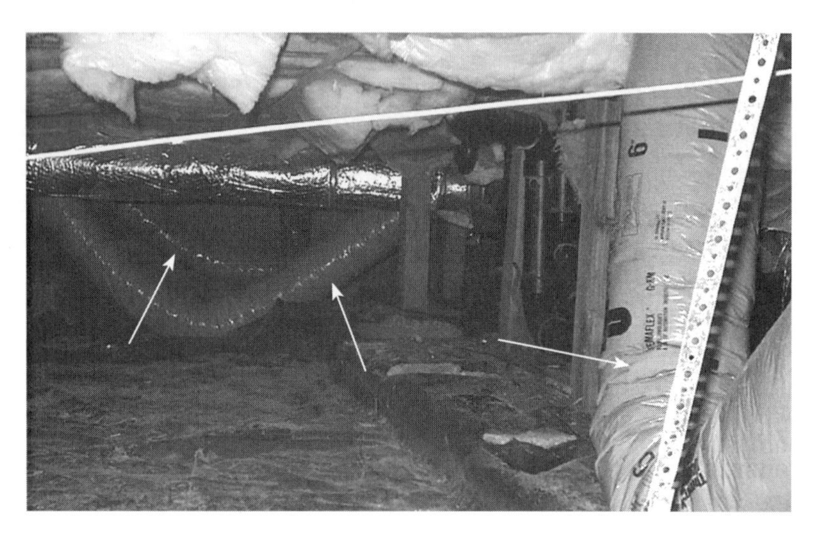

Figure 4-12 The three flexible heating ducts in this photo (one to the right and two in the background) are full of water, and noticeably sagging under the weight. These ducts need to be drained and then replaced before the heating system can be reactivated.

Attic Insulation — If attic insulation is soaked with water, you need to remove it right away. The weight of the wet insulation can cause the ceiling drywall to fall, which is potentially damaging to the house and dangerous to anyone in it. Leaving wet insulation behind will also promote mold growth in the attic, especially on drywall and OSB sheathing. Loose-fill insulation generally needs to be removed by hand. If you have an insulation vacuum (more on those in Chapter 9), you won't be able to use it with wet insulation — it's too dense and heavy.

Heating and Cooling Ducts

When a water loss occurs, the water almost always runs across the floors. If the house has ductwork in the crawlspace or in the basement, you can be assured that the ducts will be one of the first things to fill up with water.

For that reason, the ducts are one of the first things you'll want to inspect. If the house has flexible ducts, the damage is fairly easy to spot. The ducts will be sagging from the weight of the water, and may even tear loose from the boots. If you see sagging flex ducts, like those in Figure 4-12, carefully slit them open with a utility knife and drain all the water out. Be careful to position yourself so that you don't get a shower!

With hard ducting, you have a couple of options. If the weight of the water is substantial enough, you may find a few of the fittings have come apart, allowing the water to drain. If not, you can dismantle some fittings to drain the ducts. Or, you can extract the water from above by working an extraction hose down through the ducts.

A trash pump is better for this operation than your regular water extraction equipment, since ducts tend to accumulate a lot of debris from both the water loss and the original house construction.

This is important! — Never turn on the heating system until you've evaluated the furnace and the ducting! Pushing heated air through a wet duct system can aggravate mold growth. If mold or other pathogens are already present in the ducts, turning on the heating system will circulate them throughout the house. Instead, provide temporary supplemental heat until the heating system can be checked, repaired, and reactivated by a trained heating professional.

Wet Trim

The trim in the house may or may not be salvageable, based on how wet it got and the type of material. Try your best to save trim wherever possible. It minimizes matching problems (which we'll cover in Chapter 23), and speeds up reconstruction.

If you think you can save the trim, remove it carefully from the wall. To avoid damage to the drywall — *especially* drywall that's soft from the moisture — use a wide drywall blade between your pry bar and the wall. The metal blade will spread the pressure from the pry bar over a larger area, and will help prevent drywall damage and patching later.

As you remove each piece of trim, pull the nails through from the back side using a pair of end cutters. That minimizes damage to the face of the trim.

Start removing the trim at a logical spot, such as next to the main door, and work around the room in a clockwise direction. Label each piece of trim consecutively with a number as you remove it. By always working in a consistent and logical manner, and numbering everything as you go, you'll greatly speed up the reconstruction process.

Figure 4-13 Salvageable materials stacked on a job, awaiting transport back to the warehouse for cleaning and storage. Notice the doors stored vertically, and the trim banded with tape (circled).

As the trim is removed, find a dry place to store it. Usually the garage makes a good, out-of-the-way spot. Bundle the trim from each room together to minimize the possibility of damaging or losing any of it. You can use blue painter's masking tape or small rolls of plastic-wrap banding material to hold everything together. *Don't* use duct tape, regular masking tape, Scotch tape, wire, or anything else that might damage the trim.

It may be helpful to remove the interior doors. That will allow the air to circulate easier and you can move contents and equipment around with less risk of damage. Pull the hinge pins, remove the door, *and then put the pins back in the door hinge for storage.* Put a piece of blue painter's tape on the edge of the door, and label it with the room it came from.

When you store the doors, find a spot where they won't be in your way and won't get damaged. Lean them against a wall, as vertically as possible (see Figure 4-13). This will help prevent warping as they dry. You can place the doors one in front of the other to save space. Use mover's blankets or scrap pieces of dry carpet pad between the doors for protection.

Step 7 — Decide On a Drying Plan

As you inspect the house, think about where to set up your drying equipment. Your decision will be based on your understanding of where the water is, how wet the different materials are, and what your different pieces of equipment can do. You also need to know exactly what pieces of equipment you currently have available and how they'll interact to best advantage in this situation. Your training classes will be your guide in making these decisions.

In the training classes, you'll learn about the science of psychrometry. You'll study water, and how moisture moves and reacts in different situations. You'll learn formulas to calculate the equipment size you need for different humidity and temperatures. You'll also receive guidance in the positioning and moving of equipment. All of this is important to your understanding of your profession.

That information is far too involved for me to cover here. Instead, we're just going to look at the basic procedures you'll go through in setting up your equipment. After a short time on the job, doing the calculations and coming up with a plan for how much equipment to use and where to place it will become second nature.

Classifications of Water Losses

In an effort to help standardize water-loss restoration, several industry groups have agreed on four broad classifications of water losses. As you'll see later in the chapter, these classifications will help you decide how much equipment you need on a job, and where to set it up.

- **Class 1:** This is the lowest class, and considered the easiest to dry. It primarily involves relatively dense materials that don't absorb large amounts of water. These include wet plywood, concrete, and framing lumber, but little or no wet carpet or carpet pad. The water loss is generally confined to one room, or a part of one room.
- **Class 2:** This is a moderate water loss, involving an entire room of wet carpet and pad. It also involves drywall that has wicked moisture up to a height of 24 inches. In a Class 2 loss, at least some of the materials in the affected area are porous enough to retain moisture. See Figure 4-14.
- **Class 3:** This is a severe water loss, involving several rooms with carpet and pad. Drywall has wicked up water in excess of 24 inches above the floor, and the insulation is saturated. Class 3 losses often involve water that comes in from overhead, as in Figure 4-15.
- **Class 4:** This is considered a specialty drying situation. It involves water concentrated in deep pockets that are relatively difficult to dry with standard air movers. This might include hardwood floors, saturated subfloors, ceramic tile with a saturated mortar base, logs, crawlspaces, and other specialty situations. We'll look closely at Class 4 losses in Chapter 6.

Figure 4-14 An example of a Class 2 water loss, involving one room with porous materials — carpet, pad and drywall. Water wicked up the drywall to height of less than 24 inches, and has been cut out. The higher drywall cutout was made to repair a broken pipe. You can also see an air mover with a carpet clamp, set up to blow air underneath the carpet. This method dries both the carpet and the subfloor.

Figure 4-15 An extensive Class 3 water loss. The water came in from overhead, and involved a number of different rooms and different porous materials.

Open and Closed Drying Plans

For the most part, you'll be dealing with what's known as a *closed drying plan*. That means that the building is closed off from the outside, allowing you better control of the temperature and humidity levels during the drying process.

If the circumstances of the water loss are favorable, it may be possible to speed up the drying process by also using outside air. That's known as an *open drying plan*. For an open drying plan to be effective, the humidity in the outside air needs to be lower than the inside humidity, and the outside air temperature needs to be above 70 degrees. Under those conditions, you can speed the drying by opening windows and turning on interior ventilation fans.

When considering an open drying plan, you also have to be aware of security. Don't leave windows or doors open when you're not there to monitor the situation, or when other security risks might be present. Also, you have to be sure that there won't be any weather-related problems, and that you have the homeowner's approval.

Step 8 — Set Up Air Movers

You need to move air across wet areas in order to draw out the moisture. So, on a typical water restoration job, you'll begin by setting out your air movers. Here are some guidelines for the number of air movers you'll need:

- **Class 1 loss:** One air mover for every 150 to 300 square feet, depending on the amount of water and the condition of the area.
- **Class 2 and Class 3 losses:** One air mover for every 50 to 60 square feet.
- **All affected areas:** Regardless of square footage calculations, you need at least one air mover set up in each affected area, including each room, hallway, closet, and alcove.

Blow Air Across Surfaces

Some of your air moving equipment needs to be set up so that it's blowing a stream of air across the wet surfaces, such as across a wet floor. To make it easier to extract moisture from the air, position the air movers so that the air is moving *toward* your dehumidifiers. You also want good air circulation, so avoid placing them near any obstacles.

Since what you're pushing air at now is invisible moisture vapor, as opposed to visible standing water, sometimes the positioning of your air movers is hard to visualize. Let's look at an example of how to set things up.

> ***Tip!*** *When drying wide surfaces, such as vinyl or tile floors or subflooring, you can use an air mover with a 48-inch boot attached. This spreads the air stream over a wider area and speeds the drying process. But because the air stream is wider but not as concentrated, this technique works best for moderately porous materials.*

Imagine that you have a room that has a floor drain in it, and there's a lot of water on the floor. Your job is to take a big squeegee and push the water down the floor drain. You'd place yourself at the far edge of the standing water, and then push it toward the drain.

The floor drain represents your dehumidifiers, the squeegee represents your air movers, and the standing water on the floor represents the moisture vapor in the air.

To clean up the standing water, it wouldn't make sense to be between the drain and a wall, and then push the water toward the wall, right? It's the same concept with the air moisture in the room. You'd place the dehumidifier in the room, then set up your air movers at the far side of the wet area to push the moisture vapor toward the dehumidifier. You want to make sure to place the air movers so they don't blow at the wall, or at some other obstruction.

Now visualize that you have two rooms with water on the floor. The floor drain is in the next room, and there's only one door between the two rooms. Again, you wouldn't push the water toward a wall in the hopes that it will find its own way through the door. Instead, you'd position yourself at the far side of the first room, opposite the door, and push the water through the door and into the room with the drain.

Again, think of the dehumidifier as the drain and the air mover as your squeegee. The air movers should be blowing a stream of air across the floor and at the dehumidifier, not across the floor and into a blank wall. If the dehumidifier is in the next room, you'll get the most efficient moisture extraction by setting up the air movers to push the air toward the doorway leading to the room with the dehumidifier.

In some instances, the air mover setup may not be quite so obvious. Let's go back to the water and the floor drain example. Suppose the room you need to squeegee is one of several rooms with standing water in them, and the floor drain is a couple of rooms away. Now suppose that you have a helper with another squeegee. You could be in the first room — the one farthest from the drain — pushing water through the door into the next room over. At the same time, your helper in the second room will continue pushing the water along toward where the floor drain is located.

It's the same concept with the air movers. You can set them up so that they push air along a wall and toward a doorway. At the doorway, set up another air mover to push that stream of air toward the dehumidifier.

Here's one more example using the squeegee and the floor drain. Suppose you've pushed the water from your room to the room where your helper, John, is located. Now suppose that John has done the same thing, but he's pushing the water along to Ted, who's standing right by the floor drain. Ted also has a squeegee, and he's giving the water a final push into the drain. The three of you, working together, will get a nice flow of water across all the affected rooms and into the drain.

Remember, for a refrigerant dehumidifier to work properly there has to be a flow of air over the coils. So, just like Ted, who's standing right in front of the floor drain with his squeegee, you can also improve the dehumidifier's performance by setting up one of your air movers to push air directly toward the dehumidifier.

Blow Air Under Carpets

To effectively dry carpeting, you want to introduce a stream of air beneath it (refer back to Figure 4-14). This air stream will evaporate the moisture from both the subfloor and the underside of the carpet.

First, use your water extractor to remove as much water as you can from the carpet and pad. Next, starting at the corner of the room where the greatest amount of moisture is, disengage the carpet from the tack strip and roll it away from the wall. Remove the wet pad from under the carpet, and dispose of it. If the tack strip is saturated and rusting, remove and dispose of that as well.

> ***Tip!*** *Never disengage carpeting from the tack strip by pulling up on it with a pair of pliers or other tool, as this can damage the carpet backing. Instead, use a knee kicker to push the carpet toward the wall. This will disengage the backing from the angled nails on the tack strip. Then use a sharp awl to pull the carpet up and off the strip.*

Set up an air mover with a carpet clamp in the corner. Unroll the carpet back toward the air mover, so that it's draped over the air mover's snout. Clamp the carpet to the air mover. If the tack strip is still in place, press the edges of the carpet lightly back down onto the tacks.

Start the air mover on its lowest speed setting. You'll see the carpet lift up as the air starts to move under it. This is called *float.* How much float you get out of a piece of carpeting depends on how large the room is and how wet the carpet is. Adjust the speed of the air mover until the carpet appears to be lifted up several inches, but not so much that it's flapping up and down against the tack strip (that can ruin the backing). In large rooms or with carpet that's very heavy due to excessive moisture, it may be necessary to use more than one air mover.

Using TES Equipment for Drying Under Carpeting

Another option is to use a TES drying system to push heated air under the carpet. The drying concept is essentially the same, but with the addition of heat.

> ***Tip!*** *When using the TES or E-TES system, be sure you have complete training in the proper method for setting up and operating the equipment.*

After extracting the water and disengaging the carpet, set the TEX box in place in one corner of the carpet in the same way that you would a conventional air mover. If you're using the propane version of TES, connect the supply and return fluid hoses to the boiler and activate the system as per the manufacturer's instructions. If you're using the electric E-TES, connect the power cords to the proper electrical circuits.

Install a conventional air mover into the slot on top of the TEX box. Be sure the carpet is properly secured or weighted adjacent to the box, then activate the heat and the air mover. Use enough air flow to get a soft float out of the carpet to facilitate rapid drying.

Blow Air into Hidden Cavities

You also want to make sure that you direct the air flow of the air movers into any wet hidden cavities. These are the areas that you discovered with your moisture meter during your initial investigation of the house.

Figure 4-16 An air mover equipped with a snout adapter and mini-boots. This configuration allows air to be blown into a wall cavity (left center), under the hardwood floor (center), as well as to another location down the hall (left front).

The most effective way to direct air into small, enclosed areas is to use an air mover with a splitter system, as shown in Figure 4-16. A splitter divides the air stream into thirds, converting the wide snout into three round ports. Each port fits a 4-inch flexible hose, with a clamp to hold the hose in place. If you don't want to use all of the ports, you can close off one or two of the openings with a blank cap. The more ports you close off, the more air flow you'll have through the remaining openings. So if you need a highly concentrated stream of air, close off two of the ports and use only one hose.

The ports can accept a standard 4-inch dryer hose. However, those hoses aren't very sturdy and puncture easily. Also, they're not designed for the high pressure coming out of the air movers. It's better to use the more rugged, black flexible hose. They're made from a thicker material, but still have the flexibility to get into tight spaces. You can buy this type hose through many of the same suppliers that sell restoration equipment. You'll also find it at some appliance stores and plumbing retailers, as well as woodworking outlets. (It's commonly used for connecting tools to a dust collection system.) The hose can be cut to any length you need, which makes it very useful.

There are a wide variety of ways to utilize air movers for pushing air into hidden or inaccessible cavities. Here are a few suggestions you can try:

- **Open-ended hose:** Leave the end of the hoses open, with no attachments on them. Simply run the hoses into any openings or locations where they're needed, as shown in Figure 4-17. You can direct a hose into the cabinets to dry them from the inside out. You can run hoses between floor or ceiling joists. You can drill a 4-inch hole in a stair riser to direct air into a closed bay under stairs. You can tuck hoses up into the ceiling grid of a suspended ceiling and hold them in place with the ceiling tiles, allowing you to direct air into the dead space above the ceiling. Drill a 4-inch hole in a wall or a subfloor to push air into enclosed joist cavities, soffits, chases, and other inaccessible areas. You can dry all these areas without having to do extensive demolition. The possibilities are endless.
- **Mini-boots:** You can also install a mini-boot on the end of the hose. Mini boots are similar to the big 48-inch boots, but they're quite a bit shorter and narrower. That spreads the air stream out so it's wider

Figure 4-17 A snout adapter without mini-boots allows the open hoses to direct air into cavities between the joists of this wet ceiling.

and flatter for more effective drying. Mini-boots are ideal for working in confined areas such as behind toilets and pedestal sinks, as well as standard cabinet toekick spaces. They're also very effective for drying hardwood floors. (We cover that in Chapter 6.)

- **Drying under cabinets:** To dry the space underneath a cabinet, use a mini-boot setup. Drill a series of holes in the toe kick, install the boots, and blow air under the cabinets. When you're done with the drying, the holes can be covered with a new strip of thin wood, called a toe-kick cover, and stained to match the cabinets. This method of drying eliminates the need for the very costly removal and reinstallation or even complete replacement of the cabinets.

> ***Tip!*** *Save all your short hose pieces. You'll find places to use them on different jobs. You can also make longer hoses by joining two flex hoses together with a 4-inch sheet metal splicing sleeve and a couple of clamps. The sleeves are available at home centers or any other place that sells sheet metal fittings.*

- **Wall drying with boots:** Wet wall cavities are very common in water losses. The water either flows into the walls through pipes inside the cavities, or water wicks up from the floor. Once in the wall cavity, the

(Photo courtesy of TES Drying System)

Figure 4-18 A TEX box, part of the TES heated drying system, equipped with an Octi-Dry™ Wall Dryer. This setup is pushing heated air into a wet wall cavity to speed evaporation and drying.

water wets both the face and the back side of the drywall, as well as the framing. If the walls are insulated, then you have the additional problem of wet insulation.

To dry wall cavities, first remove the baseboard. Drill a series of 1½-inch holes in the drywall, parallel to the floor. Drill the holes as low as possible, but above the line of the wall plates. Install a 48-inch boot on the end of an air mover, and place the boot in contact with the wall and over the holes. The air mover will force air very effectively up into the wall cavities.

You can also use the TES or E-TES system, combined with an Octi-Dry™ attachment, as another alternative for drying wall cavities. Attach the Octi-Dry Wall Dryer to the front snout of the TEX box, then insert the individual Octi-Dryer jets into holes drilled in the wall, as shown in Figure 4-18. Heated air from the TEX box is pushed into the wall cavities, speeding the drying process.

If you drill the holes low enough, they can be repaired with drywall tape and mud without having to do texture matching. The taped area will then be covered by the baseboard.

- **Drying crawlspaces and other large areas:** If you need to dry out a crawlspace, attic, or other large space, one option is to simply put an air mover into that space. However, since the fan motor on the air mover pushes a very large volume of air, it can create a couple of problems. For one, it'll blow a lot of dust and dirt around. But a greater problem is that the dust and debris will be drawn into the intake side of the air mover. That's really hard on the fan motor, which can be damaged.

  A better alternative is to use a single 10-inch or 12-inch duct, either off the snout of an air mover, or with an axial fan like the one in Figure 4-19. This will push a very large volume of air, and still allow you to direct the air stream effectively. Most importantly, it keeps your air movers out of the dirt and dust.

As we discussed in the last chapter, in addition to charging to rent your air movers and TES system, you can also charge rent for all of the attachments. Splitters, mini-boots, wide boots, axial fans, Octi-Dry, and large-diameter hoses can all be billed as separate items.

Figure 4-19 This axial fan, equipped with a 12-inch collapsible duct, has been set up to direct air into a wet attic. By using the duct and keeping the fan out of the attic, insulation and other debris aren't drawn into the fan's intake port.

Having the knowledge, skills, and equipment to do this type of drying is what sets you apart from the crowd, and the end result is beneficial for everyone: You speed up the drying process, getting the home back to normal faster. You salvage materials, not only saving money for the insurance companies, but supporting the environment as well. And, best of all, you make a nice income while growing your company's reputation as the one to turn to in an emergency.

Some Cautions About Air Movers

An air mover is one of the most effective tools that you'll use for drying a building. However, there are a couple of things you need to be aware of when setting them up for use.

Always keep their air speed in mind. These pieces of equipment move air at speeds up to 3,500 CFM. That's about 50 times the speed of an average bathroom ventilation fan — we're talking a lot of air movement here.

One problem you'll have with all that air speed is dust. When you turn the air mover on, especially since it's directing its air stream along the floor, a lot of dust, dirt, and small bits of debris will be pushed along in front of that air stream. You can fill a room with a dust cloud pretty quickly. And all that dust will settle out over all the furniture, window coverings, cabinets, counters, and anything else it can land on. If you're not careful, you can even blow the ashes out of a fireplace — and that really makes a mess!

The high velocity pushing out through an air mover means that it's drawing an equal amount of air into the intake, located on the side of the machine. Air movers suck dirt and debris toward themselves, along with papers, plastic, drapes, and anything else that isn't nailed down. So, be sure and do some cleaning up before activating your air movers. Do a good sweeping to get the bulk of the dirt off the floor. If you've been drilling or cutting drywall and there's drywall dust on the floor, sweep or vacuum it up — drywall dust makes quite a mess.

Be careful how you position air movers. Make sure they aren't facing directly at drapes, blinds, wall hangings, or other items that can flap around in the air stream. All that flapping will not only damage those items, but they can also damage the wall surfaces behind them.

Be sure that you don't set the movers too close to drapes. Fabrics will be drawn to the intake of the air mover. The intake has a screen over it for protection. But the air mover has enough power to suck the fabric tight up against the screen. That will not only damage the drapes, but it will also put a strain on the air mover's motor.

Also be aware of the location of pilot lights on appliances powered by propane, natural gas, and other types of combustion fuels. If improperly positioned, the high speed of the air can blow the pilot lights out. This can allow gas to enter and build up in a room, with potentially deadly results.

If your air movers have adjustable speed settings, always start them out on the lowest possible speed. Check the drying action on that speed setting. If you don't feel that it's generating a sufficient amount of air for the specific drying situation, increase it to medium speed, and then high speed as necessary.

One more consideration in your air mover setup is to be sure that the air stream of one air mover isn't competing with another. In other words, don't set one air mover where its air stream is blowing directly into the air stream of another unit.

Get into the habit of setting up all your air movers so that they generate airflow in only one direction around the room, either clockwise or counterclockwise.

Step 9 — Set Up Dehumidifiers

The set up for your dehumidifiers is determined by the class of the water loss and the size and layout of the rooms. Use the following information, based on the *Guide to Restorative Drying, Revised Edition*, from the Dri-Eaz Education Series, to help you with your set up.

Required Number of Refrigerant Dehumidifiers

1: Determine the class of water loss.

2: Determine the cubic feet of space to be dried.

3: Determine the AHAM (Association of Home Appliance Manufacturers) rating of the dehumidifiers you have available to use. (This is the dehumidifier's water removal capacity, in pints of water per 24 hours at 80 degrees F and at 60 percent relative humidity. It's provided by the manufacturer, as part of the specifications and instructions.)

4: Determine the dehumidifier factor (see the chart in Figure 4-20).

5: Divide the cubic feet of the space by the dehumidifier factor to determine the number of AHAM pints required.

Dehumidifier Factors				
	Class 1 water loss	Class 2 water loss	Class 3 water loss	Class 4 water loss
Conventional refrigerant	100	40	30	N/A
Low-grain refrigerant	100	50	40	50

Figure 4-20 Factors for different types of dehumidifiers

Let's do an example: You have a Class 2 water loss, and it's affecting two adjacent rooms. One room is 12 x 14 feet, and the other one is 14 x 16 feet. Both rooms have 8-foot ceilings. Find the total cubic feet of space affected:

$$(12 \times 14 \times 8) + (14 \times 16 \times 8)$$

$$1{,}344 + 1{,}792 = 3{,}136 \text{ cu. ft.}$$

The total cubic feet of affected space is 3,136.

Now refer to the Dehumidifier Factor chart in Figure 4-20. If you use conventional refrigerant dehumidifiers, under the Class 2 heading for that type of dehumidifier, you'll find the factor of 40.

Following the instructions in Step 5, divide the cubic feet of space by the dehumidifier factor:

$$3{,}136 \text{ cu. ft.} \div 40 = 78.4 \text{ pints}$$

That gives you the number of AHAM pints of dehumidification required.

So, if you use conventional dehumidifiers rated at 64 pints, you'd need to set up two dehumidifiers. Whenever you're setting up more than one dehumidifier, they should be spaced evenly throughout the wet areas. In this case, one in each of the affected rooms would work best.

Remember, this is just a rough formula, and it's only for the initial drying phase — typically the first 24 hours. After that, you'll have to carefully monitor the drying process to see how the humidity levels in the building are responding, and adjust your equipment accordingly.

Set Up Pump-Out Tubing

When you use refrigerant dehumidifiers that have a pump-out system, you also need to set up your discharge tubing. You can route the tube to a sink, bathtub, shower, or other area that continually drains. *Don't route the tube to a toilet bowl*, as the bowl can overflow. Also, be sure to run your tubing in such a way that it doesn't present a tripping hazard.

Another, although less preferable, option is to run the tube to a location outside the house. If you do that, make sure the water coming out of the tube won't damage anything, and that it can't run back into the building. Also, make sure that the tubing can't be restricted or pinched closed by an exterior door.

> ***Tip!*** *When routing the tubing to a sink or bathtub, tie the tubing in a loose knot around the faucet. That prevents the tubing from accidentally being pulled out of the sink and spilling all the water back onto the floor.*

Limitations to Refrigerant Dehumidifiers

Refrigerant dehumidifiers are by far the most common type used in our profession. They're dependable, easy to operate, and less expensive to purchase than desiccant dehumidifiers. But there are some limitations to their use.

- **Low Temperatures:** Because a refrigerant dehumidifier relies on condensation to work, the temperature of the air in the room needs to be warmer than the temperature of the evaporator coils. If the air in the room gets too cold, there won't be enough difference between the air temperature and the coil temperature for condensation to occur. Refrigerant dehumidifiers work best in temperatures between 70 and 90 degrees F. If the heating system in the house is working, use that to keep the air temperature up. If the heating system isn't working, you may need to bring in some portable heaters.

- **Low Humidity:** Refrigerant dehumidifiers work best in a very moist environment. The higher the humidity in the room, the better a refrigerant dehumidifier works. A conventional-refrigerant dehumidifier will work well in conditions down to around 45 percent humidity. Low-grain refrigerant (LGR) dehumidifiers will work down to about 30 percent humidity. Below those points, their drying abilities drop off considerably.

  You can compensate for this to some degree. As the relative humidity in the room drops, increase the dehumidifier's effectiveness by lowering the room's temperature down toward 70 degrees F. Since cooler air holds less moisture, cooler temperatures will cause some of the additional moisture to be released from the air.

When conditions are cool and relatively dry, or in some Class 4 situations, you'll need to switch from a refrigerant to a desiccant dehumidifier. This is also the case when you're trying to dry materials with low moisture content, such as wood and paper.

Step 10 — Monitor and Document Progress

After your equipment is set up, you can't just leave it and forget about it until the house is dry. You need to monitor, move, and add and subtract equipment on a regular basis. The house should be checked at least once a day. During each monitoring visit, test the moisture levels of all the wet areas identified during your initial inspection. In all but the wettest of houses, you'll see significant improvement within the first 24 hours.

As you monitor the house, look for places where you need to adjust the equipment. For example, you might need to reposition some of your air movers in order help the dehumidifiers work more effectively. Or, you may need to relocate an air mover to better access pockets of remaining moisture. Also, air movement can cause moisture to migrate to other areas. That may mean moving some of your wall boots, or the hoses feeding them into cabinets. See Figure 4-21. As the structure dries out, you can reduce the number of both air movers and dehumidifiers.

When you're going through the house, *remember to document, document, document!* Keep track of the equipment being moved, the equipment being added and the equipment coming out. You'll need all of this information for billing purposes later on. It also helps to ensure that none of your equipment gets lost.

(Photo courtesy of Dri-Eaz Products)

Figure 4-21 A typical water-loss drying situation, with air movers and dehumidifiers in position.

There are liability issues to be considered as well. That's another reason to be thorough in your documentation. It's a very good idea to keep your files for several years. When a house is sold, there are laws that dictate what the seller has to disclose about the home's condition. Those disclosures usually include any water damage that was covered by insurance. Buyers may want to contact you with questions regarding the drying process and repairs. If this happens, talk with the seller (your former client), and if necessary, your attorney. You'll find it's easiest just to provide the requested information. It keeps everyone happy and you out of potential legal arguments later on.

Equipment Use, Monitoring Time, and the Insurance Companies

As with many aspects of insurance restoration, you have to walk a fine line with restorative drying. On one hand, the insurance companies want you to do everything necessary to mitigate damage to the home after a water loss. On the other hand, they don't want you do *more* than what's necessary. It's a constant balancing act — trying to set up enough equipment to do quick and effective drying, without going overboard on expenses.

The same can be said for your monitoring time. In order to dry a structure effectively, you need to check moisture levels and move equipment around. That gets the job done quickly, and minimizes the time the equipment's on the job. But the monitoring process involves trips to the jobsite charged to the insurance company. Again, you need to find the right balance between making enough trips to effectively monitor and move equipment, and making so many that you're wasting time and money.

Stay in communication with the adjusters. Learn what their personal preferences are, as well as their company's policies. You'll soon discover that what's fine with one adjuster will be irritating to another, and what you can charge one insurance company without problem will be continually questioned, even objected to, by another.

Whenever you're questioned by the insurance company about why you had certain equipment in place for certain time periods, refer to your documentation to explain your decisions. And, be ready to compromise. If an adjuster really feels that your air movers could have come out a day earlier, or that you only needed three dehumidifiers on the job instead of four, work out an arrangement with him. Discount part of the equipment cost, or decrease how much labor you're charging.

Always show a willingness to keep an open mind and to answer questions. If you discuss disagreements calmly, it goes a long way toward cementing relationships with your adjusters.

REAL STORIES:

That's About as Wet as a House Can Get!

It was a beautiful house, nestled in an upscale subdivision with sweeping views of the town below and the distant mountains. The owners had been transferred to another state, and forced to move before their house sold. It was now sitting vacant, waiting for a buyer who could meet the $1 million-plus asking price.

Late one Thursday afternoon, the real estate agent was closing up the house after a showing. The temperatures were dropping, and low, gray snow clouds were piling up in the distance. She checked the thermostats and the alarm system, locked the front door, and headed home. Her next visit would be on Monday, when another showing of the 5,800-square-foot home was scheduled.

Unbeknownst to her — or anyone else — there was a time bomb lurking in the attic. Just above one of the third-floor bedroom closets was a single stub of ¾-inch copper water pipe. The pipe had been installed as part of a residential sprinkler system the builder had planned as an added feature. But the system had been abandoned at some point during the construction process, and never completed. This lone stub of pipe was left uninsulated, unprotected, and vulnerable to the freezing temperatures in the attic.

That night, the snow hit as forecast. Temperatures continued to drop, and ice-cold air entered the attic through the vents. The attic temperatures got cold enough to freeze that unprotected water pipe. The frozen water in the line expanded, overwhelming the poorly-soldered cap on the end of the pipe. Finally, the cap popped off. When the daytime temperatures rose a bit, the ice plug in the end of the pipe melted slightly, and was eventually driven out by the water pressure. A ¾-inch water line was wide open in the attic of the three-story home!

On Monday afternoon, the real estate agent returned to the house to open it up for her next appointment. She was momentarily stunned to see long rivers of ice coming out of the attic vents and down the sides of the house. The landscaping next to the front porch was completely encased in ice that had worked out from under the lap siding. The front porch itself was a skating rink. Forcing open the front door, she was greeted by a torrent of water coming down the main stairs like a set of river rapids.

We'd done work for her previously, so she immediately gave us a call. The first technician who showed up had never seen so much water in one house. He found the main shutoff valve, stopped the flow of water, then wisely called into the office to ask for as much backup as we had available.

A tour of the house revealed the extent of the damage. Drywall had been overwhelmed by the weight of the saturated fiberglass insulation in the attic and had been ripped off the trusses. Cabinet doors had swollen and cracked off their hinges. Wood windows were completely inoperable. Doors were so badly jammed in their frames that some had to be pried open with crow bars in order to enter the rooms.

The bottom floor of the house, which was a daylight basement with carpet over a concrete slab floor, had become a swimming pool filled with several inches of standing water. Ceramic floor tiles had already begun popping up from the pressure of the swollen subfloors. Walking on the severely cupped hardwood floors was an exercise in balance. The only fortunate thing about the loss was that the owners had taken all their contents with them.

The drying and demolition work alone took over two weeks. When things were dry enough to estimate accurately, we found that many of the rooms had to be completely gutted. Even the electrical panels required rebuilding due to the amount of water they had in them. Repairs took several months — all from one small forgotten water pipe in the attic.

(Photo courtesy of Dri-Eaz Products)

Figure 4-22 A top-of-the-line mobile furnace for providing temporary heat to a building. This diesel-powered unit is thermostatically controlled and designed to be set up outside the building.

Winterizing a Loss

If you live in a cold climate, one situation you're going to face is what to do with a house where the heating system is inoperable because of the loss. You run the risk of having pipes freeze, toilets crack, washing machines damaged, and similar problems that freezing water can cause.

One solution is to provide temporary heat. There are propane-fired heaters you can rent or buy that'll keep a couple of rooms heated, at least enough to protect pipes from freezing. However, propane heaters come with a few problems: There's the danger of the open flame, the potential for fumes, and their need to have the fuel bottles replenished. Also, burning propane releases additional moisture into the air, which isn't helpful when you're trying to dry a house.

There are very effective portable furnaces, like the one in Figure 4-22, that are made specifically for use in restoration environments. This one operates on diesel fuel and is designed to be set up outside the building. Heat is provided to the building through a duct, eliminating the problem of fumes or additional moisture. Some of these units are large enough to heat an entire building for extended periods of time.

Temporary electric heat offers a better solution for some jobs. You can also purchase a small electric furnace (you can sometimes find used ones through your heating contractor), and have it wired with a cord and a plug. You can then plug it into a dryer or range circuit at the job, and have a safe and continuous flow of heat. The electric furnace method also gives you the option of adding a thermostat, and even a duct or two to better direct the heat. You can also use it when painting, installing drywall, and doing other types of repair work.

If you opt for a temporary electric furnace, make sure that the cord is wired in by an electrician, and that the cord and plug are compatible with the amperage of the furnace and the amperage of the outlet you'll be using. *Remember that this is only temporary, and should never be used as a permanent or semi-permanent solution to the building's heating needs.*

Even if you're able to provide temporary heat, and especially if you're not able to, you still want to winterize the house. This is a three-step process, typically done by a plumbing contractor. Here's what's required:

1. Shut off the main water supply, then drain all of the water lines.
2. Use compressed air to blow out any moisture remaining in the lines.
3. Add recreational vehicle-type antifreeze to toilets and other areas where there may be standing water.

Winterizing a house protects it against extensive freezing damage and costly repairs. It's in the insurance company's best interest to have this done, so they're usually willing to pick up the cost.

Other Winter Protection Problems

When working on a water loss job in the winter, never lose sight of the simple fact that water freezes. No matter where the water is or how it got there, it's going to freeze when the temperature drops below 32 degrees. The result can be both a safety hazard, and a hazard for the building itself. Keep these things in mind during those cold winter jobs:

- **Storing toilets and washing machines:** Never leave a toilet sitting outside. Even if you're sure that you've removed all the water, there may still be some left in the trap. A toilet left outside can collect rain water, and once that water freezes it's going to expand and crack the porcelain, ruining the toilet. The same goes for washing machines, which can sustain a lot of damage if the pump freezes.
- **Protect walking surfaces:** As you're moving wet materials and contents from the house to the disposal area, you'll be dripping and spilling water. That water will freeze along walkways, truck ramps, and other walking surfaces, forming a patch of ice that can easily result in serious injuries. When working in freezing conditions, carry some ice-melter with you (available at any hardware store, home center or retailer if you live in snow country), or at least some sand to provide traction in icy conditions. And remember that slipping hazards aren't just confined to ice. A wet walkway is a slip-hazard even in warm climates.
- **Select storage areas carefully:** It does no good to move contents and building materials out of the house and into storage, if that storage area is going to freeze as well. Be very careful about storing things in garages, sheds, and other areas where temperatures may drop below freezing. If necessary, talk with your adjuster about renting heated storage for the duration of the job.
- **Debris freezes as well:** As you remove wet carpet, drywall, and other saturated materials from the house, be careful where you put them. See Figure 4-23. If you dump everything in the driveway with the good intention of cleaning it up later, you may come back to find a frozen mass that's blocking your access to the house. Same goes for your truck; don't leave materials that can freeze in the bed of your truck. Try and get it unloaded at the landfill as soon as possible.
- **Don't overlook animal safety:** Pet owners are very attached to their pets, and part of your responsibility as a restoration contractor is to help your clients with the health and safety of their animals in a cold-weather loss situation. Remember, food and water bowls left outside can freeze, as can some types of livestock feed. Aquariums also need protection against power loss and freezing temperatures.

Figure 4-23 Where you place piles of wet debris is an important consideration, especially during the winter. Wet debris can freeze solid, creating potential safety hazards as well as limiting access to the building.

Check with your clients about their animals' needs, and help them coordinate with their adjuster as well. If pets or livestock are to be left at the property, the owners may need emergency electrical power or heat, temporary fencing or shelters, or other assistance for their animals. In most cases, this temporary assistance, including the cost of temporary kenneling or other unusual expenses, will be covered as part of their outside living expenses. That's something they'll need to clarify with their adjuster.

- **Don't forget the plants:** House plants can freeze and die in cold conditions. Don't leave them in the house, and don't store them in cold areas. Instead, move them to a place that's sufficiently heated to keep them protected. Use care while transporting them as well. Moving them in the back of a cold truck for a prolonged period will kill plants just as easily as storing them in a cold warehouse.

Some Thoughts on the Drying Process

Here are a few final thoughts on structural drying. You're going to be bombarded with information during your classes. Then, you'll get even more from suppliers, manufacturers, salespeople and even other contractors. Some of it will make sense, and some of it won't. Some will even be contradictory. Your work will be questioned by the insurance company and by your clients.

In the end, just do the best you can. Stay informed, be honest, and make sound decisions based on what your professional knowledge tells you is the best way to dry the building.

5

Mold Remediation & Contaminated Water Losses

There's certainly no other hot-button issue in insurance restoration that's bigger or presents more problems than mold. In recent years the media has covered story after story of horrific mold infestations in homes. I'm sure you've heard about people getting so sick they've had to abandon their homes, and others who've gone to every extreme, even burning their homes to the ground in a last ditch effort to clear it of mold.

So how much of this is hype, and how much is a legitimate concern that you may have to deal with from clients? There's no easy answer. But one thing is for sure: when dealing with any mold loss, you must respond quickly, cautiously, and thoroughly. You have to keep an open mind, listen to what your clients tell you, and take on each job as an individual case.

In this chapter, we'll look at mold remediation basics, as well as how to deal with losses involving contaminated water.

What You Need to Know

First of all, mold and contaminated water losses *should never be taken lightly!* They represent some very serious health risks for you, your employees, and your clients.

Dealing with these losses requires special protective equipment, and very careful and methodical procedures. They also require specialized testing to determine how badly a structure is contaminated, and when it's safe to reoccupy. What we'll be covering in this chapter is strictly an overview of the subject of mold remediation and contaminated water losses. Every job is different, and every job is potentially very dangerous. See Figure 5-1.

Figure 5-1 Mold remediation losses, like the one at this home, need to be handled with great care. They require planning, safety precautions, and the assistance of an industrial hygienist trained in proper procedures. They also require coordination between the contractor, the insurance company, and the building owners.

Mold remediation and contaminated water losses can represent a serious liability problem for your company. While you're new to the business, you may want to use experienced specialty subcontractors to handle these losses for you. Later, if you decide that you want to undertake this area of remediation work, make sure everyone on your crew has the proper training, certification, equipment, and protective gear.

These areas of work demand that you always keep up with industry changes and improvements. Mold remediation in particular is a rapidly changing field, with advances in both the methods for dealing with infestations and the tools and materials available. You, or the designated mold specialist within your company, will need to keep up with continuing education classes to stay current in the field.

Keep in mind that we're offering only a general overview of the subject of mold remediation and contaminated water losses in this chapter. Just reading this chapter won't prepare you for the work in this field. *Do not attempt to undertake any mold or contaminated water loss without:*

1. The proper training, and
2. The assistance of a trained and experienced industrial hygienist or other trained professional (as described later in this chapter).

Occupant Safety

Your primary concern when dealing with a mold or contaminated water loss is the safety of the building's occupants. People who have any type of compromised immune or respiratory systems should be removed from the building as soon as possible. This includes infants and very young children, the elderly, people who are undergoing chemotherapy treatments, and people who have just had surgery or some other types of medical procedures.

Before undertaking any type of remediation work, discuss the situation with the building's occupants, and with the insurance company. It's important to advise the occupants to check with their doctor or other healthcare professional if they have any questions about their health or the advisability of remaining in the building.

IICRC Certification

In addition to the water restoration training we talked about in Chapter 3, if you're also going to be working with mold remediation, you'll want to take the IICRC's Applied Microbial Remediation Technician (AMRT) course. This specialized training is essential before you take on these types of losses. Even if you'll just be overseeing a subcontractor, you really should have this training.

Here's the official description of this IICRC course:

> *AMRT: (4-day course) The Applied Microbial Remediation Technician course covers mold and sewage remediation techniques for individuals engaged in property management, property restoration, IEQ investigations or other related professions. Emphasis will be placed on teaching mold and sewage remediation techniques to individuals who will perform these procedures in the field. Course graduates will be adequately equipped to perform remediation services, while protecting the health and safety of workers and occupants.*

Another excellent class that'll be a great help in establishing procedures for all areas of your company is the Health and Safety Technician training offered by the IICRC. Here's the description of that 2-day course:

> *HST: The Health and Safety Technician class is a must for anyone performing restoration or mold remediation! This course is provided to increase safety in your workplace and reduce your company's risk of costly fines and penalties. It is also designed to put you ahead in the restoration industry — to show insurance companies you and your staff are ahead of the competition regarding health and safety. Topics to be covered are: OSHA Standards, Inspections, Citations and Penalties, Record Keeping, Personal Protective Equipment (PPE), Hazard Communication, Hazardous Materials, Confined Spaces, and Bloodborne Pathogens.*

For more information on IICRC training classes, refer to Appendix B.

Introduction to Mold

Mold is a form of fungi (the plural of fungus). There are thousands of different types of molds, and they're everywhere around us, both indoors and outdoors. Molds and fungi are most commonly associated with the breakdown of organic materials, such as dead leaves and other plant materials. However, some larger fungi, such as mushrooms and truffles, are edible. And some molds are important to the production of foods, such as cheese, or medicines, such as penicillin. In short, mold and fungi are both useful and essential.

Unlike other types of plants, fungi don't contain chlorophyll or use light to make food. Instead, they absorb food from the things that they eat during the process of decay. With no need for sunlight, mold and fungi grow well in dark and damp locations.

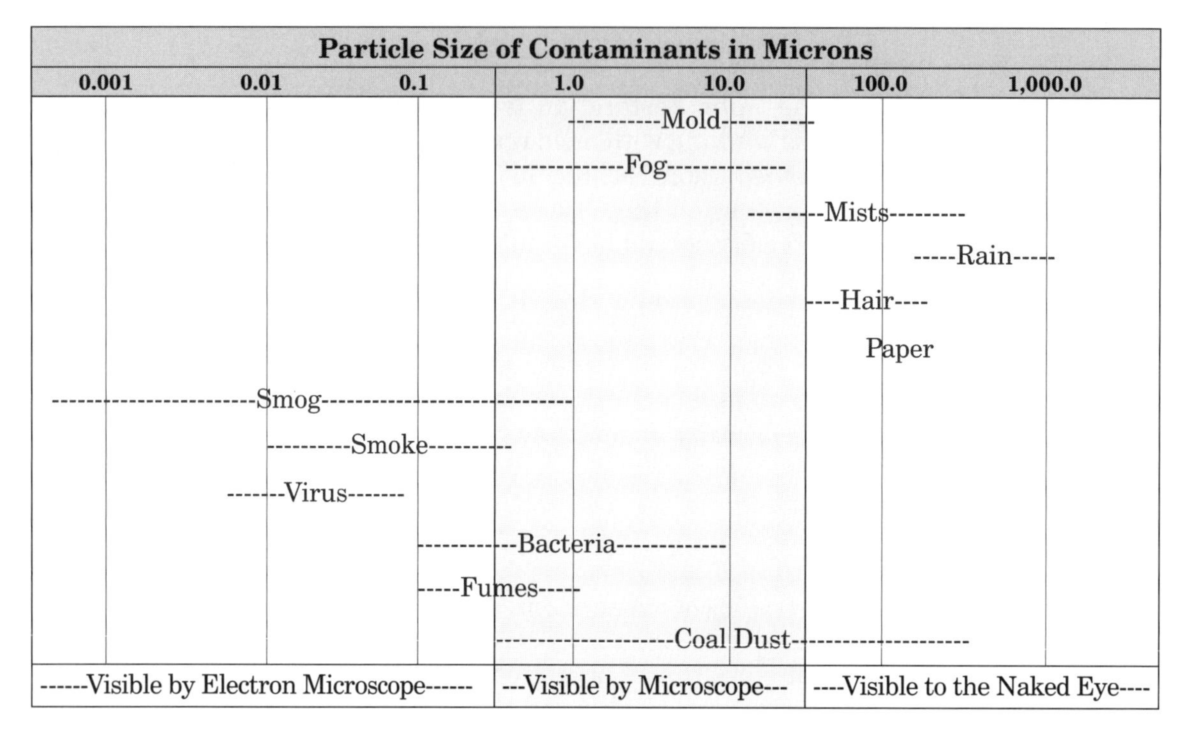

Figure 5-2 This table, comparing the micron size of common particles and other objects, gives you an idea of how small mold spores are.

Molds reproduce through the generation of millions of spores. Due to their microscopic size, mold spores are measured in microns. A micron, which is short for micrometer, is a unit of measurement equal to one millionth of a meter, or $^{1}/_{1{,}000}$th of a millimeter. In inches, a micron is 0.000039 of an inch. Common mold spores are only about 2 to 20 microns in size. As a comparison, the average human hair is about 60 to 70 microns in diameter, and a sheet of standard copy paper is about 100 microns thick. Figure 5-2 gives examples of the particle size of different contaminants and the particle size of a few common items for comparison.

Because of their small size and light weight, mold spores easily become airborne. They can travel great distances and remain in the air for long periods of time. That's how they come to be in the air around us, at all times. You can see the effects of these invisible spores in your home every day, from mold that forms on old bread, to the decay of fruit that's been left out too long.

Mold needs three things to grow:

1. an organic food source, which can be anything from paper to wood;
2. an abundance of moisture;
3. and, to a lesser degree, warmth.

In the average house, there are hundreds of potential food sources. So, when a house sustains a water loss, it's essentially creating a perfect environment for mold to

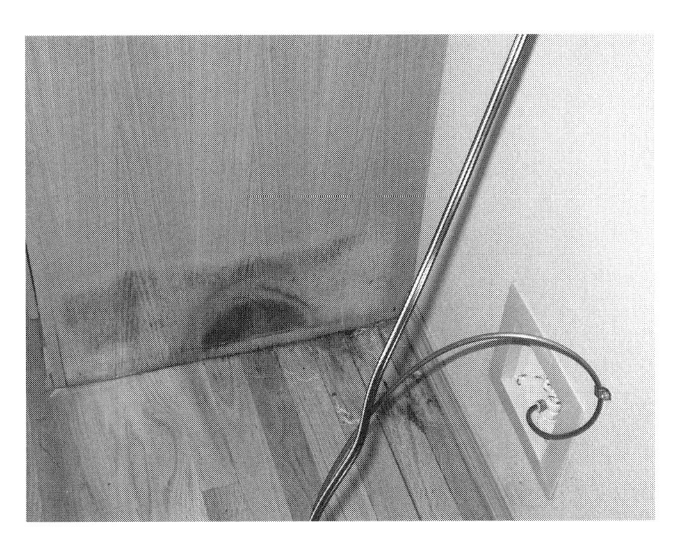

Figure 5-3 Even a small water loss has the potential for mold growth. This loss was caused by a drip in the fitting on a refrigerator icemaker line (on the right). Because it was behind the refrigerator, it went unnoticed. The water wicked up the side of the wood cabinet, causing mold to grow.

thrive. It has food just about everywhere, it has lots of moisture, and it has a nice warm place to grow. Once the mold has set itself up in its new home, it will begin to eat, and to reproduce by generating more spores.

For the most part, normal concentrations of mold spores in our homes don't bother us. Our bodies have adapted to their presence, and our respiratory systems can handle some spores without incident. But when a water loss and the resulting mold growth creates an overabundance of spores in the house, it can be more than our systems can deal with. That's where the health issue comes in.

People are affected differently by the various types and quantities of mold spores in the air. Those most affected are people with weakened respiratory systems, the elderly, infants, and individuals who already have allergies to mold. The Centers for Disease Control (CDC) offers the following information about how molds affect people:

> *Some people are sensitive to molds. For these people, exposure to molds can cause symptoms such as nasal stuffiness, eye irritation, wheezing, or skin irritation. Some people, such as those with serious allergies to molds, may have more severe reactions. Severe reactions may occur among workers exposed to large amounts of molds in occupational settings, such as farmers working around moldy hay. Severe reactions may include fever and shortness of breath. Some people with chronic lung illnesses, such as obstructive lung disease, may develop mold infections in their lungs.*

Mold Losses

Every house has the potential to become a mold situation. See Figure 5-3. And, the longer a wet condition in a house sits, the greater the chances for a serious mold growth.

There are two things you, as the restoration contractor, can do right off the bat to help reduce or eliminate mold growth. First, get the house dry by either removing wet materials or by drying the structure — or a combination of both. That will remove the moisture and food sources that the mold needs in order to grow.

Second, reduce the temperature in the house. At temperatures below about 55 degrees, mold growth slows down. However, as you saw in Chapter 4, lower temperatures also slow the drying process, so this becomes something of a balancing act. So, lower the temperature in the house to retard the mold growth until you can get the drying process started. Then, bring the temperature back up when you have your equipment in place to begin drying.

What do you do if the mold has already set up housekeeping in the building? The temptation is to immediately tear out the drywall and other materials it's living on. But in doing that you'd release millions of additional spores into the air. Unless you can do the demolition under very controlled circumstances, you run the very real risk of increasing the contamination and spreading it to other parts of the building. Once the contamination spreads, the risk to the building and its occupants spreads as well. So, if the mold is already there, you need to call in an industrial hygienist.

The Industrial Hygienist

Dealing with a mold loss not only creates a health risk, it can also create tremendous liability for your company. With a standard water loss, the extent of the water damage is visible to a large degree. What you can't see, you can detect with your moisture meters. And, you can then use the same meters to tell you when everything is dry.

But with a mold loss, detection and removal is more difficult because mold spores are always there. If you did a mold test in your house right this minute, it would test positive for a variety of different types of molds. So how much is okay, and how much constitutes a danger? And, since you can't see mold spores, how do you know when you've effectively removed enough for the house to be safe to occupy?

You need the services of an independent third party, an industrial hygienist, to perform the required mold testing for you. They have the necessary equipment to test the air and the surfaces in the house. They also have the knowledge, training, and experience to interpret the results of those tests.

The industrial hygienist, also called an occupational hygienist, is your new best friend when it comes to mold and contaminated water losses. The International Occupational Hygiene Association (IOHA) refers to Occupational Hygiene as:

> *The discipline of anticipating, recognizing, evaluating and controlling health hazards in the working environment with the objective of protecting worker health and well-being and safeguarding the community at large.*

While the primary focus of industrial hygienists is the workplace, their extensive training and education makes them ideally suited to work in residential situations as well. Industrial hygienists perform all of the necessary pre- and post-remediation tests, assess mold growth conditions, and provide you with written results of those tests.

The hygienist also provides you with a very important document known as a *protocol.* You can think of a protocol as an instruction manual for your mold or sewage remediation job. The protocol provides you with clear instructions on everything from worker protection to the specific procedures, chemicals, and equipment to use during the remediation. You'll find an example of a typical mold remediation protocol in Appendix C.

Independent Certification

Imagine that you've been hired to paint over the graffiti on the outside of a building. First you paint all the big surfaces, and then you paint all the little nooks and crannies. When you're done, no graffiti remains. If any does, then you haven't done your job correctly, and it's obvious to everyone.

Unfortunately, mold remediation isn't that simple and straightforward. Suppose you finish up a mold remediation job, and two weeks later someone in the house gets sick and files suit against your company. How can you prove in court that you're not to blame?

To protect yourself and your company, you need to have certification from an industrial hygienist that the job was done correctly, and the building was safe to occupy. You, and the courts, rely on their experience and training to ensure the safety of the occupants. Technically, you could purchase the equipment and hire the labs and do all of the testing on your own. But certifying your own remediation procedures is an obvious conflict of interest. It's much better to have an independent professional test and verify your work.

REAL STORIES:

Why We Love Independent Reports!

A contractor was hired to do a fairly extensive remodeling job on an older home. As part of the remodeling, his plumber installed several new copper water lines in the walls and under the floor. When the remodeling was just about done and the owners were checking the house before moving back in, they noticed a musty smell coming from one of the newly-renovated rooms.

An inspection revealed that a fitting on one of the new hot water lines under the house hadn't been soldered properly. It had been spraying a fine mist of hot water into the crawlspace for quite some time, resulting in a substantial mold growth covering the crawlspace framing. Openings in the floor framing had allowed the moisture to reach up into the interior wall cavities — all the way to the attic — all of which now had mold growth.

Our company was hired to do the demolition and remediation work. The original contractor was then going to come back in, make repairs, and put everything back together. Our first step was to bring in the industrial hygienist we regularly use. She performed a number of air tests, determined what molds were present and in what quantities, and wrote a protocol for us. Over a period of a couple of weeks, we did all of the necessary work according to the protocol. The hygienist came back out, repeated the air tests, and wrote a report that declared the house safe to occupy.

The contractor came back in and finished up the repair work. While he was working, he became concerned that there was still mold present, and that we hadn't done our job correctly. So he hired our hygienist to come back out and retest the house. Sure enough, she found elevated levels of mold were present once again. But she knew that the house had been safe when we finished our work, so she did a little more investigating. It turned out that the contractor had used several pieces of very wet lumber, which were supporting mold colonies, to make his repairs. He had inadvertently recontaminated the house!

We had to come back in, treat the wood and clean the air again. When we finished, the hygienist certified that house was again safe to occupy. Without the hard work and the independent certification of that hygienist, we would have had the almost impossible task of proving that we had done our work correctly in the first place!

(Photo courtesy of Zefon International, Inc.)

Figure 5-4 An air sampling being done to test for mold spores inside a wall cavity. The air pump draws air through the spore trap (far right). The pump is calibrated to draw air at a precise rate for a set amount of time.

Locating an Industrial Hygienist

To understand more about what industrial hygienists do, and for help in locating a qualified hygienist in your area, you can contact one of the national organizations that conduct the testing and certification of industrial hygienists:

- American Board of Industrial Hygiene
- American Industrial Hygiene Association

You'll find their contact information listed in Appendix A.

Ten Steps in a Typical Mold Remediation Job

In mold remediation, there are certain procedures you need to follow. It's a fairly methodical, step-by-step operation. Take your time and complete each stage — and pay attention to detail. The following are the 10 basic steps involved in a typical mold remediation job. But remember, *each job is unique. Your protocol will dictate the exact steps you need to take.*

Step 1: Eliminate the Source

As with any other water job, your first step is always to determine what caused the water to enter the building, and then to make sure that's taken care of. It won't do you much good to go through a mold remediation process and dry the structure if there's still water coming in.

Step 2: Conduct Pre-Remediation Testing

This is where your industrial hygienist makes an appearance. He or she will inspect the site, ask the cause of the water loss, determine the extent of the water damage and the mold locations, do some moisture testing independently of what you may have already done, and then perform a series of specialized tests.

The testing that the hygienist performs may vary between hygienists and jobs. However, there are three fairly standard tests that you'll see done:

- **Air Sampling:** In an air sampling test, a small air pump, like the one in Figure 5-4, draws a measured amount of air through a spore trap at a precise flow rate (see Figure 5-5). The spore trap is then analyzed in a laboratory to determine the spores present and the quantity of each

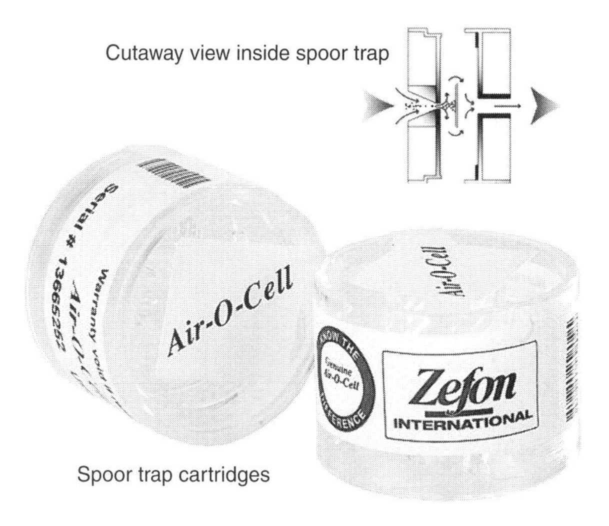

(Photo courtesy of Zefon International, Inc.)

Figure 5-5 To prevent possible contamination, the spore trap cartridges are kept sealed until ready for use. After they're used, they're resealed and immediately sent to a lab for evaluation. The cutaway view of the inside of a spore trap shows how the air is channeled into the canister. Any mold spores in the air stream are trapped on the plate in the center.

type. Since the size of the trap, the amount of air flow, and the duration of the air flow are all regulated, this provides the hygienist with a uniform test that can be easily duplicated during the post-remediation clearance testing.

The hygienist usually repeats this test in several places throughout the house. One or more samples will be taken in the rooms where the mold is visible. Samples are also taken in other parts of the house to determine if the spores have spread, even though mold isn't visible. And one test will be taken outside the house as a control sample. This one provides the hygienist with information about what mold spores are naturally present in the air around the house, and in what quantities. Figure 5-6 shows the Lab Report from three tests taken inside a basement and the outdoor control test.

- **Tape Sampling:** A special sticky tape material is used for this test. The tape is pressed on top of a horizontal surface to determine if there's an excessive amount of mold spore settling out of the air (see Figure 5-7). It can also be used on a moldy surface to identify the specific type of mold growth. The tape is placed on a microscope slide, inserted into a special shipping case, and sent to the lab where the number and type of spores are counted.
- **Mold Culture:** For this test, a specialized mold culture medium is exposed to the air using a specific type of air sampler. The sample is taken for a precise period of time at a specific flow rate. In the lab, the mold spores are allowed to grow on the culture. This test is primarily performed when the hygienist needs more specific information about the types of molds present in the building.

Step 3: The Remediation Protocol

Once the hygienist has the pre-remediation test results, he'll prepare a specific protocol for the job. The protocol will provide you with considerable information about the mold, as well as laying out the basic parameters for the remediation. Among the information in the protocol you'll find:

- **Baseline spore counts:** This establishes a series of baseline numbers for mold spore counts in each of the areas where the initial testing was performed, including outdoors. These numbers tell you the condition

Lab Report

Sample Identification	East end of basement				West end of basement				Carpet				Outdoors			
Date Analyzed	12/01/2011				12/01/2011				12/01/2011				12/01/2011			
Volume(M³)	0.0750				0.0750				0.0750				0.0750			
Percent Of Trace Analyzed	100% of Trace at 600x Magnification				100% of Trace at 600x Magnification				100% of Trace at 600x Magnification				100% of Trace at 600x Magnification			
Debris Rating	2				2				4				2			
Analyte	**Total Count**	**Count/M³**		**%**	**Total Count**	**Count/M³**		**%**	**Total Count**	**Count/M³**		**%**	**Total Count**	**Count/M³**		**%**
		Result	**DL**			**Result**	**DL**			**Result**	**DL**			**Result**	**DL**	
Mycelial Fragments	2	27	13	n/a	2	27	13	n/a	45	600	13	n/a				
Pollen	6	253	42	n/a	2	84	42	n/a	43	573	13	n/a				
Total Fungal Spores	144	6050	42	100	1966	82800	42	100	341	14100	42	100	9	321	42	100
	Fungal Spore Identification				**Fungal Spore Identification**				**Fungal Spore Identification**				**Fungal Spore Identification**			
Alternaria													1	13	13	4
Arthrinium																
Ascospores	1	42	42	0.7					4	169	42	1.2				
Aspergillus/Penicillium	135	5700	42	94.2	1950	82300	42	99.4	236	9960	42	70.6	7	295	42	91.9
Basidiospores	6	253	42	4.2	7	295	42	0.4	37	1560	42	11.1				
Bipolaris/Drechslera									1	13	13	0.1				
Botrytis																
Chaetomium					1	13	13	0								
Cladosporium	1	42	42	0.7	2	84	42	0.1	51	2150	42	15.2				
Curvularia									1	13	13	0.1				
Epicoccum									3	40	13	0.3				
Ganoderma									3	127	42	0.9	1	13	13	4
Myxomycetes	1	13	13	0.2	1	13	13	0								
Oidium/Peronospora																
Pithomyces																
Rusts																
Smuts/ Periconia																
Stachybotrys																
Scopulariopsis					3	40	13	0								
Torula																
Ulocladium					2	27	13	0	4	53	13	0.4				
Unclassified Conidia									1	13	13	0.1				

(Report courtesy of Wise Steps, Inc.)

Figure 5-6 A typical lab report showing the results of an air-sampling spore-trap test. For this particular job, four pre-remediation tests were performed: three inside the house, and a fourth, the control test, outdoors. The control test established what mold spores were naturally present in the environment around this home. Notice the high readings under the heading of *Aspergillus/Penicillium*, especially in the west end of the basement (82,300). The outdoor reading was only 295. If the remediation is performed successfully, post-remediation-testing should show levels in the basement very close to the outdoor readings.

of the home at the start of the remediation. Remember that you can't get the spore count all the way down to zero — all you can do is make a significant reduction in the count. The only way to document that reduction is to know what the levels were when you started the job.

- **The types of mold present:** Based on the testing, the protocol establishes the types of molds that are in the building. Since different types of molds are more prevalent under certain conditions, this will help the hygienist determine what may have caused a particular type of mold to grow. That's a helpful clue in making sure you've identified and removed the sources causing the mold infestation. Also, because certain types of mold spores are potentially more dangerous than others, you need to know what you're dealing with.

(Photo courtesy of Zefon International, Inc.)

Figure 5-7 A sticky tape test is used for taking mold spore samples of potentially contaminated surfaces. After testing, the tape samples are sealed in a case and sent to a lab for analysis.

Some of the most common molds found in buildings include *Alternaria*, *Aspergillus*, *Cladosporium*, and *Penicillium*. Another type of mold, *Stachybotrys*, is one that you have may have heard of as well. It creates a toxin that can make some people quite ill, so it's been in the news more than some other types of mold. However, many molds have the potential to make toxins, so Stachybotrys isn't really all that unusual. In fact, any mold can make a person sick under the right conditions. You need to treat all molds with respect when you're working a mold remediation job.

- **Worker safety precautions:** Based on the findings of the initial testing, the protocol will establish minimum standards for worker safety. This includes the personal protective equipment (PPE) required (see Figure 5-8), as well as any safety precautions that workers must follow. We'll take a closer look at PPE later in the chapter.

Figure 5-8 A worker in personal protective equipment (PPE), including a disposable suit and respirator, working in a crawlspace.

- **Remediation procedures:** Finally, the protocol covers the specific remediation procedures recommended by the hygienist. That includes the materials to be torn out, the type of equipment to be used to control emissions, the areas to be sealed and treated, and any other procedures needed.

Step 4: Erect Containment

When part of a building is contaminated with mold, it's important to protect the rest of the building so the spores don't spread. The contaminated portion of the building must be physically separated from the rest of the building in a process known as *containment.* This involves setting up plastic sheeting to seal off the contaminated rooms. The exact areas to be contained will be specified in the protocol.

> ***Tip!*** *For everyone's safety, it's a good idea to use 6-mil flame-retardant plastic sheeting for containments, clean rooms, and other remediation uses, rather than standard plastic sheeting. Flame-retardant sheeting should be available by special order from any supplier that sells standard sheeting materials.*

When you set up your containment, consider how you're going to access the contained area. For example, suppose you're working in a bathroom at the back of the house. At first, it seems like it would be easy to isolate and contain that area, since only one small room is involved. However, in order to get to that bathroom from outside, you have to walk through the living room, hallway, and bedroom. That means you'll be spreading mold spores through those three rooms as you move back and forth to the bathroom, and eventually the spores will travel throughout the entire house.

Part of your containment must include a safe, enclosed interior walkway from an exterior door to the work area. That allows workers to move from outside the building to the containment areas without contaminating anything else. Hang plastic sheeting from temporary support poles, or any other means that works for your particular situation, to create containment corridors. Don't forget to cover the floor with plastic sheeting as well.

One particularly effective means of erecting containment is a system called ZipWall® (Figure 5-9). ZipWall uses spring-loaded poles, combined with top and bottom plates that allow you to quickly and easily set up plastic sheeting wherever you need it. It also works great for dust control, as shown in Figure 5-10. You can add zippers wherever you need them for safe access to the containment areas (Figure 5-11).

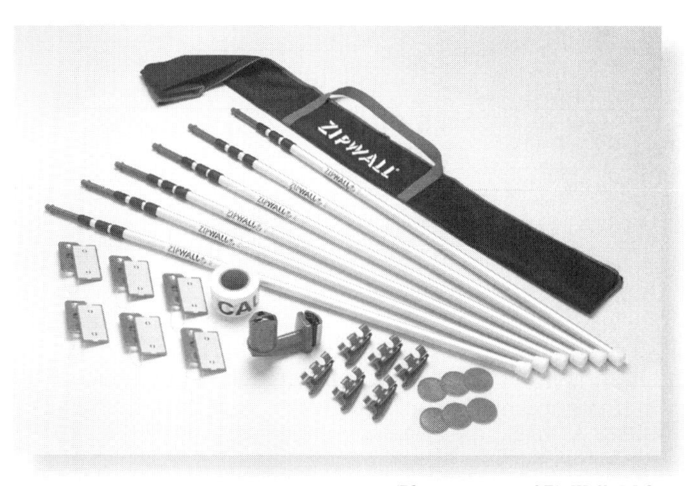

(Photo courtesy of ZipWall, LLC)

Figure 5-9 An example of a complete ZipWall kit, which greatly simplifies setting up containment for mold remediation jobs. It also works well for general dust control purposes.

(Photo courtesy of ZipWall, LLC)

Figure 5-10 The ZipWall poles adjust to different lengths, and pads hold the plastic sheeting in place as you lift the poles. This makes it easy for one person to set everything up.

(Photo courtesy of ZipWall, LLC)

Figure 5-11 Wherever access is needed to the containment area, simply cut the plastic sheathing and add a zipper.

Also included in the containment section of your protocol will be procedures for constructing clean rooms. A clean room is an area that's enclosed on all sides with plastic sheeting, and has double flaps or zippers at each end for doors. The clean room gives workers a place to put on and take off protective clothing as they enter and leave a containment area.

> ***Tip!*** *You can construct an inexpensive portable clean room out of PVC pipe and fittings. The pipe can be quickly assembled on-site, then wrapped with plastic sheeting on the sides, top and bottom to create an enclosed room. When the remediation work is complete, simply remove and discard the sheeting, then disassemble and clean the pipe and store it for next time.*

REAL STORIES:
When is a Contained Area *Not* a Contained Area?

We were once asked to step in and take over a mold remediation job that had been started by another contractor. He didn't specialize in this type of work, but the project was limited to just a wet laundry room, so he thought he could handle it. The room had fairly high concentrations of moldy materials, but they were contained within an easily isolated space. It should have been a small and pretty straightforward project.

At least, it was small and straightforward when he started. But, it didn't stay that way for long!

The contractor hadn't bothered with a protocol, or even pre-remediation testing. He just hung a piece of plastic over the door, and considered the area contained. From there, he was pretty much just winging it.

The laundry room had an exhaust fan, so he turned it on to create a little negative air pressure in the room. Normally that would be a good thing, but he hadn't checked to see if the fan was ducted to the outside of the house. Although it should have been, it wasn't; and he'd spread the mold spores into the attic, where they settled out onto the blown-in insulation.

He then borrowed a dehumidifier from another company to help with the drying. Under most circumstances, this also would have been a good thing. However, since the laundry room wasn't very big and he needed room to work, he set up the dehumidifier in the hallway outside the laundry room. And, since dehumidifiers have a fan in them, and need to draw air *toward* them in order to work, the dehumidifier pulled contaminated air out of the laundry room into the adjacent hallway.

His last mistake was to turn up the heat to aid with the drying. Again, this would have been the right thing to do in many situations, but not in this one. Why? Because the furnace was located *inside* the laundry room. So while he was working away, tearing out moldy drywall and releasing huge quantities of mold spores into the room, the fan on the furnace was happily sucking in all those spores and then pumping them through the duct system where they were distributed throughout the entire house!

All in all, it's a perfect example of how good intentions, but poor execution, quickly turned a simple one-room project into a whole-house remediation!

Step 5: Install Drying and Air Scrubbing Equipment

The next step in the sequence is installing specialized drying and air cleaning equipment. You can set up the equipment all at once, or during the various stages of the remediation. You'll be using three different types of equipment, and each has its own distinct use in the overall remediation process. Let's look at those three processes:

1. ***Drying the building:*** It takes some type of water intrusion to get mold started, so one of the first things you need to do is dry out the building. The drying procedures and equipment used for mold remediation jobs are typically the same as those used for any other drying job.
2. ***Scrubbing the air:*** In addition to drying the building, you also need to clean the building's air to remove airborne mold spores. This critical process is commonly known as *scrubbing*, and requires the use of specialized HEPA filter air scrubbers. We'll talk about HEPA (high-efficiency particulate air) equipment in more detail at the end of this chapter.
3. ***Maintaining negative air pressure:*** The sections of the building with mold contamination should be isolated from the unaffected areas. In order for those unaffected areas to remain free of mold spores, you need to keep the contained areas under negative air pressure during the remediation process.

Let's use the operation of a bathroom ventilation fan as an example of creating negative air pressure. When the fan is turned on, it exhausts air from the room. Since there's now less air in the bathroom than in the surrounding rooms, the bathroom is in a negative air condition. To fill that void and to equalize the pressure in the bathroom, air will flow toward the bathroom from other parts of the house.

You want the same condition existing in the containment areas on a mold remediation job. As long as there's negative air pressure within the containment area, outside air will flow *toward* the contaminated rooms, keeping the spores safely contained. Air scrubbers are all that's typically required to keep the containment area under negative pressure. However, in some circumstances, especially where large spaces are involved, you may need supplemental exhaust fans or negative air fans as well.

When you've erected your containment area and activated the air scrubbers, you'll see the plastic sheeting bow inward a little into the containment area. This is a good indication that you've correctly achieved negative air pressure within the containment.

Step 6: Seal Penetrations

As we discussed, mold spores are everywhere in the air, and you can't remove all of them. So, when you create a negative air flow in a room, air is going to be drawn into the room from somewhere else.

It's very important to be aware of *where* the air being pulled into the containment is coming from. Sometimes there are sources of mold unknown to you in the house. Those spores can be accidentally pulled into the containment area, so always provide *filtered* make-up air for your containment.

Figure 5-12 The penetrations around the pipes were sealed with expanding foam to prevent any spores from coming up from the crawlspace.

For example, if you create negative air in a room that's over a crawlspace, you may be drawing air from under the house into the room you're trying to dry and clean. If the incoming air contains an elevated number of spores, it can make scrubbing the air a more difficult task. Draw air first though a filter, then into the containment.

The hygienist may require you to seal gaps between the areas being cleaned and undesirable areas, such as attics and crawlspaces. You can use caulking, spray foam, plastic sheeting, or other methods to seal the areas before installing the air scrubbers. Some common areas to seal off include wiring and plumbing penetrations, gaps around heat registers, gaps around crawlspace and attic access doors, and openings in framing. Figure 5-12 shows foam filling the spaces around pipes coming up from the crawlspace in this house.

The hygienist generally takes an outdoor air sample in addition to the interior ones. In some instances, there may be elevated spore counts in the air outside the house. This occasionally occurs on farms and ranches, or during times of unusual environmental conditions, such as high pollen counts. In that case, the hygienist may require that you seal off all exterior doors and windows in the contained areas as well. You can use 6-mil plastic sheeting and blue painter's masking tape to create an effective temporary seal.

Step 7: Remove and Dispose of Affected Materials

A properly managed mold remediation, like a properly managed restorative drying project, allows you to salvage a lot of the materials in the building. But there are always going to be materials that either can't be salvaged because of their condition, or that are simply not cost effective to try and save.

When you and your hygienist do the initial walk-through on the job, you'll examine the various materials in the building and make decisions about what can and can't effectively be saved. Those decisions become part of the protocol, which gives you, the homeowner, and the insurance company a clear idea of what's going to be torn out.

Because mold spores are everywhere — especially in landfills, where it's basically a buffet lunch for them — mold-contaminated materials don't usually have any special disposal requirements. They're not like asbestos or certain other materials that have to be treated as hazardous waste. Unless the hygienist tells you otherwise, or unless the local landfill has some restrictions, moldy materials can be bagged and disposed of in much the same way as other construction debris.

Just be careful to not contaminate clean areas with the debris when you move it from a contained area out to the dumpster. For example, carrying a roll of moldy carpet or a big chunk of moldy drywall through a part of the house that isn't contained and wasn't previously contaminated will quickly spread the mold spores, and will expand the contamination area. If you've properly set up containment corridors from an exterior door to your work area, removing contaminated materials shouldn't be much of a problem. If necessary, you can also wrap carpet, drywall, and other bulky items in plastic sheeting and seal it with tape prior to carrying it outside.

Finally, *don't use standard trash bags for mold debris.* They're too easily punctured, which can release spores into areas where you don't want them. Instead, use bags made from 6-mil plastic. These heavy-duty bags are available from most janitorial supply houses, as well as through remediation equipment and supply distributors.

Step 8: Clean and HEPA Vacuum Remaining Surfaces

After all nonsalvageable materials have been removed and disposed of, the next step is dealing with what remains. On large remediation jobs, that often includes framing lumber, sheathing, and other structural materials.

One option is to sand the surfaces, using a power sander attached to a HEPA-filtered vacuum cleaner to contain the dust. Another option is hand sanding or hand scraping. In some cases, wet washing might be an option.

However, due to hidden areas and irregular surfaces, a lot of these materials are difficult to clean effectively. At the very least, it can take many tedious manhours to

Figure 5-13 This attic framing and sheathing was once covered with mold. After soda blasting, it's actually cleaner than it was when the house was brand new.

Figure 5-14 Specialized equipment for soda blasting. Compressed air is used under very controlled conditions to blast bicarbonate of soda (baking soda) against building materials, safely removing mold spores.

remove the mold spores. An alternative used by many restoration contractors is surface blasting. This method uses compressed air to blow fine sand, crushed walnut shells, or other small particles at these materials. The particles are small enough to get into and clean small irregular surfaces.

A relatively recent innovation in mold-spore removal is soda blasting — using fine particles of bicarbonate of soda (baking soda) to do the blasting. This technique has proven very effective, and does very little damage to the structural materials (see Figure 5-13). However, it requires specialized equipment (Figure 5-14) and is best left to trained subcontractors.

After the spores have been scraped, sanded, or blasted off, you need to clean them up. Like fine dust particles, mold spores will eventually drop out of the air and settle onto horizontal surfaces. You might not see them in the same way that you see dust, but they're there, and they need to be vacuumed up and disposed of.

You need a specialized vacuum that traps the spores for easy disposal, one equipped with a HEPA filter. Standard shop vacuums stir up too much airborne dust and too many mold spores end up back in the air. We'll talk more about HEPA vacuums shortly.

Remember that each job will have very specific requirements for the removal and cleanup of the mold spores. Those requirements will be included in the protocol prepared by your hygienist.

Step 9: Encapsulation and Antimicrobials

Once you've removed whatever materials you're going to, thoroughly cleaned and vacuumed the remaining surfaces, and cleaned the air, there'll probably still be a certain amount of mold spore remaining in hard-to-reach areas. The final step is either to seal the affected areas in a process known as *encapsulation*, or to treat the areas with solutions that prevent further mold growth.

- **Encapsulation:** Encapsulation can be accomplished using either clear or pigmented shellac, or other approved sealers. The objective is to encase the surfaces and prevent any remaining mold spores from becoming airborne. Encapsulation is effective for specific isolated areas, such as a section of wood framing that's difficult to clean completely. It's also very effective for large areas such as subfloors or roof and wall sheathing, where extensive cleaning would be labor intensive and not very cost effective.
- **Antimicrobials and Biocides:** Antimicrobials are products which, by definition, control, inhibit, or destroy microorganisms. A biocide is another type of antimicrobial. However, a biocide is specifically formulated to kill the microorganism, not just inhibit its growth. Biocides may be designed to kill a wide range of microorganisms, or they may be formulated to work only on one specific target.

Tip! *A product with the suffix* stat, *means it's intended to inhibit or control without killing. For example, a bacteriostat is designed to control bacteria without completely destroying it. On the other hand, a product with the suffix* cide *means it's intended to kill the organism. An insecticide, for example, is designed to kill insects.*

Antimicrobials or biocides are usually spray-applied. The fine mist covers wide areas quickly, and penetrates small gaps in framing and other areas. They're often a final step after all the other remediation has been completed. Your hygienist will list the specific antimicrobial treatment areas and products to use as part of the remediation protocol.

All antimicrobial products need to be handled with great care. Each one has very specific mixing, handling, application, and ventilation requirements. There are also requirements and recommendations for the appropriate clothing, respirators, and other safety equipment to use when applying them. *Be sure that you obtain a Material Safety Data Sheet (MSDS) from your dealer for every product you purchase. Always read and follow both the MSDS and the manufacturer's specific instructions!*

Step 10: Clearance Testing

The final step in the mold remediation process is clearance testing. Clearance testing is done at the very end of the job. It allows the hygienist to evaluate the effectiveness of the remediation work, and the current condition of the building.

The hygienist repeats the same tests that he took initially, in the same locations. These tests will be evaluated by a lab, and the results documented in writing. By comparing the spore counts after the remediation with the spore counts in the same locations prior to the remediation, as well as with the outdoor control tests, the hygienist can determine if the air quality in the building has improved enough to be considered safe for occupancy.

The importance of these clearance tests can't be overemphasized! Not only do they establish that the building is safe for the occupants, but from a liability standpoint, they also provide written, independent verification that you've done your job correctly.

Sadly, you always need to remember that where there's mold, there's also likely to be attorneys anxious to find fault and file lawsuits. Proper documentation, including clearance testing, is your best protection against them.

Mold and Insurance Coverage

Now that we've covered the basics of mold remediation, here's another problem that you're likely to run into: Mold is most likely *not* covered by insurance.

It used to be. But in the wake of several natural disasters and all the moisture and mold problems associated with EIFS (Exterior Insulation and Finish Systems, also known as synthetic stucco), which had moisture intrusion issues, that's all changed. Today, most, if not all, homeowner's insurance policies either specifically exclude mold coverage, or they have very low maximum coverage amounts.

Any mold coverage, if it's there at all, will only extend to mold that's the result of a sudden and accidental loss, as opposed to a maintenance issue. In other words, when a pipe breaks and the building gets wet and mold starts to grow, that's part of a sudden and accidental loss. But if the mold is the result of ongoing seepage or it's caused by something like a leaking roof that's outlived its useful life span, then that's a maintenance issue.

For you, the severely limited mold coverage can become a real headache. You might be called in on a water loss, which is covered, only to find mold growing inside the wall, which isn't covered. So how do you fix one issue without addressing the other?

To some degree, there's an overlap that works in your favor. Let's say that a broken pipe has caused some drywall and carpeting in a bedroom to get wet. Both of these materials are saturated to the point where they can't be saved, and both have developed mold.

Since they got wet as part of a covered loss and aren't salvageable as a result of that same loss, they can be torn out and disposed of, and the insurance company will cover it. The fact that they have mold on them isn't an issue, since they're covered as part of a water loss, not a mold loss.

What *is* an issue, however, is that as soon as you tear those materials out, you'll be releasing mold spores into the room. You'll be potentially contaminating the air, and you'll need to deal with that problem. The insurance company is almost certainly *not* going to foot the bill for that, so it's going to be up to the homeowner to pay for the additional mold-related costs. In most cases, it's an expensive undertaking, and the average homeowner will be reluctant to pay for the services of a hygienist and the other costs involved with mold remediation.

So you're stuck. If you do anything to contaminate the house, it becomes your liability. Even if the current homeowner assures you that they won't hold you liable in any way, and even if they're willing to sign a disclaimer, at some point they're going to sell the house. When they do, they have a legal obligation to disclose the fact that there was mold in the house at one time. Their buyer may have concerns about the mold and come after you for proof that the house is safe. Without the necessary testing, you simply don't have any legal proof to offer them. Even with your disclaimer in hand, a smart lawyer will argue that you're a professional. As such, you should have recognized that the situation was potentially hazardous, and you should never have allowed it to happen.

So, if you find yourself in a situation, like the one in our example, where there's a potential mold hazard and no one wants to pay for remediation, you basically have three choices:

1. ***Proper Remediation:*** You can do a proper remediation, get your clearance tests, and bill the homeowner for the costs. You can shift as much as of the cost as possible onto the insurance company as part of the covered portion of the loss, including demolition, debris removal, and reconstruction. However, the costs of the hygienist, the testing, and the air scrubbing may still be substantial, and will have to borne by the homeowner.
2. ***Take Your Chances:*** At some point, you'll probably be tempted to go ahead and undertake the water loss and bill the insurance company for all of the covered work. As far as the mold is concerned, you can take a head-in-the-sand approach. You can hope that through a combination of natural ventilation and removal of the affected materials, the mold infestation won't be a problem. But be forewarned — you're taking a tremendous risk! Not only could you cause someone to get sick, but you could also sink your company in a sea of lawsuits. *This option is simply not worth the risk to you or to the building's occupants. Don't do it!!*
3. ***Walk Away:*** It's tough to do, but there are going to be a lot of these jobs that you simply have to walk away from. If the insurance company isn't going to cover the mold remediation, and the owners either don't want to do it or can't afford to do it, you're in a lose-lose situation. It's lousy for you, and lousy for the homeowner, but that's the reality of it. The risk is too great; you just have to walk away and go on to the next job.

Contaminated Water Losses

This area is certainly not the most glamorous part of insurance restoration, but at some point you're going to encounter losses involving contaminated water, including sewage. A contaminated water loss could be something as simple as an aquarium that leaked or a single toilet that overflowed, or it could be a massive sewer backup that affects an entire house. Or worse.

Beyond the jokes and the "yuck" factor of sewage and contaminated water lies a very real health danger. Like mold remediation, the cleanup and repair of these losses requires special precautions. They need to be carefully planned, and that plan needs to be conscientiously carried out.

Water Categories

In Chapter 4 we talked about water damage classifications, ranging from a Class 1 to a Class 4 loss. Those classifications deal with the wetness of the structure and its contents. You can use them as a guide for planning how much equipment is needed to dry a structure.

There's a second set of classifications you need to become familiar with: the three water categories. These classifications include sewage losses. Even when a loss initially involves clean water, it can become contaminated fairly quickly under the right circumstances. The water categories will help you determine how potentially dangerous the loss is, what materials you can safely salvage, and what precautions you need to take during the course of the remediation work.

Water categories are based on several factors, all of which influence the water's potential danger. Whenever you enter a wet building, it's very important that you consider these factors. You can then weigh the information to decide how to deal with the loss. To determine the water category, ask yourself the following questions:

- **Where did the water come from?** Was it a water line? A toilet? A washing machine? Part of a sewer or septic tank malfunction?
- **How long was the water in the structure and in contact with the affected materials?** Did you get the call within a short time of the occurrence, or had the water been in the building for days before it was discovered?
- **What was the initial temperature of the water?** Did it come from a cold water line, a hot water line, from a water heater, a boiler, or another hot water source?
- **What are the temperatures inside the building?** Is the building warm or cold now? What was it at the time the water entered?
- **What preexisting conditions do you need to know about?** Did the water run across an area that was contaminated in any way? For example, is there pet urine or fecal matter in the carpet or pad?

After considering these factors, you can make a determination about which water category applies to the loss. Remember that these are only general guidelines. *If you have any doubts, questions, or concerns about the category of water you're dealing with, always take Category 3 precautions until you can consult with an industrial hygienist or other experienced professional and complete any required testing.*

The following quoted material dealing with water categories is taken from the *Guide to Restorative Drying, Revised Edition*, from the Dri-Eaz Education Series.

Category 1 Water

Category 1 water is *"water with no significant risk of causing sickness or discomfort."* In other words, it's considered clean water. The water has come from a potable water source, such as a frozen water pipe, a defective water pipe fitting, a broken icemaker line, or is the result of damage to a clean water source.

Regardless of the water source, the water *cannot* have been inside the building for more than 72 hours in order to be classified as a Category 1 water loss. Also, the contents and structural materials in contact with the water must be generally clean and well maintained. And, there should be no odors present in the house.

In the case of a Category 1 loss, you can typically dry the structure without any special precautions. However, if the water has been in the house for more than 72 hours, or if you detect odors, or there are materials that you don't feel are clean, then the water must be degraded to a different category.

Category 2 Water

Category 2 water involves *"water that carries a significant degree of chemical, biological, and/or physical contamination."* This classification includes water from sources such as *"aquariums, waterbed leaks, toilet bowl overflows that originate from the sanitary water supply yet contain urine from the bowl, dishwasher discharge, clothes washer discharge, and water that enters the structure from hydrostatic pressure (from below grade)."*

> ***Tip!*** *You may hear Category 2 water also referred to as* gray water. *In general usage, gray water is water that's not clean and potable, and that may contain some type of liquid contamination. However, for safety and accuracy, it's better to understand exactly what Category 2 water intrusions cover.*

"In order to remain a Category 2, water must not be allowed to dwell in the structure for an extended period of time. If the water source was originally a Category 1, the water will degrade to a Category 2 after approximately 72 hours. When water is a Category 2, it will continue to degrade and become a Category 3 after 48 hours."

Remember that these are only general guidelines. Water with a high initial temperature or in structures with higher-than-normal indoor temperatures will degrade more rapidly. Also, the higher the degree of contamination in the water initially, the faster it will degrade.

When dealing with Category 2 water, you'll need to take special precautions. These *minimally* include: protective gear for workers, removal and disposal of carpet pad, hot water cleaning and extraction of carpets, and treatment of porous materials using approved antimicrobial solutions.

Category 3 Water

Category 3 is the highest category and is generally considered the most unsanitary and potentially dangerous water. A Category 3 loss is one that *"results from a grossly unsanitary source, carries pathogenic (disease causing) agents, or when water has dwelled in a structure for more than 120 hours."*

Tip! *Category 3 water may also be referred to as* black water. *This generic term is typically taken to mean that the water is contaminated by both liquid and solid waste materials. Here again, you need to understand what Category 3 water intrusions are, to know them by that term, and to handle them accordingly.*

"Category 3 water sources include: discharge from toilets that originate from beyond the toilet trap (from the sewer or septic system), and intrusions from the surface of the ground into the structure (flood waters).

"Regardless of the category of water from its initial source, water can be deemed a Category 3 if it is allowed to dwell in the structure for an extended period of time. Typically, this period of time is approximately 120 hours, or five days. The amount of time necessary depends upon temperature and the water's content after migration through the structure."

A Category 3 water loss poses an extremely significant health risk to you, your workers, and any occupants of the building. When you encounter a Category 3 loss, for safety and liability issues your best course of action is to consult with an industrial hygienist or other qualified third-party professional.

The Industrial Hygienist

With a loss involving Category 2 or Category 3 water, you're dealing with a lot of bacteria. As with a mold loss, these are invisible microorganisms, so it's extremely difficult to be able to judge the overall effectiveness of your remediation work. You need to get an industrial hygienist involved from the very beginning.

The hygienist may take air samples, or samples on various surfaces. The exact type of testing will depend on the type and extent of the loss. Once the testing is complete, the hygienist will write a protocol for the remediation, which will be similar to those for a mold remediation.

There's a high likelihood that workers on a Category 2 or Category 3 water loss will be exposed to pathogens that could result in illness. So the remediation protocols for these types of losses will also include a section on personal hygiene, such as hand cleaning.

Contaminated Water Remediation

The methods for dealing with contamination by a Category 2 or Category 3 water loss are very similar to those for a mold situation. The primary difference is that the bacteria tends more to be on surfaces, as opposed to the easily-airborne mold spores. And, some of the cleaning and surface treatment solutions used for these losses are different from those used on a mold remediation.

Depending on the extent of the loss, you may still need a containment system to protect other parts of the building from contaminants. You'll still be removing contaminated materials and using HEPA equipment to clean the air. After the work is complete, the hygienist will perform clearance testing to ensure that the building is again safe to occupy.

Tip! *If you're unsure about the conditions you're dealing with on a contaminated water loss, treat it as a Category 3. Make sure all your workers are fully outfitted with PPE.*

Depending on what contributed to the Category 2 or Category 3 water loss, you may need to treat contaminated materials as hazardous waste. If that's the case, the hygienist and your local landfill or hazardous waste disposal facility will dictate the proper disposal procedures. For both safety and liability reasons, be sure that you follow those procedures exactly.

Worker safety is an important consideration with any contaminated water loss. Workers will be required to wear the protective clothing and respiratory protection outlined in the protocol.

Contaminated Water Losses and Insurance Coverage

Whether or not a contaminated water loss is covered by insurance depends on where and how the loss originated.

If it was caused by a problem in the section of sewer line owned by the building owner, it's probably going to be covered. Problems with private sewage disposal systems, such as septic systems, are also normally covered. If it was the result of a toilet or washing machine overflow, or if it was caused by something like a break in an aquarium or a waterbed, chances are that will be covered as well.

If the loss was caused by a problem with a municipal sewage disposal system, beyond the point where the municipal system connects to the building's sewer line, chances are it won't be covered. For example, if a sewer line backs up in the street as a result of a problem with a city's system, and sewage enters an individual building as a result of that backup, the homeowner's insurance will probably not cover the loss. The insurance company will consider it the city's fault and the city's responsibility. However, municipalities are often legally exempt from paying for those losses. Even if they do, the amount paid is often insufficient to cover the overall cost.

Tip! *Insurance on a covered loss will generally extend to the cost of the hygienist and necessary lab testing, in addition to the remediation work.*

Another contaminated water loss that's generally not covered is one resulting from a flood or other surface water, such as snow melt that's become contaminated. The exception would be if the building owner specifically carried flood insurance coverage.

With any contaminated water loss, you need to establish whether or not the loss is covered. If it is, you want to know whose insurance is covering it, and if it'll be covered in full. If it isn't covered, you need to have a clear understanding with the homeowner that they'll be responsible for covering the cost of the loss.

Even if the homeowner is planning on going after their insurance company or the municipality for the cost of the loss, *never get caught in the middle* — make sure the homeowner pays you directly. The last thing you want to do is incur a substantial expense doing the remediation work, and then have to wait until the lawyers finish battling things out to get paid.

When coverage and payment issues arise, you have the same three basic options as you did with mold losses that aren't covered. You can do a proper remediation and bill the building owner, you can take your chances and do the work without the proper testing, or you can walk away. You know my recommendation: If the homeowner can't or won't pay, walk away!

The only way to do this type of work is the *right* way! As with the mold remediation, improper or inadequate remediation of a contaminated water loss can create a situation where someone may become seriously ill. *Don't put your company and your personal finances at risk.* It's not worth it!

Other Types of Contamination Losses

You'll occasionally be called out on losses that involve the cleanup of urine and feces *not* related to a water loss. These can include losses stemming from an overabundance of animals living in a home, or even people who've been without a functioning toilet and have been using alternate means of capturing and disposing of waste, such as plastic bags, instead. These types of losses must *always* be treated with the same degree of caution as a Category 3 contaminated water loss. This includes using personal protective equipment (PPE) as a worker safety precaution, and calling on the help of an industrial hygienist on larger losses.

Some popular television programs have taken a very casual approach in portraying this type of work, showing workers doing hazardous cleanup in tee shirts and simple dust masks instead of proper respirators. *Never make the mistake of overlooking the very real health risks of dealing with these types of biological hazards!*

Personal Protective Equipment (PPE)

When dealing with many types of water losses, especially mold and contaminated water losses, you'll need to use personal protective equipment, or PPE. There are actually several forms of PPE, to fit the specific circumstances of the job. However, they all have the same goal: worker protection.

Here's the OSHA definition of PPE:

> *"Protective equipment, including personal protective equipment for eyes, face, head, and extremities, protective clothing, respiratory devices, and protective shields and barriers, shall be provided, used, and maintained in a sanitary and reliable condition wherever it is necessary by reason of hazards of processes or environment, chemical hazards, radiological hazards, or mechanical irritants encountered in a manner capable of causing injury or impairment in the function of any part of the body through absorption, inhalation or physical contact."*

You need to have a good understanding of what PPE to use for specific situations. Your hygienist will usually specify the PPE necessary for mold and contaminated water losses. If you're in doubt about whether personal protective gear is needed, use it. Always be on the safe side of caution. Let's look at some of the basic PPE items.

Figure 5-15 A partial-face respirator covers the mouth and nose, and is held in place with a pair of adjustable straps behind the head. There are two replaceable filter cartridges on this style.

Respiratory Protection

Respiratory protection devices are among the most important types of PPE. They're used to protect your lungs and respiratory system from inhaling fumes, particulates, and other potentially harmful jobsite contaminants.

Dust Masks

There's a wide variety of respiratory protection available. The simplest form is the dust mask, which fits over the nose and mouth for protection against larger airborne particles, such as sawdust. Dust masks are useful when cutting or installing drywall, doing basic demolition work, cleaning, and many other construction tasks.

Respirators

For protection against fumes, the small particles and airborne pathogens present during a Category 2 or Category 3 water loss, or during mold remediation, you'll need a respirator. A respirator can filter and trap smaller particles than a dust mask can.

There are several basic types of respirators available. The one you choose depends on the jobs you'll be doing, how well it fits and seals, how comfortable it is, and your budget. Respirators are a vital piece of safety equipment, so be sure you get the right one. Your hygienist or a representative at a safety supply outfitter can help you with the selection process.

- **Partial-face respirator with disposable cartridge:** This is probably the most common style, covering the mouth and nose. See Figure 5-15. As you breathe in, the air is drawn through a disposable filtering cartridge. Exhaled air leaves the respirator through a one-way valve. To extend the life of the cartridges and prevent early clogging, there are pre-filters available. The pre-filters act like a standard dust mask, and filter out dust and larger particles before the inhaled air is drawn through the cartridge.
- **Full-face respirator with disposable cartridge:** It's similar to the partial-face respirator, but covers the entire face.
- **Battery-powered respirator:** These respirators have a rechargeable battery pack and replaceable filters. The battery is carried in a case on the worker's belt or on another type of harness, and typically provides about eight hours of air supply. These respirators supply cooler, more-consistent air, and are usually preferred by workers who'll be wearing them for long periods of time.
- **Supplied-air respirator:** A supplied-air respirator (SAR) utilizes a small air pump to generate a clean air supply, which is pumped

through a hose to the respirator mask. This is the most expensive type of respirator. It provides a very clean air supply that also helps cool the mask and prevent fogging. There are also SARs available that utilize a single pump to supply air to two separate hoses and two separate face masks. This allows two workers to share the same pump, making the system a little more economical.

Respirator Cartridges

Most respirators have one or two interchangeable, disposable cartridges on the front, like the one in Figure 5-16. The cartridges have different uses, and need to be selected for the type of exposure hazard. For instance, there's one type of cartridge designed for exposure to certain types of paint fumes; and another for the spores present during mold remediation. There's a wide variety of cartridges available, so it's important to select the proper one for the type of job you're doing.

Respirator Fit Testing

All employees working on jobs where there's exposure to known hazards need to be supplied with respirators. You'll also have to ensure that the respirators fit properly, and that the employees have access to replacement filter cartridges.

Since the purpose of a respirator is to prevent the user from breathing contaminated air, the respirator mustn't allow air to leak in around the edges. Each employee needs to be individually fit tested with his respirator to ensure a proper fit.

To fit test, the worker puts on and adjusts his respirator. A smoke vial from a special testing kit is waved in front of the respirator. The vial generates a safe smoke with a pungent odor. If the wearer is unable to smell the odor, then the respirator is making a proper seal. Otherwise, the respirator either needs to be adjusted for a better fit, or replaced with a different style of respirator.

Figure 5-16 Disposable respirator cartridges screw onto the respirator. Different types are rated for different uses, indicated by a combination of their color and their numerical coding.

Respirator fit-testing kits and replacement smoke vials are available from suppliers that sell professional-grade respirators and cartridges. Some suppliers will also do the fit testing for you, either free or for a small fee.

One of the things that negatively affect the fit of a respirator is facial hair. Some types of full-face respirators will work with a short well-trimmed beard, but most types simply won't seal well. For safety, you'll need to develop a company policy addressing the issue of facial hair for employees who may be required to wear respirators, and make it part of your company handbook.

Respirators and cartridges are regulated by the National Institute for Occupational Safety and Health (NIOSH), which is part of the Centers for Disease Control (CDC). See Appendix A for more information about contacting either of these organizations.

Gloves

Gloves are another important part of a worker's PPE, and there are several styles to choose from.

Disposable latex gloves are appropriate for many contamination situations. They're inexpensive, come in several different sizes, and hold up well for light to moderate tasks. For people with latex allergies, there are also non-latex disposable gloves.

For tougher working conditions, you may need to wear a pair of outer gloves over the latex gloves to protect against tears and punctures. Jobs involving more extensive contamination or the use of toxic chemicals or compounds may require heavier gloves made of rubber or other protective materials. Depending on the situation, these may be worn alone or over the lighter, disposable gloves.

Body Suits

Workers need to wear full body suits for protection on most sewage and mold remediation jobs. These are typically lightweight, one-piece disposable suits made of Tyvek® or similar material, with elastic cuffs at the wrist and ankle. Most types also have hoods, and some also include booties that cover the shoes. See Figure 5-17.

On contaminated sites, tape the suit to the gloves at the wrists for additional protection. If you're using separate booties, tape the suit to the booties at the ankles as well.

Certain types of hazardous waste cleanup operations require a more specialized hazmat suit for additional protection.

Figure 5-17 Two workers on a mold remediation project in full PPE. The disposable suit has a hood, booties, and gloves. The gloves and booties are sealed at the wrists and ankles with tape for added protection. Both workers are wearing full-face respirators. Since they're working in a crawlspace, they're also wearing kneepads for comfort and to prevent their suits from being torn while they're on their knees.

Boots and Booties

Use slip-on disposable Tyvek booties to protect the worker's shoes from moderate contamination sites. They're relatively light-duty, and won't stand up to heavy traffic or rough surfaces. For those situations, there are protective boots with fully sealed seams and rubberized bottoms that provide both traction and protection against contaminates. For some jobs, you might need slip-on rubber boots as well.

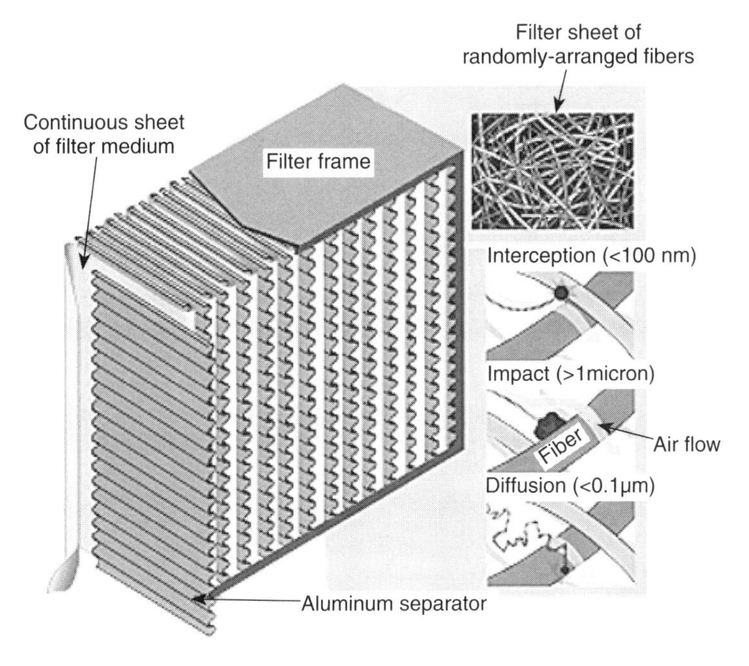

Figure 5-18 A cutaway drawing of the interior of a high-efficiency particulate air (HEPA) filter.

Splash Goggles

Splash goggles offer protection from airborne gases and fumes while cleaning and spraying. They're also an important safeguard while mixing and applying chemicals, and cleaning up liquids and other materials that may be contaminated. They differ from safety glasses in that they're designed specifically to protect all parts of the eyes, not just the front. Here again, it's very important that they fit comfortably and form a good seal.

HEPA Equipment

HEPA stands for high-efficiency particulate air filter. The exact definition of a HEPA filter is one that will remove 99.97 percent of all airborne particles down to 0.3 microns in size. See Figure 5-18. This filtering effectiveness is sufficient for cleaning the air in a mold or bacterial remediation situation.

HEPA Air Scrubbers and Negative Air Machines ($900 to $1700)

A HEPA air scrubber, like the one in Figure 5-19, is essentially a very efficient fan and filter arrangement. Room air is drawn into the scrubber, where it passes through a pre-filter. The pre-filter, which is easily removable for cleaning or replacement, traps larger particles such as normal dirt and dust before it can reach the internal HEPA filter. Some types of scrubbers also have optional charcoal filters to help remove odors.

After passing through the pre-filter, the air moves through the HEPA filter itself. Due to the sensitivity of the filtering medium, the HEPA filter is protected inside the internal housing of the machine. After filtering, the clean air is exhausted back out through a port on one side of the scrubber.

(Photo courtesy of Dri-Eaz Products)

Figure 5-19 A portable HEPA air scrubber and negative air machine. Air is drawn in through the large port, then exhausted through the smaller port. A pre-filter, visible through the intake port, traps larger particles before they reach the more sensitive HEPA filter inside.

(Photo courtesy of Bosch Tools)

Figure 5-20 A lightweight canister vacuum with HEPA filtering capabilities and an easy to clean polypropylene tank. It also has a Power Tool Activation mode, so any power tool that's attached to it will also start the vacuum. That works well when using a power sander to sand mold off building materials, drawing the sanded material directly into the vacuum.

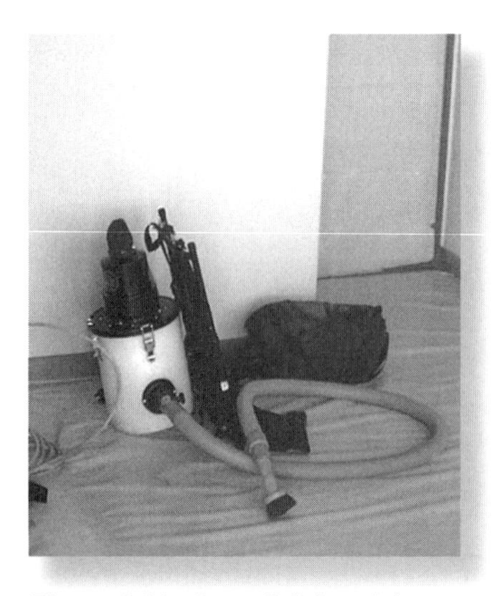

Figure 5-21 A small, lightweight HEPA vacuum with a pack and harness to be worn on the back. It has a long hose for convenience.

Most HEPA scrubbers have an adaptor that allows you to connect a duct hose to the exhaust port. You can then direct the cleaned air to a specific location, or outside the house to create additional negative air pressure within the room.

When a HEPA machine is recirculating its air within the room, it's considered an air scrubber. When it's ducting the air to the outside, it's often referred to as a negative air machine. Most machines are capable of both tasks.

HEPA Vacuums ($250 to $650)

A HEPA vacuum cleaner is very much like any other vacuum cleaner, except it has a HEPA filter inside. There are three basic designs to choose from:

- **Upright:** It's a standard upright vacuum cleaner for vacuuming carpets, with a HEPA filter cartridge inside.
- **Canister:** This is similar to a standard rolling shop vacuum (see Figure 5-20). It has a hose and various nozzle attachments you can use, depending on the job and your specific needs.
- **Backpack:** A smaller, lighter version of the canister vacuum, except instead of having wheels, it has a harness that allows the worker to wear it on his back (Figure 5-21). At also has a hose and various attachments. The advantage of the backpack style is that it's easier to use while on a ladder, scaffolding, or other areas requiring portability.

Cleaning and Maintaining HEPA Equipment

All of your HEPA equipment will be used in tough, dirty situations. When you remove it from a jobsite, it'll be contaminated with bacteria, mold spores, or other material, both inside and out. In order to prevent potential contamination of your storage facilities and your next job, all of your equipment needs to be cleaned as soon as it's removed from the job, and ideally before it's returned to your storage facility.

Clean your equipment by wiping everything down with an antimicrobial solution. Each manufacturer will have specific products and procedures that they recommend, so you'll want to read and follow their instructions. While doing the cleaning, your workers will still need to be wearing protective clothing.

The other maintenance procedure that you need to keep up on are the filters. In order to be effective, your air scrubber needs to have a clean HEPA filter in it. However, since what it's filtering is primarily invisible, it's difficult to tell when the filter is no longer working like it should.

One solution is to replace all of the filters on the machine at the end of each job. (In certain dusty situations, the pre-filters may need to be changed more often than that). By replacing *everything*, including the HEPA filter, you're assured that the machine is operating at 100 percent efficiency, and you eliminate the risk of cross-contaminating another job.

An option with air scrubbers is to have their discharge air tested. This is done with a device called a particle counter. You can purchase one to do your own testing, but they're very expensive — around $2,000 and up. Instead, ask your hygienist to test it for you. Hygienists usually have the necessary equipment available. Even if they charge you a small fee, the test is probably worthwhile. A replacement HEPA filter for a scrubber runs $150 to $300, so you don't want to replace them more often than necessary.

Remember, the filters are contaminated, so you need to dispose of them properly. After a mold remediation, place the filters in a 6-mil plastic bag, then seal it completely. As long as there are no restrictions on mold disposal at your local landfill, you can dispose of it there. With a contaminated water loss, bag and dispose of the filters in the same way you're instructed to deal with other materials from the jobsite.

Mold remediation and contaminated water losses represent both a lucrative and complex field. It's important to stay up to date with the latest information, and you need to build a relationship with a qualified industrial hygienist that you can rely on. Always work very closely with the insurance companies on these jobs, and approach each loss carefully, cautiously and methodically.

6

Specialty Drying Situations & Other Water Loss Considerations

In the preceding chapters we covered a lot of ground pertaining to water losses. But there are still a few more water loss odds and ends to talk about.

Class 4 Water Losses

Class 1 through 3 water losses have to do with how far water has spread through the building, how many rooms are affected, and how much water has been absorbed by the different materials. You're dealing primarily with what's known as *free water* — water that's contained in open areas of wood and other materials. When water is absorbed relatively easily by a material, it's also more easily released and removed during the drying process.

Class 4 losses are different. They involve pockets or areas where water has been absorbed deeply into the cells of the wood, or into dense materials such as mortar. Known as *bound water*, the moisture is locked tightly into the material, and is considerably more difficult to remove. Drying out a Class 4 loss is much harder than the other three types.

The *Guide to Restorative Drying, Revised Edition*, offers a good explanation of the problems and possible solutions involved in a Class 4 loss:

> *"Standard air movers generally will not create air movement directly across the wet surface in a Class 4 scenario. In addition, the affected material is dense, and will contain primarily bound moisture. Class 4 intrusions involve other difficult situations as well, including exotic materials and special or unforeseen situations.*

Figure 6-1 Here, air movers with 48-inch boots are being used to create a wide air flow across the hardwood floor. This drying step helps break the boundary layer, while pushing moisture toward the dehumidifier on the other side of the room.

"When determining the air movement needed for a special situation, the purpose of air movement is considered: removing the boundary layer [see below] *and distributing dry air in the structure. Enough air movement is installed to achieve these two objectives."*

We've already talked about the importance of using air movers to free up moisture so the dehumidifiers can pick it up, but let's break those two air-flow goals down a little further:

1. You want to remove the boundary layer surrounding the materials you're trying to dry.
2. You want to distribute dry air throughout the building.

Removing the Boundary Layer

The boundary layer is *"a region of high humidity near a wet material which hinders evaporation by acting as a vapor barrier."* In other words, the material you're trying to dry is so wet it's created a layer of very high humidity around itself. That high humidity is like a blanket over the material, slowing down your attempts to evaporate the moisture.

To address this, you need to move air over the wet surfaces as directly as possible. Set your air movers so that they're blowing air *across* the surfaces, as in Figure 6-1, not directly *at* them. That helps disperse the "blanket" of humidity, and increases the rate of evaporation.

Distributing Dry Air

As dehumidifiers remove moisture, they generate dry air. So your second goal is to move that dry air around inside the structure to help release the bound moisture and speed the drying process.

How you set up your dehumidifiers depends on the number of rooms affected, and the type of material you're dealing with. You may not be able to install a large number of dehumidifiers simply to generate a lot of warm air. Instead, make the most of your dehumidifiers by carefully setting up your air movers to move the dry air where it's needed.

Each Class 4 loss is unique, and requires a combination of specific equipment and specific techniques to dry it. The next couple of real-world stories are examples that illustrate some of the issues.

REAL STORIES:
Talk About Some Wet Wood!

Late one winter, a pipe froze and broke in a beautiful vacation home along the river. The house was unoccupied at the time, and so the water ran for quite a while before it was discovered. That wasn't really such an unusual problem, except for one thing: The house was built of solid logs, as you can see in Figure 6-2.

The logs were 8 to 10 inches in diameter, and had the normal checking and cracking found in any large piece of solid wood. They were interlocked at the corners in traditional log-home construction style. Some of the interior walls were also solid logs, and some were conventional stud and drywall partition walls. The plywood subfloor was water-saturated, as was the crawlspace. Moisture vapor filled the home, and condensation was running down the windows in sheets.

After ensuring the leak was stopped and then winterizing the home, our first goal was to remove as much standing water as possible, including pumping out the crawlspace. We then removed the unsalvageable materials, including the carpet, pad, and drywall. We took out all the contents — even removing the wet cabinets.

The plywood subfloor was so wet it wasn't cost effective to try and dry it. We removed as much as we could without affecting the structural integrity of the house. Having it out gave us better access for working under the floor.

After that, it was a matter of slowly removing the moisture from the logs, down to the normal moisture level for wood of that density. We installed some temporary heat to help the wood release

(Continued on next page)

Figure 6-2 The massive, solid logs in this home presented a Class 4 drying situation. There was a tremendous amount of water bound deep inside the material, and it took a long time for it to release back into the air and be removed from the building.

REAL STORIES:

Talk About Some Wet Wood! — *Continued*

the moisture, and positioned several conventional dehumidifiers in different areas of the house to deal with the water vapor as it was released.

To help create dry air, we set up a couple of desiccant dehumidifiers. Dry air from the desiccants was directed, through ducts, to the wettest areas. We also set up air movers to circulate the dry air, and to blow across the logs and break down the boundary layer.

The drying process took several weeks. We monitored, repositioned and gradually removed equipment during that period. We had to consult a log home contractor to help us determine when everything was dry enough to begin reconstruction.

The log home crew re-leveled the logs, using jacks that were part of the home's original construction. See Figure 6-3. We then blocked the joists and patched the subfloor back in, before finishing up with the cosmetic repairs.

All in all, it was a lengthy and costly drying job. But it was far less expensive than rebuilding the house with new logs.

Figure 6-3 Many of the newer solid log homes have jacking posts (center) that allow the home to be re-leveled if settling should occur. This is a real advantage after a Class 4 drying operation has been completed. However, the re-leveling operations are best left to qualified log home contractors.

Natural Wood Floors

Wet wood floors present another challenge that in many ways falls under the general Class 4 heading. Wood flooring can soak up and retain a lot of water, and may require special techniques and equipment to dry.

Before starting the job, you need to determine the type of wood flooring. Wood floors are very popular right now, and you're going to encounter a wide variety of both natural and synthetic wood and laminated floors. Some can be dried, some can't. And some of the ones that *can* be dried aren't necessarily worth the effort.

> ***Tip!*** *Class 4 losses typically require a lot of time and equipment, so work closely with your adjuster, before and during the job, to avoid any misunderstandings.*

Natural, or traditional, wood flooring consists of long, solid wood strips of just about any species, from softwoods such as pine and fir, to hardwoods, including oak, maple, cherry, and walnut. There are also more exotic species available, such as teak or bamboo. The strips are typically interlocking tongue and groove. Relief grooves are milled into the bottom edge, parallel to the board's long dimension.

Traditional wood flooring is installed as raw, unfinished boards. After installation, they're sanded, sometimes stained to add color, and then coated with two or more coats of clear finish. The finish penetrates the surface layer of the wood, and also seals the joints between the boards. However, the edges, ends, and bottoms of the boards remain raw.

When the flooring gets wet, water enters through gaps around the edges of the room where the floor meets the walls, or where it butts up to other flooring. Once it gets in, the water will travel underneath the flooring, covering a wide area. This is especially true with traditional tongue-and-groove flooring — the grooves underneath act as channels for the water to follow.

If the flooring is older, the finish may be worn, creating areas where the top seams aren't well sealed. That creates another opportunity for water to enter the flooring. The result is an uneven moisture pattern in the wood, and the cupping that you often see in a wet wood floor when the edges begin to curl upward along the length of the boards. See Figure 6-4.

Drying Wood Floors

Drying the floor traditionally requires a combination of air movement and dehumidification. Many restoration contractors are also adding in carefully-directed heat to the drying as well. However, since there's a waterproof finish on top of the flooring, you need to set your equipment up so that the air can reach the moisture layer that's down below the finish. That'll force the moisture up and out so the dehumidifiers can remove it.

There are three basic methods you can use to get out the moisture that's below the wood flooring's surface finish. Some methods work better with different floor types, and sometimes a combination of methods is needed.

Figure 6-4 The dark stains on this very wet hardwood floor are typical of moisture migrating through joints between the ends and along the edges of the boards.

REAL STORIES:
Can We Save That Tile?

Vacation homes are at high risk for Class 4 losses. They're often unoccupied, and water leaks can run for a long time before they're discovered. That was the case on a job where a frozen pipe had broken in the kitchen.

The home had ceramic tile floors that extended through the entry, kitchen, dining room, hallway, and on into other areas. The tile was set on a solid mortar base. When we arrived, some of the tiles had already begun to pop from the moisture absorbed by the mortar. The adjuster requested that we attempt to dry the tile rather than remove it. He understood it was a gamble, and that he'd be wasting money on the drying if it wasn't successful. But he wanted us to give it a try.

We started by moving air across the face of the tile. That dried the surface of the tile, but didn't impact the moisture underneath. We then tented the floors with plastic sheeting and used our desiccant to direct warm air under the tent. But that still wasn't sufficient to reach the mortar.

Finally, we tried to access the mortar from underneath. The tile had been laid over a plywood subfloor with a crawlspace below it. From below, we drilled access holes through the plywood into the bottom of the mortar, and attempted to push warm air up at the mortar to see if we could get the moisture to release. Once again, there were no dramatic signs of success.

After two days, the adjuster agreed that the drying process would take way too long. It was obvious that replacing the tile would be faster and more cost-effective than any type of drying.

Keep in mind that you can salvage just about anything if you want to commit the time and resources to it. But it has to make financial sense, and this one simply didn't.

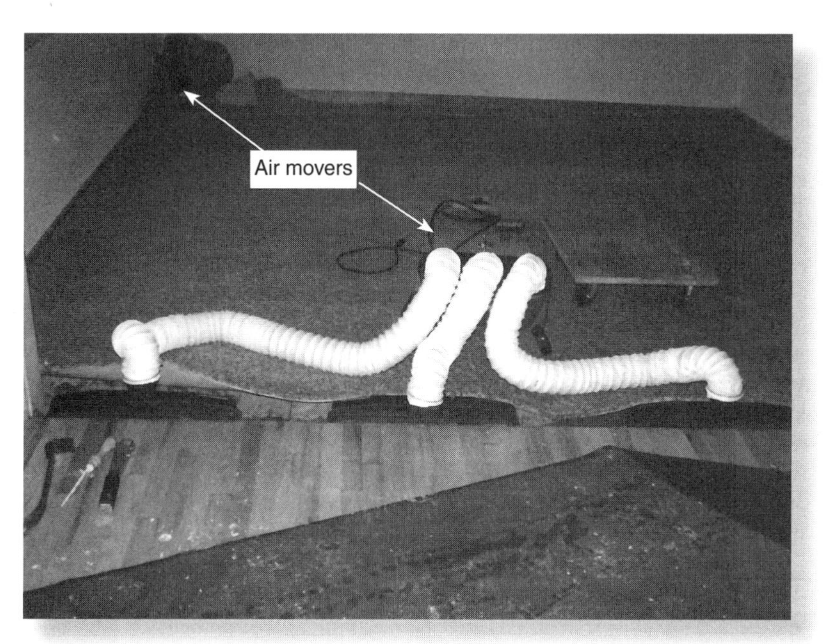

Figure 6-5 An air mover with a snout adapter and three mini-boots pushes air beneath a hardwood floor. This is an ideal situation, with carpet butting up to the ends of the hardwood flooring. By simply lifting the carpet, access is provided to the ends of the flooring, and air can be forced under the floor, parallel to the flooring. There's also an air mover in the far corner drying the underside of the carpet and the top of the subfloor.

Using Boots and Mini-boots

Use air movers equipped with wide boots to disperse an air stream (look back at Figure 6-1). This will move air across the hardwood floor for the initial surface moisture removal.

Air movers equipped with splitters and mini-boots are effective for moving air under the floor boards. You can sometimes remove the baseboards perpendicular to the long dimension of the flooring to expose a gap between the ends of the boards and the wall. If there isn't a gap, you can cut one with a saw. Then you can install a boot or a mini-boot over the gap to push air under the boards and into the grooves, drying the floor from below. Other gaps where you might find access include areas around floor registers, and where the wood floor meets cabinets or butts up against other flooring areas. See Figure 6-5.

Tenting

You can tent the floor by taping or weighting 6-mil plastic sheeting over it, as shown in Figure 6-6. Push dry discharge air from a desiccant dehumidifier or heated air from a directed heat system under the tent through a hose to dry the wood. With a desiccant system, another hose at the opposite end of the tent will draw moist air into the dehumidifier for drying and reprocessing. With a directed heat system, a lay-flat duct is often directed to an air mover to discharge the moist air to the outside.

To help speed the drying, you can perforate the waterproof finish on the flooring by rolling a special spiked roller over it. This punches thousands of tiny holes in the finish to release the moisture trapped below.

Drying Mats

Dri-Eaz has developed a Rescue Mat® Hardwood Floor Drying System. Place the mats on top of the hardwood flooring, then connect the vacuum extraction system, as shown in Figure 6-7. As air is pushed below the flooring, the mat and vacuum system

Figure 6-6 A hardwood floor tented with plastic sheeting. The hose in the foreground is coming from a desiccant dehumidifier.

create a powerful suction action that draws the water up from floor. The extracted water is then pumped to a disposal site.

Is the Floor Worth Drying?

All of the drying methods just discussed are labor and equipment intensive, and therefore costly. Once the floor is dry, you'll also have to sand and refinish it. So before you undertake drying a wood floor, make sure it's worth the effort.

How big is the floor, and how isolated is it? If it's a single small area, like an entry way or a small dining room, it might be better to remove it. If it's just a small part of a kitchen floor that also extends into adjacent dining and family rooms, then saving it may make sense. If the flooring is Brazilian cherry with a zebrawood accent stripe, even in a small isolated room, such as a study, it might well be worth the cost to dry it and save it.

Every job and every floor is different. Discuss it with your clients and your adjuster, and make a decision based on the available facts.

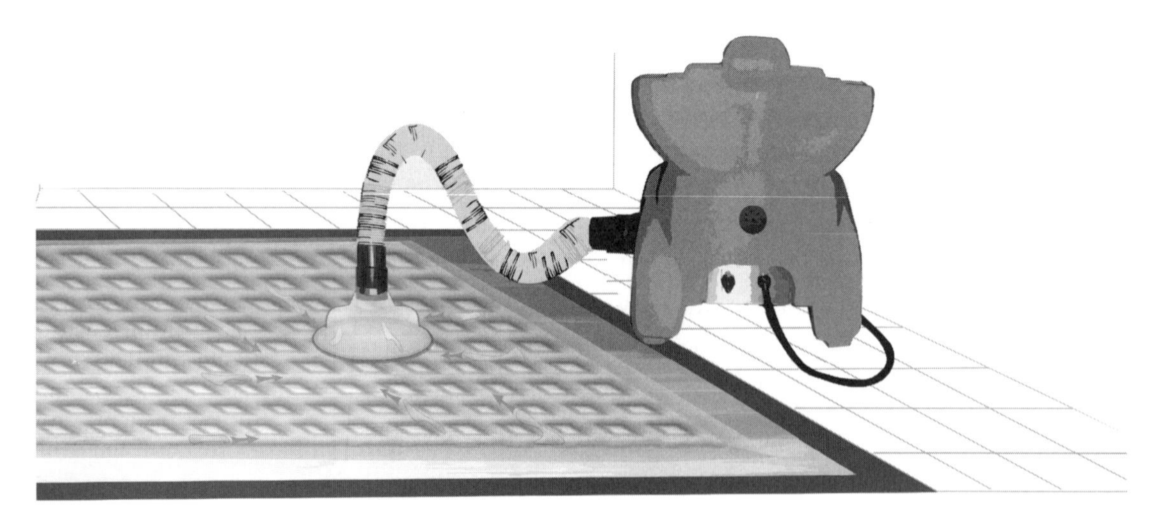

(Photo courtesy of Dri-Eaz Products)

Figure 6-7 Drying a hardwood floor from above with a Rescue Mat system. Suction is used to draw moisture up through the floor, which makes its way through channels in the mat. It's then drawn through the hose and out for disposal.

Make Sure It's Dry

Make sure the floor is completely dry and the wood is down to the proper moisture content before refinishing. Otherwise, there might be other problems down the road.

As wood flooring gets wet and swells, the edges rise up and it cups. As it dries, those edges will lay down again. If you sand a wood floor before it's completely dry, you'll sand down those raised edges. Then, as the floor finishes drying, the edges will continue to drop, creating small indented grooves down the length of each board. And, any remaining moisture that's trapped behind the finish can make the finish cloudy and cause it to fail prematurely.

Before refinishing, the moisture content (MC) of wood flooring should usually be down to approximately 6 to 8 percent. Have your hardwood floor subcontractor verify the MC of both the wood and the subfloor before he does any refinishing.

Drying Other Wood and Synthetic Floors

With most types of engineered wood and synthetic wood floors, you'll find that it's not worth the time and effort to dry them. For example, laminate flooring absorbs quite a bit of water, and is very difficult to dry effectively.

Many types of engineered, prefinished wood floors have layers of different materials sandwiched beneath the finished wood veneer. Those different materials can react unpredictably to both the water and the drying process. It's difficult to know when the material's dry, and how the restored material will hold up over the long-term. Engineered wood floors also have veneer layers of differing thicknesses, some of which are thick enough to be refinished and some aren't.

Unless you know from experience that the type of engineered or synthetic flooring you're dealing with can be successfully dried — and, in the case of engineered flooring, refinished — you're better off replacing it.

Pressure-Testing a Loss

If you have a water loss caused by a leak in a plumbing pipe, it's a good idea to have your plumber pressure test the water lines before turning on the water to the building. The test can detect additional leaks in the system quickly and easily, without running the risk of additional water damage.

This is especially important in the case of a frozen pipe. When water freezes in a pipe, it expands. It can split apart weak seams in the pipe, cause fittings to come apart, and cause damage to O-rings, cartridges, and other faucet components. Never think that the water you see leaking from one break in a pipe or from one damaged fitting is the full extent of the problem. There may be literally dozens of areas in the plumbing system damaged as well.

To conduct a pressure test, your plumber will place an air gauge on the main water line where it enters the house. He'll then use a compressor to pressurize the system, and will watch the gauge for a set amount of time. If the gauge begins to bleed down, it indicates that air is escaping from the system. The plumber will then track down the air leak, determine and repair the cause, and repeat the test until the system holds air.

Again, it's in the best interest of the insurance company to have this test done. Talk with your adjuster; explain the need for the test and the reasoning behind it, and verify that the insurance company will pay for it. If they won't, then it's a cost the homeowner will have to pick up — or in a worst case scenario — you'll have to pay. In a frozen pipe situation, you never want to close up the walls without verifying that the plumbing system is completely intact.

Equipment Power Usage

When working on a water loss, you need to keep in mind how much electrical power each piece of your equipment uses so you don't overload a circuit. An average-sized air mover draws about 4 to 5 amps of power, so a 15-amp circuit can only handle three air movers. When setting up, you'll need to split the equipment between two or more circuits. Kitchens typically have two 20-amp circuits, so you can draw a lot of power from there. Laundry rooms also usually have a 20-amp circuit for the washing machine.

In addition to all of your drying equipment, plan on purchasing a collection of heavy-duty extension cords. Avoid the temptation to purchase cheap 16- or 14-gauge cords on sale at your local hardware store. Instead, stick with 12-gauge cords that can handle the amperage draw of your equipment without problem. You should also consider at least one 10-gauge cord for longer runs. And, you'll find it a lot more convenient if some of your cords have a 3-tap end on them.

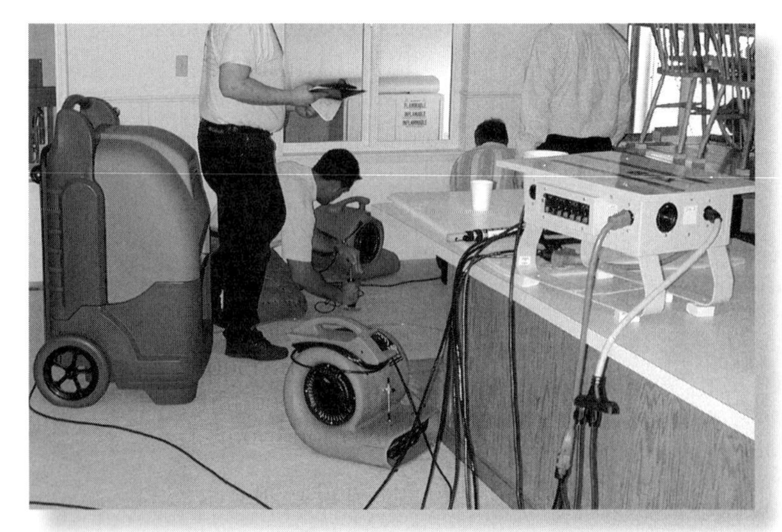

(Photo courtesy of Dri-Eaz Products)

Figure 6-8 An electrical spider box (right) being used in a training class on water restoration techniques.

Electrical Spider Box

An electrical spider box, like the one in Figure 6-8, is a particularly useful accessory to help meet your electrical needs. It's basically a UL-approved metal box with a cord designed to be plugged into a 220-volt electrical outlet. The circuitry inside the box safely splits the incoming power into several 110-volt outlets, allowing you to tap several pieces of equipment off of the one box. The outlets are GFCI-protected, and are also protected by individual circuit breakers.

A spider box has several advantages, but the most important one is safety. It's designed and tested to provide multiple safeguards for both users and equipment. It also gives you a means to get additional use out of a building's electrical system, since you're tapping into a 220-volt circuit that you wouldn't normally be able to access. Since the box provides a central location for several outlets, it reduces the number of cords you need, and also reduces the hazard of workers tripping over multiple cords coming from different directions.

Client Concerns About Power Usage

Sooner or later, while you're in the middle of drying out someone's home, and feeling like a bit of a hero for saving it for them, you're going to hear the following question: "How much electricity does all this equipment use, and who's going to pay for it?"

To calculate how much power your equipment is using, and then determine how that translates into dollars and cents for the homeowner, you can use the following formula.

Calculating Costs for Equipment Electrical Usage

Calculate each piece of equipment separately, and then total them.

Step 1: Multiply the voltage of the building's power by the amperage rating of the equipment. This will give you the number of watts being used by the equipment. (If you don't know the exact voltage, use 120 volts.)

Step 2: Multiply the watts being used by the number of hours that the equipment is in operation. This will give you the total number of watt-hours.

Step 3: Divide the total number of watt-hours by 1,000, which will convert it from watt-hours to kilowatt-hours. Electricity is measured and billed in kilowatt-hours.

Step 4: Multiply the kilowatt hours by the cost of electricity per kilowatt hour in your area. This price will be shown on your client's utility bill, or you can call your utility company and ask. That will give you the total cost to operate that piece of equipment for that period of time.

Let's use the formula to calculate how much it costs to operate one air mover, drawing 4 amps, on a two-day job. You can determine the amperage use from either the manufacturer's catalog or from the label on the air mover itself. We'll say electricity in your area costs 15 cents per kilowatt hour (which could be high or low, depending on where you live).

Step 1: 120 volts × 4 amps = 480 watts

Step 2: 480 watts × 48 hours = 23,040 watt-hours

Step 3: 23,040 watt-hours ÷ 1,000 = 23.04 kilowatt-hours

Step 4: 23.04 kilowatt-hours × 15 cents per kilowatt-hour = $3.46

In this example, it cost $3.46 to operate one air mover for two full days (48 hours). You can calculate the cost per hour for each piece of equipment and keep a record of those costs. The only variable from job to job will be the local cost of electricity per kilowatt-hour.

As far as who's going to pay for it, that's a question the homeowner needs to ask their insurance company. Typically, insurance will pick up the tab for any unusual expenses directly related to a loss. That includes the electricity to operate all the equipment, as well as expenses for such things as operating the home's heating or cooling systems for extended periods due to the loss.

The insurance company will usually request some type of documentation to justify the expense. The easiest way for the client to do that is to show them their utility bills for the same period in the prior year. For example, if their loss occurred in February of this year, they need to get last year's February utility bill. If it's $40 higher this year than February of last year, and there are no other changes to account for that increase, such as a higher rate, then the insurance company will probably reimburse them for the additional $40.

To be honest, except in the case of a very large loss, most of your clients won't even think to ask about it — it's the least of their worries. But some will. If they do, you can calculate the equipment usage for them, or just refer them to their insurance company.

Equipment Noise

One thing that you and your clients will quickly learn is that drying equipment operating inside a house is pretty noisy. Between the air movers and the dehumidifiers, getting a building dry enough for reconstruction creates a considerable disruption for the homeowner. If they're still living in the home during the drying process, sooner or later you're going to get complaints about the noise. Here are some suggestions for how to deal with it:

- **Be proactive:** Let them know up front that the drying process is going to be noisy. Explain that you understand how disruptive the noise can be, but without the equipment, the house can't be dried in a timely manner.
- **Show them the off switch:** Show them how to safely shut off the equipment if they absolutely need to. Some people want to know that they can shut off an air mover or two when they want to sleep, or when the baby needs his nap, or when they're on the phone. Ask them to be sure and turn it back on as soon as possible. If you have more sophisticated equipment that you don't want shut off, such as a

REAL STORIES:

But the Dog Needs His Beauty Sleep!

A woman had an icemaker line come loose on her refrigerator. The water poured into the kitchen, down the hallway, and into a couple of adjacent rooms before the leak was discovered and stopped. We were called out to get the house dry and put back together.

After removing the wet materials and extracting as much water as possible, we set up our drying equipment. We told her we'd be back the following day to check on the progress of the drying. But, later that afternoon, we got a call from the woman complaining about the noise our machines were making.

One of our technicians went by the house and checked the equipment. Everything was working fine, and he explained that there was nothing he could do about the noise. He showed her how to shut off the machines if she needed to do that in order to sleep that night. And he carefully explained that the equipment needed to run as much as possible to get the house dry and her life back to normal.

The following day, he went by the house to check on the progress of the drying — and found all of our equipment sitting outside! The woman had shut everything down, unplugged it, and put it all on her front porch. Several thousand dollars worth of equipment sat outside all night where anyone could walk off with it!

Our technician knocked on the door, and asked the woman what had happened. "It was just too noisy! I put up with it as best I could, but it was keeping my dog awake!"

We picked up our stuff, and sent the insurance company and the woman a polite letter saying that since the house was still wet, we wouldn't be able to make any repairs. To this day, I don't know what happened with her house. But I assume that the dog got a good night's sleep!

desiccant dehumidifier or an air scrubber, post a warning sign on the machine that tells them not to shut it down.

- **Warn them about the insurance company:** Tell them that their adjuster is well aware of how long it takes to dry a house. Explain that if they turn off the equipment for extended periods of time, and thus increase the amount of time your company bills the insurance company for, it could become an issue.
- **Suggest that they move out:** If the noise is a major issue and they can't sleep, they should discuss the situation with their adjuster. Many adjusters will authorize a couple of nights in a motel if the circumstances warrant it. If nothing else, they might be able to stay with a friend or relative until the worst of the drying is complete.

Coming in After Someone Else Has Done the Drying

You'll sometimes be faced with a situation where you come onto a job after another contractor has been in the building doing the drying work ahead of you. He may have finished up the drying, but not been hired to do the repair work. Or, he may have left the job before the drying was completely finished.

There could be a number of reasons why this happens. The homeowner may have been unhappy with the contractor and asked him to leave. Or perhaps the contractor had problems with the homeowner, and couldn't or didn't want to finish the job, and left on his own. Whatever the reason, you've been called in to finish things up.

Here are some tips for when you encounter this situation:

- **Do a little bit of homework first:** See if you can find out what happened so you don't walk into the middle of a mess. If you know the adjuster on the job, give him a call and see what's going on. If you have a good relationship with the other contractor, call him and check things out. You don't need the grief of taking on a bad job if you can help it!
- **Be professional:** Don't come in like the conquering hero and start bad-mouthing the other contractor. You don't know all the circumstances behind what happened, and there are always two sides to the story. Also, you never know when and if the situation might be reversed.
- **Be careful with your opinions:** You may be asked by the homeowner to give your opinion about something that the other contractor said or did. *Be very careful about being drawn into that conversation.* Some homeowners are looking for liability, and you could easily find yourself subpoenaed into court as a witness. It's better to stay out of things if possible.
- **Don't mess with their equipment:** If the other contractor's equipment is still in place, leave it alone! Refuse to take on the job until the other contractor has removed all of his equipment and has fully vacated the premises. Once again, there's lots of potential for liability and legal finger-pointing here. *Make sure he's completely out before you step in!*
- **Do your own inspections:** Ignore anything the previous contractor did. *Treat the job as though no one has been there before you.* This is very important! Do your own inspection of the building, and do your own metering. Even if you know and trust the other contractor, and even if the homeowner assures you that the building is dry, *do your own inspections!* If you don't, and it turns out that something is still wet and mold grows or additional damage ensues, you could very well be held liable for it.

Leaving a Job Before It's Dry

Now for the flip side of the coin: What if *you're* asked to leave a job before it's dry? Or what if your relationship with the building owner has deteriorated so badly that you simply don't want to deal with that job any longer?

First of all, tread lightly. If you can work things out, try. See if you can determine why the client is unhappy, and then rectify the problem. Of course, that's not always possible. So if you have to leave before the job is dry, here are a few pointers on what you need to do to protect yourself and your company:

- **Collect all your equipment:** Don't leave anything behind that you'll have to come back for later.
- **Send a letter to the homeowner and the adjuster:** Immediately send the letter. Don't point fingers, argue, insult, or anything else. Be clear and professional. Simply document in writing, clearly and succinctly, that as of this day and time you have left the job, the job is not dry, and you recommend that the owner of the building continue with the drying process as soon as possible to prevent the possibility of additional damage. Send the letter registered mail with a delivery confirmation, so you have proof of delivery and a receipt. If you're really concerned about liability, consider spending the additional money to send the letter by overnight mail or even by a courier service, again with a delivery confirmation.
- **Document the situation:** Immediately, while it's fresh in your mind, document all of the conversations that led up to your being asked to leave the job, or that resulted in your decision to walk away. Interview your employees and any subcontractors who might have been involved or might have witnessed anything important, and document what they have to say. Type up their statements, and if possible have them sign a copy for your file.
- **Make a complete file:** Gather all your records for that job and put them in a safe place. That includes photos, drying logs, call sheets, phone logs and emails — anything and everything pertaining to the job. Make a second copy of the entire file and store it in another location, just in case something happens to the first one.

It's a lawsuit-crazy world out there. Always hope for the best, but make it a point to plan for the worst!

7

Understanding Fire & Smoke

While water losses will probably make up the bulk of your work as a restoration contractor, you'll likely have your share of fire losses as well. Encountering a fire loss, even a small one, is a shock to the senses — especially sight and smell. No loss is as dramatic as a fire. The devastation caused by a large structural fire can be overwhelming, especially the first time you see one.

Fires, and the smoke they produce, are both complicated and unpredictable. Structural damage can extend to areas you don't expect, and smoke can permeate a building far beyond the reach of the fire. Before you can effectively estimate and make repairs to a fire-damaged structure, you need to understand what a fire is and what happens as it burns.

Fire losses affect your clients differently than water or storm losses. Even if the actual damage isn't as bad in terms of the cost or severity, a fire is somehow worse. It's sudden, more invasive, and much more frightening. It's like having an intruder in your home whose sole purpose is to do harm.

That's also what makes the aftermath so much worse. Fire changes things. A wet house might have $10,000 worth of damage, but it looks pretty much the same as it did before the water arrived. A fire-damaged house, needing the same $10,000 in repairs, looks different. It's a different color and it smells different. Familiar objects have taken on unfamiliar shapes. See Figure 7-1.

Even after years of doing fire restoration work, I still find something awe-inspiring and a little frightening about the destructive power of a fire. Don't lose sight of that — it'll help you be more understanding and empathetic with your clients, and make you a better and more thorough estimator.

Figure 7-1 A fire can do a devastating amount of damage, especially to a home's contents. Fires are also extremely traumatic for the homeowners.

Training and Safety Considerations

You'll want to have specialized training and IICRC certification to work in fire restoration. A good place to start is the 2-day IICRC-certified *Fire & Smoke Restoration Technician (SRT)* class. While not really structural in nature, the class gives you a good overview of different types of fires, including what smoke is, and how to deal with fire restoration losses.

Here's the IICRC's course description:

> *SRT: The Fire & Smoke Restoration Technician course concentrates on technical procedures for successfully completing the restoration of a fire and smoke damaged environment. Students will learn how to combine technical procedures with a practical approach to managing the job site and how that relates to pricing the job.*

See Appendix B for more information on IICRC classes in fire restoration.

Safety

Construction is always a dangerous occupation, but few construction sites have as many hazards as a fire job. This is especially true of major structure fires, but even the smallest fire can represent serious health and injury risks.

REAL STORIES:

That Mobile Phone is Intimidating!

Our crew was tearing out saturated drywall and insulation in the bathroom of a wet job, and had to remove the shower door and framework. They took it apart and stored all the pieces in the garage.

My partner and I had been in business for several years, primarily doing remodeling, but we wanted to move into insurance restoration. So, we started doing the dreaded cold calling, making the rounds, introducing ourselves to agents and adjusters and trying to get a foot in the door.

One afternoon, we got a call from an adjuster we'd recently introduced ourselves to. She had a fire loss with structural damage in the attic and smoke damage everywhere else. Another contractor was already on his way out to do an estimate. But her insurance company required competitive estimates on fire losses and, since she had our card in front of her, she thought she'd give us a try.

Excited to get the opportunity, we dropped everything and rushed over to the job, only to find that the other contractor was still there. Of course, we recognized him immediately — he was the biggest insurance restoration contractor in the area!

He was standing beside his truck, talking on his *(gasp!)* mobile phone. Mobile phones were new back then, and only the really "big" boys had them. The phone was the size of a brick, and was tethered to his truck with a thick curly cord. We were very impressed. He just continued chatting away, totally ignoring the two young upstarts invading his territory.

The next thing we saw was the big burnt hole in the home's roof. Blackened rafters and charred sheathing were visible, along with smoky, cracked windows in the dormers. Our eyes went from the burnt house to the big contractor with the impressive mobile phone and back to the burnt house. It was hard to say which was more intimidating.

After the big contractor left, our construction experience kicked into gear, and we calmed down. We met with the adjuster, and later with the homeowner, an elderly lady who was somewhat overwhelmed by the entire process. She appreciated the time we spent going over everything with her, something the other contractor hadn't done (which was an important lesson we learned that day).

We got the job! The fire repairs went fine, we learned a lot, and we established a great relationship with that adjuster, who began sending us a steady stream of work.

Sometimes it pays to be a little intimidated.

Often, floors have holes in them (Figure 7-2), or are barely supported by fire-damaged joists. Roof structures hang by a thread above you, and splintered boards grab at your clothes. Razor sharp metal and broken glass litters the floor, and wires dangle everywhere (Figure 7-3). Furniture and other contents are unrecognizable remnants, intermingled with piles of carpet and collapsed drywall (Figure 7-4). Everything under your feet is saturated with water from sprinklers and fire hoses. Insulation, soot, and dust congests the air, making it difficult to take a deep breath.

As you move through these hazards, even on the brightest of days the darkness inside is almost complete. Windows are fogged over with soot, and every light source is covered in black. Even the best flashlight won't effectively penetrate more than a few feet into the gloom.

There are dangers associated with the smoke as well. Soot particles are as small as mold spores, so they're easily inhaled into the lungs and can present a serious health hazard. Many types of soot particles are also known to be carcinogens.

Fires produce a wide variety of gasses during their combustion process. These include carbon monoxide, hydrogen cyanide, and those known as polycyclic aromatic hydrocarbons, or PAH, another carcinogen.

Figure 7-2 A fire in the second floor bedroom of this home left the floor unstable. A large hole, partially buried under debris, opened to the concrete slab of the garage below.

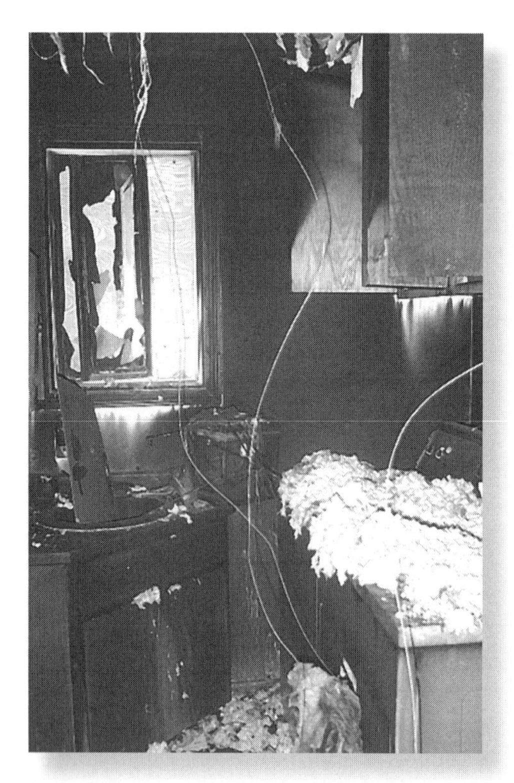

Figure 7-3 This house was a minefield of hazards. Wires dangled from overhead, and there was broken glass scattered around. Loose-fill insulation dropped down from overhead when the ceiling collapsed, covering tripping hazards on the floor.

Figure 7-4 The kitchen of this home, which was several rooms away from where the fire started, was almost unrecognizable in the aftermath of a major fire. The overhead flex ducts melted, leaving coils of wire hanging (center). This created one more hazard, in addition to broken glass, metal edges, and a floor that was severely weakened from the fire.

When you're working inside a structure fire, you're putting yourself at risk, even just doing an estimate. *This is no joke!* All workers at a fire scene should be properly equipped with PPE. This includes:

- A properly-fitted respirator with fresh, organic-HEPA filter cartridges. Because of the amount of dust, prefilters are also recommended to prevent premature clogging of the cartridges.
- Gloves and safety glasses.
- Proper foot protection. No tennis shoes or other soft-soled or open-toed shoes.
- Hard hats to protect against hazards from above.

Finally, you need good ventilation. Use axial fans or other ventilation fans to keep work areas clear of excess dust and particles. Also use HEPA air scrubbers with activated charcoal filters for these situations.

If you have any questions about the proper type of PPE and ventilation equipment to be using to ensure adequate worker safety in a particular situation, consult with your industrial hygienist or your local OSHA office.

What Is Fire?

Fire is a chemical reaction between oxygen and a fuel source. Oxygen is readily present in the atmosphere, and in a typical building there are fuel sources everywhere, including wood, paper, fabrics, flammable liquids, and other materials. So if every building has the two main elements of a fire present and ready — fuel and oxygen — what's missing? Why isn't everything on fire?

For a fire to start and sustain itself, you need a third element: heat. These three elements make up the Fire Triangle (shown in Figure 7-5):

- Oxygen
- Fuel
- Heat

Figure 7-5 The Fire Triangle: Oxygen, Fuel, and Heat. These are the three elements required to start and sustain a fire.

If you have oxygen and fuel, all that remains is to supply the fuel with heat in a sufficient quantity to reach its ignition point. Then, a fire will start. Different materials have different ignition points — it will take different amounts of heat to ignite them.

Take a simple wood match as an example. The match has a fuel source, which is the wooden stick with chemicals on its head. And of course, it's surrounded by oxygen. By itself, it doesn't catch fire. But move the head of the match rapidly across a rough surface, and the friction generates heat. The chemicals on the head of the match have a low ignition temperature, so the heat from the friction is enough to ignite them into

Figure 7-6 Fire consumes and decomposes the material it's using for fuel. Here it has left behind charred wood. This was the header over a window. You can see the damaged remains of the header, trimmer, sheathing, siding, and trim.

Figure 7-7 As long as oxygen, fuel, and heat continue to be present, the fire will continue to burn. A byproduct of that process is heat. You can see the intensity of the heat consuming this brush.

flame. The wood stick will sustain the flame until the wood is burnt up.

That's how the fire starts. For the fire to continue to burn, the three elements — oxygen, fuel, and heat — all need to continue to be present. Remove any one of them, and the fire can't sustain itself.

A good example is a campfire built with wood. If you put water on the wood, the temperature drops below the wood's combustion point, and the fire dies out. Or, if you don't put any additional wood on the fire, once the last piece is consumed, the fire dies for lack of fuel.

Now consider a candle burning inside of a jar. As long as the lid's off the jar, the candle burns, but if you replace the lid, you remove the oxygen source and the flame dies out.

Heat

When a fire burns inside a building, it does damage by physically decomposing the materials it's using for fuel, like the charred wood beam shown in Figure 7-6. It also does damage by coating surfaces with smoke, and by the effects of the intense heat it creates (Figure 7-7).

The effects of heat are often strange and surprising. Light bulbs and glass light fixtures melt into surreal shapes, even though the metal portion of the fixture appears undamaged. Windows shatter as their vinyl frames melt and twist, and fiberglass showers droop off walls. You'll encounter all types of severely heat-checked materials (Figure 7-8), and heat-shattered materials that you thought were heat resistant (Figure 7-9).

Heat can melt the insulation on electrical wiring, even in areas where the rest of the structure appears undamaged. Of particular concern is metal conduit, which can appear undamaged from the outside (except for smoke), while the wiring inside has melted.

Heat will also soften solder joints on copper water pipe fittings. In older homes, it can soften the lead and oakum or soil cement that seals the joints on cast iron bell-and-hub drain fittings. It can also melt ABS and PVC pipes in areas you thought were otherwise undamaged.

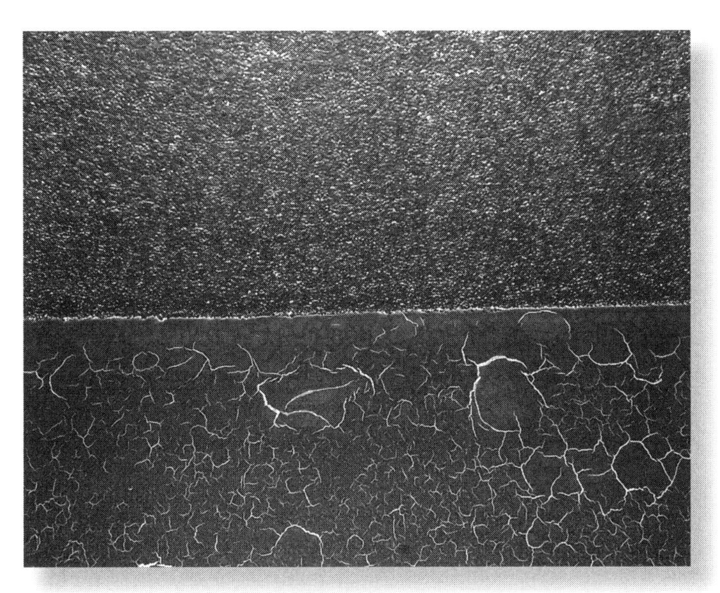

Figure 7-8 Drywall that's been heat-checked by a fire. The material expands as it's heated by the fire. When it cools and returns to its normal size, it cracks, which is why there's no smoke in the cracks.

Figure 7-9 A gas fireplace in the aftermath of a fire. Even glass that's designed for the high temperatures of a fireplace can be shattered by the heat generated during a structure fire.

Two Types of Fires

Those in the insurance restoration industry typically categorize fires into two groups. A fire is considered *hot and clean* or *cold and dirty*, based on how the fire burned, and how much smoke it produced.

Hot and Clean Fires

A hot, clean fire is one that has burned with a lot of oxygen and a lot of dry fuel. You'll see this type of fire in a home with dry structural wood and organic contents. Usually windows were open or broken, doors were open, or there were other sources of oxygen to feed the flames.

With this type of fire, less smoke is produced. The soot tends to be dryer, and it's generally easier to clean up. However, hot fires also pressurize the air to a greater degree, and can push smoke deeper into cavities and crevices.

Cold and Dirty Fires

If you've ever damped down a camp fire or a fire in a wood stove, you know that as the temperature of the fire goes down it produces a lot more smoke.

A cold and dirty fire is one that has smoldered. It often occurs in a building with very little oxygen. This could be because the building's well insulated and weather sealed with all the doors and windows closed; or it could be that the fire started very small and never really got enough fuel to flare up and draw in the oxygen it needed to grow hot. Dirty fires also take place when the primary fuels are synthetic materials such as foam or plastic, or other materials that produce thick smoke with high concentrations of volatile organic compounds.

Figure 7-10 The fire started in this bedroom, probably from an electrical short. The heat was intense here, as you can see by the melted television. Even the switches and thermostat on the wall between the closet doors have melted. Once you determine where the fire started, you can start there and work out for your estimate.

Fire Spread and Fire Damage

As you evaluate fire losses and write your estimates, it's important to understand some of the influences that affect how a fire spreads, and how much damage results.

Where Did the Fire Start?

The first thing you want to know is where the fire started (Figure 7-10). That will tell you a lot about how and where it traveled, and how extensive the damage is likely to be.

A fire that starts in a kitchen usually means someone was cooking, so they were there to quickly stop the fire's spread. A fire that originates in a chimney flue could mean structural damage in concealed attic spaces. A first floor fire will cause smoke damage and possible structural damage to the second floor, while a second floor fire usually only results in water damage on the first floor from the firefighting efforts.

How Did the Fire Start?

A kitchen grease fire will typically cause certain patterns of damage to the cabinets and appliances, while flue fires will damage the chimney, flue pipes, and adjacent framing. An electrical fire will require a careful inspection of the entire electrical system, from the meter through every circuit, and for a roof fire you'll need to check vents and flashings for damage and possible leaks.

What Is the Building's Construction?

Buildings made from flammable materials such as wood are usually going to sustain more damage than buildings made from nonflammable materials such as block. However, in a hot fire, stone, block, stucco, even concrete construction can sustain serious structural weakening of the masonry.

What Is the Building's Use?

The building's use and whether or not it was occupied at the time of the fire can provide important clues to the extent of the damage. A wood residential structure will have different materials and contents than a wood building used as an automotive garage or a retail store, and each use-type will burn differently.

Figure 7-11 This dramatic photo of a house fire shows how much smoke is generated by the fire. The fire heats and pressurizes the smoke, forcing it deep into building materials, especially in the attic.

How Was the Fire Fought?

You'll quickly learn that there are big differences in how fires are fought, and how effective different fire departments are.

In the past, fires were always fought by soaking them with water, and the water often did as much damage as the fire itself. Today, most fire departments use spray nozzles that disperse the water, allowing them to use less water in a more effective way. Some fire departments use water additives that increase the water's ability to penetrate burning materials, or foam products that both lower the fire's temperature and help deprive it of oxygen.

However, some smaller districts or rural areas, especially those with a volunteer fire department, may not have the latest equipment and technology. Those fire departments may still fight fires using just lots of water.

Another factor in fire spread and the resulting damage is how the firefighters attack the fire. Remember that their primary mission is to extinguish the flames and prevent the fire from spreading to adjacent structures. To accomplish that, they may punch holes in walls and ceilings, break windows or doors, or even cut into walls or roofs with a chainsaw — sometimes right through wiring, plumbing, or structural supports.

What Is Smoke?

As a fire burns whatever material it's consuming as fuel, it releases a combination of gasses, vapor, and fine particles of unburnt carbon into the air. This combination is what we know as smoke (Figure 7-11). The unburnt carbon particles suspended within the smoke are called soot, or technically, PICs — *particles of incomplete combustion.*

How much visible smoke a fire produces is dependent on two things:

1. The fuel being burned
2. The amount of available oxygen

Figure 7-12 Smoke residue is readily apparent on this refrigerator. It's relatively clean where the smoke was blocked by paper or a magnet, but there's smoke residue everywhere else.

You can see common examples of smoke production in a home fireplace. To start your fire, you place some newspaper, kindling, and firewood in the fireplace, then strike a match and ignite the paper. Newspaper has a low ignition point and doesn't require much heat, so it starts burning right away. The heat from the burning paper rapidly evaporates moisture from the small, dry kindling. That moisture is released in the form of volatile compounds. What you see as smoke is the visible evaporation of those compounds.

Heat from the kindling starts the same process in the firewood. Again, the volatile compounds in the wood, which are hydrocarbons, evaporate off and you see smoke. Eventually, the wood catches on fire. As it burns down to embers, less smoke is produced. That's because the organic compounds in the wood are now gone. Toss a fresh piece of wood on the embers, and the smoke starts up again until the wood starts burning, and the process repeats itself.

Here's another common home example — a hot wood fire in a wood stove. There's very little smoke as long as the door of the stove is open and the fire is burning hot. That's because the combustion is rapidly consuming the hydrocarbons.

However, if you close the stove door, you begin to starve the wood of oxygen. The combustion temperature drops, and the burning process slows. The combustion process is now far less complete, and there are a lot more unburnt particles in the vapor. In other words, you're seeing more smoke.

Three Types of Smoke Residue

Smoke, as the result of incomplete combustion, leaves tiny particles of unburnt material in the air, and different materials produce different kinds of smoke. As smoke cools, it leaves behind a thin film called a smoke residue (see Figure 7-12). This residue produces both odor and staining. How much odor and staining depends on the type of residue, and how long it's in contact with a surface before it's cleaned off.

There are three general classifications of smoke residue, depending on the type of burning material that produced the smoke: natural, protein and synthetic.

REAL STORIES:

Easily the Worst Smell in the World!!

Donna put the roasting pan on the stove, on medium heat, to brown the large beef roast, then went out front to get her mail. She ran into a neighbor, then another neighbor showed up. One thing led to another, they walked over to the neighbor's house, and the big pot roast was forgotten.

Several hours later, Donna returned home, opened the front door, and the smell was nauseating. She instantly realized what she'd done, and ran into the kitchen. The heat was still on. The pan had not caught fire, but the roast was just a sad little wisp of black charcoal.

It was, to put it mildly, a smell like nothing I've ever smelled before — and hope to never smell again: A thick, cloying, somewhat sweet, totally overpowering odor of burnt meat that really defies description, and it had permeated the entire house. The kitchen and all the adjacent rooms were coated with a sickly yellow film. The kitchen floor was like walking across fly paper, and everything you touched in the kitchen, you stuck to momentarily. The owners wisely took their kids and moved out.

The adjuster showed up early the next day. The odor, having lingered all night, seemed to have gotten worse. The adjuster was fighting the flu, and we warned him that going in might not be such a good idea. But he was a conscientious guy, and needed to see the damage for himself. He was back out in a heartbeat, and his breakfast ended up in some bushes near the front door.

Never has a house smelled so horrible and had so much damage, while looking so undamaged. Other than the yellow film (and the odor), it really didn't look like there was anything wrong with it. But there was, and it was a lot more than we bargained for.

This job involved some of the most intensive cleaning and sealing we've ever had to do. The odor had permeated the carpet pad, and even the underlayment. We tore out the carpets and vinyl, and applied a clear encapsulant over the subfloor. All the walls were washed, sealed, and then painted. We took down kitchen cabinets, and cleaned and sealed the unfinished backs. Switch plates were washed with a commercial degreaser. Door hinges and door knobs were taken off and washed, and the bare wood behind them sealed. Every porous material within range of that darn pot roast had to be cleaned, sealed, or replaced.

The house was finally made livable, but it was weeks before any of us could look a hamburger in the eye again.

Natural Material Residue

Natural material residue is also called organic material residue, and is the result of burning natural materials, such as wood, paper, cotton, wool, cork, feathers, etc. The residue from natural materials is gray, gray/black, or black in color, and has a dry, powdery consistency. The odor tends to be a little milder than some other smoke odors, depending on the exact source. For example, many types of wood and plants produce a pleasant, aromatic smoke when burning.

Protein Residue

A protein residue results from the burning of meat, fish, poultry, or any type of skin. It's the least visible type of residue, with a brownish or yellowish color that's mostly transparent. It has a thick and sticky texture that adheres to just about anything, including vertical surfaces and ceilings. Protein residues have a heavy, sickly-sweet odor that's often very difficult to get rid of.

Synthetic Residue

Synthetic materials, such as plastic, foam, carpeting, synthetic textiles, and many other materials, produce a black residue that has a rather oily feel to it. It adheres fairly well to different surfaces, and tends to smear easily when wiped. The odors can vary widely, from relatively mild to very strong, depending on the material or combination of materials that burned.

Smoke from burning synthetic materials often results in *smoke webs*, sometimes called *streamers*. These are fine black webs, closely resembling spider webs, which cling to the corners of rooms, sometimes to cabinets, or even along curtain rods. These hanging chains are the result of smoke particles with opposite charges that are attracted to one another.

A Fourth Type of Smoke Odor

In addition to the smoke odors produced by natural, protein, and synthetic smoke residues, there's a fourth type of smoke odor. You may not find it in a physics class or a fire science book, but you'll certainly hear about it in your training classes. It's known as *psychological smoke odor*, and it's sometimes the hardest one of all to get rid of.

A psychological smoke odor is one your clients *think* they smell. And it's very real to many people. Sometimes, they're right, and it *is* the result of something you missed along the way. But if you've done all your restoration and deodorization steps correctly, and neither you nor anyone else can smell it, then it's probably a psychological smoke odor.

So what do you do about it? First of all, listen to what your clients have to say, and take it seriously. Ask the person to rate the smoke odor, on a scale of 1 to 10. That helps them evaluate how serious and lingering the odor is, and also gives you a benchmark to start with so you can determine if you're making progress eliminating it — real or not. Investigate their complaint. You may need to do a little extra cleaning, or perhaps take one more deodorization step.

After that, if you're convinced the odor is a phantom one but your client's equally convinced that it's real, you need to sit down and talk about it. Explain all the steps you've taken up to that point, and discuss any remaining options. You might, for example, suggest a week or two of natural ventilation.

If they're withholding a substantial amount of your final payment, that should be discussed as well. Ask that most of your payment be released, and a smaller amount be held. Or ask that the disputed amount be held in escrow. (We'll discuss that option at the end of the book.) Request that a neutral third party, that you both agree on, make an inspection, do a *sniff test*, and see if they can help resolve the situation.

How Smoke Behaves

During a fire, the smoke produced by the various burning materials will behave in different ways. That behavior can have a huge impact on what you'll be facing when it comes time to do an estimate or make repairs.

REAL STORIES:

Why Can't Any of You Smell That!?!

We had just finished up a long and rather difficult fire restoration. The job involved structural repairs and the restoration of a lot of contents and clothing. The clients had been a real handful throughout the job, and we were looking forward to getting things wrapped up.

It was the final day. We were finishing up the last of the pack-in, along with some typical end-of-job cleanup chores. Everything looked great; and even the man of the house supervising the final details was pleased with how everything turned out.

Then his wife came home. She stopped at the door, made a face, and immediately turned and glared at me. "I still smell smoke."

We had been down this road before — a couple of times. The house had been thoroughly cleaned and deodorized, painted, then deodorized again. Everyone agreed that there was no smoke smell. Everyone but her. We had all talked to her about it. Even her husband had tried, but it simply wasn't doing any good.

"I'm sorry," I said patiently, "but I don't smell anything. We've been all through the house, including the attic and the crawlspace. None of us can detect an odor anywhere."

She turned to her husband, who for the most part had been equally demanding throughout the restoration process. "You smell it, don't you?" But to his credit, he had to admit that he didn't smell it either.

Just then, she spotted her neighbor's car pulling into the driveway across the street. Her face lit up. "Finally," she said, referring to her neighbor, "Ben's home. He has the best nose of anyone I know. He'll smell the smoke!"

We all waited nervously as she went and corralled poor Ben. She led him into the rooms where she was convinced the odors were coming from. He followed behind her, sniffing patiently. Finally he looked at her and smiled. "I'm sorry," he said, as nicely and as gently as he could, "but I don't smell anything. I was here right after the fire, and I think these folks did a wonderful job getting your house put back together."

She never said another word, just disappeared upstairs and never came down again as we finished up and said our goodbyes. For all I know, she's still wandering the neighborhood, looking for someone to come over and smell smoke with her.

There are five basic things that affect how smoke behaves during a fire. Depending on the size and severity of the fire, you'll be able to identify most or all of these influences as you inspect the damage.

Heat

Heat is a primary influence on smoke in a building. Heat causes building materials to expand, opening up natural cracks and fissures that allow smoke to move deep inside the material. Heat also causes fasteners to expand inside wood fibers. As the fasteners cool and shrink, they leave voids in the material that smoke can enter.

Tip! *The most likely place to have a residual smoke odor problem in a house you've deodorized is in the attic during the summer. High attic temperatures during the hot summer months cause wood to expand and release any smoke molecules that weren't adequately cleaned or deodorized during the restoration process. The second most likely spot for odor to reoccur is in the bathroom, where hot, humid air from bathing will open up the pores in building materials and release the smoke odor.*

Heat naturally moves toward cooler surfaces. For example, imagine standing in front of a campfire on a cold night. The front of you is warmed, as heat from the fire moves toward your body, which is cooler than the flames. But your back feels cold, as your warm body gives off that heat to the cooler night air behind you.

You might think that all of the inside of a burning building would be the same temperature, but there's actually a lot of variation in the heat. That's why you'll find more smoke residue on windows, window frames, toilet bowls, mirrors, and other cooler areas. You may even see smoke streaks around the bottom of a door, indicating a cool air leak where the weatherstripping didn't seal well.

Pressure

As the air within the building is heated during the fire, it expands greatly, which drives the smoke deep into every available nook and cranny. With heat and pressure working together, the smoke is pushed deeper into wood and other materials than with heat alone.

Try to visualize this effect, using the following example. Imagine you have a piece of wood with a crack in it, and you also have an uninflated balloon that's coated with black soot. The wood is cold, and you push the uninflated balloon into the crack as far as it will go, then pull it out again. When you look at the wood, you see that the soot from the balloon hasn't gone very far into the crack.

Now imagine you could heat the wood, raising its temperature by a couple of hundred degrees. As the wood heats, it expands, the cells and fibers open, and the crack softens and enlarges. Now, if you pushed the uninflated balloon into the crack and removed it, you'd see that the soot had penetrated farther. That's the effect of heat alone.

Next, imagine you could push the balloon into the crack and then start inflating it at the same time the wood was heating. You would heat the wood and inflate the balloon in a continuous, simultaneous process. The crack opens up more and more from the heat and the pressure of the inflating balloon. That allows the balloon to go deeper into the crack, depositing soot much farther down into the wood. That's the effect of heat and pressure working together.

Finally, imagine the balloon is deflated and the wood is cool again, which is the same as a fire being put out. The crack decreases in size, but you're left with a deposit of soot deeply imbedded in the wood.

Magnetism

A strange effect of heat during a fire is the magnetization of metal surfaces. The smoke particles themselves have an electrical charge, due to their rapid excitement in the air as they move during the fire. The result is that the smoke particles become magnetically attracted to metal surfaces.

Examples of this magnetic attraction after a fire include plumbing pipes and metal conduits coated with smoke, as well as metal window frames, faucets, towel bars, and other metal surfaces. One of the more bizarre things you'll encounter is

Figure 7-13 Due to the effects of magnetism, the nail heads in this drywall are clearly visible, because smoke particles adhere magnetically to them.

drywall fasteners that become visible (see in Figure 7-13). This happens because the smoke particles adhere magnetically to the heads of the screws and nails holding the drywall in place.

Ionization

Ionization occurs when opposite charges in the smoke particles attract, forming a bond. This is what creates the smoke webs described earlier. You may also see it in closets, where smoke is heavily attracted to plastic dry-cleaner bags due to the static electricity present in the plastic.

Impingement

In the restoration world, impingement is just another way of saying *splatter*. When we talk about smoke impingement, it means that smoke or another burned object has struck a surface with sufficient force to impinge, or splatter, and remain on that surface.

Smoke and Soot Particles

The solid material suspended in smoke is called *particulate matter*, or PM. The size of these tiny particles depends on what material was burned, how hot the fire was, and how completely the fuel material was consumed by the flames.

REAL STORIES:
That's a Good Way to Lose Your Business

One of the larger insurance companies once hired me as a consultant to inspect a job in a small town a couple of hours away. The homeowners had experienced a relatively small flue fire with some attic and roof damage, and a local remodeling contractor had done the repair work.

A couple of months had passed, and the homeowners were smelling smoke. The contractor had inspected their complaints, but said it was just smoke from their fireplace, and refused to do anything about it. The insurance company had already settled the claim, so they hired me to go out and take a look.

The first thing I noticed when I entered was the smoke smell. It was noticeable in every room, not just the living room, so it wasn't coming from the fireplace. There was also a slight brownish tint in the corners of their freshly painted ceilings.

As soon as I lifted the attic hatch, I was bowled over by the smoke smell. For a moment, I actually thought there was another fire up there. I grabbed a flashlight and began an inspection.

There was a fresh layer of blown-in pink fiberglass insulation. I rooted around in it in a few places, and found that it had been blown in over batts — which was unusual. Since it had been an attic fire, I would have assumed the old insulation was removed. If that had been the case, why replace it with both batts and blown-in material? Brushing aside the loose fill, I found that the batts underneath weren't new. The contractor had left the old smoke-filled batts in place, and covered over them with new blown-in insulation. You can see the smoke layer in Figure 7-14.

(Continued on next page)

Figure 7-14 Smoke has clearly permeated the old batt insulation in this attic. These batts are retaining smoke odor, and should never have been left in place.

REAL STORIES — *Continued*

That's a Good Way to Lose Your Business

I moved back more of the insulation, exposing the back side of the ceiling drywall. A simple swipe with a white cloth picked up a fine layer of black soot deposited from the still-smoky insulation. A later conversation with the homeowners revealed that the contractor had never cleaned the ceilings from inside before painting them. That lack of cleaning, combined with this soot from the attic, accounted for the smell as well as the mysterious brown tint in the corners of the rooms.

Above my head in the attic, all of the trusses and roof sheathing were a shiny white. I knew from reading the insurance company's report that the contractor had sealed all the framing and sheathing with sealer, so the white coating was no surprise. What *was* a surprise was all the black char showing through (see Figure 7-15). The contractor had replaced a few of the worst of the fire-damaged sheets of plywood sheathing, but he'd left several in place that were still structurally damaged, and had just sealed over them.

My final report to the insurance company laid out over 30 different areas where the contractor had made mistakes. Some were minor, but most of them were pretty severe. All of them combined to create the smoke smell in the house — a smell that the contractor had attributed to the homeowner's fireplace, and refused to anything about.

Whether the contractor was doing fraudulent work or simply didn't know how to do the job correctly, I'm not sure. Either way, the insurance company had to pay someone else to come in and make additional repairs. I did hear that the insurance company was planning on suing the original contractor to recover their money — a suit that could easily put that contractor out of business!

Figure 7-15 A charred rafter and charred roof sheathing were left in place and covered over with white pigmented shellac. Restoration contractors need to be very careful about what materials they choose to leave in place after a fire. Anything left behind must be carefully and correctly treated, which wasn't done in this case.

Wood smoke particles, for example, average 0.2 – 3 microns in size. Other types of smoke particles are as small as 0.01 microns. In general, smoke particle sizes range from 0.1 to 4 microns (a micron is one millionth of a meter or 0.000039 of an inch). Smoke particles are commonly even smaller than mold spores!

You'll find smoke residue and soot particles in varying degrees all over a house after a fire loss. Different areas will require different cleaning methods:

- **Loose soot:** Loose soot, particularly from natural materials, settles out on horizontal surfaces. This is relatively easy to remove by vacuuming.
- **Magnetized smoke:** Magnetized smoke particles coat metal surfaces, both horizontal and vertical. These require more direct cleaning methods (which we'll discuss in Chapter 9).
- **Smoke in cracks:** Smoke that has penetrated into cracks requires a combination of cleaning, deodorization, and encapsulation.

Fire restoration is a fascinating and challenging field. Admittedly, it's not for everyone. But like any other facet of the construction industry, if you're going to do it and get paid for it, then you have an obligation to do it right.

8

Structural Fire Losses

As we saw in the last chapter, a fire requires oxygen, fuel, and a heat source. And since the oxygen and fuel are all around us, all that's needed is an appropriate heat source. It could be an unattended candle, an overworked space heater, an unwatched frying pan, an electrical short circuit, a carelessly-discarded cigarette, or a child playing with matches. There are a thousand simple everyday sources of heat that can set an unfortunate chain of events in motion.

The fire situations you'll be called on to estimate and repair will range from simple to complex. Some may take an hour to estimate, some can take days. You may complete the repairs on one job in a week, while others might have your crews and your subs tied up for six to eight months or more. Some fires may have a comic twist to them, some may be heartbreakingly tragic.

Fire losses can be hard on your crew, and hard on your equipment. Dissecting and reassembling a partially-burnt building is demanding, and can also be very dangerous. The repair process is stressful for everyone involved, from the homeowner and the adjuster to your crew and your subs. If you take on fire work, be prepared for a challenge; and be prepared to get dirty.

Typical Fire Loss Situations

Most fires fall into one of several categories. Let's look at some of the fire loss situations you're most likely to encounter, the particular items to be concerned with in those situations, and special problems to be aware of when inspecting and estimating those losses.

Keep in mind that even though we're discussing generalities here, every fire loss you encounter will be unique, and you'll need to evaluate each situation carefully, on an individual, job-by-job basis.

Kitchen Fires

Kitchen fires are among the most common fires, and usually the least severe. They often start with a careless cooking mistake, either leaving something unattended on the stove or spilling grease or other flammable material on a hot burner (see Figure 8-1). Here's what to look for during your inspection:

- **Cooktops and ranges:** The fire will almost always originate on the range or cooktop and travel upward, so you'll have damage to the appliances. This might mean simply cleaning or replacing the elements, or it may be severe enough to warrant complete replacement. When in doubt about an appliance's condition, ask the adjuster to authorize an appliance repair company to inspect it.
- **Range hoods:** Range hoods are generally directly above the source of a stovetop fire. Since they're likely to sustain a lot of damage, and are relatively inexpensive, it's best to just replace them. Remember that the hood vent runs through the attic and up through the roof. Depending on the heat and severity of the fire, that could result in damage to the attic framing or even to the roofing, especially if the fan was running at the time of the fire. Look for damage to the ducting as well.

  If you're dealing with a downdraft unit, check and see if the fan was on at the time of the fire. If it wasn't, it may only need cleaning. If it was, flames could have been drawn into the fan, damaging the wiring. If the heat was severe enough, it could damage the ducting itself, or it may have been drawn down under the house where it could do additional damage.

Figure 8-1 Like many kitchen fires, this one started on the range-top. There was damage to the hood (now removed), drywall, wiring, and upper cabinets.

- **Cabinets:** The path of the flames will usually damage the upper cabinets, even in a minor fire. It doesn't take much heat to melt or bubble the finish. The damage can range from cleaning and/or refinishing, to charring that requires replacement. See Figure 8-2. Because the burn pattern in a kitchen fire usually goes up, the lower cabinets are often not affected, other than perhaps for cleaning off soot.

  But, if the upper cabinets have to be replaced, you'll have to deal with the

Figure 8-2 This is typical of the damage done to upper cabinets by a kitchen fire. There's scorching on the side panel, face frame, door edge, and hinge. The area above the panel, where the cabinet was protected by the hood, is relatively undamaged. Notice the damaged hood wire (left), and scorched drywall. Because of the drywall opening for the wire, there's probably smoke odor inside the wall cavity.

issue of matching — a potentially complicated issue that depends on the policies of the adjuster and the insurance company. You might need to have one cabinet custom-matched; you might need to replace all the upper cabinets; or you might have to replace all the cabinets in the entire kitchen due to matching issues. We'll cover more on matching materials in Chapter 23.

- **Counters:** Pay particular attention to the countertop areas around the range or cooktop, where high heat could have transferred to the counter. Just about every type of countertop material is susceptible to heat damage, including natural stone counters such as granite, solid surface materials such as Corian®, quartz products such as Silestone®, and ceramic tile counters, especially if there's a plywood substrate under the tile. If you have questions about the condition of the counters, have a countertop dealer evaluate the material.

- **Flooring:** Depending on the severity of the fire and how it occurred, you may also have damage to the flooring. This can range from simple cleaning of smoke or spilled oil to complete replacement. Hardwood flooring with light burn marks can probably be refinished, while deeper burns may require repair or replacement.

REAL STORIES:
I Thought Water Put *Out* a Fire!

A teenager came home from school, and — big surprise — he was hungry. Dinner was still a couple of hours away, so he decided to make himself some French fries. He put a pan on the cooktop, poured in some oil, and turned the burner on high. Then he went to grab a package of frozen fries from the freezer in the garage.

When he returned, the oil was good and hot, so he dumped the ice-coated frozen potatoes into the hot oil. The reaction was instantaneous! The ice reacted with the overly-hot oil and splattered all over the top of the stove. The oil ignited on the burner, and the flames traveled up the pan and in turn ignited the oil inside.

Confused and panicked, all the young man could think to do was put out the fire with water. He grabbed the burning pan with a pot holder, swung it around, dropped it into the sink, and turned on the tap. More chaos ensued as the water splashed the burning oil into the sink and up onto the window curtains! At that point, he spotted the pan lid on the counter and had the presence of mind to toss it over what was left of the oil in the pan, smothering the fire.

The final tally of his after-school snack? The hot oil seeped into the cooktop and burned out the wiring, requiring a completely new cooktop. The flames burned the hood, the hood wiring, and the upper cabinets. Splatters of oil also burned holes in the laminate counters. Some of the oil dripped down the front of the sink cabinet and scorched a large section of the finish. The insurance company agreed to replace the countertop and all of the cabinets rather than deal with matching issues.

As the boy grabbed the pan and swung it toward the sink, an arc of burning oil spread across the vinyl kitchen floor and the carpet of the family room. The flooring in both rooms had to be replaced. The kitchen and the family room had to be cleaned and painted. The curtains above the sink had to be replaced. Even the rubber seal on the garbage disposal was damaged from the hot oil and needed replacing!

And the teenager's only comment? "I thought you were *supposed* to use water to put out a fire." At least *he* wasn't hurt.

Problem Areas

Most kitchen fires are relatively straightforward, but there are some potential problem areas:

- **Damage to gas lines and electrical wiring:** The heat from the fire may damage electrical wiring and gas lines concealed in the walls, attic, and even under the floor. Before reactivating any electrical or gas systems, always have them inspected and tested by a qualified expert.
- **Splash patterns:** During a kitchen fire, it's not uncommon for someone to try and carry a burning skillet or pan to the sink, outside through an exterior door, or to some other location away from the kitchen. This can result in splatters of burning material or scorching in unusual places. Talk to the homeowner about exactly what happened, then check the floors, adjoining counters, windows, even walls and draperies — any areas you think might have been affected by the fire.
- **Attic insulation:** Not all range hoods and exhaust fans are vented all the way to the outside like they should be — many are just vented into

Figure 8-3 An older wood stove and flue damaged in a flue fire. This particular stove was set directly on top of a hardwood floor without a protective hearth pad under it.

the attic. This can result in smoke, soot, or burning material traveling up the flue pipe and getting onto the attic insulation.

- **Fire extinguisher damage:** Many home fire extinguishers utilize a dry chemical such as sodium bicarbonate (baking soda) in a very fine, powdered form. It's very effective in extinguishing the fire, but it also makes quite a mess. If you know that a fire extinguisher was used to put out the fire, be sure and consider the extra cleanup time you'll need in your estimate.

Flue Fires

Flue fires are caused by soot and creosote, and are also fairly common. Creosote is a thick, oily material that results from the distillation of wood smoke, which solidifies as it cools. Soot is basically the particles of partially burnt material that builds up in chimneys, metal flue pipes, and flue caps. The soot and solidified creosote can eventually clog the interior of the flue, and if the temperature in the flue gets high enough, the creosote will ignite.

When inspecting a house that's had a flue fire, here are some things to be looking at:

- **Damage to masonry fireplaces:** The high temperatures of a flue fire can break down mortar. It can also weaken the bricks inside the chimney and even the firebox, leading to a very dangerous situation where the fireplace appears structurally solid, but during subsequent use flames can escape through cracks in the masonry. After any flue fire, have an experienced masonry contractor or structural engineer completely inspect the entire fireplace and chimney. And make sure that your client *does not* use the fireplace until the inspection's been completed.
- **Damage to wood stoves:** If the flue fire was in a wood stove with a metal flue, like the one in Figure 8-3, the entire flue system needs to be replaced. Also, the wood stove itself needs to be carefully inspected for damage.

Figure 8-4 This fire began in the wooden chase surrounding the masonry chimney. It then spread to the attic and did quite a bit of damage to the roof framing. Due to the amount of heat generated by this type of fire, it's important to carefully examine the concrete blocks and the mortar in the chimney for possible damage.

- **Attic damage:** The fire can easily break through any weakened masonry or loose flue pipe joints, and from there enter the attic. That can cause structural damage to ceiling joists, rafters, trusses, braces, and other framing. See Figure 8-4. Pay particular attention to the areas of framing and blocking directly adjacent to the chimney and the flue pipe. Look carefully at the roof sheathing. Move the insulation and look at the top side of the ceiling. Inspect the ductwork for both HVAC and exhaust fans (Figure 8-5).
- **Roofing:** Sparks and flames can easily get outside of the flue cap. They can ignite wood roofing materials, as well as dry leaves or needles on the roof or on the ground around the house.
- **Insulation:** Soot from a flue fire will often settle out over the top of the attic insulation. This can be very difficult to clean, and often leads to an ongoing odor problem that only gets worse in the summer when attic temperatures climb. The only effective solution is to completely remove the insulation and replace it with new material. (We'll cover more on this process in Chapter 9.)

Problem Areas

Flue fires tend to be fairly localized and therefore fairly straightforward. Here are potential problem areas to watch for:

- **Roof vents and localized roofing damage:** Whenever you've had a fire that's reached the attic, inspect it carefully for heat damage. Even small fires can do structural heat damage to plastic and other soft materials such as roof vents (Figure 8-6).

  If you have to replace roof vents or flashings in an older roof, be very careful with the old shingles. The insurance company will usually

Figure 8-5 Heat from a flue fire caused this nearby flexible heating duct to melt. All areas in an attic need to be inspected for this type of hidden damage after a fire.

want you to replace only a handful of shingles right around the damaged area. But if you do that, you'll often find that the client holds you responsible for future leak problems elsewhere on the old roof. Evaluate the situation carefully. Sometimes the only proper solution is complete replacement of all the shingles on that exposure of the roof. Discuss this with the insurance adjuster and the homeowner before you proceed with repairs.

Figure 8-6 Heat from an attic fire melted this plastic roof vent. Notice the discoloration of the composition shingles surrounding the vent. These were also heat-damaged.

- **Environmental Protection Agency (EPA) wood stove certification:** Many states have laws stipulating that new wood stoves have to be certified by the EPA. Those laws often state that if a wood stove has to be moved for any reason, it can't be reinstalled unless it's a certified stove. That includes having to take a stove out during the course of repairs to a building.

If you have an undamaged wood stove that has to be moved in order to do fire repairs, first determine how these laws apply in your state. If the undamaged stove needs to be replaced under state law, then that's going to become a code upgrade issue. (We'll discuss code upgrades more in Chapter 22.)

REAL STORIES:
But We've Had Thousands of Fires in that Fireplace!

An older couple owned a home built in the 1920s that they'd lived in for several decades. One winter night they had a fire in the fireplace, and then went to bed as the embers were still dying down. They awoke in the middle of the night to the awful sound of a smoke alarm, and saw the glow of flames outside the windows.

A neighbor had seen the flames as well and called 911. Help arrived quickly, and the damage appeared to be limited to one side of the house, part of the upper floor, and the attic. Due to the age of the house and the condition of the chimney, we wanted to call in a structural engineer before we estimated the damage. We felt that might save some unnecessary structural demolition and repairs, so the insurance company agreed.

After a lengthy inspection, the engineer declared the fireplace unsafe to reuse, except for the concrete footing that it rested on. So we set up scaffolding, and began removing the chimney by hand to avoid any further damage to the house. The fire department came by during the dismantling process, and made a careful inspection of the wood. That was the first time I ever heard the term *pyrophoric carbon.*

The house had been built before building codes and established standards for clearances to combustibles. The fire inspector pointed out how the brick chimney had been built right up against the wood sheathing boards and wood roof framing. All of the wood in that area was black and crumbly, as though it'd been roasted over a low campfire.

"That's pyrophoric carbon," he explained. "Thousands of fires in this fireplace have turned the wood to carbon, until its ignition point was well below that of normal wood. It just took that one last fire to reach the tipping point and ignite it. I guess that's what happened the other night."

We had a really good adjuster on that particular job. He agreed that even though the cause of the loss could be considered ongoing damage as opposed to sudden and accidental, the house couldn't be put back together as it was. It needed to be brought up to code to prevent a repeat of this incident. It became a judgment call on his part as to what was damaged in the course of the fire and what damage already existed. This is a good example of the discretion adjusters have in how they view and handle certain losses. He decided that because the homeowners had code-upgrade insurance as part of their policy, all the carbonized wood had to be replaced as covered damage.

So, we replaced all the damaged wood and rebuilt the chimney — with the proper clearances to combustibles. Now that house will have cozy fires for another 100 years!

- **Pyrophoric carbon:** This is an unusual phenomenon, and even somewhat controversial among firefighters. Basically, it means that if wood is continually exposed to a heat source over a long period of time, the wood will slowly carbonize, lowering its ignition temperature.

  An example is an older house with wood framing very close to a fireplace. Over the years, heat from the fireplace will carbonize the wood. Eventually, the wood's ignition point gets so low that just the heat from the fireplace will set it on fire, even without direct exposure to the flames. See Figure 8-7.

  When you work on an older house with a flue fire, be aware of the potential for this situation. Look closely at the wood surrounding the fireplace, including framing, siding, and sheathing. Look for discoloration or darkening of the wood, charring, blistering paint, or other alarming conditions. Bring anything you see to the attention of your client and the adjuster.

Figure 8-7 A fire occurred in the wall cavity of this older home and spread to the attic. It was thought to be caused by pyrophoric carbon, resulting from heat exposure over the years affecting the wood adjacent to the chimney. Notice that the framing, wall sheathing, and siding all butt directly up against the chimney.

Garage and Outbuilding Fires

Garages and outbuildings such as barns and storage sheds are mostly unoccupied and packed with equipment and flammable materials, so in many ways they're perfect places for a fire to start.

With these fires, you'll be looking at many of the same things you would with any structural fire, including the condition of the framing, the wiring, and other components, as in Figure 8-8. But there are some special conditions to check into as well.

- **Open framing:** Many of these buildings have open framing, so there's no drywall or other covering over the framing to offer fire protection. This can mean greater structural framing damage, and the need for more extensive smoke cleaning and sealing.
- **Insulation:** Here again, if you have uncovered walls and ceilings, any batt insulation between studs and joists will be subjected to direct exposure to the smoke. It's best to simply remove and discard exposed insulation.

Figure 8-8 A fire in a wood-frame storage shed behind a home.

Figure 8-9 You can see the spalling and blackening of the concrete slab after the fire in this detached garage. This slab damage was caused by the intense heat from a piece of equipment that burned and melted in this spot.

Figure 8-10 This foundation bolt is rusted and bent as a result of a fire. You can also see damage to the slab around the bolt.

- **Foundations and concrete slabs:** If the outbuilding had flammable liquids, such as gasoline, that caused the fire to burn hot enough, you may encounter heat problems with the concrete. That's often evident by a slightly pinkish discoloration of the slab. It can also show up as scorching and spalling, as shown in Figure 8-9.

  Usually, the problem is one of just cleaning and sealing the concrete. A lot of burnt structural material and contents pile up on the slab, and that, combined with water from the firefighting efforts, forms a thick black sludge that requires a lot of cleaning. If it isn't removed, that material will work its way into cracks in the slab and give off an odor later on.

  If there was a lot of heat and structural damage to the open framing of the walls, that damage may extend down to the stemwalls as well, showing up as discoloration, spalling, and structural cracks in the concrete. With both stemwalls and slabs, you may also see damage to and around the anchor bolts, so check them carefully. See Figure 8-10.

  When you suspect concrete has been structurally damaged in a fire, you need the help of an expert. Talk with a structural engineer who can either do the necessary testing or recommend a specialist.

- **Contents:** Garages and storage sheds are usually crammed with contents, which is a real headache when it comes time for reconstruction.

REAL STORIES:
That's Taking Recycling a Little Too Far!

A young couple rented a vacation house for a week at a local ski resort. They enjoyed the snow, the slopes, and all the pleasures of a winter getaway, including a glass of wine in front of a cozy fire in the evenings.

When their vacation was over, the helpful couple tidied up the place in preparation for heading back home. They packed, loaded their car, and took out the trash. Just before leaving, they sorted out their wine bottles and other recyclables, and set them aside for collection. Then, in an effort to be as nice and accommodating as possible, they went one step further. They took the partially burnt log from the night before out of the fireplace, and set it back on the pile of firewood in the home's attached garage. No sense wasting a perfectly good log, right?!

Late that afternoon, a housekeeping company arrived at the rental house to clean up in preparation for the next arrivals. As they entered, they smelled a very strong smoke odor. Tracing the odor to the garage, they found the wood pile smoldering and a thick layer of smoke in the garage. Opening the door from the house provided the smoldering wood with a fresh shot of oxygen, and it burst into flame.

The housekeepers called 911, and the fire was quickly doused. Fortunately, the structural damage was relatively light. But the heavy smoke that had billowed unchecked for hours, coupled with the garage's open, uninsulated framing, made for a lengthy cleaning and sealing job before the odor was finally tamed.

Sometimes recycling just isn't worth it!

Everything has to be moved out in order to do the necessary cleaning and restoration work on the structure. You then need to clean the contents, so you don't reintroduce odors back into the building.

- **Smoke damage:** In the case of an attached garage, it's likely that there's smoke damage inside the house, even if the fire didn't breach the wall between the house and the garage. Inspect the crawlspace, basement, attic, closets, and other enclosed areas, as well as all of the rooms, and check for visible signs of smoke in addition to any smoke odor.

Problem Areas

When inspecting and evaluating a fire that occurred in a garage or outbuilding, you want to keep these things in mind:

- **Contaminants and accelerants:** Be very aware that outbuildings often contain materials you wouldn't find in a house. These materials include gasoline, mineral spirits, solvents, pesticides, herbicides, cleaners, and other potentially toxic substances. Some of these are accelerants — materials, such as gasoline, that cause a fire to burn faster, longer, and/or hotter than it otherwise would. In similar fires, one where accelerants are present and one where they aren't, you'll see greater damage in the one where accelerants are present because the heat is more intense and more concentrated.

  The other issue is contaminants. Some materials produce toxic fumes when they burn, and others become toxic when they combine during a fire. Until you're sure what you're dealing with, wear a respirator while inside the building, and use an axial fan or another type of ventilation device to provide a fresh air flow. Also, consult with the fire department or other trained professionals for information on what materials may have been in the building, and what special precautions you should take.

- **Insurance coverage issues:** An attached garage is part of the house, and is covered under the house policy. Typically, outbuildings that are on the same property as the main structure have coverage extended to them as part of the main policy. The maximum amount of that coverage is usually a percentage of the main policy's value, but that's not always the case. Be sure to verify the coverage and policy limits.

Exterior Fires

An exterior fire is one that starts somewhere outside the structure itself. It may be the result of an accident or carelessness, such as a problem with a gas barbecue or a discarded cigarette, or it could be from arson, vandalism, a wildfire, or even from a fire in an adjacent building. Here are some of the challenges:

- **Siding damage:** In addition to structural damage, siding damage can present a real challenge. You have to determine how far the damage extends, and how best to remove the damage and patch in new material, including accessing and replacing any damaged sheathing underneath the siding. See Figure 8-11.

  Wood siding: The issue with siding is twofold. You have to first match the material. That may be easy, as in the case of T-1-11 plywood or a newer composite lap siding. Or it may be quite difficult, as with vintage clapboards or patterns that are no longer made. The second issue is where to make your transitions. When patching in new horizontal siding, for example, you can't simply make a vertical cut and then patch everything back in a straight line. You have to consider how to stagger back the joints.

  Vinyl siding: Damage to vinyl siding is another issue altogether. Rather than scorching, vinyl siding melts and makes an incredible mess. See Figure 8-12. It's very difficult to patch in, so the best plan is usually to replace the entire wall of material.

  Masonry and stucco siding: Repairs to stucco, brick, and other masonry depend on the amount of heat sustained, so look closely at mortar joints, and look for discoloration of stucco. Also, don't forget that even when siding of any type looks OK, there may be damage behind it that needs to be accessed.

- **Attic and crawlspace damage:** Depending on the fire's location, how it traveled, and the prevailing winds, flames may be drawn underneath the house through the foundation vents. Cross ventilation in the crawlspace, along with leftover construction debris and flammable dryer lint, can spread the fire. Flames and smoke are also commonly drawn into the attic through eave and soffit vents.

Figure 8-11 An exterior fire caused extensive damage to the plywood siding and porch overhang on this house. Almost all of the siding on the front of the house had to be replaced.

Attic lumber tends to be very dry and highly flammable, and attics are common storage areas as well.

➤ **Soffit and sheathing areas:** Because of the upward sweep of the flames, soffit damage is common.

Closed soffits: The damage may be as simple as replacing some trim or cornice boards. However, closed soffits can also act as a chase, funneling the fire to other parts of the attic, so look at the pattern of the smoke and charring to see where the fire traveled.

Open soffits: One of the main problems with open soffits is the exposed rafter tails, especially with manufactured trusses. See Figure 8-13. You need to be very careful with how you repair the

Figure 8-12 This older house had vinyl siding installed over wood siding (visible at right). It also had replacement vinyl soffits and trim. Heat from a fire melted the vinyl into a glob, and scorched the old paint on the wood siding. All of the vinyl material needed to be removed, and the paint stripped. Then the wood had to be cleaned and primed before new vinyl siding could be installed.

Figure 8-13 The tails on these manufactured roof trusses were burned, as was the top plate of the wall. Depending on the amount of charring and how much structural support is left in the trusses, these may or may not be salvageable. It'll require certified plans from either a structural engineer or a truss company to obtain the necessary building permits to make repairs.

Figure 8-14 A fire on the outside of this home swept up the wall and scorched the underside of the open soffit. The rafter tails, fascia, and roof sheathing were all damaged.

tails. Simply cutting them off and patching in new ones can alter the overall structural integrity of the truss or rafter. The other option is to cut off burnt tails and add new ones next to them. But that can throw off the spacing of the tails, which can affect the spacing of the sheathing and also the visual appearance of the open soffit. These are all things to consider.

Roof sheathing: If the sheathing is burnt from the underside, as in Figure 8-14, it'll need to be replaced. That means removing roofing. This gets you into a patching situation, which of course can be difficult since you're working from the bottom up.

- **Decks and fences:** Exterior fires often cause damage to decks, ranging from minor to quite extensive. An updrafting fire can take all the structural support out from underneath a deck. A fire that burns hot enough can do a lot of damage to concrete and masonry, as well as damaging porches, walkways, fences, railings, and other exterior hardscape.
- **Smoke damage:** Even if the fire didn't breach the exterior shell of the building, it's very likely that the interior suffered smoke damage, as in the attached garage we discussed earlier. Here again, make a careful inspection of the inside of the house to look for visible signs of smoke as well as any smoke odor.

REAL STORIES:

Now That's a Strange Place to Burn!

It was a warm summer evening, and Bill fired up the gas barbecue to grill some burgers for dinner. He didn't notice that his kids had pushed the barbecue closer to the house earlier that day during a backyard ball game. Neither did he notice the grease that had been building up around the grill. Through an unfortunate chain of events, the barbecue caught the grease on fire, and the fire spread to the side of the house.

It wasn't an extensive fire, and since Bill was right there, he was able to put it out quickly by himself. We thought the damage was confined to some siding, a few pieces of trim, part of the deck, and of course, the barbeque. The insurance adjustor authorized Bill to replace his old barbecue with a cool new one, which was some consolation for his embarrassment.

As I was examining the siding damage and writing my estimate, I noticed something black on the ground, right at the edge of the deck. I had seen it before, but assumed it was just a piece of charred decking. Looking closer, I realized it was the cap for the sewer cleanout. The deck had been built right over the spot where the main drain line exited the house on its way to join the city sewer.

I got out some tools, and pulled up a burnt deck board. The fire had burned through the decking and part of the deck had landed right on top of the sewer cleanout! Grease from the barbecue had been dripping through the deck and had built up around the cleanout fitting. That allowed the fire under the deck to get hot enough to melt the ABS fitting and the pipe beyond.

We ended up having to remove a good portion of the deck to get to the piping. We then sawed out part of the concrete garage floor and excavated under the stemwall in order to access the main sewer drain line. We spliced into that, extended a new pipe back to the outside to replace the damaged one, and then installed a new cleanout.

I have to say, of all the things I expected to see on that estimate, a fire in the sewer line underneath the garage slab certainly wasn't one of them! Now that's hidden damage!

Problem Areas

Here are a few of the potential problem areas associated with exterior fires:

- **Hidden damage:** There's potential for hidden damage with all fires, but that's especially true with exterior fires. Heat, smoke or cinders may have caused damage inside the exterior walls, the attic and crawlspace, underneath shingles, or behind siding. It can be just about anywhere, so always keep an open mind.
- **Wind-driven embers:** With exterior fires, the wind may distribute burning embers that can cause widely scattered damage, so try to determine what the winds were doing at the time of the fire. Examine the general pattern of the ash and debris, and look for additional areas of damage. For example, you may find scorched fencing or deck boards in an area that you assumed was untouched by the fire.
- **Outbuildings:** Outbuildings are often overlooked, but they can easily sustain smoke, heat or even fire damage. This is another place where you may discover damage from wind-driven embers.
- **Landscaping:** Exterior fires typically mean damage to landscaping, which can be tough to estimate because you don't always know exactly what the extent of the damage is. For example, you may see that the lawn's scorched by the fire, but it'll probably come back to life in time. So is it really damaged? Same thing goes with trees and shrubs that are not killed outright by the fire. You may need to have the insurance company authorize a consultation with a landscape expert to determine what will survive and what needs to be replaced.

Also under the heading of landscaping are underground sprinkler systems, including pipe, sprinkler heads, valves, valve boxes, timers, wiring, and other components. Again, your best bet is to have a landscaping contractor make a site visit to evaluate everything from the plants to the sprinklers, and give you an estimate.

When it comes to coverage, insurance companies differ in how they handle the issue of landscaping, especially if the landscaping will grow back. Rather than trying to second guess any of this, just do your evaluation and then follow your adjuster's guidelines.

➤ **Neighboring buildings:** An exterior fire can often damage the buildings or homes around it, especially in a tightly-packed neighborhood. This can range from minor smoke damage to major structural issues. See Figures 8-15 and 8-16.

The liability for this damage can be a real headache, and one that you definitely need to leave up to the adjuster. For example, suppose a fire originates in House A, and also damages House B next door. It's usually the responsibility of House A's insurance company to pay for repairs to House B. However, getting the insurance companies and the lawyers to agree on that takes time. So what usually happens is that House B's insurance company will step in and get House B repaired for their policyholders. Then they'll go after House A's insurance company to recover their expenses.

Figure 8-15 A fire occurred in the rear of this home (the one with horizontal siding). Due to its age, there were code-related issues that needed to be addressed with the building department and the insurance company prior to reconstruction. Repairs took a while to be completed.

Figure 8-16 Flames from the home fire shown in Figure 8-15 jumped across the alley, damaging the shed (with the plywood siding), and totally destroying the adjacent garage (right side of photo) and the truck parked inside. These structures were handled as separate claims, which was fortunate for the owner of the garage. His insurance company quickly authorized repairs — and didn't wait for the neighbor's insurance company to settle the claim.

You don't ever want to be caught in the middle of this situation. And, since House B has a different owner, your contract with the owners of House A doesn't extend beyond House A. If the owner of House B wants to hire your company to do repairs, that's fine. *But that's a totally separate job.* It's a separate client, a separate estimate, a separate insurance company, and a separate contract.

Major Structure Fires

Amidst all the fire, water, storm and vehicle damage losses that you'll be dealing with on a regular basis, there will be the major structure fires (Figure 8-17). These are the ones that make the 6 o'clock news; the ones with miles of bright yellow caution tape and crowds of curious neighbors.

These are, in a word, intimidating. They're the kind of job where you stand outside all by yourself for a moment, looking up at the charred and dripping structure, and wonder about the owners and what this must have done to their lives.

But then you remember that you have a job to do, and that's why you're there. So your next thought will almost certainly be, "How in the world can I ever take that apart and put it back together?" Or, "How in the world can I ever estimate anything this complicated?"

You may have heard the old story about the man who sits down to dinner, and is served an entire elephant. "How in the world can I eat all of that?" the man asks. "It's easy," his host tells him. "Just do it one bite at a time."

So if you're ever intimated by an overwhelming fire job, remember that advice: Just take small bites of the elephant. Find a logical place to start, and a logical way to proceed. When you do that, you'll get a comfortable rhythm to your estimating and your reconstruction efforts that will allow you to bid and then rebuild even the largest of structural fires.

Figure 8-17 Major structure fires can be both challenging and intimidating. Fires like this one take a long time to estimate accurately. They also take several months to repair, and require a lot of attention to detail to prevent the recurrence of smoke odor.

Initial Board-Up and Stabilization

The first undertaking on any major structural fire is to stabilize and protect the building and contents. You want to keep anyone from coming in and getting injured, and you want to keep the building from

Figure 8-18 Piles of debris are left after a structure fire destroyed part of this home. To prevent possible injury and limit liability, this needs to be cleaned up as soon as possible. Notice that the open end of the home has been temporarily covered with plastic sheeting for immediate weather protection, until a more secure covering can be installed.

sustaining any additional damage. You also want to protect the contents against weather and theft. In some instances, you may even be securing a crime scene.

Get the Necessary Authorizations

Your very first step in a major structural fire is to meet with the insurance adjuster and the homeowner and get the necessary authorizations to enter. With larger fires, insurance companies often use a cause-and-origin inspector to determine where and why the fire started. This is done to ensure that the fire wasn't deliberately set, and also to see if there's any negligence or liability that can be subrogated against; for example, did an appliance or a heater malfunction and cause the fire.

In some cases, law enforcement agencies such as the police, the FBI, or the DEA may be involved, and the fire may be considered a crime scene. In that case, you'll also need their approval before you're allowed to enter.

Board Up the Building

To secure the building against intruders and the curious, install plywood or OSB sheets over all broken windows and doors. Ideally, these should be installed from the inside so that people don't have easy access to the screws.

Install a plywood door at a convenient access point. Cut a sheet of ¾-inch plywood to size, attach some heavy-duty T or strap hinges, and install a hasp and a padlock. This lets you get in and out of the building with less effort, and saves you a lot of unscrewing and rescrewing of panels.

Clean Up the Outside

Big fires make a big mess. There are often piles of insulation blowing around the neighborhood, and all kinds of sharp and dangerous materials on the ground outside the house (see Figure 8-18). As soon as possible, get authorization from the insurance company to do an exterior cleanup. Remove broken glass and other materials, contents that have been discarded outside, and any other exterior debris. This is just a general exterior cleaning for safety and liability purposes, and is something that needs to be done even if the building is a total loss.

Figure 8-19 A temporary electrical meter base and meter was installed for this fire-damaged home. One section of siding was stripped to allow for connections to the existing house wiring, then temporarily replaced. The inside connection is shown in Figure 8-20.

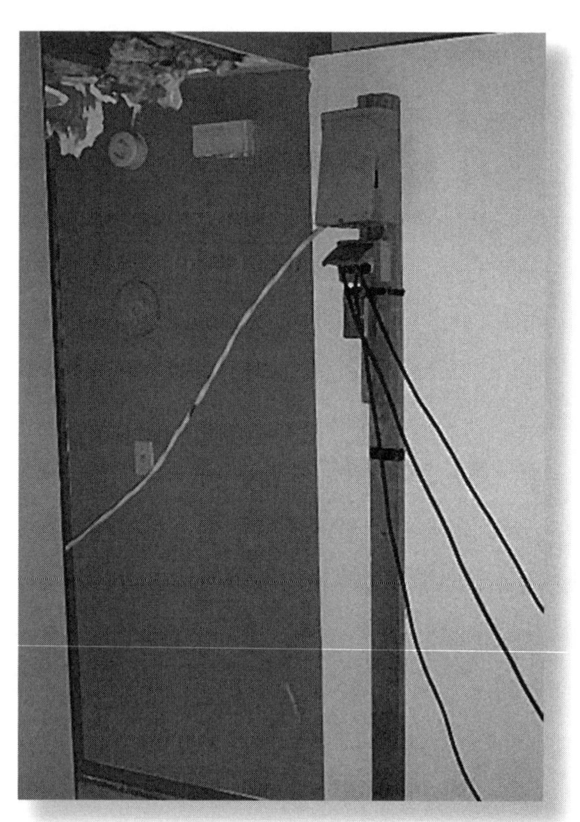

Figure 8-20 This temporary installation, inside the house shown in Figure 8-19, allowed for quick resumption of power for lighting, drying equipment, and construction equipment. All installations of this type are inspected and approved under a temporary electrical permit application, the same as for new home construction.

Establish Power and Lighting

When a major fire occurs, the power company responds and shuts off power to the building as a safety precaution until the wiring can be examined. If the service panel and meter base are undamaged, they may simply remove the meter and install a locked protective cover over the base. If the meter and meter base are damaged, then they'll disconnect the service drop or the underground service feeds.

Tip! *Never, under any circumstances, enter a fire-damaged building until you've been authorized to do so!*

It's going to take some time before power is reconnected to the house. You can either have your electrician set up a temporary power panel in the same way you would for the construction of a new home (see Figures 8-19 and 8-20), or you can set up a temporary generator. You need to get power so you can establish temporary lighting, which can be in the form of temporary work lights on poles, or a string of heavy-duty emergency lights on a long cord. Until you can provide adequate safety lighting, keep as many people out of the building as possible!

Inspect the Building's Stability

Once things are cleaned up and you have a reliable light source, inspect the building to see how stable it is. Shore up any structural parts that are in danger of collapse. See Figure 8-21.

Weatherize and Winterize the Building

Once you've established that the roof is stable enough to stand on, use heavy-duty plastic sheeting to cover any holes. Cover any damaged wall openings as well, so that the building is as weathertight as possible. Look back to Figure 8-18.

If necessary, either protect the building's plumbing system by providing temporary heat, or winterize the system.

Dry and Dehumidify

In the initial rush to deal with the structural issues of a fire-damaged building, including the power, stabilization, and winterization issues, drying the building is something that often gets overlooked. Water was used to fight the fire, and rain or snow may have gotten inside before you were able to protect it. So, in addition to the fire damage, you also have to deal with a wet building.

After you've taken care of the other essential steps, see what areas need to be dried in order to salvage them and prevent additional structural deterioration. You'll need to set up air movers and dehumidifiers, in much the same way as you would for any wet building. This may involve renting a generator to provide adequate power until you get temporary electrical power established.

Figure 8-21 Temporary supports under fire-damaged trusses are used to stabilize the roof so that tarps can be installed for winter-weather protection until reconstruction can get underway.

When drying a fire-damaged building, there are a couple of extra things to keep in mind. In order for the drying to be effective, you need to keep the moisture vapor you're generating contained in one place so the dehumidifiers can pick it up. In a fire-damaged building, you may have larger open areas than you might otherwise have, so you may need to enclose or contain spaces by tenting them with plastic.

Be careful when you position your air movers. You don't want to blow a lot of soot and debris around and make your cleanup job worse. You also need to guard against sucking a lot of soot and ash into the pickup side of your air movers and possibly damaging the motors.

You'll have to clean all your drying equipment at the end of the job, so keep track of all this time. You can bill for the use of the drying equipment, as well as for equipment cleaning and deodorizing.

Initial Walk-Through and Assessment

The evaluation process is something that's best done with the adjuster. The two of you can work together to assess the condition of the building, and can jointly set some parameters for the estimate. This simple process will save both of you a lot of time and misunderstanding.

One method we've found effective is to set up a meeting time with the adjuster, then show up about an hour early. Walk the job through on your own, take a few general notes, and get an idea of some of the issues you'll be facing. Get a feel for where the problem areas are going to be, so you can bring them up and discuss them with the adjuster while you're there together.

On your initial walk-through with the adjuster, there are two things you should avoid if you can:

1. ***Building owner involvement:*** Ideally, the best and most effective walk-throughs are with just you and the adjuster. Sometimes, the building owner wants to be there as well, and that's obviously his right. But since the building owner is going to be somewhat confused and traumatized by the fire and the entire insurance, bidding, and reconstruction process, having him there initially will greatly slow the walk-through process. It also makes it more difficult to stay focused simply on the technical aspects of reconstruction and estimating.
2. ***Multiple contractor walk-throughs:*** Of far greater impact is the very irritating tendency of some adjusters to do a walk-through with two or even three competing contractors simultaneously. They think this saves time, or helps generate more ideas, or fosters healthy competition, or — who knows what? It does none of those things. All it does is make everyone uncomfortable and edgy, and greatly slows the whole estimating process.

If an adjuster suggests a joint walk-through with you and a competing contractor, politely ask if it would be possible to do the walk-through with just the adjuster. If he refuses, and you're new enough in business that you're trying to get some work from this particular adjuster, then you may just have to put up with it. But if you can, avoid it.

Cost Reserves and Time Estimates

In addition to discussing reconstruction issues, most adjusters are going to want to know two other things from you — a cost reserve and a time estimate.

> ***Tip!*** *The reserve amounts and estimated completion times are important numbers for insurance companies. Try to be as accurate as possible, but always err on the high side — damage that may not be visible on your initial walk-through will always add to the cost and the time.*

A cost reserve — known in the industry simply as a *reserve* — is a very rough cost estimate of how much the loss is going to cost. The insurance company will then set that amount aside, at least on paper, to cover their anticipated expenditures for this loss. Guess on the high side. If you're really uncomfortable coming up with a number on the spot, tell the adjuster you'll call him later in the day, after you've had some time to take a better look. But don't forget to call him.

The adjuster will also want to know about how long it will take you to complete the job. Again, this is a financial consideration that the insurance company must deal with, since they'll be paying outside living expenses for the homeowners.

Walk-Throughs with Others

Depending on the size and nature of the job, part of the walk-through and assessment process may require the expert opinions of others. You might need the advice of a structural engineer, or an industrial hygienist if you suspect that the fire has somehow contaminated the building. Some fires may require a licensed architect or an environmental consultant. And of course, you'll need consultations with your own subs.

Total Losses

No matter how badly fire-damaged, just about any building can be rebuilt. However, that doesn't necessarily mean that it *should* be. And that's one of the elements you and the adjuster need to determine.

There are a lot of things that come into play when making the decision to declare a building a total loss. The first, obviously, is financial. Let's say that a fire-damaged house would cost $100,000 to repair, or it could be torn down and rebuilt for $150,000. The decision is obvious from the insurance company's point of view — they'll pay the $100,000 to have it repaired. If the same house would cost $200,000 to repair, the decision is equally obvious. The insurance company will declare it a total loss and pay out the $150,000.

Now take the same example of a house that could be repaired for $100,000 or torn down and rebuilt for $150,000. The insurance company is willing to pay the $100,000 for repairs, but the homeowners want a new house. The homeowners are under no obligation to have the house repaired. Likewise, the insurance company is under no obligation to build a new house if the old one can be repaired for less money. What happens is that the insurance company will get at least one additional estimate, confirm that the house can indeed be repaired for $100,000, and then they'll settle the claim with the homeowners for that amount.

At that point, the homeowners can use the $100,000 settlement however they please. They can opt for building a new smaller house for the $100,000 they have, they can add $50,000 of their own money and build a house the same size as the house they used to have, or they can add in even more money and build a mansion if they choose. Of course, if there's a mortgage on the home, the bank or mortgage company will have a say in the reconstruction of the property. They won't, for example, allow the homeowners to build a $20,000 cabin on the property to live in and spend the rest of the money however they choose. The value of the reconstructed home must exceed the value of the mortgage. The amount depends on the conditions of their mortgage agreement.

Sometimes, the opposite situation comes up. It may be less expensive to build a new house than to repair or rebuild the old one, but the owners want the old one reconstructed. The house may have a lot of sentimental value for them, or it may have significant historical value. In that case, the decision may be made to rebuild the house, even at the higher cost. If the cost differential isn't significant, the insurance company may agree to simply pay for it, usually capping the settlement amount within a small percentage of the value of the total policy. In some cases, the homeowners will have to pay the difference.

Reconstruction and Building Permits

When it comes time to begin reconstruction, you're going to need to determine if you'll need building permits, and if so, what kind. Never assume that you don't need a permit, and never try and skate by without one. The small amount of hassle you might save is nothing compared to the grief you could potentially be letting yourself in for.

Remember that when the homeowner eventually sells the property, he'll have to disclose that there was a fire in the building. If you worked on the house without the proper permits, you could be opening yourself and the homeowner up to some serious liability. You'll also open up the insurance company to liability, and if you're ever the cause of a lawsuit against an insurance company, the impact on your company will be severe, to say the least.

Always check with the building department to see what permits are required, and then obtain them. If you're told that none are required, jot down the name of the person you talked to and document that conversation for your file.

Working with the Building Department

Building permits for the repair of fire damage aren't nearly as straightforward as permits for a room addition or a new home. They require a lot of clarification regarding what's damaged, what's being removed, what's being replaced, what's to be reinforced, and other structural details not found on a typical set of building plans. They can get confusing — both for you and the building officials.

A simple way to lower your stress level and speed up your jobs is to make it a point to introduce yourself to the people who issuc the building permits. See if your can arrange a short meeting with them, and explain the type of work you do. Ask them what information they need, and what will make their job easier when it comes to understanding and processing your permit requests. Discuss their system of processing requests with them. The more information you have about them, and they have about you, the easier the process becomes.

Your purpose isn't to ask for favors, and you don't want them to break any rules for you. What you want is the inside scoop on how to get things done efficiently, and these are the people who know how to do it. You might get a tip like, *"You know, if you bring me an 8½ x 11 drawing with just a cross section of the roof, I can probably slip it in front of the plans examiner and get it processed a lot faster than if you bring in a whole set of roof drawings."*

When you apply for building permits for a fire repair, let them know that you're working with homeowners who've been forced out of their home. If they understand that you need the permit so that you can get a roof on someone's home and their kids out of a motel, they'll be a lot more sympathetic. Most jurisdictions will expedite permits when they know it's an emergency situation.

What Permits Do You Need?

As with any type of construction, the required permits will vary from job to job, so first determine if you even need a permit. For example, a small kitchen job may not have damage that warrants a permit. Neither will fires that result in only smoke or cosmetic damage.

If the fire did any structural damage, or damaged any of the building's major mechanical, electrical, or plumbing systems, then a permit will probably be needed. The following sections cover some of the permits you'll most likely be required to obtain for larger fire repair jobs.

Repair Permits

Some jurisdictions have what's known as a repair permit that covers many types of general repairs, usually up to a small dollar amount, like $2,500. If this type of permit is available, it may be the easiest kind to obtain for smaller jobs. Repair permits should cover minor repairs to plumbing, electrical wiring and other systems as well.

Structural Permits

Permits for structural repairs are the most confusing. Most building codes require a permit if damage has occurred that affects a building's structural shell, or if you're altering the shell in any way. With a fire-damaged building, it's not always clear how to apply that regulation.

Figure 8-22 Manufactured roof trusses that have been in a fire require careful inspection. The extreme heat can cause the steel connector plates to swell, separating the fibers in the surrounding wood. When the steel cools again, the joint is often structurally loosened.

You'll probably have to explain to the building department exactly what you're doing. For example, suppose the fire burned through six rafters and four ceiling joists. Typically, you'll have to present them with drawings showing the roof in both section and plan views, indicating exactly where the damaged areas are. Within the drawings, you'll have to indicate how extensive the damage is, and what you intend to do.

If the fire damaged any manufactured roof trusses that you'll either be repairing or replacing (Figure 8-22), then you'll need an engineered drawing to submit with your permit request. These drawings are available from whatever company you get your trusses from.

Other Permits

You'll also need permits to cover the other types of repairs you'll be making. These include:

- **Electrical:** You'll need to know the number of lighting circuits being replaced, and the number of specialty circuits. You'll also need to know if the service panel is being replaced, as well as the meter base. Low-voltage wiring, such as that for alarm systems and doorbells, is usually permitted separately.
- **Plumbing:** You'll usually need permits if you're making additions to the plumbing system. Repairs usually don't require a permit, unless you're doing something that adds to the city's sewer or water load. Repairs or alterations to a septic system may require a separate inspection by that department.
- **Mechanical:** Simple repairs, such as the replacement of a flue pipe or heating duct, usually don't require a permit. The replacement of a fireplace, furnace, wood stove, heater, water heater, or similar heating appliance will almost always require a permit.
- **Nonstructural repairs:** The replacement of cabinets, flooring, drywall, insulation, counters, and other nonstructural components typically won't require a permit. Replacing windows typically doesn't require a permit, unless the size of the structural opening is altered. Replacing roofing doesn't usually require a permit, although some areas do have reroofing permits. The exception to this would be if you're changing to a heavier roofing material that requires reinforcing the roof structure, such as some types of cement tile.

REAL STORIES:

Sometimes *You* Have to be the Expert

Fire losses can be a new and unusual occurrence for everyone involved. Sometimes, that even includes the building inspector.

I'll always remember one building inspector who unexpectedly dropped by a big fire job we were working on. It was early in the demolition process, and no rebuilding had begun. We hadn't yet called for any inspections — in fact, we were a long way away from needing any.

"I was just curious," he explained when he walked in and introduced himself. "I saw the fire right after it happened, and then I saw that you guys had pulled permits for it. I've never been in a big fire situation like this before. How in the world are you going to take this apart and put it back together?"

I spent about half an hour walking him through the job, explaining what was going on and what we were planning to do. He asked a lot questions, and seemed genuinely interested in what we did and how we did it. At our invitation, he stopped by a couple of times during the rebuilding process as well. Needless to say, our framing inspections passed pretty easily!

He was a nice guy, but there are jerks out there as well — the ones who don't know something, but are afraid to admit it. We ran into one like that on another fire job.

We were dealing with a very smoky fire. In one of the bathrooms, a lot of the framing and ABS plumbing was smoky but otherwise undamaged, so we were going to leave it in place. We cleaned everything; then sprayed it with white pigmented shellac, a common fire-restoration sealer, as we usually do.

The inspector came through on a rough framing and plumbing inspection. When he saw the pigmented shellac on the ABS, he said he didn't think that was an allowable use of shellac. *In his opinion*, it would degrade the ABS and lead to leaks in the future. He refused to sign off on the job until we either cleaned the pipes or replaced them.

I never like to go over an inspector's head unless I have to, but this was one of those times. I went to his boss, explained the situation to him in private, and told him there was nothing wrong with using the product in this manner. He agreed, but refused to override his inspector without the ever-popular *documentation* for the file.

I contacted the manufacturer of the shellac, who got quite a kick out of the whole thing. (They clearly thought it was funnier than we did!) They drafted a letter, faxed it to our office, and I hand-delivered it to the inspector's boss. He presented it to the inspector, who then grudgingly agreed to come back out and sign off on the job.

The end result was that the job sat for almost a week with nothing being done because one inspector refused to acknowledge that he simply didn't know something!

- **Demolition:** Most demolition and debris removal doesn't require a permit. However, some jurisdictions require a permit for the demolition of an entire building.

The Insurance Company and Permits

The insurance company typically covers the cost of just about anything related to obtaining the necessary repair or rebuilding permits, including:

- The permit fees.
- The preparation of any plans required by the building department. It doesn't matter whether you prepare the plans yourself or have someone else prepare them.

- Your time spent in obtaining the permits.
- Fees for engineers and other experts.
- Draftsman, designer, or architect charges to prepare specific drawings for an architectural review committee.
- Fees for special permits, such as for a temporary parking permit to allow a Dumpster to be placed in a commercial zone during demolition.

Structural Engineers

Structural engineers are qualified through a combination of training and experience to assess many different types of damage to a building, including fire damage. They can be a tremendous help, especially on large structural fires where you hope to salvage part of the structure. They'll also provide you with the documentation you need for the building department.

Here's an example of how you might work with a structural engineer on a large fire job. Suppose you have a fire that's scorched the outside of a building's large main beam, like the one shown in Figure 8-23. That main beam supports several smaller beams, which in turn support a run of joists.

Figure 8-23 You need a structural engineer to help determine whether the scorching on the top of this main support beam is enough to warrant replacing it, or whether it can be salvaged and left in place.

Tip! *Structural engineers are expensive, but insurance companies are usually willing to pay for their services if it helps minimize the overall repair costs. But don't ever hire one on your own and assume the insurance company will pay the bill* — always *get the adjuster's preapproval first.*

Removing and replacing the main beam would be a complicated undertaking, and you'd prefer to save it if possible. You know that you can effectively scrape the beam to remove the charring, and then seal it and paint it to deal with the odor. But how do you know that the beam is still OK from a structural standpoint, and how do you convince the building department?

The structural engineer will determine what type of wood the beam is made from, and its grade. He'll measure the beam's overall span and size, and then he'll consult with you to determine how large the beam will be once all of the char is removed. Finally, he'll determine exactly what live and dead loads it currently supports.

From there, he can calculate whether or not the beam will still be adequate in its reduced size to meet the load requirements. If it is, he'll generate either a drawing or a letter indicating the beam is fine. You'll take this to the building department as documentation to get your building permit.

If the beam's reduced size won't be sufficient to support the loads, he'll work with you on what to do next. He may suggest an intermediate support, a reinforcement bracket, a means of transferring some of the load off it, or some other way to keep the beam in place. If it turns out it needs to be replaced, he'll suggest the most structurally sound and cost-effective way of doing it.

One of the most important things the engineer provides is his certification stamp. That stamp, whether on a letter or a set of plans, gives you and the building department the assurance that so long as the engineer's specifications are followed exactly, the repairs to the building will be structurally safe.

However, that certification stamp applies *if and only if the engineer's specifications are followed exactly. Never* deviate from what the engineer has designed, even slightly. You put the building and its occupants at risk, and open yourself up to a huge liability suit.

Occasionally, jobsite conditions, hidden damage, the unavailability of materials, or some other reason will prevent you from doing what the engineer specified. If that happens, let the engineer know immediately, and have him design and certify an alternative. Keep a copy of the addendum on site with the plans, and show it to the building inspector when he makes his site inspections.

There are few jobs in insurance restoration that are as challenging and rewarding as fire repair, especially the complete reconstruction of a home after a major structure fire. From those first overwhelming days in the blackened shell of the house, to the final walk-through of the shiny-new home with its delighted and grateful homeowner, you really gain perspective into why this is such a great field to be in!

9

Structural Cleaning & Deodorization

In a fire loss, the devastating damage caused by the flames is obvious from the moment you enter the building. What isn't as obvious is the smoke damage. However, the time and effort you'll spend cleaning and repairing smoke damage may equal or even exceed the actual fire repairs.

Smoke cleaning and deodorization is part art and part science; utilizing both high-tech equipment and cleaning products, and low-tech cleaning techniques along with everyday supplies.

Smoke is only one of the many cleaning and deodorizing situations you'll face as a restoration contractor. In this chapter, we'll look at some of the many techniques and products for cleaning and deodorizing smoky buildings, as well as dealing with other types of odor problems.

Training

There are a countless number of possible combinations of materials, contaminants, stains, and odors that you can potentially encounter. To acquire some level of comfort in dealing with these, you definitely need some specialized training.

The IICRC is once again a good place to turn for training. They offer both general and specific cleaning-related courses. Many cleaning product and equipment suppliers also offer excellent training classes that'll provide you a broad base of both theoretical and practical knowledge.

IICRC Certification

The Fire & Smoke Restoration Technician course mentioned in Chapter 7 is especially helpful in dealing with smoke damage. Here are some of the other IICRC certification courses, with their course descriptions, that you should consider for cleaning and odor control:

- **Odor Control Technician:** *"OCT (1 day course): The Odor Control Technician course covers olfaction and odor, odor sources, detection process, theory of odor control, equipment, chemical options and applications. The student will learn how to address odors caused from biological sources such as decomposition, urine contamination, and mold; combustion sources such as fire and smoke damage; and chemical sources such as fuel oil spills or volatile organic chemicals."*
- **Carpet Cleaning Technician:** *"CCT (2 day course): The Carpet Cleaning Technician course teaches the fundamentals of carpet cleaning. Topics include fiber types and characteristics, fiber identification, carpet construction and styles, dyeing at the mill, soil characteristics and the chemistry of cleaning. This course will also introduce the technician to the 5 methods commonly used in carpet cleaning. The differences between the CCT and CMT courses are, basically, a focus on either general or residential versus commercial applications."*
- **Upholstery & Fabric Cleaning Technician:** *"UFT (2 day course): The Upholstery & Fabric Cleaning Technician course covers upholstery fiber categories, fiber identification and testing, manufacturing of the fiber & fabric, chemistry of cleaning, upholstery cleaning methods, protections, spotting and potential problems. A student will have a specific knowledge about fabric and fiber content, as well as furniture construction. This enables students to identify limitations and potential cleaning-related problems on a given piece of upholstery."*

Safety, Safety, Safety

Safety simply can't be overstressed with any of the cleaning situations we'll be discussing in this chapter. That includes safety measures for yourself, your workers, your clients, and also for the buildings and contents entrusted to your care.

The restoration and cleaning chemicals and equipment available today are highly effective, and in the right hands they can make a tremendous impact on the successful restoration of a building and its contents. But here are some important points to remember:

- Always get the proper MSDS (Material Safety Data Sheets) for every product you have. OSHA also requires that MSDS material information be properly organized, stored, and available! We'll discuss this in more detail in Chapter 19.

Figure 9-1 You can see the surface discoloration on the Melamine shelves in this kitchen cabinet. The areas where objects were sitting on the shelves weren't smoke damaged. The middle shelf and a small part of the upper cabinet (left) have been test cleaned to determine if the discoloration can be effectively removed.

- Make sure all your crew members get the necessary training.
- Work closely with your suppliers and the manufacturers to ensure that you have the right products and the right equipment for each specific job.
- Always *have* and *use* the proper safety equipment and procedures, as recommended by the manufacturers of the equipment and products you're using.
- Measure correctly. If the manufacturer's instructions tell you to mix one ounce of their product in one gallon of water, that's *exactly* what they want you to do. Don't pour *"about"* a gallon of water into a five-gallon bucket, and then *"glug"* in some product and stir it up. These are cleaning products for professional use. A *"close enough"* attitude can do some serious damage.

Your Primary Smoke Objectives

With any smoke restoration job, your objectives are to remove the stain and remove the odor — both are equally important. If you achieve only one without the other, you won't have a satisfied client.

Smoke discolors all kinds of objects, and the lighter the object, the more obvious the discoloration. It's especially noticeable where a surface has been protected from the smoke. A perfect example is the inside of the cabinet shown in Figure 9-1. During the fire, the smoke discolored the exposed surfaces of the white Melamine shelves. But when objects sitting on those surfaces were later removed, the protected areas underneath were clean. Depending on the severity of the smoke, you may be able to remove the smoke residue and clean the surfaces.

The second objective is to remove the odor, and that can be very difficult. For instance, a fiberglass tub/shower unit has a smooth gel-coat surface on one side that may clean up well and not absorb any odor. But the back side of that unit is raw fiberglass, with a very rough and porous surface that you can't clean. If the smoke was heavy enough, especially in a situation where the fire pressurized the smoke, then it's likely that the back of the shower will retain the smoke odor.

Figure 9-2 Pitting from heavy smoke is clearly evident in the aluminum frame of this shower door.

Time Is of the Essence

With any smoke loss, the faster you can deal with the affected materials, the better. If you have contents or building materials in the house that you think are salvageable, get working on them as soon as possible.

Smoke is very corrosive, and it's generally on the acidic side, so it can quickly damage metal surfaces like aluminum-framed windows and shower doors. See Figure 9-2. Also at great risk are electronics, such as computers and televisions.

Figure 9-3 Test cleaning was done on the sinks and laminate counters in this bathroom to determine how well they'd clean up. The results of the test cleaning were documented for the insurance company and the building owner, and became part of the decision-making process regarding which materials to salvage and which to replace.

Smoke can permanently discolor many lighter-colored surfaces, and the longer the smoke is in contact with the surfaces, the more pronounced the staining becomes — and the harder it is to remove.

Despite the need for quick action, don't get too far ahead of yourself. Start your work by carefully documenting everything, and doing some test cleaning, as shown in Figure 9-3. This will help you determine what's salvageable, and what's not worth cleaning.

Before you begin work in a room, take a series of digital photos. Remember that much of the light is absorbed by the smoke, so use a camera flash or some supplemental lighting to ensure that you get good, clear photos.

The *Fire & Smoke Restoration Technology Student Workbook*, used as part of the Restoration Sciences Academy offered by Dri-Eaz and Unsmoke, suggests taking *swipe samples* at different sites where you're performing cleanup operations. These will document exactly how smoky the area was prior to working on it. You can use white cosmetic wipes, available in bulk at big box retailers.

REAL STORIES:

Sometimes You Have to Walk Away

We were called out to a fire loss in a two-story townhouse. The fire had started on the first floor, and had smoldered for quite awhile. A lot of contents had burned, creating fairly heavy smoke. Finally, a front window shattered, letting in enough oxygen to feed the fire and get it going. This heated the air, pressurized the smoke, and pushed a lot of hot, smoky air upstairs.

The adjuster assigned to the job was, unfortunately, an ambitious young man eager to make a name for himself with his company. The company was one with a well-deserved reputation for cost-cutting in paying out on claims. It was definitely a bad combination.

The two upstairs bathrooms had builder-grade fiberglass tub/showers on exterior walls that were heavily coated with smoke. We test-cleaned the front fiberglass and determined there was a chance that the units could be cleaned and salvaged. The problem was that the smoke would have been attracted to the cold exterior walls behind the showers. We knew we had to get behind the units to replace the smoky insulation, treat the raw fiberglass, and clean the exposed plumbing. If we didn't, a recurring smoke odor would result, especially in a hot and humid environment like a bathroom.

The adjuster refused to pay for any of that work. He said that *in his experience* we could do all the cleaning from the front, without removing the showers. We adamantly disagreed. We explained our position to the homeowner, and told her exactly what to expect in the future. We explained her rights to her, as well as what recourse she had with the state insurance commission.

She wanted us to do the job, and she wanted it done right. However, she wasn't an assertive person, and this was her first experience with an insurance claim. The adjuster saw this and put pressure on her to settle the claim at the bargain-basement price he came up with, with the shower as well as other issues in the house. He assured her that *his contractor* could get the job done for the amount he was offering.

We really felt bad for this woman and how she was being treated. But there's only so much you can do, and only so much you can explain, before you finally have to just walk away. Don't let your sympathies allow you to be bullied into doing something that you know is wrong! And working under those conditions was wrong!

Their suggestion is to swipe a representative surface prior to cleaning. Then, on an index card write the location where the sample was taken, the date, and the person taking the sample. Staple the wipe to the card, seal the card and wipe inside a small zip-lock freezer bag, and store the samples from that particular job in a labeled box for future reference. It's inconvenient when you're in a hurry to get started, but it's an important step in ensuring you'll be fully paid for your work.

Value Determinations

Just about anything can be cleaned and deodorized given sufficient time and effort. But you need to evaluate whether an item is worth the time and expense to salvage, or whether it's better to replace it. Let's use the fiberglass tub/shower unit as an example:

- **How effective was the test clean?** Is it obvious that it'll clean without residual staining? If not, then you need to try a little more aggressive test cleaning. If you're still not convinced, then don't try and salvage it.

- **Can you control the odor?** How bad was the smoke behind the shower? Is there insulation that needs to be removed or wiring or pipes that need cleaning? Is there soot on the framing, or on the subfloor?
- **Can you slide the unit forward?** In some instances, you can easily unfastened the unit, disconnect the plumbing, and slide the whole thing forward into the room for cleaning, which makes salvaging it more cost effective. In other cases, it could be a major — and expensive — undertaking.
- **What's involved in replacing it?** One-piece tub/shower units are usually set in place when the house is built. You can cut it up to remove it, but getting a new one into a small space can be a real problem. You may need to remove a window or door, or even some framing, so that becomes a factor in your decision.
- **What does your client want?** For some people, cleaning is fine, and if the finished product is a little yellowed, they don't really care. Other people won't settle for anything less than exactly what it looked like before the fire, and still others will demand a complete new unit. Your client's wishes aren't always the final deciding factor, but what they want will certainly weigh heavily on your decision.
- **Remember like kind and quality?** If the client had a one-piece tub/shower and it can't be salvaged, they're entitled to a new one of the same type and quality. That's regardless of what's involved in getting it installed. If you have to remove framing or a window, then the labor to do that needs to be included in your estimate. Some adjusters may want to substitute a two- or three-piece unit that can be installed without framing alterations, but the client doesn't have to accept it. Also, that style of tub/shower is more time-consuming for you to install, so you'll need to charge more. They're also more prone to leaks than a one-piece unit, so you'll be taking on more future liability if you agree to that installation.
- **Look for creative alternatives:** A new bathtub with a shower surround of ceramic tile or some other material might be a more cost-effective compromise than having to do extensive reframing to bring in a one-piece unit.

Five Basic Steps in Smoke Odor Removal

There are five basic steps you'll need to follow in order to achieve both the cleaning and odor removal objectives on a smoke and fire restoration project.

1. Source Removal
2. Chemical Containment
3. Vapor Phase Containment

4. Penetrating Deodorizers
5. Surface Sealants

Remember, no two jobs are alike. You're always dealing with different buildings, materials, and types of odors, as well as different quantities and intensities of soot and smoke, ranging from light to severe. So you may need to modify these steps to meet specific jobsite conditions, or you may need to take them in a different order or do some of them simultaneously.

> **Tip!** *Your nose is obviously your best tool for finding odor sources. However, it will quickly become desensitized to the smell of smoke when you're inside a fire-damaged building, so be sure and wear a respirator. This will not only protect your lungs, but will also protect your sense of smell. Remove the respirator only when trying to locate a specific odor source. Also, go outside into the fresh air periodically to help clear your nose. Sniffing fresh coffee grounds also helps, so you might want to keep some in a sealed jar or plastic bag.*

A few of these steps will be undertaken as part of the emergency services, and will be done almost immediately in order to stabilize the building and prevent additional damage. Other steps may be performed initially to mitigate additional damage, and then done again later to help with additional odor removal.

The important thing is to constantly assess and evaluate your jobs, and compare each new job to previous ones. If you remain proactive on a fire restoration job, you'll find yourself better able to adapt to situations as they come up.

Remove the Source of the Odor

The first and most basic deodorization step is to remove the source of the odor. That seems obvious, but in a fire situation, it's not always that simple.

Let's say you have a home with a roof fire. When you walk in, you see a charred board lying there, so you pick it up and throw it away, eliminating some of the smoke odor. But that's only one of 50 burnt boards in the roof, and the other 49 have to stay in place because they're still temporarily holding the roof up. All you can do is the best you can.

Content Removal

Begin by removing the contents. This will require a loss inventory (which we'll cover in detail in Chapter 14). The owners may be doing the loss inventory on their own, or that may be something you're helping with.

Loss inventories can be tough. Going through personal items that have been destroyed can be very emotional for your clients, but try and get them through it as quickly as possible. Contents are fairly easy to remove, and you'll not only be getting a lot of odor-causing items out of the way quickly, but also helping to clear the way for the structural demolition to begin. See Figure 9-4.

Figure 9-4 Whether the contents are salvageable or not, they're holding a lot of smoke odor. One of your first objectives is to get them inventoried and out of the house.

This step also applies to salvageable contents. If items are going back to your shop for restoration, get them out of the building as soon as you can. If you haven't been hired to restore them, ask the owners to move them into a storage facility.

Structural Demolition and Debris Removal

Start working with the adjuster and your clients to agree on what isn't salvageable from a structural standpoint so you can begin demolition. Floor coverings are often the first step. After that, concentrate on things like cabinets and fixtures, and then remove drywall and insulation. Strip out as much as you can. Pay close attention to worker safety, and to the structural stability of the building.

Tip! *Cover your truck or dumpster with tarps or with 4- or 6-mil plastic sheeting before you start pumping insulation into it. That'll contain the insulation, and will keep you from having a huge mess blowing all over the neighborhood!*

While you're working on the demolition, make an ongoing effort to remove debris from within the building. Don't just drop a ceiling or tear out a wall and then move onto the next room. *Remember that everything is contributing to the odor in the house.* So as soon as it's been torn down, get it bagged up, swept up and out of the house.

Removing Blown-In Insulation

In many fire situations, you'll be faced with the removal of blown-in fiberglass insulation. One way to remove it is to knock down the ceiling and then shovel up the insulation. However, that's both messy and time-consuming,

(Photo courtesy of Wm. W. Meyer & Sons, Inc.)

Figure 9-5 An insulation vacuum, powered by a gasoline engine. Insulation is drawn into the impeller through a hose attached to the inlet pipe (center). It's then discharged through another hose attached to the outlet pipe (top left), or into disposal bags. Powerful vacuums such as these are excellent for the fast removal of smoky loose-fill material from the attic or other areas.

and it obviously doesn't work if you're trying to salvage the ceiling material.

A better alternative is to use an insulation removal vacuum designed specifically for this job. The unit consists of a gas engine and a steel impeller blade, mounted on a wheeled cart. See Figure 9-5. Attach one end of a long flexible hose to the inlet side of the impeller, and use the other end of the hose to suck up the insulation in the attic.

At the other end of the impeller, you can either attach a collection bag to catch the extracted insulation, or use another flex hose to direct the insulation straight into a truck or a dumpster for disposal. Using bags to collect the insulation is both more expensive and less inconvenient, so I prefer the direct-to-the-dumpster method whenever possible.

You can rent insulation vacuums at some larger rental yards. However, you'll use this piece of equipment fairly often, so it's well worth investing in one. Also, what you charge for the extraction services will pay for the equipment in a fairly short period of time.

Crawlspaces and Unfinished Basements

Don't forget to clear away the debris in unfinished basements or under the house, as in Figure 9-6. In crawlspaces, carefully roll up and dispose of the plastic vapor barrier that covers the soil. Try and contain as much debris within the vapor barrier plastic as you can. Then do a thorough raking to remove all burnt material. This isn't an easy task in a crawlspace, but take your time and remove as much of the ash and soot as you can. Use short leaf rakes, or rakes with the handles cut off. Wide dust pans and 5-gallon buckets with lids work well for scooping up and transporting the debris without spilling it.

> ***Tip!*** *In a crawlspace, use lime to counteract the odor left behind by any remaining soot and small pieces of burnt debris. It's available in bags at home centers and other retailers that carry sand, mortar, and other masonry supplies. Just open up the bags, spread the lime out on top of the dirt, and then use a metal garden rake to thoroughly rake it into the soil.*

Structural Cleaning

Removing the odor also includes cleaning all visible sources of smoke and soot. This needs to be part of your initial odor suppression efforts on any fire and smoke restoration project.

Your test cleaning will determine what materials in the house you're going to salvage. Those materials need to be cleaned to remove the soot and smoke residue.

Figure 9-6 A crawlspace has the potential to hold a lot of smoke odor. There's smoky batt insulation, melted foam insulation, melted ductwork, melted vapor barrier, and debris that's drifted down into the soil. All of this burnt and melted debris needs to be cleaned up and removed. And then, the concrete stem walls need to be washed and possibly sealed.

That'll eliminate the odor, get rid of the smoke staining and renew the material's appearance, or prepare it to receive other treatments, such as painting, at a later time.

Boundary Lines

Within almost all fire situations, you'll notice certain boundaries. These can occur between rooms, between floors, between hot and cold areas, and even between sections of rooms. The boundaries are sometimes sharply defined, and other times are somewhat vague. There are two basic types of boundaries, and they can impact how you decide to clean.

The first type of boundary is a heat boundary, or heat line. This is usually fairly clear, and marks a line between the higher and lower temperatures in a fire. For example, in a kitchen fire there may be a clear heat line somewhere between the upper and lower cabinets. On one side of the heat line, damage to the cabinets isn't that bad, and cleaning will be very effective. On the other side of the line, the damage is more noticeable and more severe, and cleaning alone may not be enough.

The other type of boundary is less clear. In a relatively cool fire, it may be an area where you notice that damage from the fire simply tapers off. It may reach part way into a room, or part way up a set of stairs. Or you may have a fire that occurs in one room but the room next door had the door closed, so the damage in there is relatively light.

(Photo courtesy of Absorene Manufacturing Company, Inc.)

Figure 9-7 A dry chemical sponge, used for removing soot and other materials from surfaces without water or other liquids.

Cleaning Methods

There are two basic options open to you when cleaning in a fire situation, dry cleaning and wet cleaning. Try different test cleaning techniques and solutions, and see what's effective.

Dry Cleaning

Dry cleaning is typically the easiest and least invasive, but has varying degrees of overall effectiveness. It isn't the same as dry cleaning your clothes — in this situation it refers to any method of cleaning smoke and soot without the use of liquids.

Dry cleaning methods include simply vacuuming off surfaces, or using something like a feather duster or one of the dust-attracting cloths now on the market. Another option is a dry chemical sponge, like the one shown in Figure 9-7. This soft sponge is treated with special chemicals that remove soot and smoke from a variety of surfaces without the use of liquids. You can see the results on the ceiling in Figure 9-8.

Figure 9-8 Chemical sponges work especially well on flat surfaces, like this ceiling, where wet washing isn't a good option.

There are also crumbly cleaners, dough-like materials that work well for removing soot and smoke residue from delicate surfaces. There are special types of these cleaners formulated for use on wallpaper, books, paintings, tapestries, papers and other documents.

For some materials, dry cleaning is the only option available. For example, some types of wallcoverings can't be cleaned with liquids. Unfinished wood is another example, as is most flat-painted drywall. In other cases, dry cleaning is the first step in a two-step process involving both dry and then wet cleaning.

Wet Cleaning

Wet cleaning methods are usually the most effective. In a wet cleaning method, you're typically using water mixed with some type of cleaning additive, such as a degreaser.

In addition to the cleaning additives, you have the option of adding a water-based odor counteractant to the wet cleaning mix. There are general deodorizers that are formulated for a wide range of odors, and other specialized cleaning compounds that are formulated to work on heavier odors. There are also deodorizers made for specific situations, such as for protein fire odors, or the odors associated with mold and mildew.

The *Fire & Smoke Restoration Technology Student Workbook* breaks the wet cleaning process into six important categories:

1. ***Solvent Action:*** *Most smoke residues encountered after a fire are acidic. Therefore, most water-based cleaning products used in fire restoration are alkaline.*
2. ***Chemical Reaction:*** *For example, rust spots on carpet are corrected by use of this method. In this case, an acid combines with the oxidized metallic residue and the resulting material is rinsed away.*
3. ***Lubrication:*** *Reduces the amount of friction between the surface and the smoke residue. By reducing the friction between these two surfaces, the residue can be removed easily without causing scratching.*
4. ***Agitation:*** *Utilizing some type of physical force to suspend soot residue. The most frequent method of applying wet agents is with a soft towel. A different result will be achieved if the same solution was applied with a brush.*
5. ***Temperature:*** *Higher temperatures speed chemical reactions, open surface pores on the finish being cleaned, and decrease drying times.*
6. ***Dwell Time of the Solution:*** *Dwell time is the amount of time allowed for a cleaning product pretreatment to remain for breakdown of soot contamination.*

Wet cleaning can only be done on surfaces that are moisture resistant. That includes cabinets, woodwork, most hard furniture, counters, fixtures, most metal surfaces, concrete and masonry, and other similar surfaces. Wet cleaning can also be effective on drywall that's been painted with semi-gloss paint. If the smoke isn't too bad, it can also be used on satin paint.

Other wet cleaning methods can be more aggressive, depending on the situation. For concrete slabs and foundations, steel framing, and other areas, wet cleaning methods may include cold water sprays, hot water sprays, pressure washing, and even steam cleaning.

Figure 9-9 A tanker aircraft dropping a fire retardant chemical during a forest fire. These retardant solutions often contain a red, water-soluble dye, and should be cleaned off buildings and other surfaces as soon as possible.

Cleaning Fire Retardant Drops

You may occasionally find yourself faced with the aftermath of a forest fire. Since these fires occur outside the home rather than inside, they can leave you with some unusual cleaning and deodorization situations.

Drifting or even pressurized smoke from the fire may coat the outside of the building. If the building's windows were open, you may have to deal with smoke odor or even smoke residue inside. You can generally handle these situations using the same techniques described elsewhere in the chapter.

A more unusual situation is the cleaning of fire retardant chemicals. These chemicals are dropped by aircraft from overhead in order to slow the fire's advance or to cool and help smother it. See Figure 9-9. Different compounds are used for different situations, but almost all of them contain a red dye so that fire crews can easily spot the location of the drop.

Occasionally, these fire retardants are dropped on buildings in the path of the fire. Later, the material needs to be cleaned off the exterior of the buildings. That includes roofing, siding, windows, decks, and other surfaces. You can usually do that with a pressure washer, either with straight water or a water and soap solution. Sometimes other cleaners might be needed.

If you find yourself in this situation, try and find out exactly what fire retardant material was being dropped. Ask your local fire department if they can put you in touch with the agency that was handling the fire retardant drops, so you can get information on the mixture. Your cleaning chemical supplier can then help you select the proper materials and equipment to handle the situation.

Dry Blasting Methods

There's a method that falls somewhere in between the wet and dry cleaning methods, and that's dry blasting. The more common name is sand blasting, but that's not as accurate. It's not something you'll do very often, but it can be very effective on certain jobs.

Dry blasting methods involve using compressed air to blast abrasive materials such as sand, glass beads, walnut shells, or even baking soda. The materials are blasted at various pressures, and work well for removing soot and smoke from materials such as concrete, steel, masonry, and other materials that won't be harmed by the abrasives.

Be Thorough!!

Here's a simple tip that's not stressed nearly enough in classes and workbooks and seminars: *You've got to be thorough!*

Ignore it or forget it, and you'll *constantly* have problems with this field of restoration. So, at the risk of being redundant, let's review the basics:

- Odor comes from a source. In the case of a fire, that source is a piece of burnt material, or it's a film of smoke or soot on some material or surface.
- Get rid of that burnt material or that film of smoke, and the odor will go away.
- Leave that burnt material or smoke film in place, and the odor stays behind as well.

If you're coming into insurance restoration from the remodeling field, this may be a hard lesson to learn. When you tore out a wall during a remodeling job, it really didn't matter if a little bit of insulation or a couple of pieces of old blocking got left behind. *But with fire restoration, it matters a lot.* If that insulation or blocking smells of smoke, the odor will remain when you close up the walls.

The smell will become less and less noticeable over time, that's true. But before that happens, you'll probably have to tear out new drywall, or rip up new flooring, or do any of a hundred other things to try and eliminate the odor because your clients are unhappy and you're not getting paid!

The bottom line is, when you're doing cleaning and demolition work, 90 percent doesn't work. Neither does 95 or 99 percent. If you leave burnt material behind, it's going to come back to haunt you.

Chemical Containment

Chemical containment of odor is done to reduce its overall intensity and to prevent the odor from spreading. This helps to mitigate the loss somewhat, and also makes the overall work environment easier to deal with. Ideally, chemical containment is done early in the deodorization process, often as part of the initial emergency service and board-up.

This initial deodorization helps the building owners by lessening their sensitivity to the odors, and often has a positive psychological effect in reducing the stress of walk-throughs and loss inventories. It also helps establish you as a knowledgeable contractor who's taking the proper steps to protect their home and their property.

The chemical containment application is done with a mixture of water and a water-based deodorant especially formulated for this type of job. Apply the mix with a simple pump-up compression sprayer, at a low volume and low pressure, following the manufacturer's specific application and safety instructions. Workers need to be in protective gear, including eye protection and a respirator.

You typically spray the mixture on flooring and along baseboards. You can also spray the lower parts of the walls, as well as other low horizontal and vertical surfaces. For safety, be careful about spraying high vertical surfaces, and anything overhead.

Vapor Phase Containment

Vapor phase containment involves placing granular deodorizers at the site, and is usually done following the application of the chemical containment. Granular deodorizers are time-release, acting more slowly than the liquid chemical containment, but for a longer period. The granules are made up of a small grainy material, resembling kitty litter, treated with an oil-based deodorizer that releases a vapor which helps mask and contain the smoke odor.

There are a few things you need to be careful of when placing the granules. They can stain or damage certain materials, including some painted surfaces and hardwood floor finishes. They're also crunchy. If spilled, they could be stepped on and ground into surfaces, leaving scratches. So don't just sprinkle them around. Instead, pour a small amount into an unbreakable container with a wide base and fairly high sides that won't tip or spill easily. Avoid anything that's glass and could break.

Tip! *Old plastic margarine tubs make great containers. Or, use a can or other container as a mold, and press aluminum foil into the bottom to form small disposable aluminum dishes.*

Put the containers anywhere that can use some additional deodorization during the early stages of cleaning and odor removal. Some suggestions include: under and behind appliances, in sinks and tubs, inside or on top of cabinets, and in closets and other enclosed areas. Different scents are available, depending on the odors that you're dealing with.

Thermal Fogging with Penetrating Deodorizers

Your next step is to use deodorizing chemicals to penetrate deep into surfaces, through a process called *thermal fogging.* This is one of the most unique and highly-specialized operations in the fire restoration process.

Remember the balloon in the wood crack example from Chapter 7? It showed how heat and pressure combine to drive small particles of smoke deep into wood and other materials. Once the heat and pressure are relieved, the crack diminishes in size, leaving the smoke particles deeply embedded in the material. To deal with those embedded particles, you have to recreate what happened during the fire, and that's where thermal fogging comes in.

Thermal foggers, like the ones shown in Figures 9-10 and 9-11, utilize a small gasoline-powered pulse-jet engine to heat a nozzle. A solution tank is filled with a water- or oil-based liquid deodorizer, and at the touch of lever, the liquid deodorizer is sprayed

(Photo courtesy of Curtis Dyna-Fog, Ltd.)

Figure 9-10 A thermal fogger.

A The white tank holds a specially-formulated liquid deodorizer. When the machine is running and warmed up, the black lever (center) is depressed, injecting the liquid through a nozzle into the hot barrel. The liquid is atomized into very tiny particles, producing a fog.

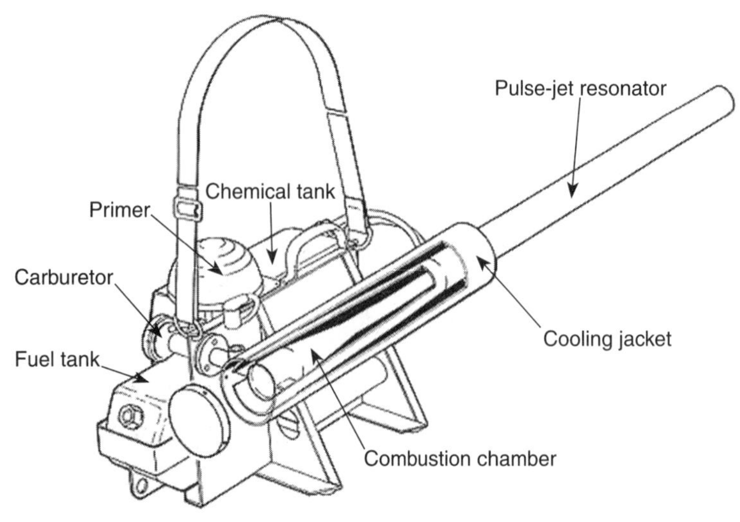

B Cutaway drawing of the thermal fogger in A. The chemical tank is at the top, and the fuel tank at the bottom (left). The half-round ball is used for priming, and batteries provide the necessary spark to start the engine.

(Photo and drawing courtesy of pulsFOG)

Figure 9-11 Another style of thermal fogger.

against the hot nozzle. This instantly atomizes the liquid into a dense fog closely resembling smoke.

The particle size of the fog ranges from 0.25 to 50 microns. That means that it's in the same size range as the original smoke particles. Coupled with the heat and pressure produced by the thermal fogging machine, this dense fog will fill the building with amazing speed. The hot deodorant droplets are pushed along the same pathways that the smoke originally followed, deep into the same tiny cracks and voids.

Repeat for Best Results

Ideally, thermal fogging is done twice, especially on smoky fires. However, you should clear this operation with the adjuster before doing it, to ensure that you won't have any problem with billing.

The process should be done once early in the initial deodorizing phase, as part of the mitigation process. In order to contain the fog within the building, you'll need to close up as many openings as possible. That means repairing or covering broken windows and holes in the roof, and sealing other penetrations.

Do the second fogging after all of the demolition is complete, and the structural reconstruction has been done, but prior to enclosing the building with insulation and drywall — about the time that you pass your rough inspections. This will ensure complete deodorization while the structure is still open.

Safety and Operating Tips

The thermal fogging process shouldn't be taken lightly! There are several very important safety precautions that you need to take while using a fogging machine.

Tip! *The liquid deodorizers used in thermal fogging machines are available in different scents; the one you select to use will often depend on the type of odor you're dealing with. Some scents are more effective than others for certain odors. Those with specific uses include:*

Cherry — for smoke odors from burnt wood or paper;

Neutral — for smoke odors from burnt wood or paper when adding another scent isn't desired;

Citrus — for odor resulting from protein fires, or for tobacco smoke;

Kentucky Blue Grass — for smoke odors from burnt synthetic materials, such as plastic and rubber.

Understand the Equipment and the Process

Your thermal fogger will come with complete operating instructions, and usually a video as well. Be sure you fully understand everything about it, from how to start and operate it, to how to shut it down and maintain it.

Ask your dealer for a complete demonstration of the machine when you buy it. If you don't understand something, or if you or any member of your crew isn't comfortable with the operation of this equipment, *wait until you and your workers get proper instruction before using it!*

You also need to know which fogging solutions to use in which situations. Your dealer should be able to advise you on this as well.

Wear Safety Equipment

You *must* be in protective gear when fogging. It's particularly important to use a respirator with the proper cartridge. Refer to the MSDS and specific instructions provided by the manufacturer of the deodorizing material you're using.

Operate Using Only Approved Materials

Never attempt to use a thermal fogger to fog any liquid that's not specifically approved for that use. *The results could be a fire or an explosion!*

Turn Off All Ignition Sources

When using oil-based fogging solutions, it's possible for the fog to build up in sufficient quantities to become explosive. *Be sure that all sources of ignition in the building are off,* including pilot lights for the furnace, water heater, and appliances.

Use a Two-Person Crew

Thermal fogging is *always* a two-person operation. The fog fills the inside of the building with incredible speed, and it's very easy to become disoriented! You need one person to operate the fogger and a second person to act as a guide. See Figure 9-12. Before starting to fog, the two crew members should carefully plan their route

Figure 9-12 Two workers prepare for a thermal fogging operation. Both have already inspected the building, and know the exact the route they'll follow. The man on the left will be using the fogger and the man on the right is the guide. The guide will also carry a fire extinguisher and be ready for any emergencies. Both are wearing full-face respirators.

through and out of the building. They should determine where their entry and exit points will be, and then clear the path between those points to ensure there are no obstructions.

Always start the fogger outside the building, and make sure that it's operating correctly. The operator and guide should proceed to the starting point, making certain to doublecheck that the path is clear. As the operator begins the fogging operation, it's up to the guide to ensure that the operator is quickly and safely guided to the exit door and out of the building. Fogging an entire house only takes a few minutes.

Don't Over-Fog

Understand how much volume of fog to put out, based on the size of the house and the manufacturer's guidelines. Over-fogging is dangerous, and offers no additional deodorization benefits.

Advise the Fire Department

The thermal fogging process duplicates the smoke from the fire, so you'll be producing what looks like visible smoke. It can leak out of gaps in the house and may cause the neighbors to become alarmed, especially since the house had just recently experienced a fire!

> ***Tip!*** *Consider extending an invitation to the fire crews to come by and watch the fogging operation sometime. It's kind of cool to see!*

Before beginning your fogging operations, call the local fire department on their non-emergency phone number and let them know what you're doing. This may save them from a false alarm call, which they'll greatly appreciate. It also gives you a marketing opportunity to get your name out there, and to be known as a conscientious restoration company that takes their responsibilities seriously.

Thermal Fogging Equipment

If you live in a large city with a lot of resources for restoration suppliers and equipment, you may be able to rent a thermal fogger. But if you plan on doing much fire restoration work, you should purchase one. They cost around $2,200. In addition to the convenience of having the equipment available whenever you need it, it'll pay for itself quickly. Once the equipment is paid for, thermal fogging is actually a pretty profitable operation.

Billing

Thermal fogging is billed by the volume of the area fogged, in cubic feet. You can easily calculate the volume as long as you know the square footage of the building and all of the ceiling heights. If you're fogging the crawl space, basement or attic, be sure and include the volume of those spaces in your billing calculations as well.

Encapsulants and Surface Sealants

In some lighter smoke situations, a combination of some or all of the first four steps alone may be sufficient to remove the smoke odor. But for moderate to heavy smoke losses, or for situations where odor or staining remains, you may need to include one additional step. Depending on the situation, you may need to apply an encapsulant or a sealer to seal off the surfaces.

Encapsulants

An encapsulant is a clear liquid additive that's mixed with water. It isn't a deodorizer; instead, its purpose is to enclose and seal surfaces by forming a thin, clear film on them. The application provides a barrier to prevent the further release of tiny soot particles or other particulate matter. It can also seal the surface to prevent the outgassing of odor vapors. There are different encapsulant materials designed to produce different results.

Encapsulants are sometimes used for exposed wood framing in areas such as attics and crawlspaces. Due to the porous nature of wood, it absorbs a lot of smoke odor during a fire, which is very hard to get out. After cleaning and thermal fogging, encapsulation may be a worthwhile step, especially in a situation where you want to leave the wood looking unfinished.

Another common use for encapsulants is in heating and air conditioning ducts. After the duct cleaning process is complete, apply an encapsulant to help seal the inside of the ductwork against possible odor release.

It's important to remember that encapsulants are water-based, so they have a high permeability. That means that the film they produce when dry allows more moisture to pass through than an oil- or solvent-based material. Stains and smoke odors can pass through an encapsulant more readily than through a sealer. An encapsulant is intended for lighter smoke situations or for situations where you'll later cover the material with an actual sealer.

You can apply encapsulants using a variety of methods, including brush, roller, low-pressure pump-up sprayer, ultra-low-volume cold fogger, and by any other method approved by the manufacturer. Be sure and obtain an MSDS, and follow all of the manufacturer's specific mixing, application, and safety instructions.

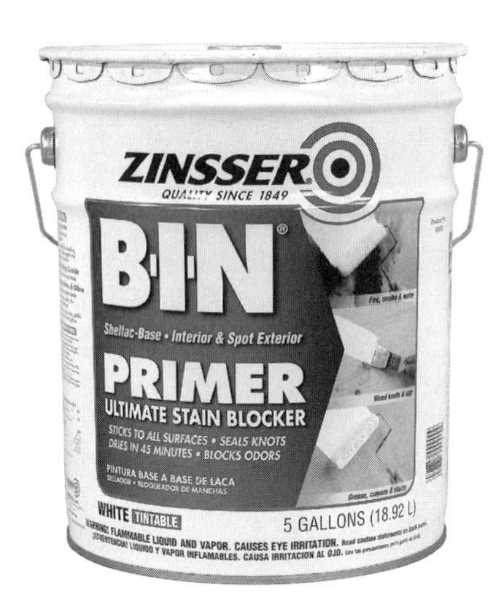

(Photo courtesy of Zinsser)

Figure 9-13 Shellac is a very effective sealer for a wide variety of restoration tasks. It's used for smoke and other stains, as well as for mold remediation, sewage losses, and other applications where you need stain and odor sealing. It's available in tintable white (pigmented) and clear (unpigmented).

Sealers

Sealers are one of the most important materials available to the restoration contractor. They're used in fire and smoke restoration, mold remediation, odor control, and many other areas of restoration. Sealers go a step beyond encapsulants, and the proper sealer, correctly applied, will help you achieve the two main objectives discussed earlier in the chapter: blocking stains, and blocking odors.

There are basically two types of sealers commonly used in restoration work — alkyds, and pigmented or unpigmented shellac. Of the two, shellac is the more effective, and the more commonly used.

Making the Case for Shellac

Zinsser is one of the leading manufacturers and suppliers of sealing products for the restoration and construction industry (Figure 9-13). They recently released a Fire Damage Restoration Information Bulletin entitled *Shellac for Fire Damage Restoration*, which offers an in-depth explanation of why shellac is so effective and important in a fire restoration environment. That bulletin is reprinted here, courtesy of Zinsser:

Shellac for Fire Damage Restoration

What is so different or unique about fire-damaged surfaces? Controlled fires — such as those in a fireplace, a furnace or a barbecue pit — usually involve the combustion of one particular kind of material, such as firewood (oak, hickory, maple, etc.), coal, fuel oil or charcoal. A house fire is an example of an uncontrolled fire that involves the combustion of many different materials, such as wood, plastic, rubber, paint, varnish, synthetic fabrics, etc. The smoke from a house fire is often very thick and laden with oily soot, with a powerful, acrid odor that permeates adjacent areas not yet consumed by the flames.

In most cases the fire is extinguished before it can completely destroy the structure, so the premises are further damaged by the water used to put out the fire. The water dissolves much of the soot and other by-products of combustion and carries them into any porous surface, such as wood, drywall, plaster, concrete, masonry, etc. The end result is that many of the surfaces in the structure that the flames never touched are affected by the fire in one way or another, whether it involves odor, water stains, soot deposits or mildew. Fire damaged surfaces and structures are among the most difficult and challenging to successfully restore. In recent years an entirely new kind of professional — the Fire Damage Restoration contractor — has emerged as the most knowledgeable and reliable person to whom homeowners and insurance companies turn when fire has damaged a home or a business.

What Is the Best Primer or Sealer for Fire Damage Restoration?

In addition to the direct damage caused by flames and heat, fire damage includes stains from soot and smoke deposits; stains from water used to extinguish the fire, as well as mildew that often grows elsewhere in the structure due to the residual dampness and moisture. The most common element of fire damage is the acrid, pungent odor that permeates every porous surface. Although several methods of treatment (including ozone saturation) are used, none is completely effective at removing the odor itself. This can be a big problem if the odor-saturated surfaces have not been completely sealed, since the odor will eventually leach through any new drywall as humidity in the restored building increases.

The most effective fire damage primer or sealer is one with all the following properties:

1. Blocks all odors and all stains completely and permanently with just one coat.
2. Exhibits excellent adhesion over glossy or dense surfaces.
3. Has maximum square-foot per gallon coverage.
4. Dries quickly with minimal or no lingering odor.
5. Is not a food source for mildew growth immediately after application.
6. Cleans up easily, preferably with a minimum of wasted solvent.
7. Can be used in colder temperatures for wintertime restoration projects.
8. Is available in a transparent version for application to rafters and other wood surfaces.
9. Relatively affordable and economical to use so that the contractor can solve the stain and odor problem in a way that allows him to minimize his material costs.

What Kinds of Primers and Sealers Are Used Today?

These days the two most common products used to prime and seal fire damaged surfaces are *fast-dry alkyds* and *shellac*, which can be either pigmented or unpigmented.

Fast-Dry Alkyds

These oil-base primer-sealers are made using a blend of natural oils and synthetic alkyd resins thinned with a combination of solvents that can include mineral spirits, naphtha, toluene, xylene and others. For most other applications, fast-dry alkyd primer-sealers are dependable products. They are easy to apply using a roller or sprayer; they will block most stains from water, smoke and soot; they will develop good adhesion over a variety of surfaces. Generally, they are fairly inexpensive, usually ranging in price from $9 to $18 per gallon, depending on the quality and the brand.

Unfortunately, for fire damage restoration purposes, fast-dry alkyds have several drawbacks:

Odor — Fast-dry alkyd primers usually have a powerful, pungent solvent odor that often lingers for days after application.

Poor coverage — Their relatively low rate of coverage — about 275 to 350 square feet per gallon — offsets their low retail cost.

Rough pigments — The higher viscosity and larger pigment particle size of fast-dry alkyd primers tends to result in the need for higher airless sprayer pump pressures which, in turn, result in faster spray tip wear and greater product loss due to overspray.

Limited stain blocking — Solvent-based alkyd primers are unable to completely block oily or solvent-sensitive stains from grease or oil fires and have a limited ability to permanently block complex or concentrated water and smoke stains.

Poor odor blocking ability — The resins used in fast-dry alkyd primers are not inherently odor-blocking, so two and sometimes three coats may be necessary to completely and permanently seal odor-saturated surfaces.

Mildew-prone film — When dry, oil-base and alkyd-base primers are an excellent source of food for mold and mildew, which often begin to infest fire damaged structures days after the fire is put out.

Difficult to clean up — Application equipment must be cleaned with mineral spirits or paint thinner, which must then be removed from the job site as hazardous waste.

Not for low temps — Although oil-base products will not freeze like water-base primers, they still cannot be used in cold temperatures (40 degrees F and below) because they will dry much more slowly (or not dry at all in freezing temperatures).

Pigmented only — Fast-dry alkyd primer-sealers are only available in pigmented form, precluding use as clear sealers over uncoated wood.

Shellac

Shellac is an alcohol-based solution of pure lac, a natural resin secreted by tiny insects on certain trees in Asia and India. After it is harvested, the dark, reddish-brown resin is crushed and then processed into flakes that are dissolved in denatured grain alcohol. The resulting liquid has a pronounced reddish-orange color. If the harvested resin is bleached beforehand, it will create a coating that has a very light straw color. Bleached shellac solutions are mixed with clays and titanium dioxide to create white-pigmented shellac. Shellac-base primer-sealers retail for between $18 and $25 per gallon.

(Photo courtesy of Zinsser)

Figure 9-14 For small areas, pigmented shellac can be applied with a brush or a roller.

(Photo courtesy of Zinsser)

Figure 9-15 Apply shellac to larger areas, such as open wall framing, with an airless sprayer. Clean up using denatured alcohol.

Both clear and pigmented shellac have numerous advantages that make them well-suited for fire damage restoration:

Easy to use — Shellac is easily applied with a brush or roller (see Figure 9-14). Because the stain-killing and odor-blocking power lies in the resin — not in the fillers — pigmented shellac has a very small particle size that gives optimum coverage and hide at lower spray pressures, thus reducing over-spray, pump stress and tip wear (Figure 9-15).

Excellent spread rate — The lower viscosity of shellac gives it a much higher spread rate than fast-dry alkyd primers — in many cases a single gallon of shellac-base primer-sealer will cover 75 to 100 percent more surface area than a single gallon of alkyd primer.

Super-fast dry time — Shellac dries to the touch in *minutes* and, in most cases, can be recoated in 45 minutes.

Clean, alcohol odor — Shellac is dissolved in denatured ethyl alcohol. It has a familiar antiseptic odor that dissipates very quickly as the product dries. When dry, shellac has no residual odor whatsoever.

Biocidal effect — Alcohol is a well-known biocide and shellac resin is not a particularly tasty food source for mildew. This makes shellac an excellent choice for application to mildew-stained and mildew-prone surfaces in an environment where mold and mildew spores are rampant.

Sticks to glossy surfaces and finishes — Shellac has incredible adhesion and sticks to just about anything, including glass, ceramic tile, stainless steel, Formica®, melamine, etc.

Impervious to odors — The unique molecular structure of shellac resin renders it impervious to odors. Just one coat applied at container consistency will completely and permanently seal any kind of odor in any porous surface. No other coating can make this claim.

Seals all types of stains — The nature of the resin coupled with an incredibly fast dry time enables shellac to effectively block almost every type of stain, including stains from water, oil, grease, soot, smoke, nicotine, mildew, mold, marker, pen, graffiti, tannin bleed, wood sap, etc.

Cold temperature application — Unlike other coatings, shellac can be applied in cold or even freezing temperatures (40 degrees F and below) without concern over proper drying and curing.

Easy to clean up or remove — In addition to alcohol, shellac can also be dissolved in ammonia. This means that contractors can clean their application equipment with a solution of ammonia and water and simply pour the wash water down the drain, as it is not considered hazardous waste. Shellac drips, spills and overspray can be removed by washing the surface with alcohol or with a strong solution of ammonia and water. Even commercially available cleaners such as Fantastik® and Formula 409® will easily dissolve and remove splatters and drips.

Does Shellac Have Any Drawbacks?

Shellac is made from a natural resource that is harvested in only a few places in Asia and India. As a result, the price of the manufactured coating will rise or fall with the cost of the raw material. In general, shellac tends to cost more than synthetic alkyd primers — sometimes twice as much per gallon. Whether or not this is a problem will depend on the contractor.

If Shellac-Base Primers are More Expensive than Alkyd Primers Why Use Them?

Experienced contractors and contracting firms understand that the cost of materials for any given restoration project represents a minor percentage of the total cost; the lion's share of the cost is in the labor.

Fire Damage Restoration companies concerned with the quality and reliability of their materials will generally use shellac-base primers for most of their fire damage projects for two primary reasons:

1. They know that they can depend on the permanent stain-sealing and odor-blocking properties of shellac-base primers.
2. They will save a lot of time and labor since only one coat of shellac is needed to properly prime the surface. Furthermore, it will be easier to clean the equipment afterward and the contractor will not have to worry about disposing of any hazardous waste solvents.

Figure 9-16 A small portable duct-cleaning machine, manufactured by Rotobrush®. The tank houses a powerful vacuum, as well as a motor that rotates the brush on the end of the hose. The rotating brush removes dirt and soot from the duct, which is drawn by the vacuum into the tank for later removal.

Duct Cleaning

Part of any fire and smoke cleaning process is cleaning all of the heating and cooling ducts. Ductwork, due to its metallic surfaces and generally cooler temperatures, naturally attracts smoke and soot. The inside of ductwork will almost certainly be coated with a black layer of material, even in relatively minor fire losses.

Rotary duct-cleaning equipment, which uses rotating brushes to gently dislodge soot from the inner walls of the ducts, is very effective for the initial cleaning of the duct work. You can interchange the different sizes of brushes to accommodate different sizes of ducts.

The brush attaches to the end of a long vacuum hose. As the brush rotates and dislodges the soot, a vacuum located in the main duct cleaning motor unit draws the soot through the hose, and back into a filtered storage tank where it's stored for later disposal.

Small rotary duct-cleaning equipment, such as the Rotobrush®, is available; they're portable, affordable, and easy to operate (see Figure 9-16). They can be wheeled directly into the house, and work well for many fire restoration applications. This type of equipment is also good for post-construction cleanup and other types of restoration duct cleaning. Many full-time restoration contractors invest in duct-cleaning equipment, both for the convenience and the potential profit. But you can also have a subcontractor handle duct cleaning.

You may be dealing with even larger losses involving heavier contamination of the duct system, including puff-backs, heavy smoke losses, and mold remediation, or commercial losses with much larger air ducts. For those situations, you'll need truck- or trailer-mounted duct-cleaning equipment, like that shown in Figure 9-17.

Truck-mounted equipment utilizes whips or other tools powered by compressed air to remove contaminants from the walls of the ducts. An extremely powerful vacuum mounted on the truck or trailer draws the loose material out through a large diameter hose and into a holding tank for disposal. This type of equipment requires a much larger investment, and is something that's usually subcontracted out.

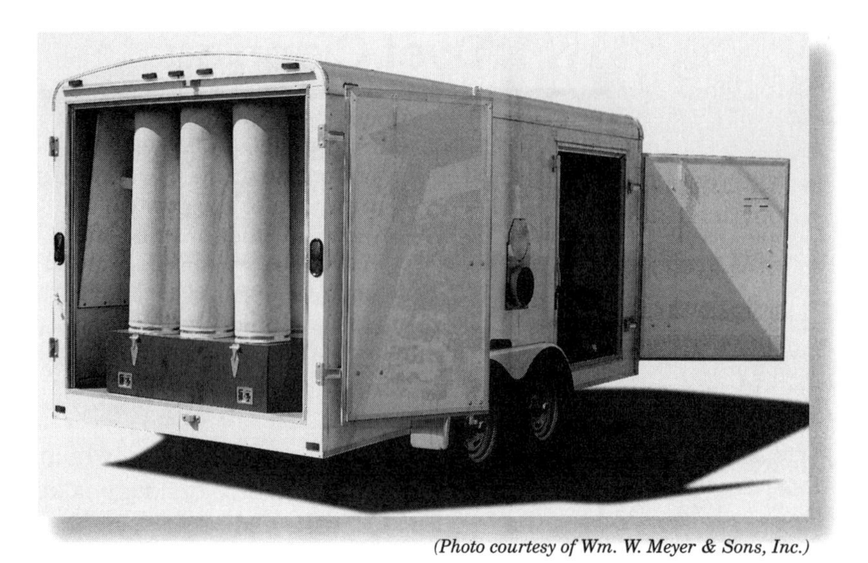

(Photo courtesy of Wm. W. Meyer & Sons, Inc.)

Figure 9-17 This large trailer-mounted duct-cleaning machine offers considerably more suction and cleaning capability. That's especially important for heavier losses, puff-backs, mold remediation, sewage losses, and other losses with greasy smoke or intense duct contamination.

The Furnace

In anything other than a light smoke situation, the furnace needs to be inspected by a trained furnace technician. Remember that smoke is very corrosive, and today's furnaces and thermostats have a lot of electronic circuits in them that need to be thoroughly tested for proper operation.

The inner and outer furnace housing needs to be cleaned thoroughly, including the blower, blower motor, coils, combustion chamber and all other components. The filter should be changed immediately as one of the initial odor removal steps, and then changed again as part of the final cleaning, just prior to occupancy.

Puff-Backs

A puff-back is a very specific type of smoke situation, caused by a malfunction inside an oil furnace. Inside the furnace's combustion chamber, oil vapors rapidly build up and explode, sending odors and a thick film of oil-based particles back through the duct system and then throughout the house. And while the term *puff-back* sounds pretty soft and mild, this oily film sticks to everything it encounters, creating quite a mess.

Puff-backs are treated in much the same manner as other fire situations, except there's no structural damage. First, make sure the furnace itself has been repaired. That's something that's typically *not* covered by insurance, so it's an out-of-pocket expense for the homeowner.

Handle the cleaning and odor restoration using the same methods you would for a petroleum or synthetic material fire. Much of the soot residue is oily, so be careful not to smear it while you're cleaning. Use degreasing cleaners with an odor counteractant. You may also need to add an encapsulant into your solution. The building will need a thorough duct cleaning. This should be handled by a professional duct cleaning company with specific experience in puff-backs. After cleaning, the ducts should be sealed with an encapsulant. Interior walls and ceilings will also need to be sealed with a stain-blocking sealer prior to painting, and you'll need to clean all the carpets and upholstery as well.

Other Odors

Up to this point, we've been discussing smoke odors. But as a restoration contractor, you'll be called upon to deal with a wide range of odors, such as those from mold and mildew, wet carpet, refrigerators that have been without power, skunks trapped in a garage, wet laundry stranded in a washing machine after a storm, and even chipmunks or other critters that have died in heating ducts. You name it, it can happen!

You'll still take the same five steps outlined in this chapter. Once the odor source has been identified, such as a dead animal or a washer full of mildewed laundry, removing that source is fairly straightforward compared to smoke odors.

After removing the source, you'll have to tailor your cleaning and deodorization steps to the specific odor. A spray from a skunk, for example, will require a specific set of cleaning products, which will be different from those required for the odor left behind by mold, or a refrigerator full of rotten food. You might need a suppression spray, thermal fogging, an encapsulant, or some combination of those methods.

It doesn't matter if you're dealing with odors from smoke, skunks or salamis — when the call comes in, you'll find plenty of help out there. From suppliers and manufacturers to trade associations, reference books, and on-line research, there's a wealth of available information. Even though you're going to run into a lot of different cleaning and deodorization situations, chances are pretty good that somebody has done it before you. So don't be afraid to ask.

10

Wind, Snow, & Other Storm Damage

Water and fire losses are certainly not the only problems you'll be hearing about when your phone rings. Variety is what makes insurance restoration so interesting, and in the next few chapters we'll look at some of the other situations you're likely to encounter.

This chapter and the next one are organized into major loss categories, with each type having four basic elements to it:

1. ***The loss:*** You'll get a good idea of the damage associated with each type of loss, and what to watch for while estimating.
2. ***Common problem areas:*** We'll look at potential problems, such as areas of hidden damage or other things you might not think to look for. You'll also get a heads up on situations where insurance companies may have coverage issues, or where you may run into problems with clients.
3. ***Permits:*** We'll look at permits and building code issues.
4. ***Examples:*** This section has a real-world example of each type of loss to help you better understand what you and your clients will be facing.

Safety Issues

Some of the losses we'll be talking about can be highly risky, such as trees that have blown down and are lying on or against a building, or a roof structure over your head that could collapse at any minute. Much of the work you'll be doing will be in harsh conditions as well, including snow, ice, high winds, or heavy rains.

To be honest, there's something of an adrenaline rush when you're responding to an emergency situation in a storm. You might feel like a hero as you arrive to take charge. But take your time. No matter what the situation, there's nothing so dire that you should put your safety or the safety of your crew in jeopardy!

Stop and analyze the situation before you do anything. Understand where the weak points in the building are, what's unstable, and where you shouldn't step. Use caution tape to rope off areas that people need to stay away from. Always remember these two basic safety rules: Take your time, and use your head.

Temporary Supports

In many situations, your first task will be to erect temporary supports to stabilize the structure so you can safely do your estimating or make repairs. In the case of snow or ice damage, temporarily stabilizing the building may be all you can do until after the snow melts.

Temporary supports can be as simple as a couple of 2 x 4s for bracing, or you might need to install an elaborate set of walls, beams, and braces to keep an entire roof structure from collapsing. Study the situation carefully to understand where the loads are and exactly what needs to be braced. Lumber is cheap, so if you're unsure about how much bracing to use, take your best guess and then double it. If you're really in doubt about how best to temporarily brace something, leave it to a structural engineer.

Wind Damage Losses

No matter where you live, damage to buildings from high winds is a fairly common problem. It can occur at any time of year, under any type of weather conditions.

The thing about wind damage that's different from other losses is that it's rarely an isolated incident. When a wind storm comes through the area, it affects dozens or even hundreds of buildings at the same time, with damage ranging from minor to severe. An intense or prolonged wind storm can have your crews scrambling for several days, even weeks.

As you might expect, the primary damage is to roofing, although there may be structural damage as well. Wind can lift and loosen shingles, or it can blow them off completely. You may see isolated shingles ripped off, as in Figure 10-1, patches of missing shingles (Figure 10-2), or an entire side of a roof stripped bare. Because wind storms tend to be from a single direction, you'll usually see the damage concentrated on the side of the prevailing wind direction.

Ridge and hip shingles are also common points of damage, especially on wood shingle roofs. Because hips and ridges are elevated above the roof plane, you'll often see damage to those areas regardless of the wind's direction.

Figure 10-1 On this roof, several individual 3-tab shingles have been lifted by the wind and snapped off in random areas. In most cases, the other half of the shingle is still underneath the course of shingles above it. This can make for a more difficult and time-consuming patch job.

Figure 10-2 This is a good example of the type of damage often found after a wind storm. One large section of laminated composition shingles has been blown off, as well as some individual shingles at the bottom edge of the roof. Shingles were blown off on other parts of this roof as well.

If the wind occurs during a rain or snow storm, water entering the house and causing damage as a result of the missing roof material is generally covered. When called out on any type of wind loss, do a thorough investigation to check for water in the attic, and possible damage to insulation or drywall.

Depending on the intensity of the wind, other materials can get torn off the house as well, from shutters and flashings to satellite dishes and television antennas. As a general rule of thumb, if it was attached to the house, it'll probably be covered as part of the structural loss. If an item was sitting in the yard and gets damaged, like a patio table, it'll usually be considered a contents item. Content losses are part of the same claim, but are handled separately.

Wind is a big problem for open structures, such as carports, pole buildings, and other types of open-sided buildings. Strong winds can get inside and under the building's roof structure, and lift all or part of the roof or roofing from below. This can cause anything from minor roofing damage, to major damage to the structure itself. See Figure 10-3.

Wind-driven rain is another problem that can create a big headache for you. It isn't unusual to find a window or a chimney that suddenly begins to leak when the wind and rain act together, but doesn't leak when there's only rain by itself. It may, in fact, only occur when the wind is coming from a particular direction, or at a certain speed.

Remember that *sudden and accidental* is the key criteria, and wind-driven rain fits the bill. If a wind storm suddenly forces rain around a window and into the house, then the damage to the window and to anything inside is typically covered.

Another aspect to wind damage is the damage caused by wind-blown objects. Remnants of a blown-down fence or ripped-off roofing may hit a home and break

Figure 10-3 This large hay barn is open on three sides, and sustained considerable damage to one end of the roof structure as a result of high winds. The amount of metal roofing peeled off, as well as the snapped beam, gives a good indication of the force of the wind.

windows or damage siding. A house can be partially stripped of its paint by wind-driven dirt, sand or debris.

Problem Areas

A big problem you're going to encounter is with roofing repairs. It's often very difficult to patch or replace roofing materials to everyone's satisfaction — and keep yourself out of a liability situation.

When you have a wind-damage loss to an older composition shingle roof, where several shingles have blown off in random areas on one side of the roof, the homeowner will invariably want a whole new roof. The insurance company won't want to pay for that, since it would essentially constitute profiting from the loss. They'll want you to simply patch in the missing shingles.

This puts you in a very uncomfortable position. If you agree to just patch, you'll never be able to match the old shingles. More importantly, you're going to have to assume liability for that old roof, and you'll probably get a callback from the homeowner whenever the roof leaks in the future. There's no way you can prove that the rain isn't coming in from an area you patched.

Your best solution in a case like this is to offer everyone a compromise. Suggest to the insurance company that they authorize replacing all the shingles on the *damaged* side of the roof, rather than patching in random shingles. That gets the homeowner a complete new roof on one side to replace the damaged area, which is what they want; and it's less expensive than a total roof replacement, saving the insurance company money. At the same time, it protects your liability because you won't be responsible for a patch job. If you want, you can then offer the homeowner a discounted price for also replacing the undamaged side of the roof, since your crew will already be out there working.

Another issue with composition roofing occurs when the wind lifts a lot of the shingles without blowing them off. The lifting can break the tar seal that forms between the underside of one shingle and the top of the shingle below it. This makes the shingles vulnerable to lifting and tearing in another windstorm, as well as to damage from wind-driven rain.

One repair suggestion for this situation is called *tabbing*. You lift each shingle individually, apply a small dab of roofing cement, and then press the shingle back down. Once the cement dries, the roofing is once again able to act as a continuous membrane against wind and rain. Although time-consuming, it's a cost effective

REAL STORIES:
Wind Losses in the Real World

A wind storm came through our area one fall, and it hit particularly hard in a new subdivision that had just been completed. All the homes had 3-tab composition shingle roofs, and most hadn't been in place long enough for the tar strips on the underside of the shingles to have adhered well.

In the aftermath of the storm, several homeowners found that the shingles on the prevailing wind side had been raised up, and hadn't settled down again. This meant that the tar strips were not ever going to bond properly, leaving the roofs very vulnerable to another wind storm.

There were a couple of dozen homes involved overall, and of course not all of the homeowners had the same insurance company. However, several of them were represented by two of the larger companies that we did a lot of work for. We were called out to look at about 10 homes, all with similar problems.

The insurance companies faced a bit of a dilemma at this point. The roofs were definitely damaged, but no shingles were actually missing, so replacing the roofs was certainly not justified. But the adjusters were smart enough to understand that if they didn't make some repairs now, they'd be looking at lots of new roofs after the next strong wind storm!

So we proposed a compromise. We would work our way across the roof, lift each loose shingle tab, apply a short bead of roofing tar under it, then press the tab back down again. This tabbing process would adhere the shingles to one another in the same way the original tar strips were supposed to, and stabilize the shingles against future wind storms.

The insurance companies were happy with this, and so were the homeowners. We insisted on time-and-materials billing — something that insurance companies don't particularly like, but unit costing doesn't work for a project like this. They agreed, so we were happy too.

In the end, it was a good solution to an unusual problem. The only ones who were less than pleased were the crew members who were stuck with the tedious job of sitting on a roof, lifting row after row of shingles and squirting little dabs of tar under them!

alternative to replacing an entire section of roofing — so it's worth discussing with the adjuster.

Keep in mind that once a claim is settled, the homeowner is under no obligation to replace what was there before. They can always apply the settlement money toward something different, and sometimes that can be very much to their advantage. For example, a homeowner with random wind damage to a wood shake roof may get payment from the insurance company to replace half the roofing. That settlement may be enough to cover the cost of an entire new composition shingle roof instead.

Building Permits for Wind Losses

Some jurisdictions have building permit requirements for reroofing. If you have minor damage to siding, shutters, or windows, that typically won't require a permit. You'll need to evaluate structural damage repairs on a case-by-case basis, depending on the severity of damage.

If the house is located in an area with a homeowner's association (HOA) or an architectural review committee, check and see what roofing requirements they have. Some have approved selections of roofing materials, as well as palettes of acceptable colors.

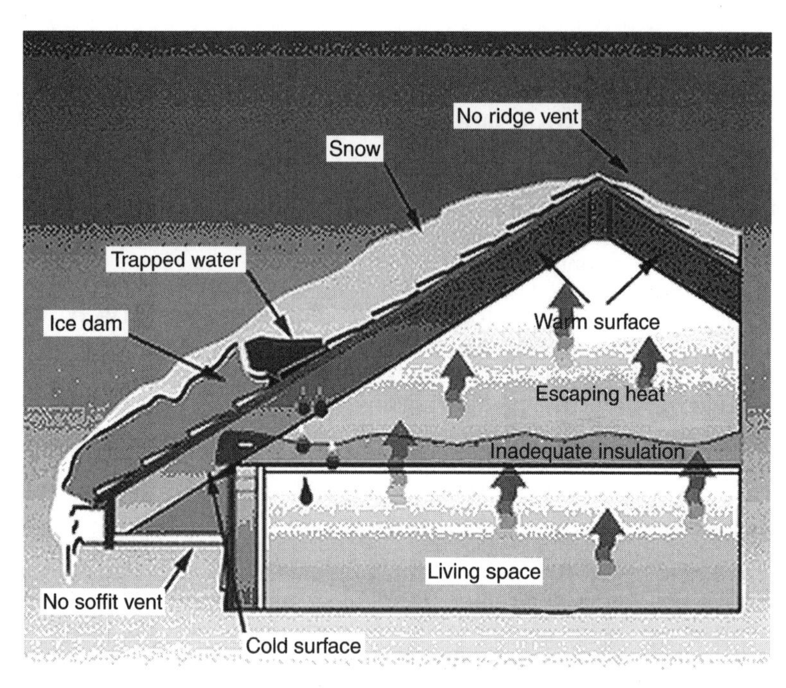

Figure 10-4 This cutaway view of an attic shows how heat loss causes an ice dam to form. The most vulnerable spot is at the intersection of the roof and the exterior wall. That's where the heat loss stops and the freezing begins.

Ice Damming

Ice losses are occasionally something of a gray area from a coverage standpoint. Under the *sudden and accidental* criteria, they're rarely sudden, but they're certainly accidental. Most ice-related problems end up being covered, but you need to work closely with your adjuster on these.

One of the most common ice-related issues is ice damming. If this is something you're not familiar with, it's important to understand the sequence of events that cause an ice dam to form.

As snow falls, it builds up on the roof. If the daytime temperatures are low, the snow won't melt, and the layer continues to grow. In the meantime, the heater is on inside the building to keep it warm. Some of that heat is inevitably lost into the attic, and once there, it rises until it contacts the underside of the roof.

With an insulating snow layer above and heat below, the snow on the roof begins to melt from the bottom, creating a film of water between the underside of the snow layer and the top of the roofing. The water runs down the top of the roof, beneath the snow, until it reaches the eaves. Once there, past the end of the attic, there's no heat from the house to keep the water warm, so it freezes. A solid dam of ice begins to form on the edge of the eaves, and the longer the cycle continues, the larger the ice dam grows. See Figure 10-4.

> ***Tip!*** *Be very careful around icicles! They can be sharp and extremely heavy. Don't walk under them, don't play with them, and don't attempt to snap the larger ones off.*

Eventually, water coming down the roof and hitting the dam has nowhere to go, so it begins to work its way back up the roof. When it gets back to the point where the roof and wall meet — where there's heat — it no longer freezes. It continues to work back up the roof as a liquid instead of as ice. Since shingles are overlapped from top to bottom, there's no protection against water coming from under the shingles, so eventually the water enters the house.

Figure 10-5 An example of ice damming. There's a solid layer of ice at the eaves of the house, covered with an insulating layer of snow. Icicles are a common indicator that an ice dam has formed.

Ice damming appears as a layer of ice at the very edge of the roof (Figure 10-5), or it may accumulate inside gutters until it's high enough to be seen from the ground. Icicles, which are part of an ice dam, are a visible sign of an ice dam, occurring as some of the dam melts, flows over the edge of the roof, and then refreezes.

Like the ice dam, the icicles simply keep getting larger as the cycle repeats. Icicles warn you that the dam is getting worse. If the ice is sufficient to be dripping over the edge of the roof, it's also moving farther up the roof and under the shingles.

With ice damming losses, the first sign of damage is usually a water stain at the joint between the ceiling and an exterior wall. That's an indicator that the ice has reached the critical intersection of the framing where the heat loss occurs, and water is now getting under the shingles and into the house. It'll drip down and wet the insulation, then pool in the corner and along the exterior walls. If enough water gets in, it'll eventually wet the drywall, and water stains will appear. Left unchecked, it'll cause the drywall joint tape to peel, and the drywall to deteriorate.

Sometimes the water will run down *inside* the wall, so you'll see water stains on the drywall at the bottom of the wall instead of the top, or above or alongside windows and exterior doors where the water encounters the headers. In some extreme cases, you may even see puddles of water coming out onto the floor at the bottom of an exterior wall.

Another unwelcome sign is the appearance of ice behind or between siding boards, as in Figure 10-6. You may also see it coming out of soffit vents, or showing up in other exterior areas where ice has no reason to be. That's a sure indication that water has gotten into the walls or into the soffits, but it's outside the layer of wall insulation.

If you show up on a job and find ice damming, there's not too much you can do until the weather warms up and everything begins to dry out. You'll have to wait to do your estimating and repairs. Sometimes you can set up drying equipment in the attic if you're able to pinpoint the source of the water, but that isn't always possible.

Figure 10-6 The ice forming behind the home's lap siding is a very unwelcome indicator of an ice damming problem. This usually means that water dammed up behind the ice on the roof has gotten into the exterior walls, and is now running out through gaps in the siding.

Problem Areas

Inside the house, look for water staining on exterior walls. Use your moisture detection equipment to track leaks. You may also have drywall damage to deal with. In the attic, carefully examine the insulation. This may be tough, because there's often water damage at that marginally-accessible joint where the roof meets the exterior wall. If the insulation is wet, you have the potential for drywall, framing, and mold problems, so it's best to replace it with fresh material.

Outside on the roof, you may find shingle problems. As the water freezes during the creation of an ice dam, it lifts the shingles, pulling the nails out of the sheathing. When the ice melts, the shingles settle back down somewhat, but the nails don't, so you'll have raised shingles all over the roof that are now vulnerable not only to another ice dam, but to wind storms and wind-driven rain as well.

There's another potential problem caused by well-meaning homeowners or incompetent snow-removal companies who try to remove the ice or the snow layer from the roof. They can do a lot of shingle damage — and that damage isn't always covered. And, while walking around on the snow-covered roof, they can do damage to other things as well, like stepping through skylights, dislodging antenna wires, snapping off plumbing vents, or damaging metal flashings.

Insurance companies usually hate ice damming claims, because the damage is difficult to see, and it's difficult to determine its extent. In some cases, the roof may look just fine, but due to the lifted shingles, the entire roof can be severely compromised. For that reason, the homeowner will often want all new roofing. Often the best thing you can do is to bring in a trusted roofing subcontractor and have him examine the roof and give you a report. Present the report to all the parties along with your recommendations, and allow them to negotiate a settlement.

Ice dams are rarely isolated incidents, so you may have multiple loss situations where the insurance companies send in catastrophe teams, commonly called *cat teams*. When that happens, you may be scrambling to write one estimate after another. If you're involved in a major cat loss and find yourself faced with more jobs than you can handle, give your loyalty and your priority to your regular insurance adjusters, agents, and companies. We'll talk about cat teams in more detail in Chapter 22.

Building Permits for Ice Dam Repairs

Due to the nature of ice damming, there usually aren't any permits associated with this type of loss. The possible exception would be if your area requires a reroofing permit.

REAL STORIES:

Dam the Ice, Full Speed Ahead!

One winter, not long after we started in business, we had a prolonged series of extremely bad storms. Day after day of snow and subfreezing temperatures created severe ice damming all over town.

My partner went out to look at an ice-damming job one afternoon, and found the homeowner up on the roof. He was all bundled up in a heavy parka and gloves — but wearing golf shoes! "Figured the spikes would give me some extra traction," he explained.

He was wielding a pick, and swinging away at the thick layer of ice stubbornly clinging to the edge of the roof. "I don't suggest you do that," my partner told him politely. "It's pretty dangerous. And with a wood shake roof, you could do quite a bit of damage to the roofing."

"So what," came the rude reply. "The house needs a whole new roof. This way the insurance company will have to pay for it, no matter what." Put off by the man's attitude, and not being able to estimate anything anyway due to the thick layer of snow and ice, my partner left. The man was still on the roof, chopping at the ice.

When the snow finally melted, we received another call from Mr. Ice-Chopper. He wanted to know if we would come out and give him an estimate to replace some random broken shakes on his roof. Replace random shakes? What happened to that whole new roof he was planning on?

Before doing anything, we checked with the adjuster who'd been assigned to his claim. Not surprisingly, the insurance company had denied the claim completely. The reason? Extensive owner-inflicted damage to the roofing!

It didn't take us long to turn this job down! Since this guy had just tried and failed to take advantage of the insurance company, it wasn't hard to see that we'd be next in line for the same treatment!

Snow-Load Damage

Damage from the weight of snow, while accidental, isn't sudden. It occurs over time, sometimes as the result of months of accumulation. A building owner has an obligation to take steps to mitigate damage, but insurance companies don't really expect people to shovel off their roofs, or even to pay someone else to do it. So in most cases, if a building is damaged by the weight of accumulated snow, it'll be a covered loss.

Snow loads can be fairly substantial. Freshly-fallen powdery snow might be only around 10 pounds per cubic foot, but as more snow falls and the layers become compacted, the weight increases. See Figure 10-7. When moisture in the snow freezes, that adds more weight. An average snow weight is more in the range of 15 pounds per cubic foot, and compacted snow can weigh 25 to 30 pounds per cubic foot. So if a roof has a buildup of 2 feet of snow on it, you could be looking at a very substantial load on the structure — upwards of 60 pounds per square foot.

Depending on wind conditions, the design of the house, and other factors, the snow load may or may not be evenly distributed on the roof. An evenly distributed load, with a roof of equal pitch and a snow layer of equal thickness, exerts the downward force of the snow more uniformly and puts less stress on the structural members. An unequal snow load can greatly increase the stress on the roof, adding lateral pressure to the

Figure 10-7 A buildup of snow exerts a considerable amount of weight on a home's roof.

already-existing downward stress. This can occur on a roof with unequal pitches, or in an area where the wind causes the snow to drift to greater thicknesses on one side of the roof.

Older homes with stick-framed roofs often have more snow-weight problems, since these structures typically don't distribute the load forces as efficiently as a well-designed manufactured truss does. Also, many older homes were built with structural framing members that are considered undersized by today's standards.

Older homes do have one advantage — if you want to call it that. They're usually not very energy efficient. They lose a lot of heat, especially through the attic, which melts the snow faster and helps keep the roof weight down. Perhaps the most vulnerable home to snow-weight damage is an older home that's been retrofitted with good-quality attic insulation.

Vacation homes are also vulnerable, since no one monitors the snow buildup or takes any steps to reduce it. The heat is kept low for long periods of time, so heat loss is kept to a minimum, allowing the snow to remain on the roof all winter.

How Snow Affects a Building

You'll see the effects of snow weight on a structure in a couple of different ways. An evenly-distributed load, with the weight generally pushing straight down, may cause the rafters to separate at the ridge. The rafters may also pull away from the wall plates. If the rafters are well-secured to the plates, but don't have good collar ties, the force may actually bow the exterior walls out.

Figure 10-8 This collapsed shed roof is a good example of how snow weight can affect a structure. The roof offered a broad expanse of area where snow could accumulate, and was poorly attached to the fascia of the building.

With non-manufactured trusses that utilize plywood gusset plates, the chords of the truss may pull apart from the gussets. In a manufactured truss that exceeds its manufactured load rating, the chords will typically tear loose from the steel gusset plates that secure the joints together.

Shed roofs may sag in the middle, or the rafters can crack or snap completely. In the case of a shed roof that's attached to a building with a ledger, like a carport roof or patio overhang, the snow weight will simply rip the ledger off the building, taking the roof with it (see Figure 10-8).

Unevenly-distributed loads exert more of a lateral force, in a direction opposite from where the force is being applied. You may see separation of structural roof members, or wood that bows, cracks, or snaps. The walls may bow, but in this case, it'll often be with more of a racking movement. Both walls will want to move in the direction of the force, so one will bow in and the other will bow out.

With snow damage repairs, be sure to examine the roof structure very carefully. Separation and spreading of the various members can be a real problem, and it's not always obvious. Digital levels or digital angle gauges are useful tools for checking rafters. They'll give you an accurate reading of the angle of the rafters, and will allow you to compare each one against the others.

If possible, take your reading at the gable end. Since rafters are supported by a wall at that point, they have a lot more loadbearing strength and are less likely to be pushed downward. Then compare that reading with readings you get on the other rafters. If the roof has flattened out from the load, variations in the digital readout should be a good indicator.

Use a long level, at least 6 feet, on the exterior walls to check for plumb and for walls that have bowed in or out. Check these in several areas, and compare your readings. If the walls are bowing, the direction of the bow will also give you a clue as to the direction of the load on the roof.

Problem Areas

Most people discover their roof problems while the snow is still on the roof, often while it's still at its deepest. All that snow is going to prevent you from lifting and repairing the roof right away — and it may even prevent you from accurately seeing and estimating the damage.

REAL STORIES:

Bringing the Household Together

In the middle of a very snowy winter, a local property management service was called in to spruce up a vacation home in a nearby resort area. The out-of-town owners were coming for a visit, and they wanted the place freshened up before they got there.

The service arrived and shoveled their way to the front door. Once the snow was cleared away, they immediately noticed that the front wall of the house was bowing outward! They called one of their handymen, who came out with some lumber and put braces against the front wall.

We were then called in to take a look, and found a very unusual situation. The house was basically just a simple rectangle, with a ridge that ran parallel to the front and rear of the house. The rear of the house was two stories, with a conventional attic and a 4/12 roof. The front of the house was one-story with a soaring vaulted ceiling. The roof on this half was a 10/12.

More of the snowfall had accumulated on the flatter 4/12 roof than on the much steeper 10/12 section. Compounding the problem, the 4/12 side had a well insulated attic that minimized heat loss, while the 10/12 vaulted side was a poorly insulated open-beam design that allowed a lot more heat loss, and so, much less snow accumulation.

The result was a dramatically unbalanced roof load — a couple of feet of wet, compacted snow on the back side of the roof and almost no snow on the front. The back of the house was pushing laterally against the front, and if the handyman hadn't put those braces up, the owners could very well have been vacationing in an open-air home!

After consulting with our structural engineer, we came up with a unique plan. The two end walls of the house were relatively stable, and the rear wall was also stable. So what we really needed to do was pull the center of the front wall back into place and lock it there. To do that, we had some 1-inch steel rods custom-made to length, and threaded on each end.

We drilled holes in the front and back walls, then passed the rods through the house, from front to back, anchoring them to the front and back walls with steel plates and lock nuts. In the middle, the rods were connected with turnbuckles, which we used to pull the front wall back into plumb. Placing the rods and plates and getting everything ready took a couple of days, but the actual pulling was surprisingly easy and effective. Within a couple of hours, the walls were back up where they needed to be!

The rods were permanent, so we finished the job by building decorative wooden beams around them, matching the beams that were already in the living room. Probably the hardest part of this whole project was figuring out how to explain to the building department what we were doing!

You'll need to get the roof temporarily stabilized and shored up, as mentioned at the beginning of the chapter. If you can do it safely, get as much snow off the roof as possible to relieve some of the weight. It may be necessary to remove snow from around the building itself to gain access for the temporary repairs. Talk with your adjuster about all this. Since these steps are necessary in order to protect the building and mitigate damage, the insurance company should be willing to pay for them.

Building Permits for Snow Damage

With snow damage, you're often dealing with structural problems, usually requiring building permits. It's up to you to document the nature and extent of the damage, explain it to the building department, and see what types of permits are required. On some of these losses, you may need a structural engineer to inspect the damage, and design and certify a repair for the building department.

For repairs or replacement of manufactured trusses, you'll need engineered truss drawings from your truss company. Repairs to chimneys and woodstove flues will most likely require a permit, and in some areas you may also need a reroofing permit.

Hail Damage

Hail creates a chaotic pattern of damage all over the outside of a house. It's time-consuming and difficult to estimate, and even more time-consuming and difficult to repair. Hail is probably the least liked of all the different types of insurance restoration losses. Adjusters and insurance companies often have a hard time figuring out what to do with the loss, and homeowners are equally baffled about what to expect in the way of repairs.

The issue with a hail loss is the randomness of the damage. When a tree falls or a vehicle hits a house, you have an obvious single point of damage. With a fire or a water loss, you have a consistent pattern of damage, even if it extends over several rooms. But when a hail storm hits, the damage is everywhere, in little random bits and pieces, over wide areas of the exterior of the house — sometimes even *inside* the house.

A hail storm can happen anywhere and at anytime, often in the summer months as part of a thunderstorm. Individual ice pellets, or hailstones, can range in size from only about 5 millimeters (about $\frac{3}{16}$ inch) in diameter to massive softball size (about $3\frac{13}{16}$ inches), and even larger. But for the most part, hailstones average approximately $\frac{3}{4}$ to 1 inch in diameter. You'll often hear reports of "golf-ball size hail" (a golf ball is a little over $1\frac{5}{8}$ inches in diameter), which is considered large. See Figure 10-9.

Figure 10-9 Hailstones are solid ice, and they can do considerable damage when they pound a building. This hailstone, photographed several hours after the storm had passed, is still quite large.

Hail storms usually last only a few minutes. But those few minutes can seem like hours when you consider the non-stop, high-speed pounding of countless 1-inch

Figure 10-10 The front of a garage after a hail storm. The vinyl siding has numerous cracks and holes in it, and there are dents and holes in the metal garage door. There are also a number of splintered areas and paint damage on the fascia and the trim around the door. All of the siding on three sides of the house had to be replaced, along with this garage door.

balls of solid ice raining down on a building for 3 or 4 minutes, you get an idea of the substantial damage that can occur even in that short time. See Figure 10-10.

The degree and location of the damage depends on the direction of the storm, its duration, and the direction and severity of any accompanying winds. The hail may come straight down, uniformly pounding all of the shingles on a roof, or it may be blown at an angle against both shingles and siding. Slow-moving storms often do concentrated damage, remaining in place and pounding the building for longer periods of time. Sometimes hailstones fall with relatively little velocity, while at other times they fall with greater force and do much greater damage.

Figure 10-11 The hail damage on this house included, among other things, a number of dents and dings in the stucco, a cracked window, and a shattered downspout.

When estimating a hail-damage claim, you need to be very thorough. Take your time and walk the entire perimeter of the house. Examine everything, including roofs, siding, decks, fences, landscaping, and patios. See Figure 10-11. And don't overlook outbuildings — they'll suffer the same damage as the main buildings.

If the hail broke any windows, as in Figure 10-12, then you'll need to extend your inspection inside as well. Make a note of everything you see, no matter how small it might seem (see Figure 10-13). It all adds up to present a bigger picture of the overall loss. The big picture becomes an important consideration when you sit down with the adjuster and the client to figure out how to approach the repair portion of the job. Hail losses usually involve some compromising, and as you'll see in the next section, that also involves some creativity.

Problem Areas

If you've never seen a building that's been hit by hail before, and you'd like to get some idea of the

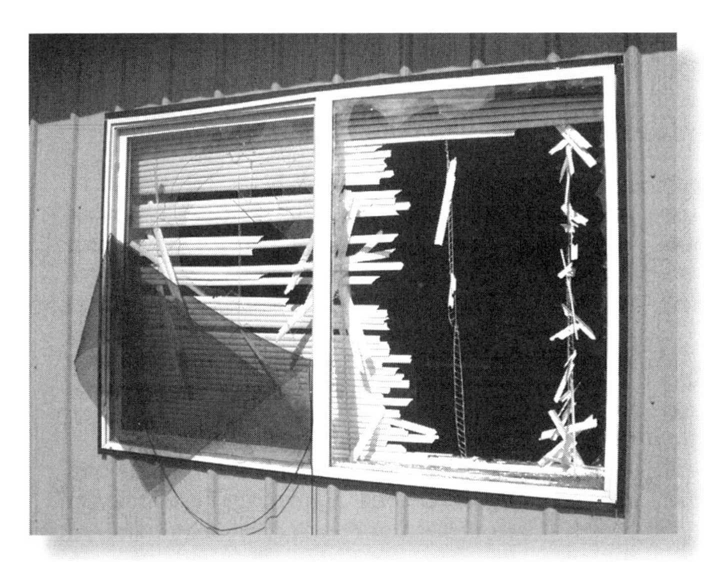

Figure 10-12 The window in this shop building took a direct and sustained hit during a hail storm. It shattered the window, and completely shredded the screen and the interior window blind. The hail damaged some of the contents inside the building, saturated the carpets, and left a big mess to clean up.

extent of the damage, here's a way to visualize it. Take a garden hose with a wide spray attachment, aim it at the side of your house, and trigger a quick spray of water. Now imagine that each dot of water is a dent in the siding, ranging from 1/16 inch up to ½ inch in depth. Some dents are smooth, others have shattered and splintered the wood around them. If your house is stucco or brick, imagine all those dots are cracks and chips.

If you also trigger a quick spray of water on your deck, deck railing, fence, and different parts of your roof, and understand that each water dot represents an area of damage, you'll start to get a real appreciation for how difficult and frustrating hail damage is to estimate and repair.

Here's the part that requires some creativity. Suppose the house has lap siding. One board has 12 dents in it. Another board 2 feet away has one dent in it. Another board a foot down has six dents. Do you fill the dents and paint the wall? Do you replace individual boards? Or, is it easier from a labor standpoint to replace the entire wall of siding?

Figure 10-13 This is an indication of some of the details to look for on a hail storm loss. The aluminum window frame was heavily pounded and dented by the hail, and can't be repaired. Damage like this also represents a negotiating point for the owner and the insurance company.

REAL STORIES:
There's Nothing Quite Like a Hail Storm!

Ironically, we were in a meeting at our office with one of our favorite adjusters when we saw the clouds building in the distance. Thunder started to rumble, and we could see jagged cracks of lightning. Then the pounding started on the roof.

It was heavy enough to drown out further conversation, so we moved to the windows to watch, as the hail cascaded into the parking lot outside. Within a couple of minutes it was over, moving off to the east.

"Damn. It's hail", my partner said.

"Yeah. Tomorrow will be a busy day," said the adjuster with a shake of his head.

We got the first call early the following day. A two-story home on some acreage just to the east of us, right in the path of the storm, had taken a sustained pounding. We went out to survey the damage.

Two exposures of laminated composition shingles were beaten beyond hope. All of the hip and ridge shingles were ruined, as were the shingles on the porch overhang. Metal gutters and downspouts on three sides of the house were severely dented. A skylight was cracked, and the flashing was dented.

Dozens of pieces of lap siding were dented or cracked. Two windows were shattered, and flying glass damaged the kitchen counters, vinyl flooring, and window coverings. Two other windows and a sliding glass door were cracked, and five vinyl window frames were cracked. Window trim was cracked and chipped in several areas, and every window screen on three sides of the house was shredded.

Two metal exterior doors were dented. Even the door knobs were dented. The fascia was splintered, and the paint in some areas had been stripped as though it had been sandblasted.

The cedar decking and deck railings had been splintered and cracked. Decorative lattice that covered the area below the deck was beaten into kindling, and an outbuilding suffered a substantial amount of damage.

Though not part of the structural claim, the hail did other damage as well. It cracked a glass patio table and shattered several pieces of plastic patio furniture. It stripped the foliage off countless trees and shrubs, cracked flower pots, broke trellises, even dented the barbecue.

And finally, it severely dented a pickup truck, and dented a motor home so badly that it was eventually declared a total loss.

We had our work cut out for us. Yep, there's nothing quite like a hail storm!

Hail does such widespread random damage that you could literally be facing the complete replacement of everything. So, you and the client and the adjuster need to meet and discuss what's reasonable to replace and what's reasonable to repair. That's why you need to have the big picture in mind. Trade-offs will be very important for everyone. It might be best to replace the entire roof, and skip doing anything with the deck. Or you might want to re-side half the house, and just patch in a few shingles here and there. Suggest a compromise that's fair to all parties, including your company, and then leave it up to them to decide.

Building Permits for Hail Damage

Hail rarely does structural damage to a house that's severe enough to require a building permit. Just about everything you'll be dealing with will be cosmetic repairs. The possible exception might be a reroofing permit if they're required in your area.

Figure 10-14 Lightning is one of nature's most powerful forces, and can do extensive damage in an instant.

Lightning Damage

Lightning is an interesting and unpredictable area of insurance restoration. A lightning strike is basically an enormous discharge of atmospheric electricity (see Figure 10-14). It can reach estimated speeds of 130,000 mph, with temperatures approaching 54,000 degrees F.

When lightning strikes a building, or even near a building, it can cause a number of different types of damage. It can affect a building's electrical and electronic systems; it can affect a building structurally; and of course, it can start fires.

When lightning actually hits the structure, there's often direct damage to the building. This will usually occur high on the building, such as at the peak of the roof, at a gable end (see Figure 10-15), or at a chimney. Wood isn't a very good conductor of electricity; the sap in the wood is turned instantly to steam, which expands the wood and causes it to shatter. You may see damage to roofing materials, roof sheathing, fascias, and soffits as well.

Metal stove flues and vent pipes are often targets. So are roof-mounted heating, air conditioning, ventilation, and other air handling equipment and ducting. Masonry chimneys are often hit, due to their height, with related damage to the masonry and mortar. Exterior air conditioning compressors and heat pumps units on the ground may be hit as well.

As discussed earlier, a fire only needs oxygen, fuel, and a heat source. The extremely high temperatures associated with a lightning strike easily provide the heat source. So, during lightning storms you'll often see fires starting as well. Building materials, dry grasses, leaves and needles, flammable liquids and vapors, and many other materials in and around the average home provide plenty of fuel.

The massive discharge of electricity from the lightning can be carried inside through the building's electrical wiring, or even through phone lines, TV cables, and other wiring. Due to the high levels of electrical energy, the path that the current takes and the damage it can do inside the building is unpredictable and often even a little bizarre.

Figure 10-15 Lightning struck the gable roof end of this home, near the metal gable vent. It shattered the fascia, blew through the siding, and did a significant amount of damage to the home's electrical wiring.

Tip! *Immediately test all of the home's smoke alarms to be certain that they're operational, and that they weren't damaged by the lightning strike. It's crucial for the safety of the building's occupants that all smoke alarms work correctly. If they've been damaged or aren't working for any reason, they need to be repaired or replaced at once.*

In addition to the electrical wiring, the lightning will wreak havoc on electronics. Since so much of today's homes are run by computers and electronic circuit boards, this damage can be both widespread and difficult to track down. It can affect appliances, thermostats, light timers, filter controls, furnaces, and a number of other components. It can also damage home electrical content items, such as computers, televisions, entertainment systems and phones.

The electrical charge from a lightning strike can be conducted along pipes and other metallic elements within the building as well, affecting odd things like plumbing joints, fixture cartridges, gaskets, seals, and O-rings. Lightning has also been known to create small holes in metal plumbing pipes.

Problem Areas

In addition to your normal walk-through of the building, you'll need to do a little more detective work with a lightning strike than you might with other types of losses. That's due to the odd nature of how lightning behaves, and the widespread and unpredictable damage it can do.

First and foremost, look very closely for fire and smoke damage. Lightning can start fires in small concealed areas that have very little oxygen. These fires can smolder for long periods of time before they're discovered. Be sure that you make the homeowners aware of this possibly so they'll remain alert for any sight or odor of smoke. If the fire department, your electrician, or any other qualified personnel are concerned about how safe the home is, the owners should strongly consider leaving.

You need to work closely with the homeowners, and listen carefully to what they have to say. Some of their stories may

AS YOU CAN SEE *from some of the examples in this chapter, insurance restoration means never having to say "I'm bored!" In the next chapter we'll look at even more major restoration losses.*

seem a little bizarre. They may tell you that the furnace only works if the thermostat is set at 48 and no higher. Or that only two burners on the electric range will come on, and one will only work on medium, while the other only works at the high or the low setting.

You'll also need to work closely with your electrician, plumber, appliance and HVAC service techs. Every system in the home will have to be closely inspected and tested. In some cases, you may never really track down the strange behavior in some of the appliances and circuit boards; they may simply have to be replaced.

Some potential problems to look for are basically structural. Lightning can cause all kinds of structural members to split, shift, or even completely shatter. It can pop boards loose from their fasteners, and can actually blow the nails out of joist hangers and other hardware. Look closely at the condition of the lumber, as well as the integrity of any connections.

REAL STORIES:

You Wouldn't Have Wanted to be There!

Late one summer afternoon the skies turned an ominous dark black. Thunder boomed, and the rain began to fall in heavy sheets. Bolts of lightning streaked across the sky, one after another. Unfortunately, one of those bolts found the gable peak on a two-story house. The impact was just a fraction of a second. But the result of that strike was the most bizarre assortment of damage that I've seen.

The lightning bolt shattered the gable end as though a giant hand had grabbed it and crushed it. The fascia split and splintered. Pieces of lap siding were blown off, and several fist-sized holes were punched through the OSB sheathing. The vinyl gable end vent melted, and the paint on the siding and trim around the gable was scorched.

Inside the attic, rafters and braces were splintered. Several batts of fiberglass insulation had been ripped out, tossed across the attic, and lay strewn in a pile, displaying black burn streaks. A small fire started smoldering in some storage boxes. The fire was discovered and put out before it burst into flames, leaving only smoke damage.

Below this section of the attic was a bathroom, with an exhaust fan in the ceiling. A length of flexible aluminum ducting led from the fan to a vent in the roof. The duct had melted, as had the plastic roof vent. The roof sheathing and composition shingles around the vent were scorched.

The aluminum duct conducted the lightning to the fan, which housed a light fixture with a glass cover. The glass shattered, and shards of glass were imbedded in the drywall of the bathroom. As the current passed through the electrical wiring of the fan, it caused such an intense arc that the electrical cable in the wall ripped free of its staples. The cable whipped violently against the back of the drywall, ripping a long, narrow strip out of the drywall from the ceiling all the way down to the fan's switch box.

There was a large mirror over the vanity, and while the mirror hadn't cracked, there were several round areas where you could see right through it to the drywall behind. The electrical wiring in the wall behind the mirror experienced an intense surge of current and the wires reacted with the silver backing on the mirror, burning through the backing and leaving little open holes in the mirror's reflective surface.

All the damage was repairable, but as the homeowner said, it was sure a good thing no one was using the bathroom at the time!

Building Permits for Lightning Damage

In order to obtain building permits, you'll first have to determine the extent and nature of the damage. Structural damage has to be outlined in the permit application, including what structural members are affected and whether they're being repaired or replaced. Cosmetic damage, including the replacement of siding and roofing, typically doesn't require a permit.

You'll need permits for repairs to the electrical wiring, including 120- and 240-volt circuits, electrical service equipment, and low-voltage wiring. Replacing appliances may or may not call for a permit, depending on local requirements.

11

Impact Damage, Vandalism & Trauma Scenes

In this chapter, we'll look at some of the other types of building losses you're likely to encounter, from tree, vehicle, and explosion damage, to issues involving vandalism, including those stemming from home foreclosures. And finally, we'll take a brief look at the more unpleasant jobs: cleaning up trauma scenes, from murder to death by natural causes.

Tree Impact Losses

Among planning departments and other agencies, homes in areas with an abundance of trees are known as *urban-forest interface zones*, and they've grown tremendously in popularity in recent years. But if one of these trees topples over and interfaces with a house, it can do a substantial amount of damage.

Tree impacts occur for a couple of different reasons. All or part of the tree may blow over during a wind storm, or a buildup of ice or snow can cause the branches to snap off due to the extra weight. During periods of prolonged rain, the ground can become saturated and a tree may topple over on its own, root ball and all. Lightning can strike a tree, causing all or part of it to fall, or even to explode, raining tree-shrapnel everywhere.

Sometimes the tree gets a little manmade help. It may get hit by a car and knocked over. Or there's always the well-meaning guy with the chainsaw who was certain the tree was going to fall the *other* way.

Trees that go over all at once can quite literally cut a house in half. If it's tall and heavy enough, it can even come down with enough power to damage one or two adjacent houses as well. And it doesn't need to be the entire tree that falls to cause damage. A single limb can often be the size of a small to medium tree (see Figure 11-1). If a

Figure 11-1 Many of the limbs on large trees can be the size of trees themselves. That was the case with this limb, which snapped off from the side of a massive pine tree. It fell across the rear of this home during a snow storm and did a considerable amount of roof damage.

large limb snaps off 50 or 60 feet up in the air, it will plummet to earth with the force of a small missile.

On the other hand, a huge tree can come over with a gentle touch as well, which sometimes happens when a tree is uprooted due to saturated ground. The tree may go over so slowly that its impact against the house is negligible. It's not uncommon for homeowners to wake up one morning with a massive 100-foot-tall pine tree leaning against their house — having slept through the entire event!

Figure 11-2 This towering pine tree was uprooted and fell over during a wind storm. It fell across the roof of the house and crushed the chimney chase. It also did structural, siding, roofing, and cosmetic damage.

Damage from Tree Impacts

The most obvious damage of tree impacts will be to the roof structure (Figure 11-2). This can include broken fascia and soffit boards, and snapped rafters and joists. It can also include snapped or separated roof trusses (Figure 11-3), and splintered or punctured roof sheathing.

Anything that's on the roof will almost certainly suffer damage as well, including chimneys and chimney chases, woodstove flues and other flue pipes, plumbing vents, attic vents, television antennas, satellite dishes, and even solar panels (Figure 11-4). And, of course, there's the roofing itself.

As the tree comes down, the limbs will also cause damage, and this can occur all over the house. The primary limbs are long and very strong, and these have a lot of smaller protruding limbs. They'll break siding and trim boards, as well as windows. Limbs do a lot of scraping, gouging, and other damage on their way down.

Figure 11-3 The weight of a falling tree caused the intermediate supports of these manufactured roof trusses to separate from the top chords (center). Because these are engineered trusses, repairs have to be designed and approved by a truss company or a structural engineer.

Figure 11-4 This solar collector was hit by a limb as the tree fell. It was a total loss. The adjacent collector also suffered some damage.

The force of the impact often opens up the building's envelope to the outside. If the tree came down during a storm, you're probably going to have a lot of water-related damage inside, in addition to damage to drywall and flooring. If a tree limb hits the kitchen or a bathroom, it can take out a massive amount of wiring and plumbing. It can also destroy cabinets, fixtures, and other interior components.

No matter how it comes down, you'll also have the tree itself to deal with. Often they're perched precariously on or against the house, where the slightest miss-cut with a chainsaw can send it crashing down or through the building, risking additional damage or, worse yet, injury to your crew.

Your best bet is to use subcontractors for the tree removal. The pros have the right equipment and the expertise to deal with a fallen tree correctly — and the proper liability insurance if something goes wrong. The tree may have to be stabilized with cables before it can be cut, or it may have to be lifted off with a crane and then cut up. In these situations, the cost of removing the tree from the house so that repairs can be done should be fully covered by the insurance company.

Problem Areas

From a coverage standpoint, there usually isn't a lot of gray area with a tree impact, since it's obviously both sudden and accidental. So if the tree did the damage on the way down, it's typically covered.

The challenge is finding and documenting all of the damage. You need to take your time, and understand what the tree did on the way down. Trace the direction

Figure 11-5 This huge tree damaged gutters, flashings, and roofing when it hit the edge of this garage. If you look closely, you can also see that it cracked the siding right below the upper trim board, and also separated the siding boards at the corner. Tree impacts require careful inspection for damage.

Figure 11-6 Even though the tree hit the roof of this home, the impact shook the house hard enough to cause a crack in the brick veneer.

of the fall, and look at where the tree impacted. See Figure 11-5. Remember that trees don't always drop straight down; they often twist and rotate as they fall.

The primary damage will obviously be from impact, but it may surprise you to find that a tree can actually bounce. As a result, that impact damage may be magnified, or it may show up in more than one spot.

A tree may fall onto the peak of a roof, bounce, and then come down again a few feet away. Or it may come down directly on a chimney, bounce off, and then crash down onto the roof. The important thing is to not make hasty assumptions about where all the damage is. A tree can cause tremendous momentary downward and even lateral movement in framing members, which will loosen joints between boards, snap fasteners, and shift connecting hardware. It can create nail pops and cracks in drywall, and do a lot of other damage that isn't always easy to spot (Figure 11-6).

All of the bouncing and movement is hard on roofing, so you'll want to have your roofer thoroughly inspect the entire roof. Look at the condition of the shingles, especially wood shakes. Look for areas where fasteners may have come up from the impact, causing the shingles to loosen from the sheathing. Also, carefully check flashings and chimney masonry (Figure 11-7). Look for damage caused by the limbs, including scrapes and scratches to siding, trim, window and door frames, window and door hardware, gutters, fascia, and other areas.

Tree limbs can cause scratches on glass that are hard to see, especially on coated glass. They can damage window films, and even low-E glass coatings. Also, the impact

Figure 11-7 The roof of this house was hit by a tree, and a branch hit the chimney. That impact broke one of the flue liners (right) and chipped another one (left). It also did quite a bit of damage to the mortar cap on the top of the bricks, especially at the left front corner. The chimney required a complete inspection and repair by a qualified mason before it was safe to use.

can damage the seal on double-pane insulated glass units, which is particularly hard to detect — especially during the warmer months. If you suspect window damage, have a glass company do an inspection of the windows and see if they can determine anything. Also, let the homeowners and the adjuster know of your concerns. Document the possibility that additional damage may become obvious in the winter months, when condensing moisture between the panes will make a blown seal obvious. All of this damage applies to skylights as well as windows.

Fallen trees have a tendency to damage landscaping, including sprinkler systems. Direct damage to things like fences, decks, and sprinklers are typically covered. But, like the exterior fire damage we covered back in Chapter 8, landscaping can be a gray area, so discuss the specifics of any landscaping damage with your adjuster.

Tree Cleanup

After the tree is off the roof, there's the mess to deal with. The insurance company will cover normal cleanup, but here's one of those areas where it can get tricky.

If the wind blew the tree over, cleaning up all the debris related to cutting up and hauling off the tree should be covered. However, the wind no doubt blew leaves or needles all over the roof and the yard as well. Raking up the yard isn't covered, since that's a normal maintenance issue. So you'll have to find a happy medium in there somewhere. See Figure 11-8.

Also not covered is cutting up the tree for firewood. The insurance company will usually pay to cut the tree into whatever manageable-sized pieces are necessary in order to haul it away. They'll also typically cover the labor to cut up the limbs for grinding or hauling. What they usually *won't* pay for is to cut the tree into rounds, or anything having to do with splitting or stacking the wood. If the homeowners want that done, they'll have to pay for that out of pocket.

Who Covers What?

Another potential problem area with tree damage is that it's not always confined to one particular piece of property. The tree may be growing on Property A, but may fall

Figure 11-8 Quite a bit of debris still remains after a fallen tree was cut up and the pieces removed. The insurance company will usually pay for the *reasonable* removal of the remaining needles or leaves, the removal of small limbs, and the cleaning of decks, patios, and roofs.

Figure 11-9 This tree came down across a fence shared by two neighbors. Damage like this can become a confusing issue from the standpoint of coverage.

over and damage Property B. Or a limb from the tree on Property A may crack off and fall on Property B. Perhaps the tree is actually straddling the property line. Or maybe the tree is on Property A, but it falls over and damages a fence that's on the property line, and that fence is shared equally by Property A and Property B (see Figure 11-9).

All of these situations can really muddy the waters from a coverage standpoint. Whenever you're asked to estimate damage from a tree-related loss, be sure and discuss the situation with the adjuster. Understand who has coverage, and which companies will be handling which claims.

Tip! *As much as you'd like to be able to trust what people tell you, you also need to develop a bit of healthy skepticism. With complex claims, such as a tree that's fallen from a neighboring property, never accept the assurance of the homeowner that the claim will be covered by the insurance company. Verify it first.*

Tree Impact Building Permits

Tree impact permits are similar to structural fire permits, so you might want to review that section in Chapter 8. All structural repair work requires a permit, but cosmetic repairs don't. Damage to trusses, rafters, and other roof or wall framing will require a building permit. Here again, structural engineers can be very helpful in a number of different damage repair situations.

You'll also need an electrical permit for repairs to damaged wiring. Repairs to fireplaces and woodstoves will most likely require a mechanical permit. If you're replacing siding, windows, or roofing, you probably won't need a permit, but be sure and check.

REAL STORIES:
Next Time We'll Take the Treat!

Tree impact damage is unquestionably sudden, almost always accidental, and generally related to some type of storm. The following story relates just how sudden and devastating the damage can be.

Early one Halloween night, a strong wind storm was starting to make itself felt in our area. A young couple had just finished handing out treats to a group of assorted ghosts and goblins at the front door of their house. They were headed back into their living room to sit down, when they heard a massive BOOM! from the next room.

The sound was so sudden and intense that both of them later said they thought something had exploded. Their entire two-story house shuddered momentarily, as though shaken by a giant hand!

They ran into the dining room, where the sound seemed to have come from, and found a massive tree limb speared right through their dining room floor! The limb had cracked off, very high up, from a huge tree next to their house, coming down point-first through the roof like a missile.

It had shattered a couple of roof trusses on the way down, passed through an upstairs bedroom, went through the subfloor, cracked a floor joist, then came through the dining room ceiling and finally lodged into the floor.

But what really caught their attention was the shattered dining room chair. It was exactly where one of them had been sitting, just an hour earlier, eating dinner! They started to rush outside to see the damage, only to find the front door blocked. Using the rear exit, they went out and came around to the front, where another shock awaited.

As if the horrible image of that shattered dining room chair wasn't enough, smaller limbs and debris were all over the front porch and front walkway. It was right where the little trick-or-treaters they'd given candy to had been standing just moments before.

The structural damage was pretty extensive. But thankfully, no one was hurt. It was definitely the wrong kind of trick for a very scary Halloween night.

Vehicle Impacts

A vehicle impact can take several different forms, and can occur for several different reasons. It may be caused by a person pulling into their garage who hits the gas instead of the brake (see Figure 11-10), or someone who's pulling out of the garage who forgot to open the door — yes, that does happen!

It could also be the result of someone pulling their car into the driveway and unexpectedly hitting ice (see Figure 11-11). Or, there's always the drunk driver, with no business being behind the wheel, driving too fast and losing control of his car. Whatever the reason, the damage to the building can be anywhere from very minor to very severe.

While most jobs you'll be doing as a restoration contractor primarily involve homes, when it comes to vehicle impacts, you'll find it's typically about a 50-50 split between residential and commercial losses.

Figure 11-10 The wall of this home was damaged when a driver pulling the car into the garage didn't stop in time. Note the proximity of the electrical meter, the telephone connection and the cable television boxes to the damage. All had to be checked for possible damage.

Damage from Vehicle Impacts

Whenever a large solid mass strikes a building, it's going to cause damage. If that solid mass is a vehicle traveling at a high rate of speed, the impact and resulting damage can be substantial.

Vehicle damage tends to be localized, occurring directly along the path that the vehicle traveled. Since a vehicle doesn't bounce, the energy dissipates relatively quickly. In a typical vehicle impact loss, the car or truck runs into the side of a building (Figure 11-12). There's generally a section of broken wall framing (Figure 11-13), along with damage to the siding on the outside and drywall on the inside. Depending on the location of the impact, you may have electrical wiring and plumbing damage to deal with as well.

There's usually a combination of framing damage, with some studs completely shattered, while others are pushed forward off their nails but otherwise undamaged. The sill plate may be tipped off the floor, and may or may not be intact. Depending on the height of the floor above grade, there can be damage to the rim joist or the subfloor. At the top of the wall, the studs will be pulled forward off the nails.

Siding cracks or shatters inward on impact. You'll need to carefully study the siding to see what type it is, whether it can easily be matched, and where the natural joints are for patching in new material. In the case of masonry materials, such as stone and brick, matching is especially tough. For stucco, consider either replacing or skim coating the entire wall.

Inside the building, the extent of the primary damage to the drywall will be obvious. After that, you'll need to look at cracks and nail pops. Factor in repairs and retexturing, as well as painting. You'll also need to consider what the impact did to the flooring.

Finally, don't forget the cleanup. You'll have debris to pick up, both inside and outside. There tends to be a lot of small debris, including everything from wood splinters to drywall dust. One of the bigger problems can be shards of glass from broken windows. Carpets will need to be carefully vacuumed, and then cleaned.

Look for fluid spills, stains, and tire marks from the vehicle itself, which can damage carpets and other flooring. Look closely for scratches on hardwood flooring, ceramic tile, and other hard-surface flooring. If there are floor registers in the area of the impact, look for debris that may have found its way into the duct system.

Figure 11-11 Both of these garage doors were damaged when a car slid down the icy driveway and struck the post between the doors. The doors had to be replaced, and the framing realigned and secured. When dealing with roll-up doors of this type, it's important to remember that the overhead springs have a considerable amount of tension on them. *They are potentially very dangerous to handle*, and should be removed by trained professionals who know how to unwind them.

Figure 11-12 This house was struck by a drunk driver who drove straight across the front lawn after missing a turn. In addition to damage to the wood framing, the impact also caused cracking in the masonry veneer.

Figure 11-13 A vehicle impact pushed this section of wall framing inward. The impact broke the bottom plate, pushed the studs off the nails on the top plate, and did other structural damage.

Problem Areas

One vehicle impact problem is hidden damage. Hitting a building with something as heavy as a car shakes things up. This can create issues with the framing or other components that you won't immediately be aware of.

With the framing, carefully examine the path the vehicle took. Look for signs of damage in areas radiating out from the impact location. Depending on the height of the foundation in relation to the height of the vehicle, there may be damage to the stemwalls or the footing. In the case of a basement, there may be damage to the basement wall that extends below grade. Damage can even affect below-grade waterproofing or drainage systems.

Impact from a vehicle can literally rip the electrical wiring out of adjacent boxes, and tear circuits loose even a distance away from the impact site. Have a qualified electrician make a careful inspection of the entire system. If any plumbing was impacted, you need to do a complete pressure test of both the water and waste systems to be certain that no other joints were loosened by the impact.

The spray of debris from the crash will often cover a lot of the floor, so you may need to wait until all the emergency cleaning and board-up is done before you complete your estimate. Factor in enough time for the cleanup of a lot of dust and debris.

Figure 11-14 Hardscaping that's damaged by vehicle impacts is typically covered as part of the loss. That includes deck railings and walkways, like these.

Vehicle damage to fences, decks, and outbuildings is typically covered (Figure 11-14). However, there's always a potential gray area with landscaping, so discuss this with the adjuster before proceeding.

Jobs like the one in the following house-moving story don't come along all that often. But when they do, they challenge everyone, from the engineer to the estimator to the construction crews. They also make this a pretty cool occupation!

REAL STORIES:
A House-Moving Story

A lot of new construction had been going on in our area, including the opening of some new home sites high on a hillside above an area of well-preserved older homes.

One afternoon, a mobile construction crane was headed up the hill to one of the new home sites to assist with setting trusses. When the crane reached the building site, the driver applied his brakes and found that they had completely gone out!

The crane began to roll backward down the hill, on a direct collision course with one of the homes below. It picked up speed on its way downhill, slowed only by the curb in front of its targeted house.

Jumping the curb, the crane tore through the landscaping, ripped out an underground sprinkler system and some landscape timbers, then hit the driveway. The driveway sloped down toward the house, which was actually fortunate. As the crane bounced over the landscape timbers, one of its stabilizing outriggers dropped, digging down and ripping a furrow into the concrete driveway.

The outrigger snagged some wire mesh embedded in the concrete, just like a fighter jet's tail hook grabs the arresting cable on an aircraft carrier's deck.

The crane, slowed by the wire mesh, slammed at an angle into the corner of the garage, and stopped. There it hung, suspended by the outrigger caught in the concrete, narrowly saved from continuing downhill and into more houses below. The terrified driver was unhurt.

After the crane was untangled and towed off, we were called in to examine the damage. Because of the hilly location, this garage was fairly unique. It sat at grade level in front, but the rear was almost five feet off the ground, perched on high stem walls like a daylight basement. Joists supported the garage slab, and the area underneath was used for storage, which was fortunate, because it gave us a lot of access for working.

The impact from the crane had shifted the garage several inches off its foundation. Some of the anchor bolts holding it to the stemwall had been sheared off, and others bent. The house had underground electrical power, and the supply conduit was cracked. There was structural damage to the garage framing and siding, and of course, to the driveway.

With the help of a structural engineer, we developed a plan and pulled the necessary permits. We decided to apply the same rod and turnbuckle idea that we'd used to repair the snow-damaged home described in Chapter 10.

We had several 1-inch diameter steel rods custom cut and threaded, and called in a concrete drilling company to bore holes in specific locations in the stemwalls. We then bolted the rods in place, and joined them together with large turnbuckles.

We had a house-moving company come in and place beams under the garage floor and then set up hydraulic jacks under them to lift the garage. To begin, they applied just enough upward pressure to barely take the weight of the garage off the stemwalls. That allowed us to use reciprocating saws to sever the remaining anchor bolts and free the garage from the stemwall. The house movers then used the jacks to lift the garage about half an inch above the stemwall.

Next, we very carefully used the turnbuckles to adjust the rods in different sequences. Slowly, we were able to pull the entire side of the garage back around until it was once again aligned correctly over the foundation. We jack-hammered out the broken electrical conduit and patched in a new one, after verifying that the wiring inside was undamaged.

We released the turnbuckles and removed the rods. The house movers released the jacks slowly and allowed the house to settle back down perfectly onto the stemwalls.

We then installed some engineer-approved tie-downs to replace the old anchor bolts, tying the house back to its foundation.

After that, it was simply a matter of repairing the remaining structural, cosmetic, and landscaping damage. A new driveway was put in, and the house was good as new. Jobs like these really make this work interesting!

Overlapping Claims

Vehicle damage losses usually involve overlapping coverage claims, with both the homeowner's insurance company and vehicle driver's insurance company becoming involved. Since there's an immediate need to get the home repaired and livable, what typically happens is that the homeowner's insurance company steps in and takes over responsibility for getting the repairs done. They will then subrogate against the vehicle driver's insurance company to recover those expenses, including any deductible paid by the homeowner.

Subrogation actions happen in the background, and rarely involve the restoration contractor. However, you do need to be very clear about the coverage and who's paying for the claim. When you're called out on a loss of this type, it's best to clarify everything with the homeowners and their adjuster.

Vehicle Damage Building Permits

Vehicle damage building permits are completely job-specific. You'll need a building permit that covers the structural repairs, so you'll have to describe the damaged components you're going to remove and replace, or any repairs you intend to make to components that are damaged, but which will remain in place. Again, for extensive structural damage, you may need the services and documentation of a structural engineer.

You'll need an electrical repair permit that describes the number of circuits being repaired, as well as any repairs to the service panel and meter base. A plumbing permit probably won't be required, unless the water heater is damaged and your jurisdiction requires permits for water heater replacements.

Explosion and Blasting Damage

Explosions aren't an everyday occurrence. They may happen as part of a fire loss when a gas can, propane tank, or some aerosol cans explode in the house. Or an explosion can happen in a car or motor home, again as part of a fire loss.

Occasionally, an explosion will occur that isn't fire related. A gas line may get punctured during construction work, and then explode when ignited by a spark. A welder's torch may set off an explosion in some fumes; or fans or other electrical sparks can set off fumes in paint shops or home woodshops.

Damage from Explosions

When an explosion occurs as part of a fire, it's unlikely that you'll really have much to deal with as a separate issue. Instead, whatever structural damage the explosion did will be estimated and repaired as part of the fire damage.

When an explosion occurs on its own, without an accompanying fire, then you have structural issues that you'll need to deal with. Always begin your evaluation at the point where the explosion occurred, which is where the damage is most severe. That'll give you the most clues about the direction the blast traveled and the force involved.

Problem Areas

Because explosions are so unusual, you'll want to tread lightly at first. Talk with the insurance adjuster to be sure the cause-and-origin people investigating the incident are done. They may feel that some other party is at fault for the explosion. For example, it may have been caused by a faulty product of some sort — in which case, they may be subrogating against the manufacturer of that product to recover the cost of the repairs. In that case, you may be asked to salvage some specific components when doing your demolition work.

Explosions can also be the result of criminal activity, either directly or indirectly, so again, you may be asked to salvage specific evidence for the investigation. If you'll be doing demolition work on a crime scene and need to remove materials that'll later be used in court as evidence in a trial, the police will need to maintain what's known as a *chain of custody*. That's a very strict, legal procedure in which witnesses have to watch and verify the removal and tagging of an item for evidence, in order to ensure that it hasn't been tampered with in any way.

When dealing with either cause-and-origin or law enforcement officials, be sure you have permission to access the site before you begin any demolition or repairs.

Depending on the severity of the explosion, the building itself may be structurally unstable. You may need to install temporary supports and bracing, and apply the same safety precautions that you'd use in a structural fire (refer back to Chapter 8).

Shrapnel is always a potential hazard in an explosion. Small bits of metal and other debris can be driven into wood, drywall and other surfaces, and if you hit any of this concealed metal with a saw blade, it can throw off small splinters that do a lot of damage. *Always wear gloves and eye protection when doing demolition work, and when transporting debris.*

Certain types of explosions can also leave dangerous residues behind. Explosions caused by chemical interactions, especially those in the manufacture of illegal drugs such as methamphetamine, can be extremely hazardous. *Be sure you understand exactly what caused the explosion before you enter the building.* You need to know what gases, residues, and other compounds may be remaining in the building before you expose yourself and your crew. Wear PPE when doing explosion cleanup work until you're confident that you're not at risk of exposure to any type of contaminants.

Explosions create some very unusual damage situations. The pressure can separate joints between structural members. Lumber can get pushed off of fasteners or out of hangers, and not settle completely back into place again; walls can shift out of plumb, drywall seams separate, and fasteners pop. Ceramic tile can loosen from the substrate below it; doors and windows can be knocked out of alignment; and dual-pane window seals can blow. The list goes on and on.

REAL STORIES:

Now *That's* an Unusual Explosion Loss!

You probably won't run into anything quite like this explosion loss, but you never know. It certainly had an impact on us!

The home was a spacious two-story farm house, newly constructed on several acres of land. It was far enough out of town that it wasn't served by natural gas, so the owners had a large propane tank installed to supply fuel for the fireplace, range, and water heater. The propane tank was full and pressurized, providing gas to the various appliances in the home.

The fireplace was a large zero-clearance unit, built into a wooden chase along one of the living room's exterior walls. Most of it was metal, with a base of fire brick and a propane-fueled log lighter. It was fall, not really cold enough for a fire just yet, so the fireplace had yet to be used for the first time. Instead, some decorative candles had been placed in the hearth opening.

The candles had been lit for the evening, and slowly began to burn down. Wax flowed, and formed small pools in the bottom of the fireplace. Eventually, one of the pools of molten wax spread across the bottom of the fireplace, and dripped over the edge.

It found a tiny crack in the sheet metal frame, and dripped down below the firebox. There, the hot wax met up with the pressurized propane that had been slowly leaking out of a badly-sealed fitting, building up fumes inside the enclosed chase.

The chase was perfectly positioned on the outside of the house like the exhaust pipe on a jet engine. When the hot wax caused the propane to explode, it roared up the enclosed chase and into the attic with unimaginable force.

As it hit the attic, the gases instantly expanded, lifting the entire roof structure straight up into the air, then dropping it right back down onto the house! The direction of the blast, and the way the energy was dissipated, kept the damage from being much, much worse. And, quite miraculously and most important, it also prevented anyone from being injured.

The day after the explosion, we were out at the site examining the loss with the adjuster and doing the initial board-up. Two fire trucks arrived, loaded with firemen. We were at first concerned that they felt they had somehow missed something, and were thinking the house was more structurally unstable than they had at first thought.

"No," one of the senior firemen told us, "we're just back to take a tour. None of us has ever seen anything like this before. We also brought some of the guys from the other station – we thought it would be a good learning experience for all of us." Now *that's* when you know you have an unusual loss!

The subsequent repairs required an entire new roof structure, quite a bit of restoration to parts of the upper floor — and of course, a new fireplace and chase.

Be aware of the potential for a lot of hidden damage, especially with a large explosion. Arrange for a detailed walk-through with the adjuster early in the job, and possibly a second one following the completion of demolition.

Blasting Damage

Contractors, excavators, developers, and others occasionally do blasting work to clear sites of rock or to demolish old structures. If you're in an area where this type of work is going on, you may be called on to repair damage related to the blasting. Blasts can create vibrations or flying debris, resulting in damage to adjacent buildings. They can also cause underground tremors that damage buildings located blocks or sometimes even miles away.

Figure 11-15 Any time there's blasting or other construction work going in an area, some damage to adjacent buildings may result. The exact cause of cracking, like that shown here, may be especially difficult to pinpoint and document in older buildings.

When an explosives professional is about to do blasting work, he's supposed to do a pre-blast inspection of neighboring properties. This is to determine if the properties have any cracks or other structural problems *prior* to the blast taking place. In reality, these pre-blast inspections are often skipped, so after the blast, it's often difficult to determine exactly what is and isn't blast-related damage.

Blasting damage from flying debris is fairly easy to recognize. The bigger problem is damage from shaking, which is similar to what you might expect in the aftermath of a small earthquake. The damage may be cracks in foundations, slabs, or masonry (Figure 11-15). It may show up as nail pops and seam separations in drywall, or as cracked windows, or doors that won't open or close correctly. There could be cracks in ceramic tile or interior plaster work (Figure 11-16). There may be plumbing leaks, or misaligned fittings. You might see humped floors, and even dust that's rolled through buildings. Some of this damage is structural in nature, and some of it's strictly cosmetic.

Figure 11-16 The homeowner attributed these interior plaster cracks to blasting work in the area.

Blasting damage also brings with it a sense of "heightened awareness." If the homeowners become aware that there's blasting going on in the area and then they see a crack in their home, they'll begin looking for others. If you're called out to do an estimate, they'll expect you to be equally thorough and diligent in finding all the damage you can. On the other side, the blasting contractor or his insurance company may be pressuring you to discount a lot of the damage as preexisting.

You'll usually notice some clues that will help you identify if certain cracks or other damage are new or old. But in the end, what was a blast-related crack in the drywall or the foundation and what was there before and they just never noticed is very hard to say. Be honest, and give the best advice and opinion that you can to all parties. Never pad your bid, and never lean to one side of the argument or the other. If there is severe structural damage involved, seek out the opinion of a structural engineer as well, especially if there are large cracks in the foundation or slab.

Explosion and Blasting Damage Building Permits

If the explosion has occurred as part of a fire, then whatever permits you need will be part of the fire loss permit process. If the explosion was a separate occurrence, then you'll need to determine the extent of the damage and which of the building's components need to be repaired.

If there are manufactured roof trusses involved, then you'll need to have your truss company prepare engineered truss drawings for you. This is true whether you're removing and replacing the trusses, or just repairing them. Separate permits will also be needed for electrical wiring, plumbing, and mechanical systems. Certain very complex explosion losses, like the one in the Unusual Explosion Loss story, will require an engineer to design and certify the repairs for the building department.

Vandalism

Vandalism is essentially any damage intentionally done to a building. It includes relatively minor acts, such as spraying graffiti or breaking a window, to major acts of willful property destruction. Arson, the deliberate setting of a fire, can be considered an act of vandalism as well.

Remember, for a loss to be covered, it needs to be both sudden and accidental. In the case of vandalism, the damage is obviously not accidental — however, as long as it wasn't done by the building owner, vandalism is a specifically-named peril, so it's a covered loss.

Graffiti

From an estimating and repair standpoint, vandalism losses cover a lot of ground. One of the most common types of damage is graffiti. See Figure 11-17. A can of spray paint is the usual tool, but it can also be done with a paint brush, marker pen, or other materials. And, you'll find graffiti applied to all kinds of surfaces.

There are several things to consider when removing graffiti. Is the surface that the graffiti's on already painted? If not, would the owners object to having it painted? If it's already painted or the owners approve of having the damage painted over, then you can wash the area to remove dirt and loose material, apply a stain-blocking primer (refer to Chapter 9), then paint over the primer with a finish paint.

If the building owners don't want the area painted, such as if it's an ornamental stone wall, then you'll have to determine how the surface material can be cleaned.

Figure 11-17 The aftermath of a vandalism spree in an unoccupied vacation home. Most of this damage isn't structural in nature. However, it requires a considerable amount of cleanup, and the replacement of materials such as carpeting and paneling.

There are a number of different cleaning products on the market that work on paints and markers with varying degrees of success. You can also try wet cleaning or even steam cleaning methods. If cleaning doesn't work, your only alternative is complete replacement of the affected material.

Structural and Cosmetic Damage

Vandalism also involves all levels of structural damage. Sometimes the damage occurs as part of a break-in during a burglary, such as a broken door or window. Sometimes the damage itself is the desired end result. In those cases, you'll commonly see holes kicked in drywall and interior doors, broken mirrors and light fixtures, or towel bars torn off the walls.

Occasionally this damage is just cosmetic, but still difficult to repair. For example, you may find items such as decorative beams (Figure 11-18) or masonry defaced. The damage isn't really structural in nature, but it isn't easily repaired either. You'll need to work with the adjuster and the building owner to arrive at a solution that's agreeable to all parties.

Figure 11-18 This main beam has been damaged by vandals hacking at it with an ax. From a structural standpoint, it's still okay. However, it's very difficult to conceal the damage and make the beam look right again, especially since it's natural wood. The best compromise is probably to box it in with more natural stained wood.

In other situations, the damage definitely *is* structural. The home pictured in Figure 11-19 and Figure 11-20 shows extensive damage from vandals. The vandals compromised not only the structural integrity, but also the plumbing and electrical systems, and other components of the building. In cases like these, you need to treat the damage as you would any other structural repair.

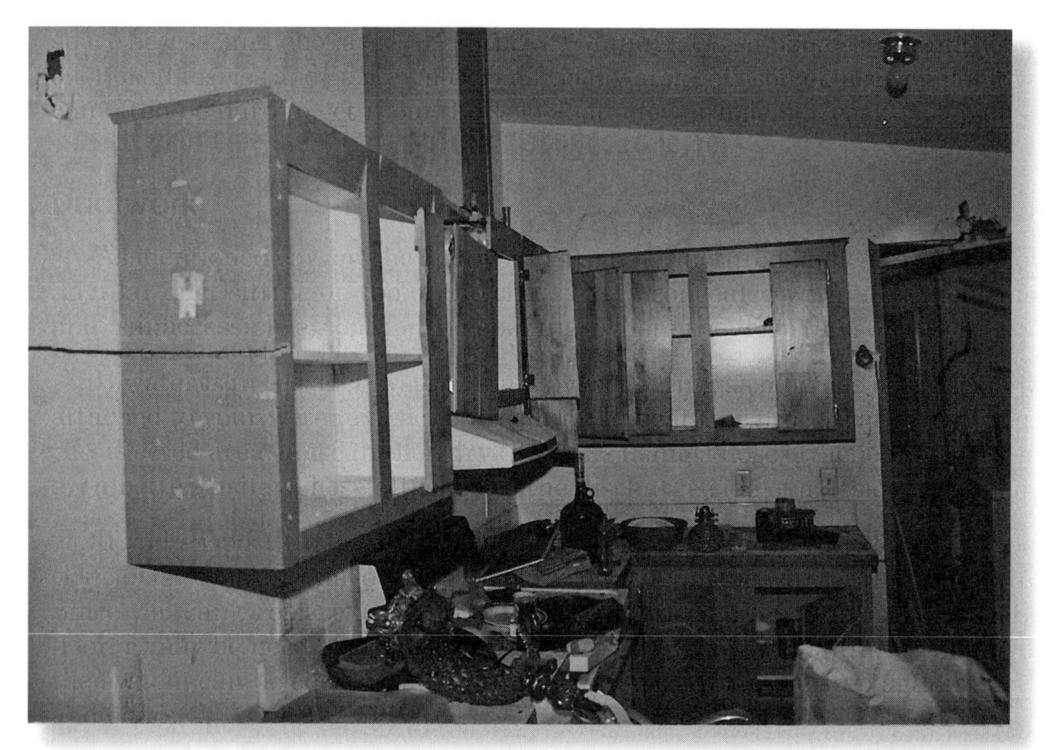

Figure 11-19 The extent and viciousness of some vandalism is hard to imagine. In the kitchen and laundry room of this home, the cabinets have been destroyed and the doors and trim have been ripped off. Drywall was kicked in, and walls throughout the house cut with a chainsaw. Broken furniture and debris were everywhere.

A Word About Foreclosures

At the time of this writing, there are an unprecedented number of foreclosed homes across the country. That's giving rise to a different kind of vandalism that will probably have an impact on insurance restoration in the coming years.

This vandalism seems to take two different forms. First there are the people who've lost their homes. Many are angry and frustrated, and some take out their anger towards the banks and mortgage companies on their home before leaving. This can range from minor graffiti or the theft of appliances and building materials, all the way up to major structural damage.

The second vandalism issue with foreclosures is that homeless people are taking up temporary residence in the unoccupied homes. The damage left behind in these cases runs from minor cleanup issues to assorted acts of willful destruction.

We expect to see more of these types of losses in the future, as banks and other lenders work through their inventory of home foreclosures.

Vandalism Problem Areas

You need to be very aware of your own personal safety in a vandalism situation. Buildings that have been vandalized may contain mold, blood or other bodily fluids, chemicals, needles, and all sorts of things that can be extremely hazardous. *At a mini-*

Figure 11-20 A chainsaw was also used on the floor of the kitchen. There was a considerable amount of structural damage to the floor joists. Rather than replace the joists, a good compromise was to install a new girder on each side of the cut area, then install a new joist to span the damaged portion. Permits were required due to the structural damage, so all repairs had to be approved by the building department.

mum, wear respiratory protection, gloves, and eye protection until you know what you're dealing with. If you suspect more severe types of contamination, take even greater precautions.

For example, if you suspect that any type of criminal activity has taken place in an empty or foreclosed home that you've been asked to inspect, *especially methamphetamine production*, do not enter the building until you've consulted with the police!

Depending on your personality and sensitivities, your own reaction to seeing the aftermath of an act of vandalism may vary. You may experience anger or outrage at what you see, or you may be offended by what you find written on the walls. Walking into a situation where someone has taken a chainsaw to a home and then defecated on the floor, can be pretty hard to stomach. Just be aware that your feelings of disgust and outrage are perfectly normal. As a professional restoration contractor, it's something that you'll no doubt be able to quickly put aside so you can move on with the job that you're there to do.

Due to the sometimes malicious and destructive nature of what's been done to the building — and the mindset of the people doing it — there's a lot of potential for odor. Rotten food, urine, fecal matter, chemicals, garbage, mildew, and other odors are all something you may need to deal with. Refer back to Chapter 9 for more about dealing with odors.

There's also potential for hidden damage in vandalism situations. Sometimes that damage is hidden because of the nature of how the building was vandalized. For example, light fixtures may have been torn off the ceiling, resulting in unseen damage to the electrical wiring behind the fixture. In other cases, the unseen damage may be the intent of the vandals. They may have left an odor-causing "surprise" somewhere, like rotten food or a dead animal or fish in or under the house.

Wear PPE, and do a thorough investigation of the building. Have your electrician make sure that all of the wiring is safe and fully operational. Have a technician check all the appliances, and an HVAC contractor test the heating and air conditioning systems. If the building's been vacant for awhile, or if you suspect any plumbing damage, you should also have your plumber do a thorough inspection and pressure test the plumbing system.

REAL STORIES:

You Want a Reference?

One of the most aggressive vandalisms I've ever encountered was the result of a tenant being evicted from a house for non-payment of rent. It made me think twice about ever being a landlord.

The tenant was several months behind in his rent, and the landlord had been unsuccessful in all his collection efforts.

The landlord finally filed a legal action against the guy, and the sheriff's department served a 30-day eviction notice. Unfortunately, 30 days is more than enough time to do an awful lot of damage — especially for someone who owns a chain saw!

The tenant packed up the few things he could take with him. What he couldn't pack, he decided to leave behind — after smashing it up and throwing it all over the inside of the house. Pieces of furniture stuck through the walls seemed to be a particular favorite of his. He then broke every mirror, and smashed every toilet and sink. Every door was kicked in, and of course, there were the requisite swear words in multi-colored paint on every wall surface.

Food was another particular favorite. Boxes of cereal and bags of flour were tossed about, mixed in with an occasional splash of canned tomato sauce or catsup. Food was left in the refrigerator, which he unplugged so it was sure to ripen to its maximum degree of odor over the 30 days. And speaking of odor, he wanted to make sure he covered all the bases, so he urinated into at least three of the heating ducts.

Then he fired up the old chain saw and walked through the entire house, cutting a nice line right through everything in his path. Drywall. Studs. Doors. Trim. Electrical wiring. Plumbing. Cabinets. Everything. He also went outside and took a few swipes at the siding. It's a wonder he didn't knock the entire house down on top of his head.

In the end, it turned out to be easier to gut most of the interior in order to make the repairs. There were so many cuts in the wires, pipes, studs and everything else that chasing it all around would have been far too time consuming, and there would have been no way to be sure that we'd found everything.

The house turned out great in the end. We even got rid of all the odors, including the ones in the heating ducts. But I doubt the guy got much of a reference for the next place he rented.

Vandalism Building Permits

Since most vandalism damage tends to be cosmetic in nature, the need for building permits is fairly limited. There's usually not enough structural damage to warrant a structural building permit. You may need a mechanical or plumbing permit if sufficient damage was done to either of those systems. And, it's not too uncommon to need an electrical repair permit covering repairs to circuits, low voltage circuits, or to the electrical panel and meter base.

Trauma Scenes

Trauma-scene restoration is essentially the cleaning and restoration of a building in which a person has been sick, injured, or died. It may be the scene of a murder, an accident, a suicide, or a death by natural causes. See Figure 11-21. You can make

Figure 11-21 Dealing with trauma scenes requires careful safety precautions, including full PPE. If you're going to make this part of your business, your crew needs to be trained on how to handle the physical and emotional requirements of this work.

trauma-scene restoration a part of your business using your own crews, or you can use subcontractors. This field may be one that you want to hold off on at first, and then consider adding it into your business later. Some people prefer not doing it at all.

The advantage to trauma-scene restoration is that it makes you a full-service restoration contractor, and so, more valuable to the insurance adjusters. Once they know that they can count on you for the really tough jobs, they'll be more likely to turn to you for every job that comes along. And while you may not want to look at it this way, this type of work is usually very lucrative. There simply aren't that many contractors who want to do it. Not only can it be distasteful, there are also some health and safety risks involved. However, because of that, you can charge top rates for your services.

Obviously, there are a number of downsides to this area of work. You'll be handling materials such as flooring and drywall that are contaminated by bodily fluids. You also may be dealing with bits of human tissue or bone fragments on occasion. Depending on the situation, there can be some fairly intense odors to be dealt with. You'll need to take serious safety precautions on every job.

When you're first called out to a trauma scene, be sure to get all of the necessary information. This can be a little difficult, depending on who's calling and requesting your services. You need to get as many of the specifics about what happened as you can gather, what the current status of the home is, and exactly what you're being requested to do.

You also obviously have to be very careful when dealing with a crime scene or in a situation where there was a suicide. Always try and verify the information you're being given. Talk to local law enforcement officials, or the coroner's office. If there's a mortuary involved, consult with them to see that the death certificate is in proper order. If the insurance company is involved, it's often as simple as just verifying with your adjuster that everything is in order, and that you have the proper authorizations to go in and do the work.

As mentioned earlier, in the case of a crime scene, be sure you have permission from law enforcement to enter and begin the cleanup process. Demolition may need to be carried out under chain-of-custody rules.

REAL STORIES:
I Think It Was Just the Breeze!

While some people learn to take trauma-scene work in stride, for others these situations are too difficult, or even a little spooky. In most cases, I can get myself prepared and do my job without incident, but even I can get spooked. That happened to me on the job in the following story, which was, thank goodness, a very unusual situation.

One day we received a call from an adjuster we worked with on a regular basis. He had a job where a woman had been murdered, and the home needed to be cleaned and repaired. Because of the nature of the crime, and the fact that it had received a lot of press coverage, he wanted to get the site secured and repaired as soon as possible.

He explained that the job was strictly volunteer, and he'd understand if we turned it down. He also said that, while he would need an estimate, we could pretty much name our own price for both the clean up and the repairs. He warned us, however, to be emotionally prepared for what we'd see.

My partner ended up being the one to take the initial call, and he went to the home with a volunteer crew to do the initial cleaning and board-up. The police were still on the scene, and the officers had requested that we coordinate everything with them. There was some blood-stained carpet and a window with a bullet hole that they wanted removed under their supervision. They had to witness and verify the removal, then tag the items, to maintain a specific chain of custody for later use in court.

After that was done, and the initial cleaning, deodorizing, and boarding up of the home completed, work stopped for a few days while the police finished their investigation. They wanted to be certain there was nothing else in the way of evidence that they needed before they released the property to us for repairs.

We were given access to the property by the police about a week later. My partner was busy with something else, so I went out to do the estimate. I hadn't yet been there, but I had convinced myself that it was just another job, and doing the estimate was no big deal.

I let myself in, set my equipment down, and started doing my measurements like I would for any job. But it was hard to ignore the fingerprint powder on the counters, or the markings on the drywall where the police had tracked the bullet trajectories, or the stains remaining on the subfloor.

I was almost done when I heard a noise, and caught a slight movement out of the corner of my eye. It was a clear afternoon – nothing eerie or spooky about it — but my hair stood on end and my blood felt like it turned to ice. The movement was just a tree outside the living room window, swaying in a slight breeze, and making a little noise as it scraped lightly against the screen. But I'm not ashamed to say — that was all it took! I'd had enough of that house for one day, and I figured the best place for me right then was back inside my truck. If I'd missed any measurements, I'd just have to guesstimate when I got back to the office!

I had to go back to the house several times after that to check on the progress of the repairs. And, I was there again to inspect the finished product. I'm happy to say that I never had that freaky feeling again — there, or on any other job. Once was definitely enough!

Trauma-scene remediation is usually a two-part process. As with a water loss, you'll usually first do an emergency response to mitigate further damage. That's when you remove stained or odor-causing materials, and begin the application of basic deodorizing pretreatments, as outlined in Chapter 9.

After that, you'll prepare an estimate for the reconstruction, which will cover whatever restoration and repair work is needed, such as drywall repairs, flooring, painting, etc. If a second round of deodorization work is needed, including thermal fogging, it would be included in the estimate at that time as well.

Tip! Under no circumstances is it ever your job to deal with a dead body or body parts! *That's a job for trained medical or law enforcement personnel — never you or your crew. If that's what you're being asked to do, you need to politely direct the caller to hang up and dial 911.*

Problem Areas

Due to the nature of this type of work, it's best to not randomly assign crew members to these jobs. If you're planning on doing a lot of trauma work, you should dedicate a specifically-trained trauma crew to those assignments. If you'll only be doing occasional trauma-scene jobs, make up a crew of volunteers who understand exactly what the job entails. Any crew members working on trauma-scene cleanup will need complete personal protective equipment (PPE). And, it's an OSHA requirement that these employees receive training, at the employer's expense, in the safe handling of bloodborne pathogens. Training classes are offered through different organizations in different areas, so check with your local OSHA representative, hospital or the Red Cross for class information. You may want to pay your crew members a higher wage, or a bonus, as compensation for handling these jobs.

It's possible that you may at some time become involved in a loss that has some local or even national notoriety. Should that occur, *never release any information to the media unless you've been given permission to do so by properly authorized parties.* Never sell or otherwise release any photos or descriptions of the scene, and make sure that your crew is under strict instructions not to do so either. Not only is this unethical, it could put you in a very dangerous position from a liability standpoint.

In all trauma situations, it's important to be very sensitive to what the family members are going through. Talk with your crew about the need to be professional and also sympathetic. These jobs are hard on everyone, and though humor can be a necessary stress reliever, there's a time and a place for it. Make your crews aware that some humor can be in very poor taste and will reflect badly on your company.

When the job is complete, try and schedule some shared downtime for the crew. A company picnic, a night of bowling, a pizza party, or some other type of lively, group activity. It will help everyone stay bonded, and keep the experience in the right perspective.

Trauma-Scene Building Permits

Most trauma-scene restoration losses are cosmetic in nature, so it's fairly unusual to need a building permit. The exception would be where a crime, such as arson, was committed that resulted in a death or a severe injury. In that case, you'd obtain a repair permit for the structural fire or other damage to be repaired, in addition to the trauma-scene restoration work.

These are just a few of the many kinds of losses that will keep your life as a restoration contractor challenging and interesting. When you head for work each day, be prepared for ... anything!

12

Mobile, Manufactured & Modular Homes

In 1950 there were around 300,000 mobile and manufactured homes in the United States. By the year 2000, the U.S. Census Bureau estimated there were nearly 8.8 million — an enormous increase during the last five decades of the 20th century. Today, you'll find manufactured homes in just about every community, and they represent another aspect of insurance restoration work that you're almost certain to become involved with.

If you're like most contractors who've never worked on a manufactured home, you probably have a lot of questions: Are they hard to work on? How do they differ from stick-built homes? Do they take special parts? Are they really as poorly built as some people say? In this chapter, we'll try and clear up some of that mystery.

A Brief History

As the U.S. highway system developed, small travel trailers first appeared. Light enough to be towed by a car, they averaged 8 feet in width, and were used primarily for camping. Their popularity grew after World War II, and a lot of them were used as temporary housing. Around the mid-1950s, the first 10-foot-wide trailers were introduced, and the term *mobile home* came into use. Larger, 12- and 14-foot-wide units soon followed.

All of these mobile homes were contained within a single unit (see Figure 12-1). But highway laws and practical considerations limited how wide a unit could be in order to be transported on the road, which also limited the size of the home. To meet the demand for bigger homes, manufacturers designed mobile homes that could be transported in two sections and then joined together. These became known as *double-wides* (Figure 12-2), and the original single-unit size, called *single-wides*.

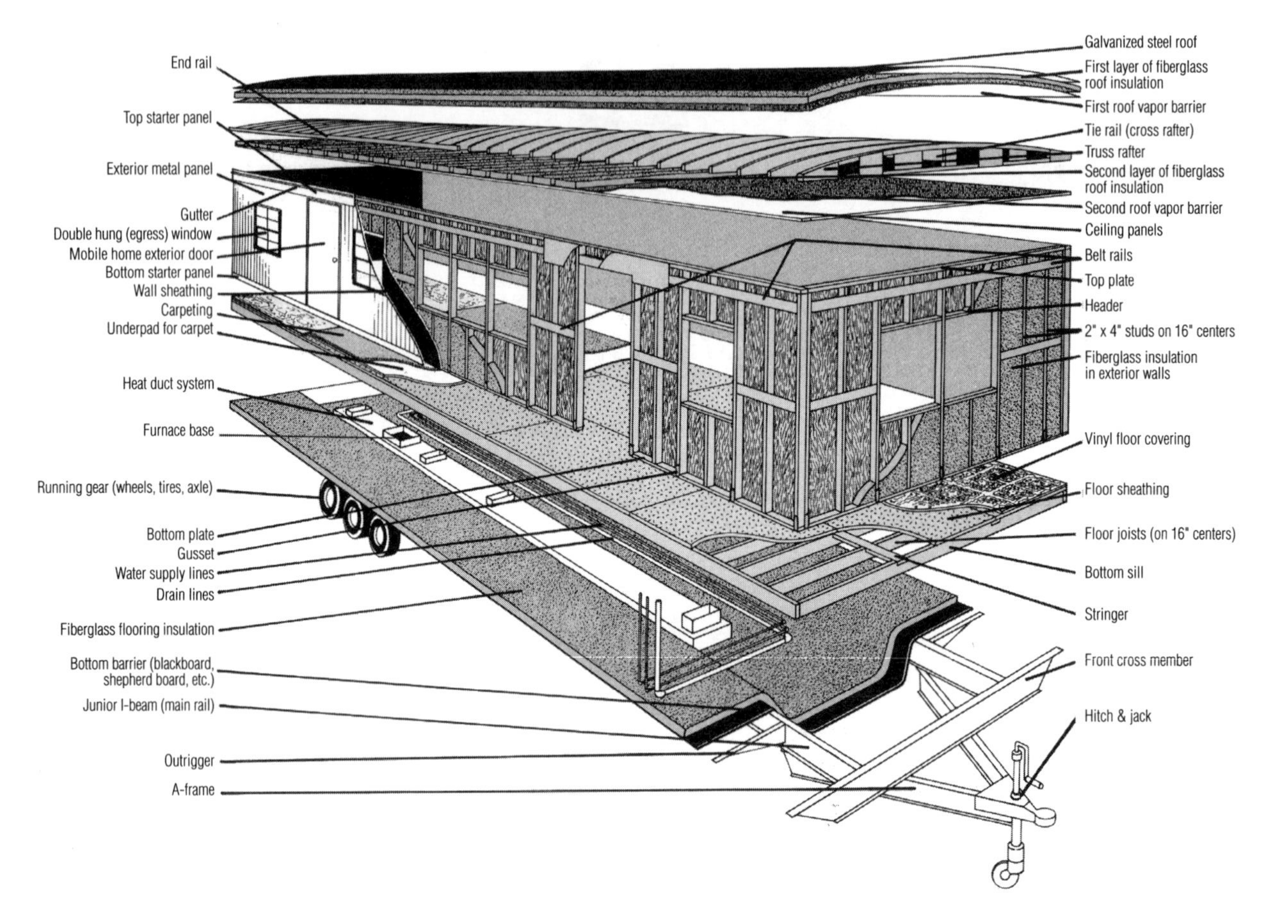

(Illustration courtesy of Foremost Insurance Group)

Figure 12-1 A typical single-unit, completely self-contained manufactured home. This home has the plumbing, heating, electrical wiring, and all other components installed at the factory. It's then towed to the site and connected to water, power, and other utilities. This type of manufactured home is most commonly known as a *single-wide*.

Figure 12-2 An older *double-wide* manufactured home. Two individual half units are towed to the site and then joined together to make a single, larger home. The point where the two halves join, the "marriage line," is covered by a vertical trim board where the siding meets. The underside of this home is enclosed with horizontal metal skirting, which can be seen below the lap siding.

The U.S. Department of Housing and Urban Development (HUD) began to regulate the construction of mobile/manufactured homes in 1976. This was done under the provisions of the National Manufactured Housing Construction and Safety Standards Act (NMHCSS). In 1994, HUD amended the Manufactured Housing Construction and Safety Standards to provide greater protection against wind damage to mobile/manufactured homes.

Understanding the Terms

You may be a little confused by the use of the terms mobile, manufactured and modular when discussing these homes. While there are many similarities between the homes described by those terms, there are distinct and important differences as well.

Mobile and Manufactured Homes

Mobile and manufactured homes are constructed inside a factory, typically under environmentally-controlled conditions. They're built on a permanent steel chassis, with removable axles, wheels, and tongue that allow them to be towed. The homes are built in accordance with HUD standards, so they aren't subject to the building codes that govern site-built homes. Instead of being checked by building inspectors, they're checked by third-party inspectors at the plants where they're assembled; when certified, they receive a HUD Compliance Label.

The home is then towed to the site and set up, usually on concrete blocks, called *piers*, and leveled (see Figure 12-3). Once the home is connected to utilities, the underside of the unit is enclosed with wood or metal *skirting* panels, which provide weather and animal protection, and improves the overall appearance. If a more permanent installation is desired, the home can be set on a foundation of concrete blocks instead, using concrete block piers for the intermediate supports.

With either type of installation, the wheels, axles, and towing tongue are removed. They're either stored under the home or, more commonly, returned to the manufacturer for reuse.

Modular Homes

A modular home isn't built on a chassis and towed to a home site or park. It's built in sections at a factory, then trucked to the site for assembly. The sections are lifted into place using a crane, and assembled on a permanent foundation to make a complete house, very similar to a stick-built home. Modular homes, like stick-built homes, are constructed and inspected in accordance with state and local building codes.

Mobile vs. Manufactured — What's in a Name?

So what's the difference between a mobile home and a manufactured home? Does anyone care? Actually, many people do care. You need to know the difference because some of your clients will be very offended if you use the wrong term.

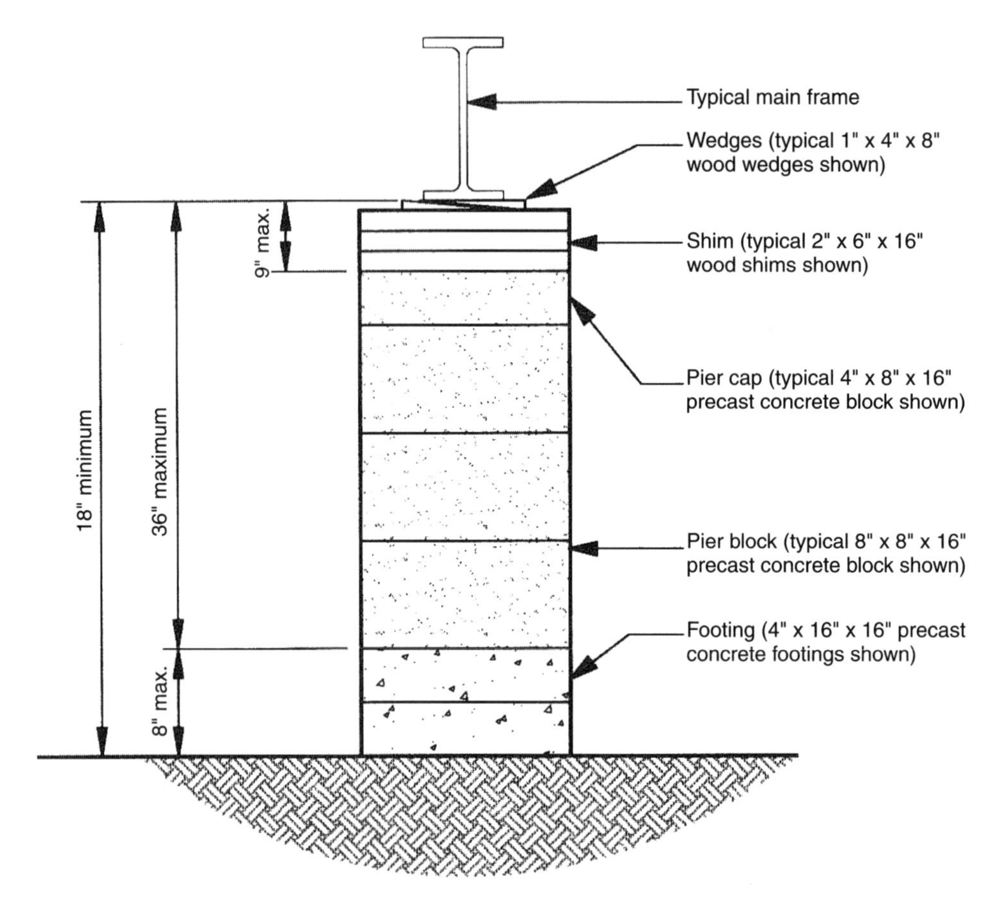

Figure 12-3 Stacks of concrete blocks are typically used to support the home once it's set up on site. Individual states or other jurisdictions often have specific setup requirements, including clearances as shown here.

Originally, mobile homes were made to "trail" along behind a car, and were called *trailers*. You'll occasionally hear mobile homes called *trailer homes*, and planned communities of mobile or manufactured homes referred to as *trailer parks*. But those terms are rarely used today when describing these homes.

The term *mobile home* is still commonly used, but the preferred name is *manufactured home*. Why? Because prior to HUD's implementation of construction standards in 1976, a lot of mobile homes were constructed using shoddy materials and workmanship. Those homes acquired a well-deserved reputation for being, to put it kindly, junk.

Unfortunately, even after the standards went into place, a lot of the homes that were produced could still be described in those terms. Though they complied with the new federal standards, they still weren't very well built. Affordability was the key selling feature, and that often meant lower grade materials and the use of unskilled or semi-skilled labor. The final HUD-inspected product wasn't held to the same standards found in building code-inspected homes.

Eventually, the manufactured home industry evolved and better homes were introduced. Common stick-built materials such as wood siding, composition shingles, and drywall replaced the old metal siding, membrane roofing, and prefinished paneling. To set these newer homes apart from the older, less desirable homes, the industry began using the term *manufactured homes.*

> ***Tip!*** *When speaking with your clients, never call their home a* trailer, *or its location a* trailer park. *That's a good way to offend them. Stick with* manufactured home, *and* mobile-home park.

Within the manufactured home industry, the 1976 enactment of the HUD standards is generally accepted as the transition point from the term *mobile* to *manufactured* home. However, there doesn't appear to be any specific legal definition to go by, so quite honestly, the terminology appears to come down to an issue of marketing. There's even confusion among building officials, lenders, HUD, and many others, so you're likely to hear both terms used interchangeably. For the sake of clarity, we'll just use the term *manufactured home.*

Manufactured Home Construction

If you're going to work on manufactured homes, it's helpful to understand the basics of how they're built. There are some substantial differences between manufactured homes and stick-built homes that affect everything from drying to estimating and repairs.

Construction begins with a steel chassis, which is essentially a pair of I-beams that run the length of the home, and several cross members. They're welded into a rectangle that's less than the overall floor area of the unit. Metal pieces called *outriggers* extend out from the beams to support the additional width. You can see the outriggers on the double-wide unit in Figure 12-4.

Figure 12-4 Part of the chassis under one half of a double-wide manufactured home. A pair of steel I-beams run most of the length of each half of the home. Triangular outriggers extend from the I-beam to the outer edges. You can see the road barrier fabric above the I-beam and outriggers. It's stapled at the marriage line (right) where the halves meet.

A tough mesh material called a *road barrier* is installed on top of the chassis (also shown in Figure 12-4). This is typically a breathable, black fabric, installed in large sheets, that protects the underside of the home both in transit and after setup.

Wood floor joists are installed on the chassis, parallel to the short dimension of the home. Aluminum heating ducts are placed below the joists, typically running in a long straight line from one end of the home to the other.

Plumbing lines, both waste and water, run between the joists. Plumbing fixtures are often kept in as much of a straight line as possible to minimize plumbing runs. Finally, electrical wiring is installed between the joists, then the joists are insulated with batt insulation. Subfloor, typically structural particleboard, is installed using both adhesive and either air-driven nails or staples.

Vinyl sheet flooring is installed on the subfloor for the kitchen, bathrooms, laundry room, and entry. This is done before the walls are up, so that the sheets can be installed quickly, and without cutting. *It's important to remember this installation process, because it can greatly affect how you dry a wet manufactured home.*

Next, the walls are framed. Some homes utilize standard 2 x 4 framing techniques (16 inches on center), while some have 24-inch on-center framing. In some older homes you may find 2 x 3 or even 2 x 2 lumber for interior walls. Some older homes were built by framing the walls and then paneling them *before* standing and fastening them in place. That means the paneling extends behind the wall intersections, which can complicate how the panels are disassembled and dried.

Another huge difference is in the roof assembly. Manufactured homes typically have very low pitched roofs, often 3/12 or even less, and they don't have attic spaces. The trusses are set on the walls, with the wiring installed as needed. The cavities are stuffed with insulation, and then the ceilings are covered. So, in the event of a leak, you have no access into the attic cavities to see what's going on.

Manufactured homes, especially older ones, are also notoriously short on attic ventilation. What little ventilation there is can easily be obstructed and there's no way to know about it. The result can be mold and structural problems that make drying and repair more difficult.

Double- and Triple-Wide Homes

Double-wide manufactured homes are especially popular today. Basically, double-wides are made up of two matching units that are about 95 percent completed at the factory. Each half has two short exterior end walls and one long exterior side wall. The fourth wall is a long open interior wall that mates with the other half (see Figure 12-5). The two open sides are covered with protective plastic sheeting during transport.

Once on site, the two halves are towed into position as close to one another as possible, then jacks are used to complete the mating process. The seam where the two halves join, located directly below the ridge line of the roof, is called the *marriage line* (Figure 12-6).

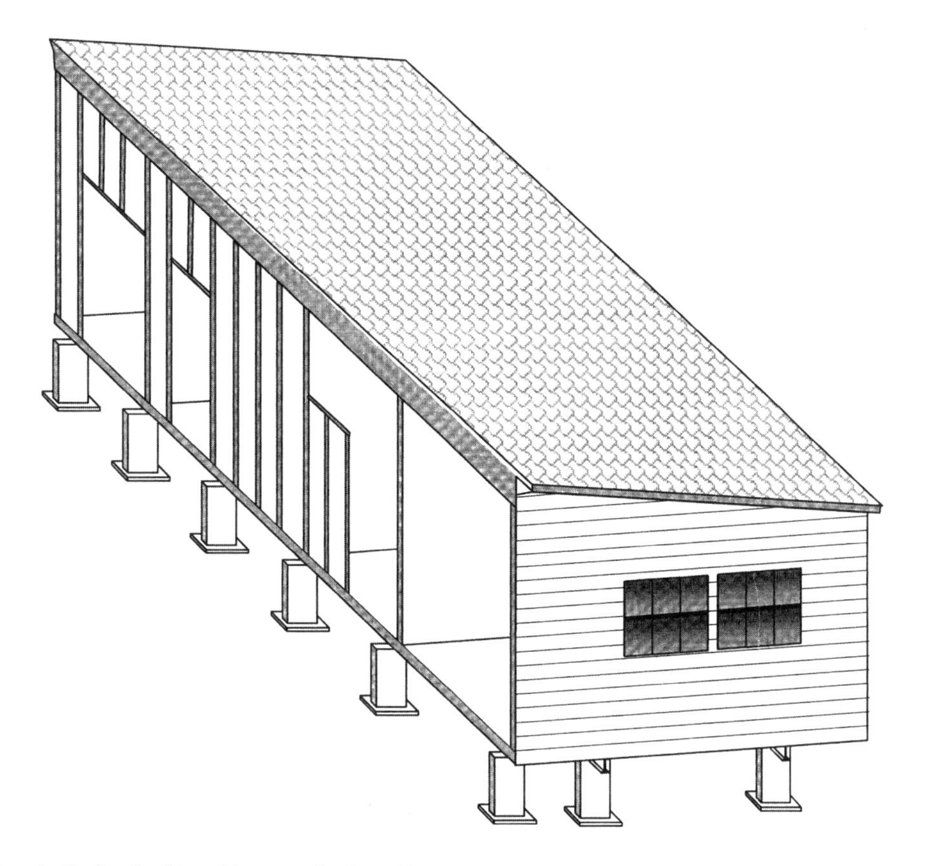

Figure 12-5 One half of a double-wide manufactured home

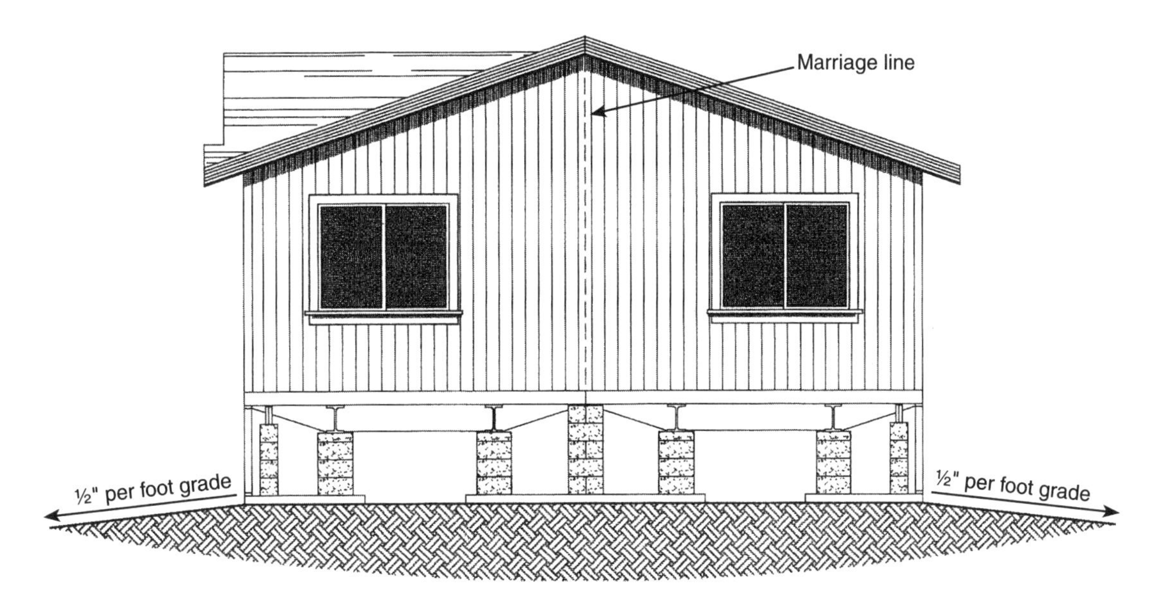

Figure 12-6 The two halves of a double-wide after connection. One row of piers is set under each of the two I-beams, and one row is set to split the marriage line. A smaller row is used under the outer supports. Wood framing supports the skirting sheets, but doesn't support the manufactured home itself. With some types of setups, decorative concrete blocks may form both the outside supports and the skirting.

From below, the two chassis halves are bolted together. Inside, the beam that makes up the roof ridge on each side is also bolted together. Trim boards or short pieces of siding conceal the joint between the two halves on the outside, and a ridge cap finishes off the joint over the marriage line at the roof.

Trim pieces cover the marriage line joint inside. If the home has a drywall interior, then the marriage line is drywalled, taped, textured and painted. Carpet covers the floor joint between the two halves.

With a triple-wide manufactured home, there are two outside units that are similar to those of a double-wide, as well as a third, center unit. The third unit has exterior walls on the two short ends only, and the two long sides are both open. The three sections are assembled in the same way as the double-wide, but with two marriage lines instead of one.

Crossover Connections

Double- and triple-wide manufactured homes are built in self-contained sections. Once they're on site, all of the utilities need to be connected to each other, allowing the sections to operate together as one home. This is done with *crossover connections*.

Double-wide manufactured homes are often designed to keep all the electrical, mechanical, and other utility equipment in one half. This simplifies both the initial construction, and the on-site connections. For simplicity, in the following examples we'll call the two halves A and B.

The home is designed so that all the utilities — water, sewer, electrical hookups, phone hookups, cable TV, and gas lines, terminate underneath the home in the same spot. That simplifies connections to the utilities on the site, especially if the home is being set up in a mobile home park. We'll assume the side with the utility connections is side A.

Electrical Crossovers

The electrical service panel is usually in the utility room, and electrical circuits for side A will be completely wired at the factory. On side B, each circuit is completely wired as well. But instead of terminating in the electrical panel, they end in one or more junction boxes attached to the chassis under the B half. Matching circuits extend from the service panel to another junction box or boxes under the A half. All of the wires in the junction boxes are labeled at the factory.

When the two halves are connected on-site, electrical flex cables are connected between the junction boxes for the two halves (Figure 12-7), and the matching wires are connected. The circuit breakers in the service panel will then be able to activate the circuits in the B half of the home.

Plumbing Crossovers

If all of the plumbing is contained in the A half of the home, then there won't be a plumbing crossover. Instead, there'll simply be a water line and a waste line connection exiting from the underside of the A half. This is usually underneath the

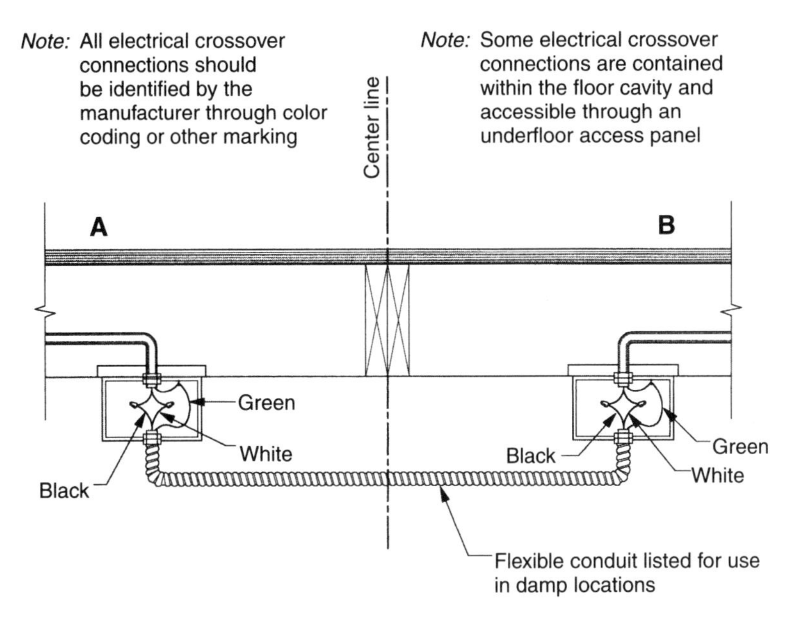

Figure 12-7 The electrical connections between the two halves of a double-wide.

utility room, or under one of the bathrooms, often near the main electrical connection conduit.

If both halves of the home have plumbing, then there'll be plumbing crossovers as well. Waste lines will slope from the B half toward the main connection point under the A half, in order to maintain the correct drainage. Manufactured homes are typically plumbed with ABS pipe for the drains, although some older models also used PVC. Either way, the waste line crossover is connected using a rubber sleeve (Figure 12-8) that forms a tight, but slightly flexible, seal between the pipes.

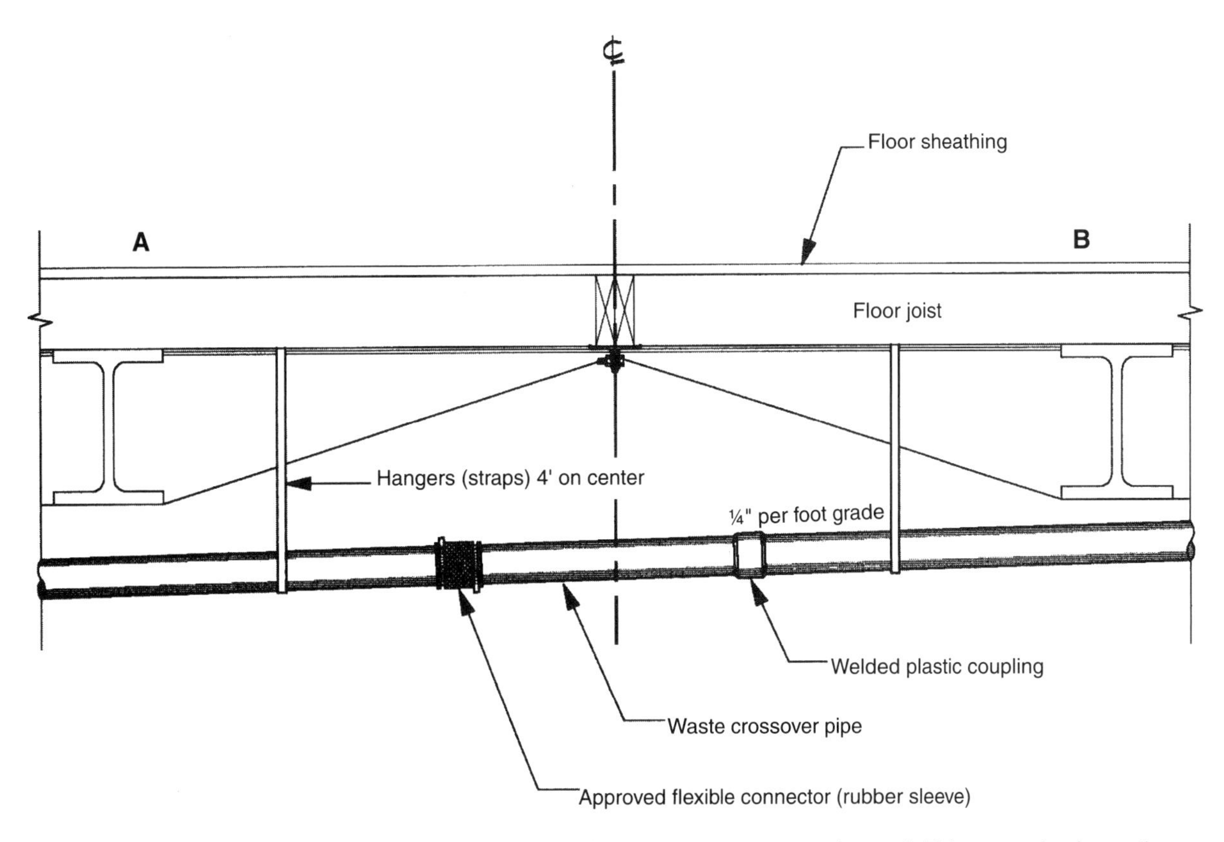

Figure 12-8 In a double-wide with waste lines in both halves, a waste crossover is used. This example shows the flexible connector and the strapping used to maintain the necessary grade.

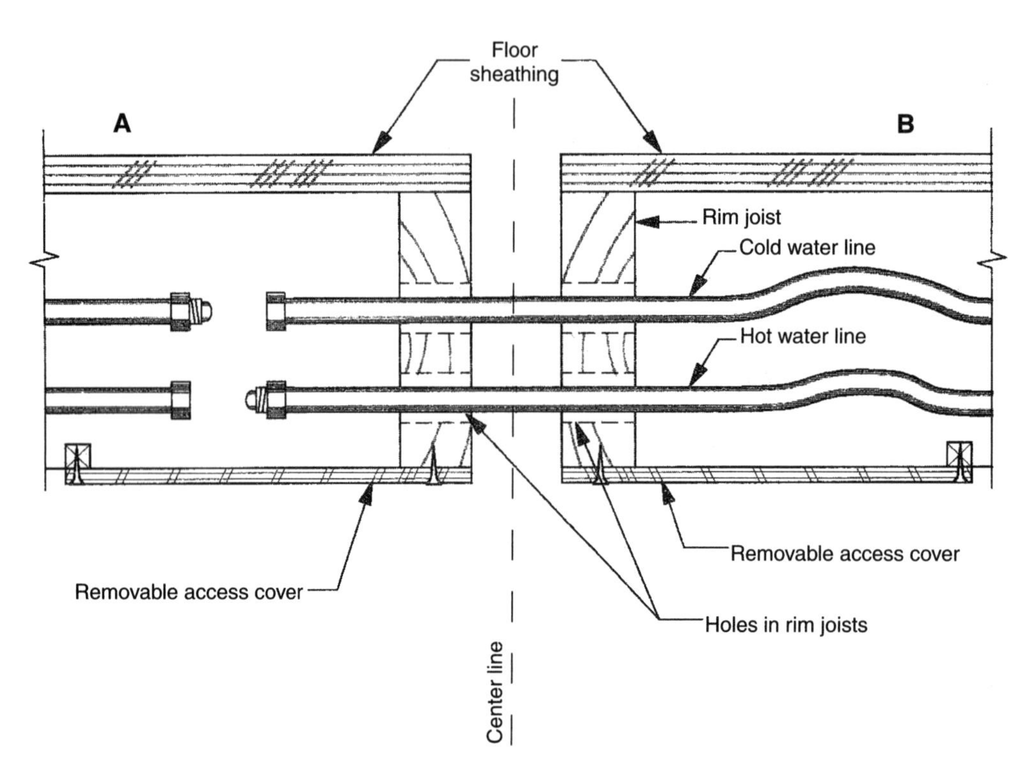

Figure 12-9 An example of plumbing crossovers for water lines. Note the access cover, which is necessary since the connections are buried up inside the floor insulation.

> ***Tip!*** *Water line crossovers are vulnerable to freezing, so be aware of that when looking for potential leak locations. Also, when doing repairs, make sure that these areas are properly insulated again when you reassemble them.*

Most manufactured homes are plumbed with plastic water lines. This makes alignment easier between the two halves, and also allows for any slight movement between the marriage lines. The crossover connections may be union fittings, or they may be metal flex connectors with threaded ends.

All of the plumbing for a manufactured home is located up inside the insulated cavities, and the water line crossover connections may be there too. If so, they'll be accessed by removing a plywood panel under the home (Figure 12-9). Otherwise they'll be left exposed under the home and separately insulated with foam sleeves, fiberglass batts, or other material.

If the home has factory-installed propane or natural gas piping that extends between the units, there'll be a crossover for that as well. These are usually standard flex pipes approved for gas line connections in exterior conditions. See Figure 12-10.

Polybutylene Pipe

Between 1978 and 1995, many manufactured homes were plumbed with polybutylene (PB) pipe, shown in Figure 12-11. PB pipe is a flexible plastic, usually gray but sometimes silver or black. It's connected with plastic or metal fittings, usually joined with metal crimp rings.

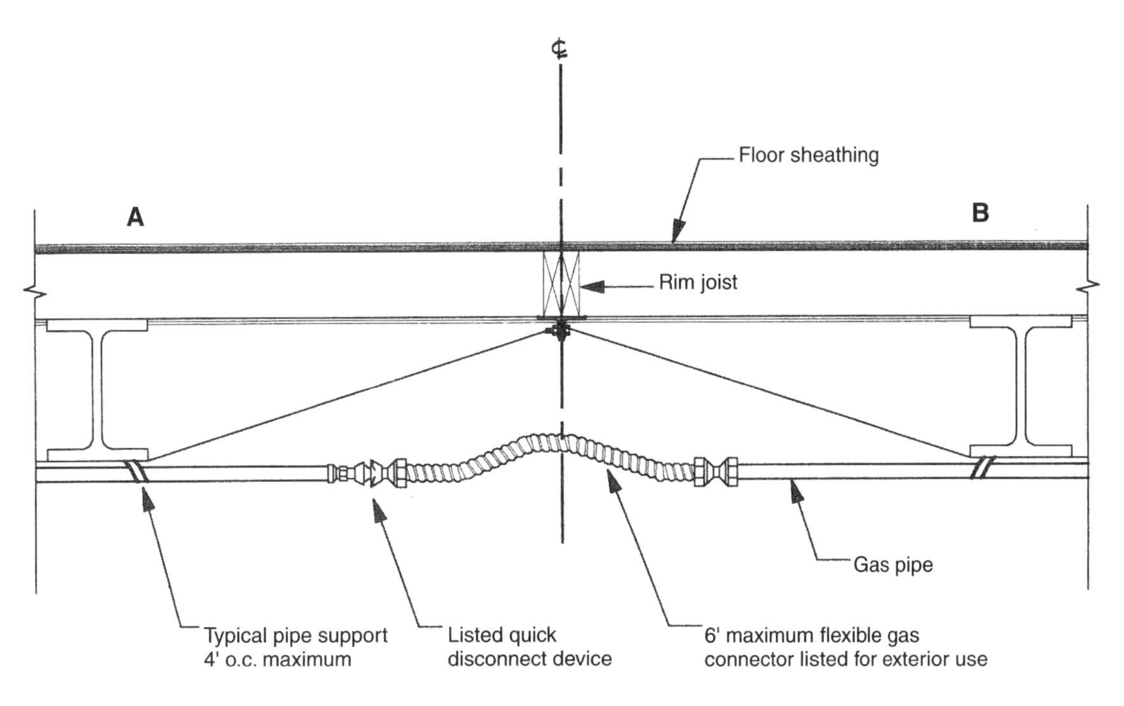

Figure 12-10 A flexible crossover connection for a gas line.

Figure 12-11 Polybutylene (PB) pipe connected with plastic elbow fittings secured in place with crimp rings.

There were a number of leak issues associated with this type of pipe and its fittings. So many, in fact, that there was a massive class action lawsuit brought against the pipe manufacturer. That suit is now closed, and most of the affected homes have been repaired. But it's still possible that you may encounter a manufactured home built within that time frame that has PB pipe or fitting issues. If you do, it's best to advise the homeowners to discuss the situation with their insurance company and their attorney before you take on any repairs.

Ducts

While there have been a lot of improvements in the design and construction of manufactured homes over the years, their duct systems isn't one of them.

In a stick-built home, a furnace dumps heated air into a plenum, where it's distributed to a series of ducts of different sizes that feed individual spaces. A return duct placed in a central location brings warm air back to the furnace for reheating.

In a manufactured home, the furnace and plenum feed the air into a single rectangular duct that runs the entire length of the home in one straight shot. It doesn't change size, and it typically doesn't change direction. (Look back to Figure 12-1). Wherever heat is needed in a room, a hole is cut in the top of the duct and a register is added. If an additional register is needed in an area that doesn't fall directly over the main duct, a lateral branch is installed, but for economy, these are kept to a minimum.

As a rule, no actual return air duct is used. Instead, a grill is cut into the wall directly above the furnace, and that's the point where room air is drawn back into the furnace for reheating. This is also where the furnace filter is usually located.

Crossover Ducts

With a double-wide, a second long, straight duct is installed in unit B, and a crossover duct is installed between the two. This is a large-diameter flex duct that simply hangs beneath the home. See Figure 12-12.

Manufactured home duct systems are very difficult to work on if they get wet or damaged. All of the ductwork is concealed in the floor, and you have no easy access to it. We'll take a closer look at ductwork and how to deal with repairs later in the chapter.

Manufactured Home Loss Repairs

Manufactured homes have fires, water losses, storm damage, and other types of problems, just like stick-built homes. But the differences in how they're built affects how you estimate, select and order materials, and make repairs.

You shouldn't shy away from working on manufactured homes, but in order to do them correctly — and profitably — there are some potential problem areas worth knowing about. So those are what we'll look at in the rest of the chapter.

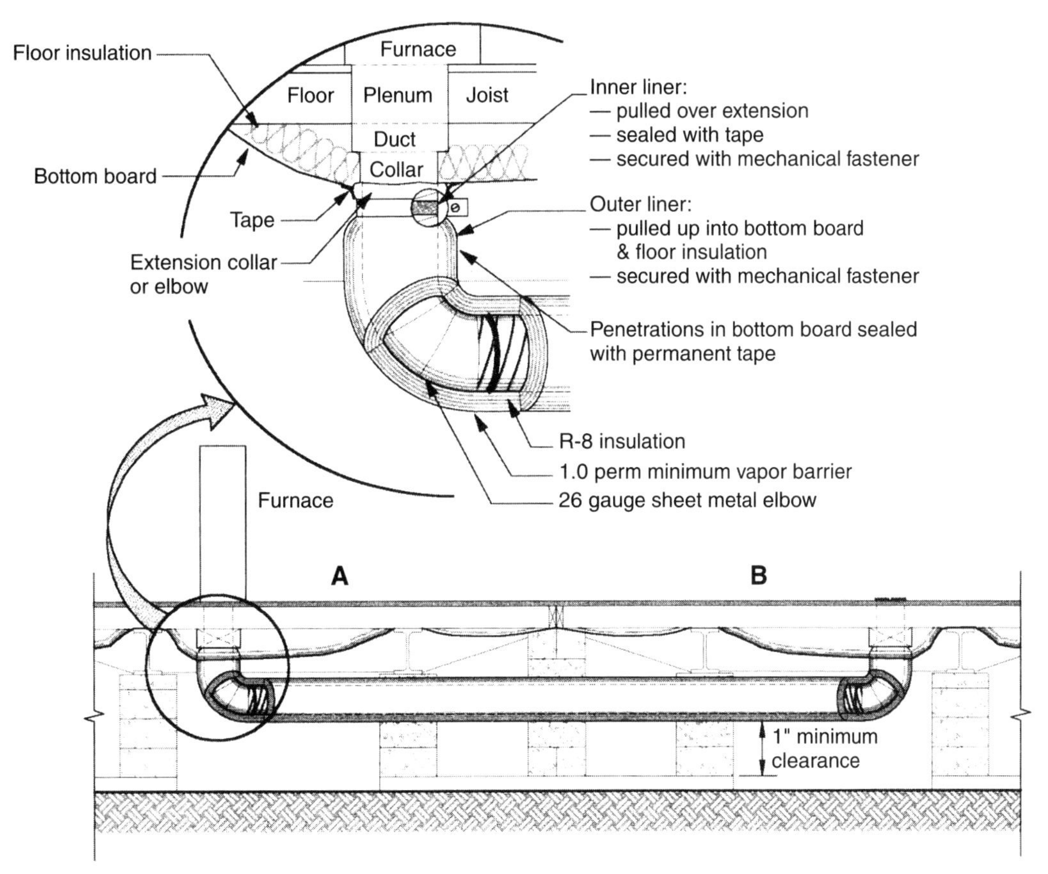

Figure 12-12 A crossover duct connecting the ducts on two halves of a double-wide. This drawing indicates that a minimum 1-inch clearance is required between the duct and the ground. In actual practice, however, the duct is often laid directly on the ground, which can lead to heat loss and moisture problems.

Trusses

As a builder, there are probably few things that will surprise you more than your first glimpse of an older manufactured home roof truss. Questions will immediately pop up, such as: "How can that support a roof? How can they get away with that?" Or, to put it more bluntly, "Are you kidding me?"

The typical roof truss on a conventional home is made of 2 x 4 and 2 x 6 lumber, held together with pressed-steel gang nail plates applied under tremendous hydraulic pressure. They're designed for use with ½-inch plywood or OSB sheathing, and will support several different types of roofing materials.

Many manufactured home trusses — especially older ones — are made of 2 x 3 or 2 x 2 lumber, or sometimes even smaller. Some are made from strips of ripped plywood. The joints are secured with pieces of scrap ¼-inch or even ⅛-inch wall paneling, held in place with a few air-driven staples. Roof sheathing may be ⅜-inch plywood on 24-inch centers.

I know you want to, so go ahead and say it: "Are you kidding me? Why is this allowed?" Because in a manufactured home, the trusses are part of the overall home

REAL STORIES:

Don't Fight City Hall!

We live in a snow area, and the snows can get heavy at times. During one particularly bad winter, we were shocked to discover first hand how many manufactured home roof trusses were frighteningly undersized for the snow loads in our area. Worse yet, when the snows cleared and it came time to make repairs, we soon discovered that our local building department wasn't prepared to deal with issuing the necessary building permits.

The inspectors had never seen trusses like these. HUD had said they were okay, but there was nothing in the building codes to cover them. The local building department had no experience with them, and they refused to assume the liability of inspecting our repairs and telling us the homes were safe to reoccupy.

There were several of these homes we needed to repair that winter, and we knew this permit problem would be on ongoing issue. So my partner sat down with our structural engineer, and together they devised a simple and effective truss repair system. The engineer ran the calculations, then they came up with a design and a drawing that he could easily alter to fit the specifics of different manufactured homes under different conditions as we encountered them.

The ideas were presented to the head of the building department, who was delighted. The homes obviously had to be repaired, and this gave the building department a way to issue the necessary permits while also giving the inspectors a clear set of guidelines for what to inspect. For us, it simplified and sped up the permit process, which also made our clients happy. For the insurance companies, the design process became more uniform, so it cut down on their engineering costs.

This is a perfect example of how to work with City Hall instead of fighting them — and everyone comes out a winner!

design package. That means every element is very carefully calculated into the design, right down to the weight of the specific shingles being used. As long as they're built and installed according to that overall design package, they're considered okay.

Today's manufactured homes are better, and the trusses are more in line with the type used on stick-built homes. But remember that there were almost 9 million mobile and manufactured homes as of 2000, and a lot of those have very weak roofs!

Truss Deflection

Once a manufactured home leaves the factory, it could end up just about anywhere. There are placement regulations regarding snow loads, tie-downs, and other issues that vary by state and even by county, but they often don't offer much assurance that a home with a roof built for a state's coastal or prairie climate won't actually end up in the mountains. It may be in the same state, but in an area with vastly different snow-load conditions.

So, a manufactured home placed in a snow area can be a bit of a ticking time bomb, with a roof that has a combination of possible problems:

- Trusses that are undersized for the potential weight of the snow that could accumulate.

- A low-pitched roof, often no steeper than 3 in 12, that, combined with composition shingles, creates a situation where there's very little chance of any snow sliding off.
- Very little roof ventilation, with often only moderate levels of insulation. As we saw back in Chapter 10, those are perfect conditions for ice damming, adding a layer of ice and that much more weight to the roof.

What can happen in situations like these is a slow downward deflection of the roof trusses, rather than a sudden catastrophic failure. The truss members start to sag due to the weight, and the joints start to separate. Visible ceiling sags, ceiling cracks, and loose moldings and trim are typically the first indications that the roof trusses are starting to deflect.

How pronounced the deflection is will be determined in large part by how many intermediate walls the home has. In a single-wide, the sagging is usually most noticeable in the living or dining room, where the trusses are only supported on the outside walls and the overall span is the greatest. In a double-wide, the same is often true in any of the open rooms, where the trusses span all the way from an exterior wall to the marriage line.

Since there's no attic space in most manufactured homes, you have to check the extent of the deflection from below. Use a long level, a laser level, strings, or other methods to check the amount of deflection over several different areas. Carefully document your findings for the insurance company.

Tip! *You can relieve some weight by removing snow with a snow rake, which is a lightweight plastic or aluminum rake head on a long handle. Working from the ground, rake as much of the snow off the roof as possible. Under no circumstances should you go up on the roof and shovel the show. The additional concentrated weight could cause more damage or even collapse!*

Push against the ceiling with your hands, in the location of the sagging trusses. If the ceiling feels solid, then the entire truss is probably deflecting down as a unit. If the ceiling has some give to it, it's usually an indication that the bottom chord of the truss has broken or separated, which is a much more dangerous situation.

You need to use your best judgment about how safe the home is. The deflection may be fairly minimal, or the roof could be sagging substantially. You might need to construct a few temporary walls to support the worst of the sagging areas, or you may feel the home shouldn't be occupied until repairs are made.

Inspecting and Assessing Truss Damage

To inspect and evaluate the actual truss damage, you'll most likely have to cut a hole in the ceiling. You may be able to do this in a closet, where you can create a temporary panel that can be screwed in place to cover the hole. However, attic space in a typical manufactured home is low and sight lines are limited, so sometimes the only option is to cut your hole in the middle of a room, close to the deflection area.

Lay plastic sheeting out on the floor, and then slowly and carefully cut the hole with a drywall saw. *Don't use a reciprocating saw.* You don't have enough "feel" with that type of saw, and you'll run the risk of cutting wires. Cut a hole about 2 feet long extending between at least two of the deflected trusses. That should give you sufficient access for an initial view of the inside of the attic.

> ***Tip!*** *The cost of the time to do the cutting and the inspection, plus the engineer's time if you're using one, are all billed separately to the insurance company.*

What you'll find is that the truss deflection will be the result of one of four things:

1. The entire truss has deflected as a unit;
2. The truss joints are loose;
3. The truss joints have separated;
4. The truss components (chords or webs) are broken.

Deflected Trusses

Sometimes the entire truss will sag as a unit. In this case, all of the components of the truss — top chord, bottom chord, and intermediate members — will move as one. The joints will still be intact, and none of the structural members will have broken.

Loose Joints

Sometimes the weight of the snow will cause the truss to deflect downward, then it will come back up again as the snow melts off. This can happen once or twice, or hundreds of times over many seasons. Each time the truss deflects down and comes back up, the poorly-supported joints loosen. You may encounter a truss where the top chord has returned to its original position, but the bottom chord is sagging down because the joints between the chords and the intermediate web members are loose (Figure 12-13).

Figure 12-13 These manufactured home trusses are made from a combination of materials. The top and bottoms chords are 1- x 1½-inch laminated plywood. The intermediate web members are 2 x 2 and 2 x 4 solid lumber. The pieces are joined by stapling through the chords into the web pieces, with no gussets or other connections. On this truss, the vertical web member is pulling off the staples. There's similar damage on the truss behind it.

Some homes may have joints that have loosened due to moisture, either from a roof leak or from inadequate attic ventilation. This is most commonly

Figure 12-14 The gusset plates on this truss were simply pieces of scrap paneling. A roof leak caused the paneling to delaminate, allowing the joint to fail.

Figure 12-15 This joint is similar to the one in Figure 12-13, except here the web member has pulled completely off the staples.

seen in older trusses, where the joints are secured with scraps of paneling or other pieces of wood (see Figure 12-14).

Separated Joints

Separated joints are similar to loose joints, and are caused by the same thing, except in this case the joints have come apart completely. Sometimes it's the result of the gusset plates coming off; or more commonly, there were no gusset plates there in the first place. Instead, the joints were pinned together with nails or staples shot through the chord of the truss and into the end grain of the intermediate member. When the truss is deflected, the members pop off the staples (Figure 12-15), leaving the joint completely unattached. Because there's nothing holding the joint together, this situation is more dangerous than a loose joint.

Broken Truss Components

The worst case scenario is broken trusses. This can be caused by a very heavy snow and ice buildup that accumulated relatively quickly, or it could be from an impact such as falling ice or a tree limb. Broken trusses require immediate reinforcement from below, and the occupants should leave the home.

Releveling, Regusseting, and Reloading Trusses

Repairing the trusses is done through a process known as releveling, regusseting, and reloading. This process, which we'll cover in more detail shortly, consists of building temporary supports to take the load off the trusses; jacking up the trusses so they're level; installing new gussets to stabilize the joints; then removing the supports so the trusses carry the roof load on their own once again.

Tip! *To make these repairs effectively, you can't have any load remaining on the roof, so work needs to wait until after all the snow has melted off. If necessary, construct temporary interior supports to prevent additional deflection or damage until the work can be completed.*

It's a somewhat time-consuming process, and the actual repairs have to be designed and certified by an engineer in order to obtain the necessary building permits. But the procedure is pretty much the same for most manufactured home

Deciding How Many Trusses to Repair

Your engineer first needs to determine how many trusses are damaged and need to be regusseted. Doing *all* of them makes the most sense, since that'll greatly reinforce the roof structure, and ensure that snow won't be an issue in the future. It's also the ideal time to make these repairs, since the home will be torn up anyway.

However, the insurance company may not want to pay to repair something that isn't actually broken yet. They would see reinforcing a truss in order to prevent possible future snow damage as *preventative maintenance*, and not something they need to pay for.

Keep in mind when planning your repairs, that manufactured homes, especially their roof systems, are very carefully designed units. When you alter that design, even by reinforcing part of it, you may be altering the way that the rest of the home's structure interacts. Even regusseting one entire half of a double-wide may stiffen that half to the point where it increases lateral pressure on the other half.

This is something you and the engineer will have to carefully consider. It's often best that all the trusses be regusseted in order to safely repair the roof. If that's the only way the engineer will certify the repair, then the insurance company will usually be okay with that.

However, if the insurance company won't cover it, the homeowner might want to pick up the cost of reinforcing the remaining trusses in order to avoid future damage and inconvenience, especially if you explain the importance of the process to them. There will never be a better time to do the additional work than while the home is already under repair.

Repairing the Trusses

Cover all of the walls with 6-mil plastic for protection. Remove ceiling moldings, take down ceiling lights, and then strip the ceiling. Remove and discard the plastic vapor barrier, then remove all of the batt insulation from the ceiling and store it for reuse.

To make repairs, you first need to lift the trusses back into their correct, level position. Build temporary 2 x 4 walls perpendicular to the trusses, then, using shims between the top of the wall and the underside of the trusses, move the trusses into position. Or, you can use temporary walls or beams fitted with screw jacks, as shown in Figure 12-16, to level the trusses. It doesn't take a lot of force to move the trusses.

Figure 12-16 A post and beam arrangement with screw jacks is used to lift manufactured home trusses back into position for repair.

You can stretch string lines between the wall plates alongside each truss to serve as temporary guides as you lift. However, a laser level set up in the center of the room is your best guide for the final leveling.

Once the trusses are where you want them, hold them in place with new gussets. *This is the most important part of the engineer's calculations.* The difficulty in doing this is that the material used in building the original trusses is often so small you can't just attach new gusset plates over the joints. Instead, the entire side of the truss needs to be covered with a gusset of plywood (see Figure 12-17). That's the only way you can achieve a sufficient connection between all of the truss members. For very low-pitch trusses like the one in the figure, it's sometimes necessary to gusset both sides of the truss.

The engineer will specify the thickness and grade of plywood to be used for the gussets, and whether it should be installed on one side or both sides. He'll also determine the size and spacing of the fasteners to be used, and specify whether adhesive is required as well.

Once the trusses are covered with plywood, there'll no longer be any way for air to circulate through the attic. To provide some air circulation, you'll need to drill ventilation holes in the plywood, above the level of the insulation, as shown in Figure 12-18. These holes will need to be in every gusset to allow for cross ventilation. The engineer will specify the size, number, and location of the holes. Be sure to drill these holes *prior* to installing the gussets. Once all of the trusses have been gusseted, remove the temporary supports.

Replace the insulation after the building department completes their inspections. Very low-pitched attics may require some type of baffles to prevent the insulation from blocking the ventilation holes. Install a new vapor barrier, then new ceiling material and trim. Finally, reinstall the cabinets, lights, and other items. Then you'll be ready for the final cleanup and final inspections.

Truss Repairs From Above

Another possibility is to make the truss repairs from above. That eliminates the need to remove the ceiling and cuts down on the interior mess, but creates a number of other potential problems. Just be aware that it's a possibility. Some contractors have done it this way, so it might be something worth discussing with your engineer.

Figure 12-17 A manufactured home's trusses are gusseted with plywood as part of an engineered repair. The engineer called for plywood on both sides of these trusses, which is being attached with both adhesive and staples. Due to the curve in the roof, a template was made for cutting the plywood. Notice the wooden block installed at the seam between the plywood sheets.

Figure 12-18 Drill ventilation holes in the gusseting to allow a continuous flow of air across the attic space after repairs are complete. Here, a template was used to locate the holes, and the drilling was done before the plywood sheets were installed. Because this attic is so shallow, baffles were also installed to prevent the insulation from covering the holes. Notice the old pieces of paneling still in place on the trusses in the back that haven't been repaired yet.

Figure 12-19 Older-style ceiling panels showing the wide-crown roofing staples used to hold them in place. Splines were used to cover the joints.

To make repairs from above, remove all of the roofing and roof sheathing so you can access the trusses. Remove the insulation and store it for reuse later. Jack the trusses up from inside the house and level them. Install the gussets from above. Replace the insulation; then install new sheathing and roofing.

There are several problems with this approach. For one, since you have to work from above, you're constantly adding and removing weight on the trusses. This is obviously dangerous, since the trusses are already damaged. But it also makes leveling them difficult, since the weight of the workers and materials shifts and puts stress on the weak trusses.

You also have the problem of protecting the relatively soft and fragile ceiling panels from damage as you lift against them with the temporary supports. And finally, you need to keep the roof covered with tarps at night while the work progresses, running the risk of weather damage to the interior of the home.

Ceiling Panels and Ceiling Drywall

Many older manufactured homes had ceilings finished with large white ceiling panels, *and this material is no longer available*. The panel material was very similar to the acoustic tiles used in today's grid ceilings. They were 4 feet wide, spanning the width of two trusses, and up to 16 feet long. The panels were held in place with wide-crown roofing staples shot directly into the underside of the truss (see Figure 12-19). Long, thin plastic strips, called *splines*, snapped into grooves in the panels to cover the staples.

Since they're no longer available, when you encounter a manufactured home with roof truss repairs, fire damage, or other losses that require repairs to existing ceiling panels, you have a definite problem. Before you start cutting into these panels, discuss the situation with both the client and the adjuster.

If you're regusseting the trusses, it might be possible to design the repair to allow you to install new drywall instead of ceiling panels. Drywall is heavier than ceiling panels, so this isn't always possible without structural modifications to the truss, but the engineer might be able to incorporate those modifications.

Another option is to place wood furring strips perpendicular to the trusses, then install standard ceiling tiles on the furring. There are other types of ceiling tiles that install in metal grids, or with metal clips. All of these installations require different expenditures of labor and/or materials than what was originally installed, so you need to discuss them with the adjuster.

Don't forget that these alterations can also change the finished thickness of the ceiling, affecting the reinstallation of moldings, cabinets, and other materials you've salvaged for reuse. Some of these items may need to be replaced, while others can be cut or altered to fit. Changes like these can affect both your profit margin and your completion time frame, so keep that in mind while you're doing your estimate.

Working with Nonstandard Materials

Older manufactured homes utilized materials that were very specific to the industry, and to a lesser degree, that's still true of the industry today. The use of nonstandard materials can create problems both in disassembling items for repairs, and when trying to match materials.

Many of the materials that you need, such as drywall, paint, and flooring, can be found through your regular suppliers. But when looking for nonstandard materials for a manufactured home, you'll need to rely specifically on manufactured home material suppliers. Many cities and towns have manufactured-home parts suppliers, and you can find them listed under Manufactured — Mobile Homes on the Internet or in your local phone directory.

These retailers usually stock everything from windows and doors to plumbing and electrical parts used specifically for manufactured homes. Many of these retailers buy overstock and discontinued items directly from manufacturers, so they're a great source when you need a couple of pieces of matching molding or paneling, or an older light fixture.

You can now order a lot of what you need off the Internet. If you know the manufacturer and the year of the home, you can also contact the manufacturer directly and see about ordering parts, much the same as you would for your car.

Let's look at some of the materials you'll want to pay particular attention to when estimating and planning repairs. You may want to consider trying to salvage and store them for reuse.

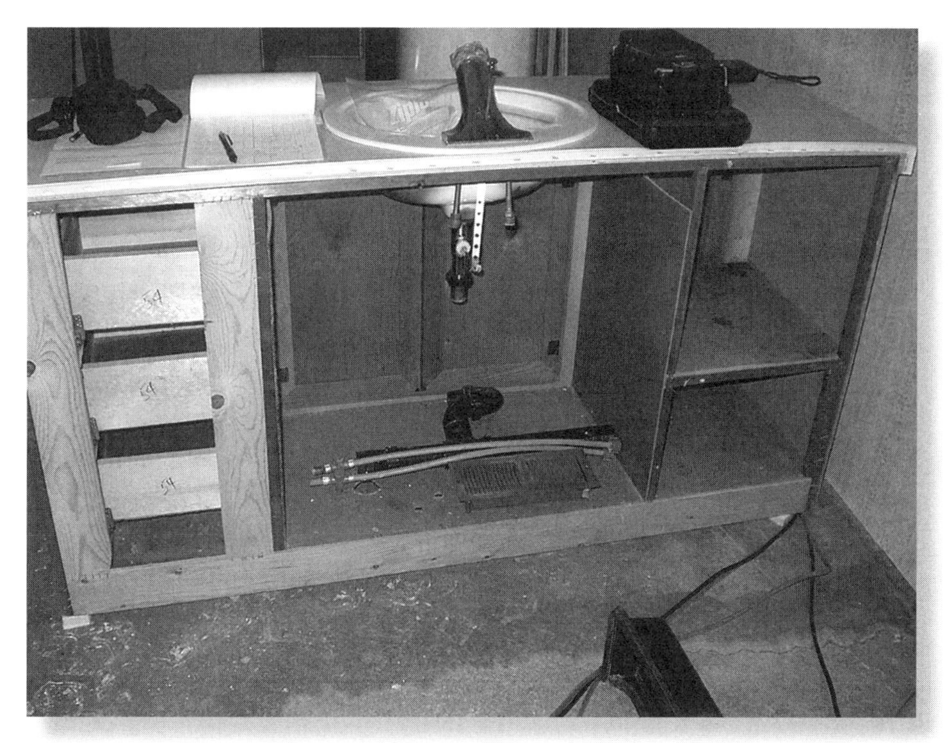

Figure 12-20 A manufactured home vanity cabinet. The cabinet is made from a framework of 1 x 1, 1 x 3, and 1 x 4 lumber stapled together and covered with paneling. The counter is glued onto the top of the cabinet.

Cabinets and Counters

Many of the kitchen, bathroom, and other cabinets used in manufactured homes are built into place at the factory (see Figure 12-20). Typically, a framework of small-dimension lumber, such as 1 x 1, is installed, then covered on one or both sides with decorative panels. Faceframes are installed on the front of the framework, followed by the doors and drawers. The cabinets often don't have backs on them, and the fasteners that hold them in place may be concealed by the panels, or by parts of the cabinet that were installed later. Because of this, they can require additional time and care to disassemble.

If you intend to save and reinstall the cabinets, study them carefully when doing your estimating. Allow sufficient time for removal, reassembly, and reinstallation. In order to simplify reinstallation, photograph and carefully label everything as it's taken apart.

If some of the cabinets aren't salvageable, your chances of finding any kind of stock cabinetry that will match are extremely slim. And, since manufactured homes utilize a lot of photo-finished cabinets, you probably won't be able to have a cabinet shop make anything either. Your best bet is to propose a compromise; for example, if the damage is to the upper cabinets, suggest replacing all of the uppers with something that complements the existing lowers.

Counters present a similar problem. They're typically laminate and often made in place at the factory. Sometimes they're glued directly to the framework of the lower cabinets, making it virtually impossible to remove them without damaging both the counter and the cabinet.

Prefinished Drywall and Paneling

Older manufactured homes often relied heavily on prefinished wood paneling and prefinished drywall. And while most of today's manufactured homes use standard textured and painted drywall, there's still a lot of prefinished material in use. With all those sheet goods comes the need for moldings. Most of it is flat stock, but you'll also need inside and outside corners, base, casing, crown, and other molding, much of it to match the paneling.

Obviously, matching the material is a potential problem. Manufacturers tend to purchase these materials in huge quantities, but when it's gone, it's gone. Your best bet is to locate it at one of the overstock outlets mentioned earlier.

If you find you need a couple of sheets of paneling for a patch, and you can't locate any replacement material, consider improvising. For example, say you need two sheets of replacement wall paneling in the living room. Take two matching sheets out of a closet, and then redo the closet with something else. Or, suggest redoing the entire living room wall as a decorative accent wall using something new. You can also prep the paneling and wallpaper over it. The more options you can offer the homeowner and the adjuster, the better off you'll be.

You can always opt to cover or replace the paneling with conventional drywall — but be careful with the difference in thickness. Installing ½-inch drywall in place of ¼-inch or even ⅛-inch paneling can create problems with doors, cabinets, moldings, and other materials. They're not insurmountable, but you need to keep that in mind and make allowances in your estimate.

Windows and Doors

Older manufactured homes have doors and windows that aren't the standard size or design of stick-built homes. Some exterior doors use a metal lip frame that attaches to the opening with multiple screws. You'll have to replace this type of door with a similar one ordered from a manufactured-home parts supplier; a standard exterior door won't work.

There are also a variety of interior doors to deal with. Some use a piano hinge, others use a special non-mortise hinge that attaches to the face of the door jamb. Some of the doors are standard sizes, and others aren't.

The same is true for windows. While newer manufactured homes utilize the more standard types of vinyl windows, older homes have a variety of nonstandard aluminum and steel windows requiring special orders or custom replacements, or even reframing of the openings.

Figure 12-21 An example of the nonstandard tub/shower units found in some manufactured homes. The use of low-end materials contributed to the unfortunate reputation these homes had for poor quality. This shower surround was glued directly to the prefinished drywall, then held in place with stapled-on trim. Replacement parts can be a problem in these situations. It's better to try and find good-quality standard materials to work with that you can guarantee.

Other Materials

There are a variety of other materials that also fit into the nonstandard category, such as odd-sized or odd-shaped bathtubs and shower stalls (see Figure 12-21). Some older toilets and sinks aren't standard, nor are some faucets. You may even run into light fixtures with hole spacings that are specific to an odd-size fixture box used by that particular home manufacturer.

Water Losses and Drying

There are differences between stick-framed and manufactured homes that you need to be aware of when water losses occur. These differences can affect your drying processes. Probably the biggest is that your access to all the spaces and components is much more limited in a manufactured home. There's no attic access, there's no easy crawl

> ***Tip!*** *When dealing with manufactured homes, don't assume that anything is a standard size! And, as a general rule of thumb, the older the home, the more likely you are to run into problems with nonstandard materials.*

space access, and duct access is difficult. Manufactured homes are tightly built, with a lot of adhesives and gasket materials, so if water gets someplace you don't want it, it can often be hard to get it out. Let's look at a few problem areas to keep on eye on:

Ductwork

As mentioned earlier, the ductwork in a manufactured home is simply a long, rectangular duct sitting on top of the chassis. It's concealed within the floor insulation, and is inaccessible except for the openings where the floor registers are.

If water is in the duct system, start by extracting as much as possible using a hose attached to your water extraction vacuum. Repeat this process at several registers in the affected area. You can then dry the ducts using a hose attached to an air mover or, even better, using a hose attached to the discharge side of a desiccant dehumidifier.

If you suspect that water has been in the system for awhile, or that other conditions exist that may have caused mold growth, consult with an industrial hygienist. Since you can't replace any of the ducting without doing a major disassembly of the floor, it may be necessary to treat the duct system with an antimicrobial. After that, the system can be encapsulated as an additional protective step against both spores and odors.

Crossover Ducts

The crossover duct between the two hard duct cans in a double-wide is the one duct that you do have access to. It's usually a 10- to 14-inch flex duct with both ends attached to a sheet metal collar protruding down from the duct system on each side of the home. Since it's both flexible and lower than the surrounding ducts, it typically fills up with water during any type of water loss.

Rather than drying it, it's easier to just replace it. Remove the bands that hold it to the collars on each side, then discard the old duct. Fit in a new duct of the same length and diameter, and secure it with new clamps if needed. The crossover duct should be properly supported off the ground, usually with straps attached to the chassis. Be sure that the straps don't crimp or pinch the duct and impede the airflow.

Vinyl Flooring Under Walls and Carpet

In most manufactured homes, the vinyl flooring is installed in large sheets at the factory, before the walls are erected. So when water flows across the vinyl floor, it tends to follow the flooring on underneath the wall plates. This creates more water travel and more wicking than you'd see in homes where the vinyl flooring ends at the wall. Be aware of this when using your moisture meters; check closely for patterns of water migration in areas where you might not otherwise see it.

Since the vinyl flooring is laid at the factory, it isn't unusual to find vinyl under carpet. When the vinyl is installed and the walls are constructed, some of the vinyl remains outside the edge of the wall in rooms that will later be carpeted. When the carpet's installed, rather than cut that excess vinyl flush with the wall, the carpet is just laid over it.

Figure 12-22 Here you see a combination of the problems you can encounter when dealing with an older manufactured home. The vinyl extends under the wall, and you can see by the tack strips that carpet was installed over the vinyl. The particleboard subfloor has started to sag from the water. The water lines are plastic, with a combination of crimp and compression fittings. There's aluminum foil stuffed around the pipe openings, which probably means the homeowners had a rodent problem. You can also see mold starting on the drywall at the left of the water heater.

Water can get underneath that layer of vinyl and remain trapped there, even after the carpet and pad are lifted for drying. Whenever you see carpet butting up to vinyl in a manufactured home, be aware of this vinyl-under-carpet situation. Look for pockets of trapped water, and set additional air movers as necessary to compensate for them. See Figure 12-22.

The Attic

If water gets into the attic of a manufactured home, through a roof leak for example, you're going to have to create your own access for drying. If possible, cut a hole in a concealed area, such as inside a closet. You can then try and dry the space with an air mover or a desiccant dehumidifier and a hose, in the same manner as you would for the duct work.

Never ignore a wet attic! Because of the lack of access and the relatively low levels of ventilation, it isn't likely that the attic will dry out on its own. The situation could easily evolve into a serious mold problem, as well as resulting in structural damage from the moisture. You have to address it, even if it means cutting a hole right in the middle of the ceiling in one of the rooms.

Removing and Repairing the Road Barrier

One of the less pleasant aspects of dealing with a manufactured home is working underneath it. Most of these homes are enclosed with wood or metal skirting, which isn't much of a deterrent to insects and animals. And, there's not usually a lot of crawling room, especially under the main I-beams of the chassis. See Figure 12-23. While there's supposed to be a vapor barrier, a lot of manufactured homes are installed without one, so you're often working on top of bare dirt.

In a water loss situation, you really have no choice but to go underneath and check it out. Water often soaks through gaps in the subfloor, around plumbing and electrical penetrations, and openings around the ductwork. It saturates the floor insulation, and can become trapped underneath the home by the road barrier. Like the attic, this is an area where you have to make your own access to the water and then dry it out.

Using rough measurements from above, determine where the leak occurred and where you think the worst of the water is likely located underneath. Set up some

Figure 12-23 A skirting panel has been removed for access to the underside of this manufactured home. The limited clearance under the main I-beams and the potential for insects can make inspecting these homes a challenge.

portable battery-powered lights under the home — due to the water, electric lights aren't recommended. Look for bulges and sags in the road barrier around the leak area. Several sags may be there naturally, so you'll need to press on them to try and find ones that feel heavy with water. It'll be like pushing on a water balloon.

When you locate one, place a 5-gallon bucket under it. Cut a small X-shaped slit in the road barrier with a utility knife to allow the water to drain. Have additional empty buckets ready as needed, depending on how much water you think you'll be dealing with. Repeat this process in all areas that need to be drained.

Next, enlarge the slits in those areas to expose the wet insulation. You can either allow the insulation to drape down through the opening and dry, or remove it and replace it later with dry material. While the road barrier is open, use your drying equipment to dry the underside of the subfloor as needed.

After everything is dry and you've made repairs, you'll need to patch the road barrier to protect the insulation. If you don't, small animals will be drawn to the warmth of the insulated floor and nest there. For smaller repairs, such as the slits cut to drain water, you can seal the road barrier with a flexible patching tape specifically formulated to stick to the road barrier. It holds up well under a variety of conditions, including transport. It's available in rolls of different lengths, and in widths of 2, 4, and 6 inches.

For larger repairs you'll need mending fabric, a woven polyethylene material sold in rolls of various widths and lengths that you can cut to size as needed. Secure the fabric patch in place with patching tape. You can usually buy the fabric and tape together in a mending kit. Both the fabric and the tape are available from manufactured-home parts suppliers. (See Appendix A for sources.)

> ***Tip!*** Don't use duct tape, plastic sheeting, Tyvek®, or other materials that aren't approved for this use. *They can fall off or otherwise become damaged, especially when the home is being transported, leaving the underside exposed and open to damage.*

Releveling

If the water loss was severe, enough water may have flowed beneath the home to affect the piers. Since the piers are all that support the home, if their stability is undermined, the home can actually shift out of level.

Figure 12-24 Concrete blocks, wood blocks, and dual overlapping wooden shims make up a typical manufactured home pier. Leveling is done by simply driving the shims farther in.

When dealing with a substantial water loss in a manufactured home, after the drying is completed, check the home for level before you begin reconstruction. It's important that the releveling be done first. Otherwise subsequent steps, such as cabinet installation, won't be accurate.

Releveling is a fairly straightforward process. First, check to be sure that the ground has dried out, and that each stack of blocks is plumb and stable. At the top of each stack you'll find two or more wide wooden shims (see Figure 12-24). These are driven between the top block and the underside of the chassis.

You'll need two people to level the home. One person should be inside the home with a long level placed on the floor in the area that's out of level. The second person, under the home, taps the shims inward to raise the floor as needed. The inside person continues to check the level, and advises the second worker when the proper adjustments have been achieved.

Building Permits and Structural Engineers

Even though manufactured homes aren't inspected by local building inspectors when they're built, they do require building permits for reconstruction when they're structurally damaged. So, who issues them, and who does the inspections?

Regardless of who had the original jurisdiction over the home's construction, the primary focus is the safety of the occupants. The local building department issues the building permits, and inspection of the repair work is done by local building inspectors while the work is in progress and after its completion, just the same as a stick-built house.

That's the theory anyway. In practice, you may find it a little confusing to obtain a permit for a manufactured home repair due to the HUD versus building code issue.

Since a manufactured home isn't built to conform to accepted building code standards in the first place, the plans examiners don't know what to look for when issuing a permit. Building inspectors aren't trained in manufactured home standards, and don't know how to interpret a lot of what they're seeing — much of which doesn't meet code, so they don't want to assume liability for it.

Unless your local jurisdiction is accustomed to issuing these types of permits, the process can be pretty cumbersome. Some local jurisdictions turn it over to the state, and some states turn it over to the locals. You can get caught in the middle, along with the homeowner who just wants his home repaired.

Having a structural engineer design a plan for your repairs is often your best bet. Granted, many engineers don't have specific manufactured home training, but the basic principles of loads and stresses are the same for any structure, so he can design and certify repairs for a variety of different structural losses. That certification removes the liability from the building department, so they generally have no problem issuing you the necessary repair permits.

Be Proactive

If you're going to be working with manufactured homes, consider doing some advance homework. Talk with your local building department, and find out what they require for issuing permits. Talk to your engineer, and see if he'll work with you on designing structural repairs, especially truss and fire repairs.

Do some online research and see which agency in your state oversees manufactured homes. Find out if they have a *Manufactured Home Dwelling Policy*, a set of *Manufactured Home Dwelling Standards*, or other similar documents, which will provide you with information on how manufactured homes are set up in your state.

Finally, talk to some of the adjusters that you work with. Learn what their policies are with regard to engineers and building permits. If you've discovered that having a structural engineering plan is the only way the building department will let you pull a manufactured home repair permit, let the adjusters know — it might avoid problems or questions that could slow up a claim down the road.

Modular Home Losses

Once a modular home is erected on its foundation, it's fairly indistinguishable from a site-built home, at least in outward appearance. Depending on how it was built, however, the modular home may share some of the unfortunate characteristics of the manufactured home. For example, you may find that there's no attic access, or you may be in the attic or crawlspace and find that you only have access to half of it because of a massive marriage-line beam.

Modular homes aren't always recognizable. In fact, sometimes the homeowners and the insurance company aren't even aware of it. Once you identify the home as being a modular, just be aware that you may have to take a few extra steps. When drying and estimating, take the same precautions that you would with a manufactured home, and keep the adjuster in the loop if you run into anything unusual.

13

Contents

Up to this point, we've been talking about all the things that can happen to the structure, and how to fix them. But just about every building you encounter, from the smallest of homes to the largest commercial offices, will have one thing in common. They'll be packed with the owner's belongings. See Figure 13-1.

There'll be everything from televisions to toothbrushes and coin collections to paperwork and pickles — just about anything you can imagine. And probably some things you can't!

In the insurance industry, all of this is collectively known as *contents*. And, you have to deal with it, often before you can get to work on the structure. In the next three chapters we'll look at how to handle contents, from packing and cleaning to moving and storage.

The Separate but Equal Parts of a Loss

In the first chapter we broke our response to losses into three distinct elements:

- emergency
- structural
- contents

All these elements are part of the same claim, and subject to just one deductible. However, insurance companies process and handle each one separately.

Figure 13-1 Dealing with contents is a fact of life on just about every loss you encounter. Some homes are organized, and some lean a little more toward the cluttered side.

Not every loss has all three parts. A water loss might have an emergency response and some structural repairs, but not have any contents damage. Or, a small fire might have both structural and content losses, but not require an emergency response. It's important for you to recognize the distinction between these three different elements, especially between content items and structural items.

Dealing with Contents

The majority of buildings you'll be dealing with, whether residential or commercial, will be occupied. That means you'll be working with content items on a regular basis. How you deal with those contents will generally fall into one of the four following situations:

1. ***Moving contents during an emergency call*** — You may have to do this to facilitate the handling of an emergency situation. There'll be times when you have to get the contents out of your way quickly so you can set up equipment and mitigate damage.
2. ***Storing contents on-site*** — You can do this when the building needs repair, but the damage isn't extensive enough to warrant removing all of the contents from the site.

3. ***Dealing with contents that aren't salvageable*** — On many losses, at least some of the contents may be ruined by water or smoke, or be damaged beyond repair. Before you can dispose of items, they need to be assessed, sorted and inventoried with the help of the owners.

4. ***Transporting and storing contents off-site*** — If the damage to the structure is extensive and there's no safe place to store the contents on-site, they'll have to be moved elsewhere. Rather than just being stored, some contents may need to be cleaned, repaired, or deodorized, which is common after a major fire loss. In either case, the contents need to be assessed, sorted, and inventoried before being moved.

Each of these four scenarios requires different packing methods and different methods for inventorying and handling. We'll look at each of these in much more detail in the upcoming chapters.

What Are Contents?

It might seem obvious, but what exactly *are* contents? Or more importantly, what are contents as far as the insurance companies are concerned?

In general, contents are those items you'd take with you if you moved. They're usually not attached to the structure. For example, kitchen cabinets are attached to the wall; when you move, they stay with the house. So they're considered structural. But the food and dishes stored in the cabinets would go with you, so they're considered contents. Another example is a closet. The shelves and rods attached to the walls are considered structural, while the clothes, shoes, tie racks, and the hamper bag hanging on the door are all contents.

The difference between content items and structural items is clear in those examples, but that's not always the case. What about in the garage? Let's say a homeowner built some shelving out of 2 x 4s and plywood and fastened them to the garage wall with screws. That's fairly permanent, and would probably be included with the structural estimate. On another wall the homeowner has a portable metal shelving unit, also secured to the wall with screws. Despite the screws, it would be considered a contents item.

Where it becomes a gray area is when the homeowner takes a metal shelving unit and some lumber and combines them to make something semi-permanent. It has both permanent and portable elements. As an estimator, you pretty much have to take a guess. Does it look like it would stay with the house if the owner moved, or would he take it with him? Make your best judgment call, then let the adjuster know whether you included it with the contents or the structural part of the loss, and why.

Three General Contents Categories

Contents can be broken down into three general categories. There's actually no hard and fast rule for this, but you'll often hear these terms, so it's helpful to understand what they refer to.

1. ***Hard goods:*** A general term for furniture, appliances, electronics, tools, hardware, kitchenwares, etc.
2. ***Soft goods:*** An equally broad and generic term that covers clothing, shoes, bedding, rugs, linens, and similar items. *Soft goods* can also sometimes be applied to upholstered furniture.
3. ***Perishables:*** Anything that can spoil or die. This includes all types of food; anything that becomes unusable due to extremes of heat or cold, such as cosmetics or some types of toiletries; and living house plants.

Appliances, Spas, Hot Tubs and Pools

Certain appliances have become accepted within the industry as being structural items and others contents items, at least in the United States. In other parts of the world, where conventions differ as to what appliances a person takes when moving, this might not be the case.

Most kitchen appliances are considered part of the structural loss. That includes built-in and freestanding ranges, cooktops, dishwashers, ovens, range hoods, built-in microwaves, and microwave/hood combinations. Kitchen appliances that would typically be considered contents items would be refrigerators, portable dishwashers, and portable microwaves. Small portable appliances, such as toasters, toaster/ovens, coffee pots, etc., are all considered contents items unless they're permanently built in.

Washing machines, clothes dryers, and stacked washer/dryer combos are typically considered contents items. Water heaters, water softeners, most types of plumbed-in water filters, wood stoves, and all vented heaters are considered structural items. In one of those odd insurance company twists, hot tubs, spas, and above-ground swimming pools are typically considered contents items.

The Personal Nature of Contents

One of the most important things for you to understand and remember is that most contents are very personal. When you tear out carpeting or cabinets, they don't mean much to a homeowner. But when you're handling dishes or clothing or a piece of art that they bought on their honeymoon, those are very personal items, so you must always keep that in mind.

First of all, ask permission to move about the home and to handle, move or pack up personal items. It's a simple thing, but it means a lot. The homeowners need to know you'll be going through their closets as you check for water damage. Make sure that's okay with them. If you encounter a closed door, knock first, or ask someone

REAL STORIES:
What's Hidden in the Attic?

We'd been hired by a gentleman that we knew to make fire repairs to the second floor of his mother's home. It was a classic older home, with a couple of upstairs dormer bedrooms that had access doors to small, semi-hidden attic spaces. One of these dormer rooms was now his mother's sewing room, but it had been the man's bedroom when he was growing up.

We were cleaning out the contents in the rooms and attic spaces in preparation for demolition and repair of the fire damage. Tucked away in a hidden recess in the attic off the man's boyhood bedroom, my partner found a dusty box. It was open, and he could see that it contained some old toy cars, a couple of toy guns — and one mild girlie magazine from the 60s.

We discretely set the magazine aside, and later gave it to the man. He got a huge laugh out of it when he saw it. "I'd forgotten all about that magazine," he said. "Thanks for saving it for me!" But I think what he was really thankful for was that we didn't share it with his mother!

if it's okay to enter. Someone may be sleeping, there may be a pet inside, or there may be personal items inside they'd prefer you not see.

Granted, you're there to help. But it's still their home — no matter what the situation — so treat them with the same courtesy you'd like someone to extend to you in your home.

Tip! *Teach your crew what's expected of them when it comes to the respectful handling of people's contents, and make it a written part of your Company Handbook.*

Contents and Your Liability

You obviously have a lot of liability when you're dealing with other people's contents. If you damage or lose something, chances are you're going to be held liable for its repair or replacement. *You'll need to have adequate insurance coverage.* But be forewarned — contents liability coverage can be a very complicated issue.

Let's say Mrs. Smith has a kitchen fire, and one of her cabinets is damaged and has to be replaced. Inside that cabinet is a very valuable antique platter, which needs to be cleaned. But during your time on the job, the platter gets broken.

Below are several different scenarios. Even though each one results in the platter being broken, you may or may not have coverage, depending on how your insurance company looks at it and exactly what policies you have in place:

- **Incidental contact:** You pick up the platter to move it because it's inside the cabinet you need to replace, and while doing so, you drop it and break it. In this scenario, you weren't working on the platter, you were simply moving it to work on the cabinet. This is typically under your contractor's liability coverage.

- **Care and custody:** You box up the platter and transport it back to your shop for cleaning, and it gets broken or lost while it's there. In this scenario, you were specifically working or intending to work on the platter itself, so you'd need specific care and custody insurance to make sure it's covered.
- **Theft:** The platter is in storage at your shop, and a stranger breaks in and steals it. That may be part of a care and custody policy, or it may be part of the insurance on the building itself. If you don't own the building, you may need to have a specific renter's policy in place to cover theft.
- **Employee theft:** The platter is in storage at your shop, and while it's there, it's stolen by one of your employees. This usually requires either care and custody insurance, or a specific employee theft policy.
- **Transit:** The platter's on its way back to Mrs. Smith's at the end of the job, but it's broken when your truck is in an accident. This usually requires specific transit insurance.
- **Interstate transit:** If Mrs. Smith lives in one state and your storage facility is just across the border in another state, the issue gets even cloudier. There are laws regarding the transport of goods across state lines, and in some cases that can have a major impact on your insurance coverage.

As you can readily see, *it's imperative that you and your insurance agent understand exactly what you'll be doing when it comes to contents.* You need to have the proper type and amount of coverage in place, and you need to update your coverage regularly.

Contents You Don't Want to Handle

No matter what type of insurance you have, there are certain contents that, from a liability standpoint, you simply shouldn't deal with. Some may even be specifically restricted by the terms of your insurance policy.

There's a list of these items below. When you speak to your clients, ask them if they have any of these items in the rooms or buildings you'll be dealing with. If they do, or if you encounter any of them as you go, ask them to please move the items for you. If they aren't able to move them, ask them to have someone they trust — a friend or family member — handle them. The other option is to bring in an insured third party, like a moving company, to handle the items *under the homeowner's supervision.* This will help take you out of the liability loop.

Avoid handling the following items:

- **Money:** Cash, coins, coin collections, antique coins and paper money, gold, silver, and even children's banks. See Figure 13-2. If you encounter a safe that needs to be moved in order to do structural repairs on a building, and the owner can't remove it, hire a locksmith with specialized equipment.

Figure 13-2 Money is something you never want to handle, especially something like this bottle of coins. There's simply no way to prove how much is in there, or whether or not money is missing. Items like this need to be left to the homeowner to handle.

- **Jewelry:** All types of jewelry, including watches, jewelry boxes, antique jewelry, and loose stones.
- **Firearms:** Rifles, pistols, shotguns, antique firearms, ammunition, and explosives. If you encounter a gun safe that needs to be moved, hire an expert with specific experience in moving them. Some locksmiths can move gun safes, or check with a local gun shop.
- **Specific personal items:** What one person considers highly personal, the next person might not. So when you arrive to move contents, just add the following to your list of initial questions: "Are there any personal items that you'd like to take care of yourself before we start moving things?"
- **Confidential materials:** What may be just a file cabinet or a box of papers to you may be highly confidential files to a business owner, or to a homeowner who operates a business from home (see Figure 13-3). Many people also feel that way about their household bills and other records. When dealing with paperwork, especially in a commercial or home office setting, ask the

Figure 13-3 Confidential files, especially on a commercial loss, need to be handled with extreme care. As soon as you arrive on-site, find a responsible person to coordinate with, then handle the files directly under his or her supervision.

owners if there are any confidential materials they would prefer to handle themselves.

- **Computers:** Computers, monitors, printers, scanners, fax machines, and other electronic office equipment. Computers aren't difficult to move, but they're highly personal items. They can also be sensitive to mishandling; if you disconnect one and move it, and it doesn't work correctly in the future, the owner may feel entitled to a new one at your expense. Ask the homeowner or business owner to disconnect and move their own equipment, or have their regular outside computer technician come in to do it.

- **Electronics:** Large screen TVs, stereo systems, projection TVs, surround sound systems, entertainment systems, and similar complex electronic systems should be handled by others.

- **Large clocks:** Large wall clocks, grandfather clocks, antique clocks, cuckoo clocks, and other large or intricate clocks are very delicate. See Figure 13-4. Incorrect moving, handling, or storing can do a lot of damage, so leave this task to the homeowner, or bring in an expert.

Figure 13-4 Large clocks, such as this antique grandfather clock, are surprisingly sensitive instruments. They have to be stored under controlled humidity conditions, and when they're reinstalled they need to be carefully leveled or they won't operate. Leave this process to a clock expert.

- **Pianos and other musical instruments:** Pianos are large and intricate instruments (Figure 13-5) that require careful handling, and you can do a lot of damage trying to move one on your own. Hire a licensed, experienced piano mover to do it for you. Pianos are also sensitive to humidity changes, so if you have one that's in a fire or a water loss, make sure it's removed from the home as soon as possible. Have it taken to a piano shop for drying and storage. Once the house has been repaired, the shop can return the piano and tune it.

  Many other musical instruments also require expert disassembly, moving, storage, setup, and tuning, and can be easily damaged by inexperienced or untrained personnel. If you're unsure of how to handle something, hire an expert to do it.

- **Pool tables:** You'll be surprised how often you come across pool tables in this work. Good-quality pool tables have a very heavy, multi-piece slate bed that's difficult to disassemble and reassemble. And, they have to be stored correctly to prevent damage. It's best to hire an expert to move them. You'll want to establish a good working relationship with a pool table company that you can trust

Figure 13-5 Pianos, organs, and many other musical instruments — including this beautiful antique — require careful handling. They need special moving equipment, controlled storage, and careful tuning after redelivery.

to handle pick up and delivery. They should also be able to level it precisely when they set it back up again so that it plays correctly.

- **Artwork:** Moving, packing, or storing artwork incorrectly can do irreparable harm. That doesn't mean you need to stay away from handling all artwork, but if you have doubts about what you're dealing with, consult with the owners or an art expert before moving or storing it. If necessary, have the object professionally crated and stored.

Equipment and Supplies for Handling Contents

Your liability when handling contents comes in several different forms. You can scratch something, drop it, damage something by stacking one thing on top of another, or encounter any of a hundred other mishaps. There's also the possibility of damaging the structure while moving the contents, such as gouging a floor while moving a refrigerator. And, of course, there's always the chance that someone could get hurt while lifting, moving, or loading contents.

To minimize this liability, you need to invest in the right equipment and supplies for handling contents. Having what you need to do the job correctly minimizes content and structural damage, improves worker safety and morale, and shows the clients and adjusters that you take pride in what you do.

REAL STORIES:
That's Worth *What*?

We were packing out the contents of a vacation home that had suffered water damage. Some of the contents weren't damaged, but there was enough structural damage that we considered it best to get all the contents into storage for protection.

In the upstairs loft, overlooking the living room, was a very large unframed painting. It was an abstract, all swirls of reds, oranges and yellows. It appeared to have been hanging up there for years, and had a noticeable layer of dust on it. We all assumed it was a cheap painting that someone had picked up in a garage sale or from a "starving artist" sale somewhere — just something colorful to fill up a big blank wall in a vacation home.

We set up extension ladders on either side of it, and lowered the big awkward canvas to the floor. We laid out a large sheet of bubble wrap and got ready to wrap it for transport. Then someone noticed a receipt attached to the back. Curious, we glanced at it. Sure enough, the painting had been bought several years before. But not at a cheap sale — at an exclusive gallery. For $16,000!

Definitely time to break out more bubble wrap!

Handling Contents On-Site

Depending on the size of your jobs and the amount of contents you're handling, you'll need some of the following items right away. Some you can purchase as you need them, and others you can add in the future.

Standard Boxes

Boxes are basic items that you'll need for handling contents, both on- and off-site. Select a couple of standard sizes, and stick with those. Having just two or three sizes simplifies inventory (see Figure 13-6), ordering, and billing. It also makes it easier to stack the full boxes inside trucks and in your storage areas (Figure 13-7).

Figure 13-6 Selecting only a couple of standard sizes of boxes will simplify your inventory, and cut down on costs. Boxes are shipped broken down in flat bundles, as shown here. That simplifies both storage and transport to the jobsite. Notice the rolled-up bundles of blank newsprint in the background. It's used for packing.

Select a medium-size box, 12 x 12 x 12 inches or 14 x 14 x 12 inches, and a larger size, 18 x 18 x 12 inches or 24 x 18 x 12 inches, or another size that you think will work well for your business. Don't forget that you have to lift the boxes once you fill them up, so choose sizes that will handle most of your packing needs without getting too heavy. Also, select boxes that are readily available from a reputable supplier.

Don't use old grocery or produce boxes from your local store, for several reasons. First and foremost, that's just downright tacky. You're a professional company, and

Figure 13-7 Sticking with standard-size boxes makes stacking them easier and safer, as you can see in this warehouse facility.

you're being hired to do a job; it's not appropriate to show up with a bunch of old banana boxes. Also, old boxes are dirty and carry odors and other contaminants. They're also unstable, they're not of uniform size and quality, and they're difficult to stack and label.

Specialty Boxes

Purchase specialty boxes as the need arises. There's an incredible variety of sizes, shapes, and configurations to choose from, depending on the types of contents you encounter. Here are some examples:

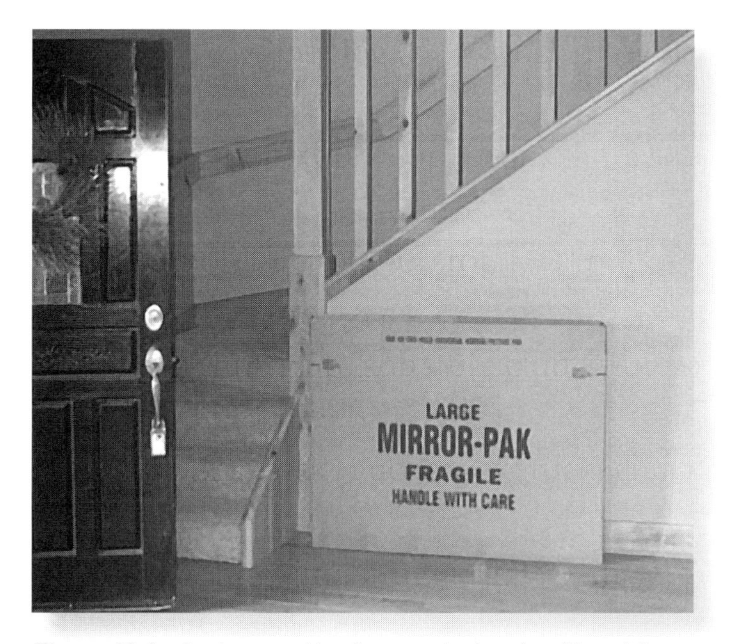

Figure 13-8 A mirror packing box, packed and waiting at the jobsite to be taken by truck to the storage facility.

- **Artwork/mirror boxes:** These accommodate large, flat items such as mirrors and paintings, Figure 13-8. There are several standard sizes to choose from, as well as ones that telescope to different sizes.
- **Wardrobe boxes:** These are designed to use for clothing. They have a fold-down front and a metal cross-rod inside that supports clothes hangers.
- **Dish and glass pack boxes:** These boxes have both horizontal and vertical dividers that can be arranged in different configurations. They separate dishes and glasses, and can be packed quickly with less need for wrapping.
- **Special use boxes:** You may find yourself called out on a one-time job with some odd requirement, such as packing up hundreds of bottles. For odd applications, you can order boxes in sizes or with dividers that accommodate anything from mattresses and surfboards to lamps and skateboards.

Figure 13-9 A carton storage rack for organizing different-size boxes. This one also has casters, to simplify moving it around the jobsite or the warehouse.

Boxes are sold, shipped, and stored broken down, meaning they're flat when you get them. Assembly is very quick — just a matter of opening it up, folding the flaps over, and taping the bottom of the box with a good packing tape.

Carton Stands

This is simply a metal rack that separates and stores boxes by size (Figure 13-9). It keeps your boxes organized, and saves a lot of space by keeping them upright instead of lying flat. Most racks have optional casters so you can roll the boxes right to where you're packing.

Packing Materials

Bubble wrap is quick and easy to use, and works well for most packing and cushioning needs. It's available in rolls of different widths, and it's perforated so you can tear it off in different sizes as needed (Figure 13-10).

An alternative to bubble wrap is unprinted newsprint paper. (Look back at Figure 13-6; you can see the paper in the background.) It's less expensive than bubble wrap and takes up less space in the box, but it doesn't offer as much cushioning. Newsprint is available in different size sheets, as well as by the roll. *Don't* use regular newspapers for packing. The ink will transfer to the contents, and then you'll have even more that you have to clean.

Figure 13-10 A bubble wrap roll in a dispenser box. This roll is perforated at regular intervals to make it easy to tear off whatever size pieces are needed. Bubble wrap is also available in larger rolls, wider widths, and with larger bubbles for additional padding.

Also, avoid foam packing peanuts. They take up a lot of storage space, and are quite messy, both on the jobsite and in your warehouse. If you do use peanuts for some of your specialized packing needs, use the biodegradable ones that dissolve in water.

Tape Guns and Tape

For assembling and sealing boxes, you'll need tape guns and packing tape, available wherever you buy the boxes. Also consider a tape gun holster, which slips onto a belt and keeps the tape gun handy while working.

Moving Blankets

Use padded, quilted blankets for protecting furniture (Figure 13-11). You can't have too many of these, and they're invaluable for protecting contents from damage. They're available in different sizes to meet different packing needs, including 36 x 40 inches, 40 x 72 inches and 72 x 80 inches. Harbor Freight Tools (see Appendix A) is a good source for these. They range in price from $5 to $8 each, but they're less expensive if you purchase them by the case.

Hand Truck and Appliance Dolly

These are two items that you must have when moving heavy contents. The hand truck is ideal for moving boxes, as well as furniture that's taller than it is wide. You can pick up a good quality hand truck for around $40 to $50.

An appliance dolly is similar to a hand truck, but it has a wide canvas strap system on it, with either a handle or a ratchet. You wrap the strap around the appliance to hold it tight against the dolly so you can safely lift and maneuver the appliance. It also has a set

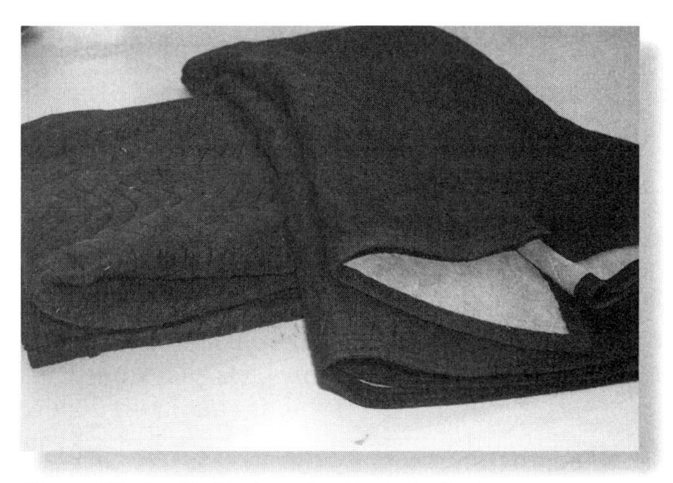

Figure 13-11 Padded moving blankets like these are essential for protecting contents from damage. They're used at the jobsite, in the warehouse, and in transit. You can never have too many of these around.

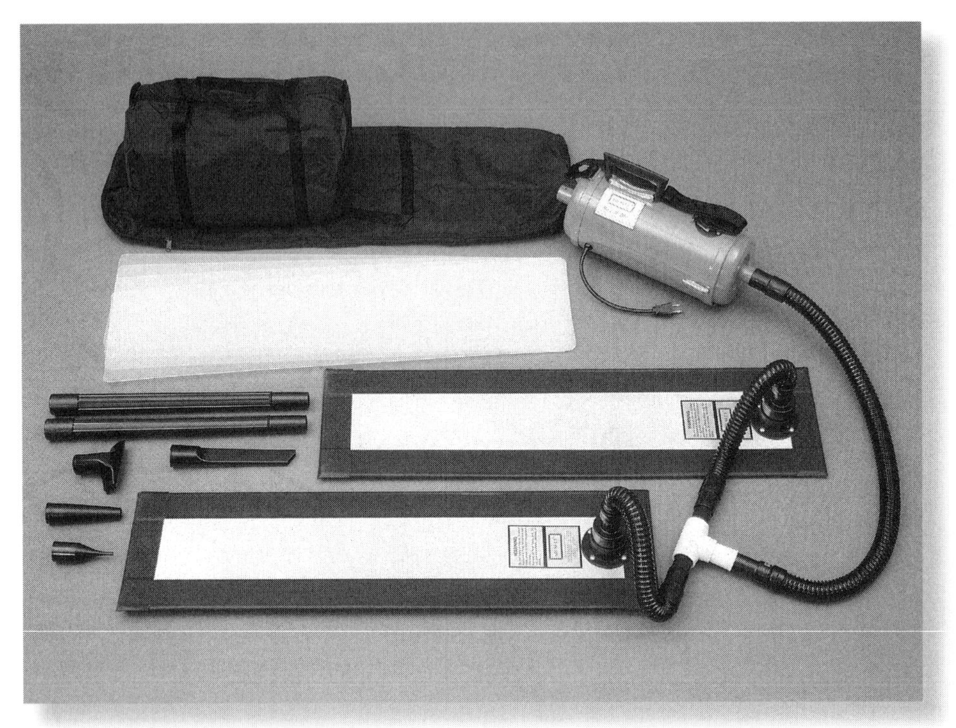

(Photo courtesy of Airsled, Inc.)

Figure 13-12 An Airsled® kit. The small electric motor (top right) blows air through the hoses into the two perforated lifting plates, which provide the lift needed to move even the heaviest appliance. The white plastic pads can be used to provide a hard, smooth surface when moving appliances over carpet. By reversing the hoses and using the provided attachments, the Airsled can also be used as a small canister vacuum.

of rollers on the back of the main handles, down near the wheels, which allows you to pull the dolly up and down stairs with more control. Appliance dollies are available for about $80 to $100.

Airsled®

If you move appliances, an Airsled is one of the best things you can invest in! At around $700, they're not cheap. But they'll save you a lot of labor, and also protect both the contents and the structure when moving heavy items.

An Airsled utilizes a small, electric-powered canister motor that blows a stream of air through a pair of hoses. Each hose attaches to a high-impact plastic plate that's perforated with thousands of tiny holes, like upside-down air hockey tables. See Figure 13-12. The thin plates slide under whatever you want to move, such as a range. The film of air lifts the plates and the range with surprising ease, and holds them just above the floor, as shown in Figure 13-13, so you can move the appliance with very little effort.

Unlike an appliance dolly, the Airsled keeps the appliance fully upright. You can move a fully-loaded refrigerator without tipping it, or having to empty it first. And since it's riding on a cushion of air instead of a pair of wheels, there's no risk of damage to expensive flooring. If the Airsled saves you the cost of replacing one kitchen floor ruined by moving a heavy refrigerator across it, you've more than recouped your investment.

(Photo courtesy of Airsled, Inc.)

Figure 13-13 Using an Airsled® to move a range. Unlike a conventional hand truck or appliance dolly that uses wheels, the Airsled uses a cushion of air for lift. Appliances move incredibly easily, and there's no chance of damaging the floor.

Mover's Dollies

Mover's dollies are small, square moving platforms with protective, non-slip carpeted tops and four casters underneath. See Figure 13-14. You can use them for moving furniture and other heavy objects around inside buildings or warehouses. They cost around $15 for a small one, and about $25 for a larger model with more weight capacity. Harbor Freight Tools is a good source for these as well (see Appendix A).

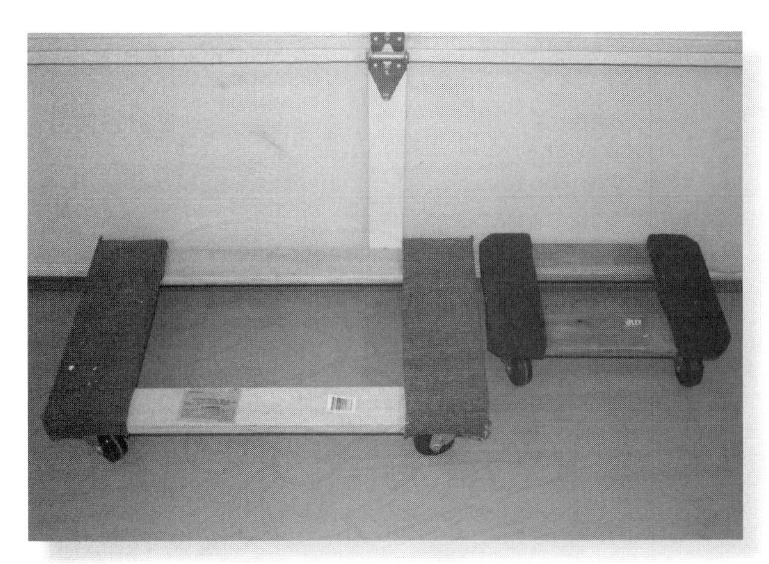

Figure 13-14 Mover's dollies in two different sizes.

Bins

Plastic bins with lids are ideal for temporary on-site storage of contents. They're sturdy, and can be moved around and stacked while still protecting the contents. They're also easy to clean, even in a fire situation. So unlike boxes, they're reusable and they nest inside each other for storage between jobs. Buy several in three sizes — small, medium, and large.

Plastic Zip-Lock Bags

Buy zip-lock bags in bulk at your local discount warehouse store — Costco or Sam's Club. You'll need the small sandwich size, the quart size, and the gallon size. All three are ideal for the storage of small contents items that would otherwise get lost in the confusion of moving and restoration.

Forms

There are two particular forms that you'll need for dealing with contents: a Pack-Out/Pack-In Inventory Form, and a Release of Contents Form. These aren't stock forms, and will have to be made up to suit your needs. We'll talk more about these particular forms in Chapter 15.

Labels and Tags

You'll need tags for labeling boxes, furniture, and other contents items. There's no perfect way of doing this, and it's something that you'll have to work out to suit your particular way of operating. You may want to use blank labels, or design and print your own. For labeling furniture, paper tags with string or wire fasteners on the end work great. You can find these wherever you buy your warehouse supplies.

Supply Bin

Plan on having one supply bin for each on-site contents job you have in progress. It's an inexpensive investment compared to the time wasted if your crew is disorganized or ill-prepared. The bin should contain pens and pencils; marker pens; labels for boxes and bins; paper tags for hanging on furniture; safety pins for attaching tags; rubber bands; a tape gun and extra tape; zip-lock bags; blue quick-release masking tape for bundling and labeling certain items; and whatever other small supplies your crew will need.

Tool Box

Dealing with contents typically requires a few basic tools as well, so put together a small tool box. It should include: a utility knife and blades; a 6-in-1 screwdriver; a small socket set, with both SAE (Society of Automotive Engineers) and metric sockets; an adjustable wrench; a set of SAE and metric hex wrenches; slip-joint pliers; needle-nose pliers; and a small hammer.

Handling Contents Off-Site

If you're going to be storing contents at your own warehouse facility, you're going to need some additional supplies and equipment. These items will make it easier to sort, organize, move, and store contents. They'll also keep everything separated and protected. All these items aren't necessary immediately, but they'll become more important as your operation grows and you have multiple households of contents in storage.

Pallets

Pallets are very handy for stacking loads of boxes and keeping them organized (and it's also why it pays to have all of your boxes in one or two uniform sizes). Pallets are also great for loading and moving larger pieces of furniture.

When you're just starting out, you can often pick up pallets for free in local industrial areas. Since they're only being used in your own warehouse, this is one area where new isn't absolutely necessary. The pallets just need to be clean and structurally sound, so repair them as needed. You can add a sheet of OSB on top to clean them up and give them a smooth, flat, uniform surface for stacking boxes and furniture.

If you choose to purchase new ones, they're available in plastic, wood, and a variety of other tough, lightweight materials. See Figure 13-15. You can purchase them through the same supplier that you use for your boxes and other packing supplies. New pallets cost from $12 to over $75 depending on their size, material, design and load-bearing capacity.

Figure 13-15 Plastic pallets are lightweight, easy to clean, and virtually indestructible.

Plastic Wrap

Once a pallet is loaded, if you're going to leave the contents on it for a while, plastic wrap is a great idea. It keeps the contents secured, and also offers some additional protection from damage. A thin, tough, stretchy type of plastic wrap, similar to the plastic wrap that you use in the kitchen but tougher, is available on wide rolls that you can just wrap around the pallet.

Mattress, Chair, and Sofa Bags

Plan on having a couple of rolls of large-sized bags in stock to fit mattresses, large chairs and sofas. Initially, you can just use plastic sheeting to cover and protect these items. But as you begin to handle more and more contents, you'll find that these inexpensive bags are faster and easier to deal with than plastic sheets. They also cover the contents more effectively, and you can include their cost in your overall contents cleaning bill.

Large bags are available in many different sizes. You can expect to pay around $150 for a roll of furniture bags — which will contain anywhere from 100 to 500 bags, depending on the size. Mattress bags run between $85 and $135 a roll, again depending on the size.

Pallet Jack

A pallet jack, Figure 13-16, is a simple and cost-effective way of moving loaded pallets around your warehouse. Pallet jacks have two long forks in front, like a fork lift, with two fixed wheels in front under the forks, and two steerable wheels in back attached to a handle. Pumping the handle activates a hydraulic pump and lifts the load. The handle is used to steer the load, and a lever releases the hydraulics and sets the load back down again.

Figure 13-16 A pallet jack. The long forks slip under the pallet, like a forklift. Pumping the handle raises the forks so the load will move. The handle is also used to steer the load, and then to lower it again.

Figure 13-17 Plywood storage vaults. These vaults are great for storing and protecting contents, especially for long-term storage. They're also designed to be lifted and moved easily with a pallet jack or forklift. Metal spring clips are used to hold the vaults together. They permit easy disassembly and storage when the vault isn't needed.

Pallet jacks are available in the $300 to $600 price range, depending on their capacity, fork length, and other features. There's also a rechargeable electric version that runs about $3,500. You can find good, used pallet jacks on the market for about half the price of new ones.

Storage Vaults

Storage vaults are big wooden crates, like those used by moving companies and warehouse operations. See Figure 13-17. They measure 7 feet long, 5 feet wide, and usually 7 feet or $7\frac{1}{2}$ feet high. They're made of $\frac{1}{2}$-inch plywood panels with a 1 x 3 or 1 x 4 reinforcing framework around the outside. They sit up on 4 x 4 or 4 x 6 risers, so that a forklift or pallet jack can lift them for moving. Most are made up of individual panels held together with steel spring clips, allowing them to be broken down for storage or transport.

Storage vaults make it much easier to organize the contents you have in storage, and to keep different households separated. It's also easier to locate items if a homeowner comes in and needs to retrieve something. It simplifies the organization of your warehouse, and makes shifting contents for long-term storage easier. Once you establish a per-vault price for the storage of contents, you'll find that it also simplifies your billing process.

Vaults can be purchased through warehouse supply sources, both new and used. Used vault prices start at around $100. You can also get quantity discounts by purchasing them in bulk, so it's worth talking to the supplier or even to one of your friendly competitors to see about splitting an order with someone else.

Warehouse Shelving

Once you set a couch down on the floor for storage, that square footage of your warehouse is used up. But if your warehouse has 12-foot or 14-foot ceilings and you have some sturdy warehouse-style shelving, you can stack and store a lot more couches in the same square footage. When you begin to handle more contents, invest in some shelving to maximize your warehouse space — and your company's profits. The price for shelving varies, depending on its size and loadbearing capacity.

Forklift

Once you begin dealing with large numbers of contents, or with vaults and warehouse-style shelving, a forklift will become a necessity. Forklifts are available in manual, electric, gasoline, diesel, and propane models. Inside a warehouse, exhaust fumes make gasoline and diesels models impractical for use around people's contents. Rechargeable electric units are a good choice, but they can be expensive, and used ones are hard to find. Your best option is probably a propane-powered forklift. Small propane forklifts are readily available, both new and used, and are easy to operate. They're also reasonably priced, with good used models available for around $2,400 to $3,000.

For a small operation, you might consider what's known as a manual stacker. It's sort of a combination pallet jack and manual forklift, allowing you to move pallets around the floor by hand, then lift them into position with a manually-operated hydraulic lift mechanism. New manual stackers cost about $1,300.

OSHA requires that all operators be properly trained and certified on forklift operations. *If you have a forklift in your warehouse, even a manual one, be sure that you check with your state OSHA office to learn the exact training and certification regulations for your employees!* Also, check with your Worker's Compensation carrier to see if owning a forklift will affect your insurance rates.

Handling contents offers you an excellent opportunity! It makes your company a truly full-service operation, and it sets you apart from contractors who only do insurance restoration work as a sideline. When you handle contents, you're providing a very valuable service to your clients and your adjusters, while taking advantage of the opportunity to maximize your profitability.

14

Contents: On-Site Storage, Loss Inventories, & Nonsalvageable Items

Contents and content losses always present something of a balancing act. Salvageable contents need to be moved so you can work, but where do they go, and who moves them? Nonsalvageable contents need to be discarded, but who decides if they're salvageable, and who inventories them?

All these decisions, including what will be cleaned and repaired, as well as what contents will remain on-site and what won't, have to be made jointly by you, the owners, and the adjuster. They'll be based on value, practicality, and the amount of content coverage available — and sometimes on a bit of raw emotion.

Moving Contents During an Emergency Call

When you first arrive at an emergency call, you'll usually need to move contents in order to deal with the emergency. You might need to move things away from the water, or out of your way so you can set up equipment, make emergency repairs, or complete a board-up after a fire or storm.

Moving contents during an emergency call tends to get hurried and chaotic, and that's something you really have to watch out for. If you're not careful, you can easily damage valuable items in your haste to get things done. Let's look at some simple steps you can take to prevent that from happening.

Step One: Evaluate the Situation

Take a moment to calmly evaluate the situation. See what portions of the house are affected, and what contents are in harm's way or need to be moved so you can work or set up equipment.

Figure 14-1 The dark discoloration on this leather couch indicates how much water it has absorbed. You can also see from the footprints how wet the carpet is. The couch should be moved out of the water and dried as soon as possible.

Step Two: Block or Move Contents Out of the Water

In a water loss, take immediate steps to get contents out of the water. See Figure 14-1. Wooden furniture sitting on wet carpet will wick moisture up into the wood (Figure 14-2), which will cause it to swell and eventually warp, crack or separate. Dyes and stains from the furniture, especially antiques, can transfer to the carpet and leave permanent stains. Metal furniture, or even metal pads on the bottom of wooden furniture legs, can rust and also leave permanent stains. If possible, move the furniture to a dry area. Otherwise, raise it up off the floor using foam blocks.

Step Three: Find a Safe Storage Area

Once you've taken care of the immediate threat to the contents, see if there's a safe area where you can temporarily store them. Maybe only half the living room is wet, so you can move everything over to the dry half. See Figure 14-3. Or maybe all that's needed is to put things in bins and move them to the garage.

For a larger loss where several rooms are affected, you may have to relocate a lot of furniture. In that case, look for an area that's away from where you'll be working, and where you won't have to move things twice. Make sure it's an area where the contents can be protected against dirt, damage, and freezing temperatures.

Figure 14-2 These wooden dining room chairs show clear evidence of how much water they've wicked up while sitting on the wet carpeting.

Step Four: Make an Orderly Move

Move the large pieces of furniture first, and keep walkways clear of tripping hazards. Keep areas clear where you'll be setting up your drying equipment, and also keep the pathways clear where you want your air movers to blow air.

Remember that this is an emergency. *The insurance company expects you to do whatever's necessary to stabilize the structure and prevent further damage, but no more than that.* So don't move any more contents than absolutely necessary.

Figure 14-3 In this situation, contents were temporarily moved out of a wet bedroom and into the dining area off the kitchen.

Step Five: Label Everything

If you're moving items out of their original room locations, label them so you know where they came from. Use a paper tag with a safety pin, or a tag with a wire attached. You can also use a piece of blue painter's masking tape, placed on a portion of the contents where you know the tape won't do any damage.

Blue painter's tape has a light adhesive that adheres to delicate surfaces without transferring adhesive, so it's a good choice for labeling contents that are difficult to put into boxes or bins. However, always test the tape first, and place it on the back or bottom of the item. *Never use regular masking tape, duct tape, Scotch tape, or other types of tape!* They leave a residue that's hard to remove.

Step Six: Cover and Protect Everything

Cover dry contents with painter's plastic sheeting. This thin sheeting is inexpensive and easy to handle, and will protect the contents from dust that's kicked up by the air movers, or by other construction activity. If you're not sure how clean the floor is in the area where you're moving the contents — for example, a garage — place a sheet of plastic down before setting the contents on it.

Step Seven: Document Potential Problems

During the initial emergency, the owners may have already hurriedly and carelessly moved a lot of the contents. You may find a nightstand sitting on top of a bed, a chair on a table without any protection under it, or a wet rug laid across the couch. It's a common thing for people to do when they're in a rush, but it can cause additional damage that you don't want to be blamed for.

Rectify any situations like this that you see, but also document that you're not the one who did it. Take a couple of digital photos and make a note in your file to establish that this was how you found things when you arrived.

Another common problem is finding furniture and other contents with scratches, dings, missing hardware, or other damage that occurred before your arrival. Here again, you need to protect yourself with photographs and notes in your file.

On-Site Storage

Storing contents on-site during the reconstruction phase of a project is different from moving them for temporary storage during an emergency, when you're just focused on getting them out of the way. When you actually begin working on the building, you need to take a different, somewhat more permanent approach toward storing contents.

Your first decision is whether or not it makes sense to keep everything on-site. This is a judgment call you'll be making jointly with the homeowners and the adjuster. There are a number of factors that affect this decision:

- **Is there room to store them safely?** You may be dealing with a home or a commercial building that's packed with contents, and there simply may not be any room to safely store them on-site. Also, if the owners are going to be staying in the building, filling up several rooms with contents may make it difficult for them to function.
- **How extensive are the repairs?** If only a couple of rooms are affected, then it makes sense to keep the contents on-site. If you'll be working in several rooms, then moving everything into the garage or into the one or two unaffected rooms might

be a good option. However, shuffling contents from room to room as the work progresses is *never* a good solution. It's time-consuming and labor intensive, and also greatly increases the possibility of damage to both the contents and the building.

- **How long will the repairs take?** You might have enough room to move the contents into the garage, or into a barn or outbuilding. But if the repairs are going to take a couple of months and winter is coming, that might not be the safest place for them. Also, a barn or warehouse on the property may not be in use right now, but the repairs could stretch into the season when the barn or warehouse space is needed for another use. Under those conditions, tying it up for content storage might not be an option.
- **Do the owners need access to the contents?** You may have room on-site to store everything, but if it has to be tightly boxed and stacked within that space, it may no longer be readily accessible. Having everything in an off-site storage facility where there's more room, and where the contents will be more readily accessible, might be a better option, especially for a business.

Establishing a Storage Area

If you've made the decision to store the contents on-site, the next step is to decide on the storage area, then prepare it.

Interior Rooms

Interior rooms that the owner can temporarily do without are the best option. See Figure 14-4. These include bedrooms, guest rooms, home offices, home gyms, family rooms, etc. These rooms have the advantage of being clean, weathertight, temperature controlled, and secure. The only real disadvantage is that they're usually already full of other furnishings, so you have to be very careful when moving items into them. See Figure 14-5.

The Garage

Garages offer a lot of enclosed space, and they're often unaffected by the loss inside the house. But they tend to be cold, dirty, and often cluttered already. Also, they're usually used to store tools, equipment, gasoline, solvents, and other materials that could potentially give off odors or possibly discolor contents they come into contact with (Figure 14-6).

When using a garage for storage, clear a space specifically for this purpose. Clean the floor, and put down a 6-mil plastic sheet or a tarp for protection against moisture and dirt. If you lean items against the walls, be sure to protect them from dirt. If the owners continue to use the garage for cars, lawn mowers, sports equipment, etc., be sure that the contents are protected against exposure and physical damage.

Figure 14-4 In this home, the owners were able to do without the dining nook off their kitchen while repairs were done to the kitchen floor. That made a convenient storage area for the island cabinet, refrigerator, and a few other contents items. Notice the caution tape hanging from the chandelier as a warning not to bump into it.

Figure 14-5 When placing things in interior rooms for storage, take great care with how they're arranged and protected. This is especially true if they're going to be there for awhile. Notice the bubble wrap under the lamps, and the foam blocks under the nightstands.

Figure 14-6 When storing household content items in a garage, be sure that everything you're storing is well protected. As seen here, garages are often storage areas for oil, chemicals, fertilizers, pesticides, and a number of other materials.

Garages aren't usually sealed and insulated like houses. Some leak around the doors, and some have drainage issues with water from the driveway, or with water and snow brought in on car tires. Check with the homeowners to see if they're aware of any water issues; if there are any, place the contents where they'll be away from the water, or raise everything up off the floor on pallets.

The Basement

A finished basement is a great option, except for the hassle of taking everything up and down the stairs. An unfinished basement offers the same potential problems as the garage, and requires the same basic precautions. As with the garage, talk with the owner about any water issues. Be sure that the contents are raised, wrapped, or otherwise protected.

Outbuildings

Barns, storage sheds, and other outbuildings are a possibility, but they also present potential problems. These include insect, rodent, animal and weather damage, as well as the possibility of theft. Discuss the outbuilding with the owners, ask them about its current use and condition, and see how suitable they feel it would be for storage.

(Photo courtesy of On-Site Storage Solutions)

Figure 14-7 If you have the room, large storage containers like this one are great for on-site storage. They're weathertight and very secure.

Renting On-Site Storage Containers

Another very good option for on-site storage is renting a storage container. These are becoming more and more available, and they come in a variety of sizes and configurations.

Some types are essentially shipping containers, of the type you see on trucks and trains, like the one in Figure 14-7. They're delivered to your jobsite by truck, and placed wherever you want them. They're secure and weathertight, with a locking door at one or both ends. A shipping container has a tremendous amount of storage space, although once loaded, the contents aren't readily accessible.

There are some containers that have roll-up doors in several locations along the sides. These are more expensive to rent, but they offer easier accessibility, and may be a better choice for a business or for people who'll need frequent access to the stored contents.

There are also national companies that rent on-site storage containers for regional locations. Some local companies that offer Dumpsters or bins for debris removal also offer enclosed, lockable storage containers for on-site storage. They're delivered by the same trucks, and can be placed at the site pretty much anywhere a Dumpster can.

For short-term storage, you can also consider renting a truck or trailer. They're weathertight and secure, and can be placed just about anywhere on the property.

Storing and Drying Wet Contents On-Site

Remember that if you have contents that are wet, you can't simply place them into storage without risking additional damage from mold and mildew, or other secondary damage to wood and metal.

Wet contents should be moved to a location where your air movers can circulate air around them, and the dehumidifiers can extract the moisture. If you cover the contents to keep them from getting dirty, the coverings need to be loose so that air can circulate. If possible, move the wet contents into a separate room so you can close off the area to protect them from dirt and dust, and set up air movers and a dehumidifier specifically for drying the contents.

Contents That Aren't Salvageable

On just about every loss, even small ones, you're going to encounter contents that need to be disposed of because they're not salvageable. The items might be as minor as couple of rolls of saturated toilet paper in a wet bathroom, or a few boxes of smoky breakfast cereal in a pantry after a kitchen fire. Disposing of these items may seem pretty logical, since they're obviously ruined. However, even though you've been hired to do the restoration work on the home, it's important that you understand this: *That toilet paper and those boxes of cereal aren't yours to dispose of.*

They're part of an important chain of contents possession and salvage rights. With toilet paper and cereal, it's pretty simple. But when it comes to valuable contents, it can get much more complex, so let's take a closer look at how it all works.

Content Losses and Salvage Rights

The contents portion of a typical insurance loss is often more confusing than the structural loss. Most of the negotiations on the value and coverage for loss items are between the policyholder and the insurance company, so that won't involve you. However, a lot of the contents will be in your way at some point, and they'll either need to be moved or disposed of so you can work. So even though you're not part of their negotiations, the final outcome will definitely affect you.

Here's how it all works. At the time of the loss, the contents inside an owner-occupied building belong to the building owner. (Toward the end of the chapter we'll talk about what happens when the occupant of the building is a tenant.) If there's damage to the contents, the owner will be asked to inventory the damaged items. He'll have to make a list of everything that was lost, and its approximate value. As you can imagine, this can be a pretty tedious process; one we'll look at in more detail later.

The insurance adjuster will review the owner's inventory list. He'll settle with the owner for a mutually-agreed-upon sum of money, up to the limit of the contents coverage portion of the policy. That means the insurance adjuster is actually buying the damaged contents from the owner. This is important to understand: *Once the insurance company settles with the owner, the insurance company owns those contents, and they can do whatever they want with them.*

In just about every case, what the insurance company wants to do with the contents is have you dispose of them. Insurance companies aren't in the salvage business, so they pay you to clean up the site and dispose of all the damaged contents.

Tip! *If you're going to buy and sell salvaged items,* always be honest about it! *If something's been water- or fire-damaged, be sure to tell the buyer its history. Never try to pass something off as new or undamaged.*

Occasionally, however, the insurance company decides that the contents still have some salvage value. In that case, they may instruct you to hold onto the items and place them into storage. At that point, you can then charge them a mutually-agreed-upon storage fee, usually by the month. If they want you to pick the items up at the damaged building, and perhaps later deliver them somewhere else, you can charge for those services as well.

Eventually, the insurance company will decide what to do with the salvaged contents — a process that could take months. They may send a specialized appraiser out to see them to determine their exact value. They may offer them for sale privately, soliciting bids among interested parties, including the people in your company. They may even put them up for public auction.

REAL STORIES:

The True Meaning of a Fire Sale

A custom cabinetmaker in our area had a fire one evening. An overheated piece of woodworking equipment caught fire, causing a pile of sawdust and wood shavings to smolder and eventually burst into flames. The damage to the inside of the building was relatively light, but the fire generated quite a bit of smoke.

The cabinetmaker was a tenant in the building, and the building owner chose to do his own repairs. However, the cabinetmaker had a renter's policy, and he filed a claim to replace the ruined piece of equipment, the plywood and raw hardwood lumber he had in stock, as well as several partially-finished cabinets he'd been working on.

His insurance company settled the claim with him, and afterward they contacted us and asked if we were interested in bidding on the salvage rights. There were several stacks of smoke-damaged lumber and plywood, and other pieces that were lightly water damaged, but other than that were in great condition. There were also several nice upper and lower cabinets, and enough parts to make about 10 more.

We knew we were in a unique position. We had the equipment to transport the lumber and to deodorize it to remove the smoke smell, so we put in a relatively low bid. Since no one else bid on it because it was too much hassle to deal with, it was ours.

We moved it to our shop, cleaned it up, ran it through our ozone chamber, and it was all good as new. We took the cabinet parts and assembled them into completed cabinets. Then we ran an ad in the local paper explaining the situation, and offered the lumber and the cabinets at a very nice discount. We sold everything very quickly, made a profit for ourselves, and also made a lot of local woodworkers very happy!

Win-Win Salvage Possibilities

Not all contents to be disposed of necessarily need to hit the trash bins. Once the owners and the adjuster have completed their negotiations, examine what's going to be discarded. Some communities have opportunity centers, thrift stores, and rehabilitation centers. They may be able to take damaged furniture, cabinets, and other items and repair and refinish them. The workers have a chance to learn and practice new skills, and the centers sell the items for a profit to benefit their ongoing operations.

You might want to check into some of these centers in advance, so you know what types of donations they take. *But remember, these items belong to the insurance company and you'll need their permission to make this donation.* It's also a good idea to clear it with the homeowner. They might not be comfortable with having their things donated, or the particular nonprofit agency taking them may be associated with a cause that they don't happen to support.

Green is Good

Another possibility is to talk with your local trash hauler. They can often set up a separate Dumpster at the site for wood and other recyclable materials. You can easily separate recyclable items as they're taken out of the house, and toss them into the two separate bins.

The insurance company won't reimburse you for the additional time it takes to separate the trash. However, most landfills charge you a lower rate for recyclable materials than they do for regular trash, so the savings should help offset the additional labor. And you'll be helping the environment at the same time!

Dumpster Diving

Realistically, the majority of the nonsalvageable content items are simply going to be thrown away. But remember: *You don't own them, and you don't have any rights to them. And neither does the owner.* To put it simply and clearly, that means *no Dumpster diving!*

Maybe you see a slightly-scorched table about to get thrown away. The thought goes through your mind that you could take that table home, sand it and refinish it, and get some good use out of it. So, you start to grab it out of the Dumpster. *But it's not yours to grab.* If the adjuster happens to pull up at the very moment you're loading it into your truck, it's going to look pretty tacky. You're essentially stealing from the company that agreed to work with you on this project. That doesn't set a very good tone for your relationship with them.

And what if the homeowner sees you or your crew scrounging in the trash after their old table? Does that make much of an impression? How do you think they're going to feel about the professional, trustworthy contractor they just hired to repair their home?

No Dumpster diving applies to the homeowners as well. If they say that they'd like to grab that table back out of the Dumpster, you're obligated to politely explain that the table now belongs to the insurance company, but if they're interested, they can talk to their adjuster about buying it back. You can also explain that for liability reasons, no one except your employees or employees of the Dumpster company should be in or around the Dumpster.

The issue of salvaging contents *will* come up sooner or later, so you need to be ready for it. *Set a firm company policy for your employees that nonsalvageable content items are never to be taken off a jobsite — by anyone, for any reason.* Make it part of your company handbook, and set some strict consequences for violations.

Odor Control

As we discussed earlier in the book, the first step in odor control is to remove the source of the odor. So if you're dealing with nonsalvageable contents in the aftermath of a fire, the sooner you can get them out of the building, the better it is.

But discarding contents isn't easy, since everything needs to be inventoried first. For the owners, there's often a sentimental attachment to many of the items. So there's always a bit of a conflict; you want to get everything out as soon as you can, but there are a lot of reasons why you have to wait.

There are a couple of ways you can handle this situation. As you'll see in the next chapter, it's recommended that *salvageable* contents remain in the house and be treated with odor counteractants. This is done as part of the process to stabilize the odors in the structure itself.

So if it looks like it's going to be a while before you can remove the nonsalvageable contents, start tearing out as much of the structural materials as you can. Then, do your initial deodorization work with the *nonsalvageable* contents still in place. It's not an ideal situation, but it's better than holding off on odor control steps that need to be done as soon as possible.

Another option is to move the nonsalvageable contents to another location for sorting and inventory. This could be your warehouse, an outbuilding, or a rented space. Weather permitting, sometimes it's even as simple as just moving them outside, as in Figure 14-8. This has the advantage of getting them out of the building and allowing you to get started on the work inside. Having contents outside where there's more air, light and space makes it easier to work with them, and it can be less traumatic for the homeowners as well.

Moving unsalvageable contents to an alternate location involves extra time and labor that the insurance company may not want to pay for. It also means that another space gets dirty and has to be cleaned. You'll also have to clean your truck. Plus there's the cost of extra boxes and packing materials to consider. This is something you'll want to discuss with the adjuster to be sure you'll get paid for the additional work and materials.

Figure 14-8 To help with the odor situation inside the house, the sooner nonsalvageable contents can be moved out, the better. If weather permits, it's sometimes helpful to simply move things outside for sorting, inventorying, and eventual disposal.

Setting Content Values

While you won't typically be involved in the content value negotiations, they can often have an impact on the job, especially with larger losses where content policy limits may be reached. It's helpful to understand how the valuation process works.

Suppose a homeowner has a wooden dining room table and four chairs that are damaged in a fire. It's of average age, quality, and condition. It was originally purchased new at a furniture or department store, and cost in the range of $400 to $600. The insurance company assumes all of that. If the homeowner has a replacement cost policy, there's no depreciation to worry about. So, if they list a value of $500 for the table and chairs on their loss inventory, chances are the adjuster won't question it.

Now suppose the dining room set was made by a high-end furniture manufacturer and the original cost was $2,000. The homeowner lists that cost on the loss inventory. If he has the original receipt, there won't be a problem. If he doesn't, then the homeowner or the adjuster will need to establish two things: The dining set really was made by that furniture manufacturer, and that it was really valued at $2,000.

Now let's go one step further. Suppose the homeowner bought the dining room set at an antique store for $10,000, or had it custom made for $12,000. One-of-a-kind items like these can't be verified with anything other than the original receipt. Either the owner will have to be able to produce the receipt, or he'll have to go back to the original shop and get some type of proof of the purchase. Even at that, many policies have limits on their coverage for antiques and other specialty or limited edition items, unless the homeowner has a special rider on his policy.

Food and Other Perishable Contents

The final decision on whether to keep or dispose of contents items belongs to the homeowner, usually in conjunction with the adjuster. However, you'll often be consulted in that decision. One area where your input is very important is with food.

There are a couple of factors involved in the decision to keep or dispose of food and other perishable items after a loss. Here are a few *general* guidelines:

- **Water losses:** If there was water on the floor of the kitchen or the pantry, but it didn't contact the food, then closed containers are probably all right. Depending on how long the water was in the rooms and where it traveled, recommend disposing of open food containers due to the possibility of contaminants carried by the water having become airborne.
- **Fires:** In a fire, it's best to dispose of all food in open containers. If the fire was sufficient to deposit soot or smoke residue on closed containers, they should be disposed of as well. Homeowners are often reluctant to dispose of bottles of wine, beer, and liquor, especially if they're unopened. It's very difficult to determine how these spirits are affected by the heat and smoke of the fire, so always suggest that they speak with an expert at a local wine shop or liquor store, and see what they recommend.
- **Mold and contaminated water losses:** If the mold loss occurs in the kitchen, pantry, or other food storage area, or the spore contamination spreads to any of those areas, then all food should be disposed of, whether it's been opened or not. If the mold or water contamination occurred in another area of the home and the food storage areas are contained from the contaminated areas, defer to the recommendations of the industrial hygienist.
- **Loss of power to a refrigerator:** Food inside a refrigerator begins to spoil rapidly once the temperature rises above approximately 40 degrees F. If there's any possibility that the power was off long enough for the temperatures inside the refrigerator to rise that high, inventory all the food and recommend it be disposed of.
- **Loss of power to a freezer:** If the power to a freezer was off long enough for any of the frozen items to show signs of melting or softening, inventory all the food and recommend it be disposed of.

- **Prescription and over-the-counter medications:** A small water loss is probably not going to affect medications in sealed containers, while a fire, mold, or contaminated water loss typically means medications should be discarded. *However, never place yourself in the position of making the final decision on this. Instead, always advise your clients to consult with their pharmacist or health care professional.*
- **Other perishables:** Open tubes of toothpaste, bars of soap, open containers of shampoo, containers of makeup, and other such items are all considered perishables. They can all be affected by the heat and smoke of a fire, or by airborne contaminants after a water, mold, or sewage loss. Many of these items aren't worth the time and expense of cleaning and packing, and should be inventoried and discarded.

Content Loss Inventories

Tip! *When dealing with food and other perishables after a loss, if you're unsure about whether something's safe, advise the owners to discard it. It's best to be on the safe side; but always let the homeowner make the final decision.*

When a loss occurs, remember that contents are separate from the structure. You can write the estimate for the fire damage to the kitchen cabinets, but you're not there to estimate how much of the food inside those cabinets was lost. It's up to the building owner to inventory lost or damaged contents.

To help you understand content losses and loss inventories, and how the three parts of the loss — emergency, contents, and structure — come together, we'll look at a couple of examples: a small loss and a large loss.

Small Loss Claims

Small losses are usually straightforward, both for you and the homeowner. Our first example is a small water loss for a client named Sally.

Sally's water loss occurs in her bathroom and extends into the adjacent bedroom. In the bathroom, some rolls of toilet paper, boxes of tissue, a can of cleanser, and a cardboard box with several partially-used makeup items were in the vanity, and are damaged by water.

In the bedroom, some books that were sitting on the floor are damaged by the water, as is a wicker waste basket. Some very expensive leather shoes are also water damaged and can't be repaired. A jacket lying on the floor will need to be dry cleaned.

All Sally has to do is fill out a simple loss inventory form provided by the insurance company. She needs to list the items in the vanity cabinet, and put down an estimated value for them. The adjuster won't expect her to list each item in the box with the makeup individually, unless they were of unusually high value. Instead, she lists something like "10 misc. makeup items avg. of $6.00 each."

In the bedroom, she needs to list the waste basket and the books, and put a value on them. She lists the shoes, but due to their unusually high value, the adjuster will want a receipt, which she has. She'll send the jacket to the dry cleaner, pay for it herself when she picks it up, and submit the receipt to the adjuster for reimbursement.

When you arrive to do the emergency work, Sally has already bagged up all the loss items except for the shoes, which the adjuster wants to take a look at. You do your water extraction and drying, and write an estimate for repairs.

The issue with the shoes is settled, and the insurance company agrees to the value, based on the receipt she provided. Sally also requests and receives approval from the adjuster to donate the shoes to a local charity. You haul the other damaged contents items away along with the structural debris.

Once your estimate is approved, you write a contract with Sally to do the repairs. You submit the bill for the emergency work directly to the insurance company, and submit a separate bill for the repair work to Sally, in accordance with the terms of your contract. Sally submits her loss inventory and her receipt for the dry cleaner to the adjuster. The adjuster reviews everything, pays your emergency bill and pays Sally for the contents and structural losses, minus her deductible. Sally in turn pays you for the structural repairs, including the deductible amount, which she must cover out of her own pocket.

Large Loss Claims

On large losses, everything is more complicated, especially the loss inventory. Until you've been in a room that's been destroyed by a fire, you can't begin to imagine how difficult it is to reconstruct what the room contained.

Look around the room where you're sitting right now. Look at all the different objects it contains. Think about all the things that are stored in the drawers, or in closets, or tucked away in other storage areas. Then try and picture it as nothing but black and gray, with everything covered in soot and familiar objects fused together into unrecognizable lumps. That's what homeowners have to face when putting together a loss inventory after a major fire loss.

Let's say that a fire starts in Bob and Lisa Smith's living room. It rages through the dining room, kitchen, and home office before breaking through into the upper floor. The flames destroy the master bedroom suite, one adjacent bedroom and bathroom, and heavily damage another bedroom. There's also heavy smoke damage in the guest bedroom, another bathroom, and in the attic.

Bob and Lisa have lived in this 3,400-square-foot home for over 30 years, and have a lifetime of contents in those rooms, as well as what's stored in the attic. Bob operates a home-based business from the office on the first floor, and all of his records are stored there.

Approximately 50 percent of the home's contents aren't salvageable, and need to be inventoried and discarded. Approximately 30 percent are salvageable, but need heavy cleaning and restoration work. The remaining 20 percent require light cleaning.

Bob and Lisa are, understandably, completely overwhelmed. They have a number of friends and neighbors who offer to help deal with the contents, but parts of the building are unstable. You need to make the building weathertight, close off and shore up dangerous areas, clean up broken materials, and make the site as safe as possible.

Once that's done, it's time to begin the inventory process. It's essentially the same as what Sally did with her small loss, only on a much, much larger scale. Everything needs to be inventoried, and everything needs to have a value put to it.

Some of the things can be lumped together, such as so many boxes of cereal or so many bars of soap. They can take average costs, such as $500 for the dining room set, or whatever the normal average is for the quality of furniture that Bob and Lisa have in their home. But any highly unusual or valuable items need to be verified and documented.

Bob and Lisa face another problem. Many of their belongings were accumulated over time, so average costs don't apply to everything. For instance, they bought a limited-edition print for the living room, which has gone up in value. They bought a new couch, but kept other older furnishings. When their son moved out, they got rid of his bedroom set and turned that room into a modern home theater. They made other changes over time, and they didn't really think that much about it.

On a loss of this size, there's the very real possibility that they'll run out of contents coverage. So they're going to be faced with some very tough decisions.

Decisions, Decisions

Let's say Bob has a favorite chair, and it's worth $200. It's probably ruined, but he wants you to try and clean it anyway. You spend an hour cleaning it, and charge the Smiths $45. They're not happy with the results, and end up discarding the chair.

If they had thrown the chair away initially, the insurance company would have reimbursed them $200. Instead, they'll be reimbursed $245, to cover both the chair and your bill for the attempted cleaning. The end result is the same — they get a new chair. But they've dipped into their pool of contents money for $245 instead of $200.

Everything that the Smiths do is charged against that pool of available content money. And remember that content coverage and structural coverage are two different things, even though it's the same house and the same policy. If they use up all of their contents coverage, but haven't used up all their structural coverage, they can't dip into the structural money to make up the difference.

Here are some of the things that come out of the contents coverage money:

➤ **Replacement items:** Everything that the homeowner buys to replace items damaged in the loss, from a new toothbrush to a new big screen TV.

- **Content cleaning:** This includes cleaning hard furniture, cleaning upholstery, dry cleaning clothing, washing dishes, and even doing laundry. It also includes both on-site cleaning of contents, and cleaning off-site in your shop.
- **Content repairs:** This could be mending clothing, or repairing a burn mark in an antique rug. It includes refinishing a piece of furniture, reupholstering a chair seat, or refurbishing and painting a favorite bicycle.
- **Specialty treatments:** This includes ultrasonic cleaning, ozone treatments, thermal fogging, drying, and other specialized treatments performed on contents items.
- **Pack-out and pack-in:** Whenever you're handling contents, everything related to those contents is charged against the contents coverage. That includes wrapping and boxing contents; moving things from room to room; loading and unloading trucks; moving contents into and out of storage facilities; and moving contents into and out of the home. The only exception is the moving of items necessary for performing emergency services, which is charged as part of the emergency.
- **Packing supplies and equipment:** This includes moving trucks, boxes, packing supplies, inventory equipment, and specialty rental equipment.
- **Inventorying:** In some cases, homeowners and business owners are completely overwhelmed by the prospect of doing a loss inventory, so the insurance company may request that you assist with the inventory. That labor is typically a charge against the owner's contents coverage.
- **Debris removal:** The removal and disposal of nonsalvageable contents are also charged against the contents coverage. This is one place where a good adjuster has a little bit of flexibility if content coverage is tight. They'll sometimes allow debris removal to be charged against the structural coverage instead of against the contents coverage.
- **Storage:** The long-term storage of contents is also charged against the contents coverage.

With that in mind, Bob and Lisa eventually complete their inventory. Due to the coverage issues, they opt to clean some items rather than replace them. Other items they have to throw away but not replace.

They take advantage of some volunteers to help out with the inventory chores. And as a goodwill gesture, you provide one employee for two days at no charge to help the Smiths and oversee the volunteers with the loss inventory. You also have a stockpile of used boxes from a previous job that you donate.

Because Bob works from home, he has a special rider on his policy that covers his home office. So all his paperwork and home office equipment is handled separately. That frees up some coverage from the home's contents, but it also requires that the office contents inventory be kept separate from the home contents. Some of the paperwork in the office requires special deodorization work. Some of it has too much odor to be salvageable, and has to be photocopied and then shredded. All of the labor and materials associated with that work is tracked and billed separately.

Bringing the Loss Together

The example with Bob and Lisa Smiths' contents is a pretty common one. But there's more to this example, which helps illustrate how an entire loss of this type all comes together in the end.

In addition to all the contents work, you contract with the Smiths to handle the structural repairs. You write an estimate for the repairs, and submit it to the Smiths and their adjuster. As you'd expect with a complex project like this, along the way you write two supplementals for hidden damage. The Smiths also decide to do some remodeling while the rest of the work is being completed, so later on you write a separate bid for that work directly to them.

In the end, the adjuster has a massive file on this loss. He has a stack of invoices from you for the initial emergency work, plus in-progress billing on the original estimate and the supplementals. He also has cleaning bills from you, and repair bills from you for restoration work done on the Smiths' furniture.

He has a pack-out bill from you for taking the contents out of the house and moving some of them into storage. He has another bill from you for moving some of the contents into the Smiths' rental home, and a pack-in bill for when you moved the Smiths back into their restored home. He has a bill for five months of contents storage, and he has a stack of invoices from the Smiths for replacement tables and chairs, clothes, and all the rest.

But in the end, it all comes down to the same three basic elements of an insurance restoration loss: an emergency loss, a structural loss, and a contents loss.

Tenant Contents

Let's take a moment and go back over something we discussed back in Chapter 2, because it's very important. If the home you're working on is a rental property, neither you, the building owner, nor the insurance company have the authority to do anything with the tenant's contents.

If the tenant has a renter's policy in place, then that policy will kick in and provide coverage for the tenant's contents. In general, it works the same way that a homeowner's policy works. There'll be coverage for items that are destroyed or need to be repaired or cleaned. There'll also be coverage for packing and moving the contents, for storage, and for debris removal.

If there's a policy in place, the tenant will have the right to choose his own contractor for the content work. He can hire anyone he wants to come in and move the contents, do the packing and debris removal, etc. The only thing he can't do is unreasonably delay the repairs to the structure.

If the tenant *doesn't* have a renter's policy in place, then you have a whole different situation. Once again, as in Chapter 2, the owner of the building may have to step in and offer to pay for the cost of dealing with the tenant's contents. Or, the damage caused by the loss to the building may be all mixed in with the tenant's unsalvageable contents. In that case, sometimes the adjuster may just tell you to toss everything and include it with your debris removal bill to the insurance company.

The main thing is to be sure that someone else gives you written authorization to deal with the tenant's contents. *Never take it upon yourself to move, dispose of, or otherwise handle a tenant's contents.* You could be opening yourself up for a lot of liability — and it's just not worth it.

15

Contents: Pack-Outs, Cleaning, & Off-Site Storage

Sooner or later you'll need to move some contents off-site for cleaning, restoration, repair, and storage. This is especially true with larger losses, when dealing with contents on-site can be challenging. Moving most or all of the contents out of a building is what's known in the industry as a *pack-out* (Figure 15-1).

We've talked previously about the fact that handling contents makes you a full-service restoration company, and that it can also have a very healthy impact on your company's bottom line. But there are some serious downsides to this part of the business, and now is probably a good time to consider them.

When you do a pack-out, you could be moving as much as $25,000 to $50,000 or more in contents out of someone's home and storing them in your warehouse. If you have contents from several homes in storage at the same time, you can begin to understand the magnitude of your responsibility!

Dealing with pack-outs and content storage requires a lot of organization and care. You have to be detail-oriented, and pay very close attention to what you're doing. You'll also need to maintain adequate insurance coverage, because you'll be responsible for any accidents or damage that occurs while the contents are in your possession.

If this phase of the restoration process isn't something you want to handle on your own, there are subcontractors who'll handle it for you. But remember, if you're acting as the project's general contractor, you still have ultimate responsibility for all those contents.

If you're going to handle contents for your clients, you'll need to know how it's done, so let's look at the correct way to move contents.

Figure 15-1 This house suffered a major fire on the upper floor. The remaining undamaged contents on the lower floor are boxed or tagged and awaiting pack-out for cleaning and storage.

Content Pack-Outs

There are several reasons why a pack-out is a good idea:

- It clears out the building, making it easier for you to do your repair work;
- It protects the contents from possible damage during the restoration period;
- It's easier to clean, repair and inventory everything when it's stored in one place;
- It can make the contents more accessible to the owners.

The disadvantage is that a pack-out takes time, which equals money. So if the contents can be handled on-site, that's usually what the insurance company would prefer. It's sometimes up to you to present your case to the adjuster and justify why a pack-out makes both practical and economic sense.

Once the decision's been made to do a pack-out, there are four basic steps:

1. Inventorying
2. Packing
3. Transporting
4. Storing

Before the contents go into storage, there's often cleaning and repairs to be done. We'll look at that part of the job in more detail later in the chapter.

Pack-Out/Pack-In Inventory

A pack-out/pack-in inventory is different from a loss inventory, and there's a very important distinction between the two:

- A loss inventory is something the *owners* put together for the insurance company. It's done to document those contents items that were ruined and they're seeking reimbursement for. The loss inventory is the owner's responsibility.
- A pack-out/pack-in inventory is something that *you* put together. It lists exactly what items you're taking from the property. Once you put something on that list, it becomes *your* responsibility, so you better keep track of it! It's also the checklist you use when the items are returned or packed back into the home, which is why it's also known as a pack-in inventory.

You'll need to develop your own pack-out/pack-in inventory sheets to meet your specific needs. You may be able to adapt an existing inventory sheet, hand write your own form, or create one using a computer program such as Word or Excel. It should contain the following basic information:

- **Identification:** The client's name, job name, job number, or other means of identifying who the contents belong to. If you have a lot of jobs in storage, to maintain the confidentiality of your clients you might want to consider using a number to identify them rather than their name.
- **Dates:** The date of the loss, the date the items are packed out, and the dates the items are packed back in.
- **Room:** The name of the room the item was removed from. This is important not only for identifying the item, but it also helps when moving things back in.
- **Item description:** A blank area for writing in a description of the item. When filling this in, the description should be as complete as possible, including color, size, shape, brand, identifying marks, etc.

- **Condition:** Use this area to note any condition issues, which is *very* important! If you see a scratch on a nightstand as you pack it, make a note of it to document that it's a preexisting condition. You may also want to have check-off boxes here, with "excellent" "good" etc.
- **Names:** A space for the name or the initials of the person who packed the item, who put it into storage, and who packed it back in.
- **Client sign-off:** A space at the bottom for the client to sign and date each sheet as the pack-in is completed and each page is checked off, confirming that everything has been delivered and returned.

Packing

Once the items are inventoried, they're ready to be packed for transport. Begin in one room, and proceed in a logical, orderly manner. See Figure 15-2. You may have multiple crews packing items in different rooms, and that's fine. Just avoid half-filled boxes, or boxes that mix contents from different rooms.

Inventory the items as you pack them, and fill out all the required information on the pack-out/pack-in inventory sheet, especially the item's condition. Smaller, non-breakable items like canned goods or table linens can go directly into boxes. Wrap small breakable items in sheets of bubble wrap (Figure 15-3) or unprinted newsprint (Figure 15-4) for protection. Fill voids in the box with crumpled newsprint to prevent the contents from shifting during transit. Don't overfill the boxes. They shouldn't be too heavy to lift comfortably.

Figure 15-2 The pack-out of contents requires care and organization, as well as careful handling. That's especially true with delicate items such as glasses, dishes and other kitchenwares.

Figure 15-3 Fragile contents items wrapped in bubble wrap for transport.

On the outside of the box, list the name or identification number of the client, and the room the contents came from. When the box is full, seal it with tape to prevent anyone from tampering with the contents after the box has been inventoried. If possible, leave the box in the room where it was packed until it's ready for transport to storage.

Large items, such as tables, chairs, couches, etc., need to be listed on the inventory sheet in the same manner. Tag each item with a paper tag that shows the client's name or number and the room, then wrap it with mover's blankets for protection while in transit. If the items are smoky, cover them with plastic first to protect the blankets.

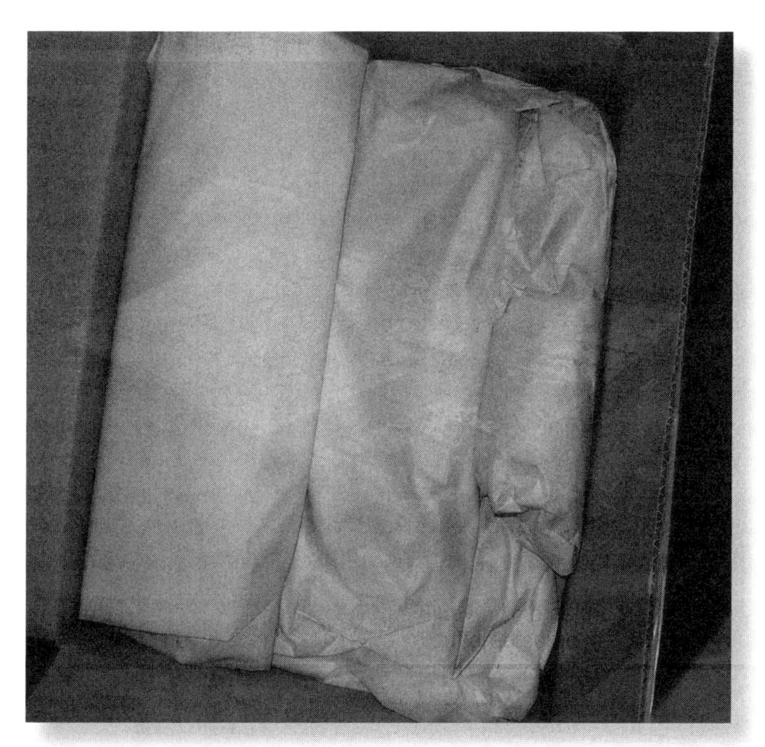

Figure 15-4 Less fragile items can be wrapped in unprinted newsprint and boxed for transport or storage.

Tip! *Don't overlook the attic, basement, and crawlspace. People often store contents in these areas, and if they've been affected by the loss, they need to be moved.*

Track all the materials you use for packing. You can charge for the boxes, bubble wrap and paper, as well as a generic additional charge that covers miscellaneous tags, tape, forms, and other packing supplies. To simplify tracking the number of boxes and other supplies you use, count everything when you first arrive at the job, then count everything remaining when the pack-out is complete. The difference is how much you used on that job.

If this is a fire situation, these boxes won't be reusable after the pack-out. If you have an area where you can store them and not risk the odor contaminating anything, then you can hang onto them for demolition use on fire jobs in the future. Remember that they're labeled with a homeowner's name and you've already charged the insurance company for them, so you can't use them for another pack-out. The other alternative is to just dispose of them by recycling them at your local landfill.

Transporting

Transporting contents can be done in your own moving truck, or in a rental truck. If you use your own truck, you can charge a fee for the use of the truck, plus the driver's time. If you rent a truck, you can charge for the rental fees, plus the driver.

Rental trucks come in a variety of sizes. Ideally, get one that fits comfortably into the driveway of the home, and will handle all of the contents in one load. The rental company can help you estimate the size of the truck you'll need based on the size of the house. If necessary, make multiple trips rather than deal with a truck that's too large, or one that's overloaded or packed too tightly.

Trucks with hydraulic lift gates that move vertically up and down, from ground level to the level of the truck bed, are the easiest to load. If one of these isn't available, get one with a loading ramp; the lower the truck's bed, the easier it is to load.

Load the heaviest items, including appliances, first, toward the front of the truck. As a general rule of thumb, if it takes two people to lift it, it should be placed toward the front of the truck. Put long items along the sides of the truck next, including rolled up rugs and carpets, mattresses and box springs, couches, and headboards. Wrap everything with mover's blankets and plastic for protection.

Fill in the rest of the truck with boxes and smaller furniture, stacking things carefully so the load won't shift in transit. Use ropes or load stabilizer bars — which you can rent — to separate the load into sections and prevent movement.

Load heavier boxes first, on the bottom, with smaller and lighter boxes used to fill in wherever there are openings. This is another reason why it pays to have all your packing boxes in just a couple of uniform sizes. It makes loading the truck in a consistent and organized manner considerably easier.

At the back of the truck, leave room for odd-shaped items. This is often a good place for things like potted plants, statues, lawn furniture, and other unusual items. Once again, be sure everything is padded and tied down.

Storage

If the contents have been in a fire, chances are everything is coming back to your facility to be cleaned. That's a multi-stage process, which we'll describe later in the chapter. If the items don't require cleaning, or if they've already passed through your cleaning process and are now ready for long-term storage, you have three different handling options: storage at your facility, storage at a rental facility, or delivery to a rental property.

Storage at Your Facility

If you have the space, storing contents at your own facility is the ideal option. It gives you complete control over everything: You know how secure and weathertight the building is; and you know exactly who has access to the contents, and you can control that access. Since you're charging rent for the storage, all of the rental income goes into your pocket.

If you live in a cold or humid climate, providing temperature- and humidity-controlled storage at your facility is an important plus. Many of the items you'll be putting into storage are wood, and they're coming from homes where they've been in a relatively controlled environment. If they're stored for long periods in cold temperatures or high humidity, there's the risk of secondary damage. Remember that you documented the condition of the contents when you took them, so if they get damaged while you have them, that's on you.

Not all restoration companies are set up for storage at their own facility, so this helps set you apart from your competition. If your competition has to rent storage spaces, you'll be able to show your clients and adjusters that with your company, their contents are in a space that's secure, climate-controlled, and organized. And, they'll have access to anything they need whenever they need it — something your competitors can't offer.

REAL STORIES:
It Pays to be Organized

There's a lot to be said for having a good crew and a good inventory system! We were once hired for a fairly extensive fire repair, including a complete pack-out of all the contents. Under the supervision of our Cleaning Division Supervisor, the subcontractor we used for all of our smoke-damaged laundry items did a complete inventory of the bedrooms and closets. They packed out all the clothes and shoes, and took everything off to their shop for cleaning.

Several days later, it all came back from the cleaners. We checked it back in, and it went into storage at our facility with everything else from the home. Since we had the entire houseful of contents, it took up quite a bit of space, and a number of boxes and storage vaults.

One afternoon, the homeowner came by. "I really hate to bother you," he said to our Cleaning Supervisor, "but is there any chance I could get my slippers? They were in the master bedroom closet." Then he looked out at all the boxes. "Oh. I guess you'll never find them in all that. Never mind."

The supervisor pulled the inventory lists from the cleaners, checked them against her own, and went right to where all the master bedroom clothing items were stored. Within a couple of minutes she was back, presenting the delighted homeowner with his slippers, freshly cleaned and wrapped in plastic!

Another time, we had a houseful of contents from a major water loss. The homeowner came by in a panic one afternoon. "I didn't think about it at the time, but all my bills and my checkbook are in the middle drawer of my desk," she said. "I need to find them right away or my bills will be late! My furniture is somewhere here in your warehouse!"

Once again, our Cleaning Supervisor pulled the pack-out inventory forms, located the desk on the form, and found where it was located in the warehouse. She took everything out of the drawer, boxed it up, and had it in the worried homeowner's hands in just a few minutes.

Slippers and a checkbook — two simple things — but because we were organized, we were able to make two clients *very* happy.

Storage at a Rented Facility

If you don't have room to store things at your own shop, then a rented facility is the only other option. Be sure the facility is secure, weathertight and climate-controlled, if needed. Make sure you know who has access to the facility, and during what hours of the day. Check with your insurance agent to see if storing contents at an off-site facility will affect your insurance coverage.

Delivery to a Rental Property

With a major loss, where the repairs are going to take several months, it's not unusual for the homeowners to move into a rental house. If they do, they may want some of their contents moved to their temporary new home. This might happen directly from one house to the other if no cleaning is needed, or from your shop to the house after cleaning has been done.

Release of Contents Form

Once you've done the pack-out inventory, *you* have control of the contents. So whenever someone wants to get something back, you need to keep track of what you're releasing. If you don't, it can cause all kinds of confusion later.

You can track items using the original pack-out/pack-in inventory form. Locate the items being released to the owner on the form, then have them initial and date the form to indicate that the item has been returned to them.

A safer and better alternative is to make up a simple release form for the client to sign. It should have spaces for the client's name, date, and the item(s) being released. Have them sign and date the form, give them a copy, and keep a copy in your file. This will eliminate any question about what's been released, when, and to whom.

Pack-Ins

When the job is finished, the final contents step is moving everything back where it belongs. This is known as the *pack-in*. Here again, you can charge for loading your truck, transport time, and truck and driver fees, including rental trucks.

The insurance company typically pays for all of the costs normally associated with moving a homeowner back into his home, within reason. That includes reassembling furniture and moving it back into the proper rooms; unpacking boxes; helping the homeowners put things back into cabinets, closets and drawers; and simple chores such as hanging pictures.

Remember that the cost of the pack-in comes out of that pool of contents coverage, so the amount available is subject to the limits of the homeowner's contents policy. Stay in close contact with the adjuster, especially as the job winds down, so you know if the owners are getting close to maxing out their contents policy limits. If they do, they'll have to pay for some of the additional labor on their own, and you may have a hard time collecting.

As you pack everything in, be sure the homeowners inspect the condition of the items and sign off at the bottom of each pack-out/pack-in sheet to acknowledge receipt of the contents. Keep a copy of these pages in your job file as proof of satisfactory redelivery of the contents.

REAL STORIES:

It Pays to Have Proof

We had packed-out a large quantity of contents from a house after a fire, and had carefully inventoried everything, as usual. After cleaning and ozone treatment, the contents were placed into storage at our warehouse. Repairs were completed on the home, and the homeowners were ready to move back in.

That day, we delivered all of their contents out of storage, checking everything off as it was brought into the house and unpacked. In the family room, there were several valuable antique artifacts that the family had collected. The owners carefully inspected each piece, and signed for them.

About a month later, we got a call from the homeowner saying that a couple of the artifacts were missing. They had looked everywhere, but couldn't find them. They were claiming that we didn't redeliver them! We took our copies of the pack-in inventory forms out to their house to show them. There in black and white was a clear description of the missing items, and their signature acknowledging delivery and acceptance.

We never heard from them again, and I don't know if they ever found the items. But without that form, and without the diligence of our pack-in crew who made sure it was signed, we wouldn't have had the proof that we needed to avoid any further accusations.

Content Cleaning and Repair

Early in the moving process, you'll need to decide whether or not contents need to be cleaned before they're moved and stored.

With some types of water losses, contents items like towels and clothes that were on the floor and got wet will require dry cleaning or simple laundering. In some types of storm and impact damage, where the shell of the building was penetrated, you may have to deal with furniture and other contents that got wet or dirty.

Homeowners may choose to handle this type of cleaning on their own. They can sometimes work out an arrangement with the adjuster to use some of that labor to offset all or part of their deductible.

Where content cleaning really becomes a major issue is with fire losses. How much cleaning is necessary — and, in fact, whether cleaning is even a practical alternative — is dependent on the type of contents you're dealing with, and the type and severity of the smoke residue.

For example, cleaning organic smoke residue off a dish is far different from cleaning synthetic smoke off a teddy bear. An unfinished piece of wood furniture that's been in a protein fire will require different treatment than a leather coat that's been in a very hot and smoky wood fire. There's virtually an unlimited number of possible combinations.

As with odor removal, there are books and classes devoted solely to the cleaning of contents, and this is an area where you'll need to invest in continuing education to keep up with the latest techniques and products.

Initial Content Deodorization

Part of the initial structural deodorization process described in Chapter 9 was a combination of chemical containment, vapor phase containment, and penetrating deodorizer, such as thermal fogging. The same concept is true for the contents.

Once you remove the nonsalvageable contents from the building, the remaining contents should be subjected to the same deodorization steps as the rest of the structure. To help minimize how much odor is taken up by the contents, do this as soon as possible. That'll help with the overall cleaning process and make deodorizing the contents easier. Since the structure is being deodorized anyway, this isn't usually an additional cost that needs to be charged against the contents coverage.

Tip! *Thermal fogging a home with the contents still in place can complicate the exit routes through the building. Plan your routes carefully, and move the contents as needed to ensure a clear pathway.*

Test Cleaning and Evaluation

As with the structure itself, you may be able to clean and salvage many content items, but not everything is worth the effort that it'll take. You'll want to do some test cleaning and evaluating, then work with the homeowners to make decisions about keeping or discarding certain items based on their age, condition, value, sentimental importance and the cost to restore them.

After a water loss, look for water stains, rust, swollen and warped wood, and separated joints. After a fire, you'll find that scorch marks from heat and floating embers, bubbled finishes, and heat discoloration are all common problems, as are joints that become weakened when wood expands and contracts and glues get heated.

Cleaning and Repair Subcontractors

As you might imagine, there's more to cleaning and repairing contents than just wiping something down with a rag, or gluing a loose joint in a piece of furniture. There are companies that specialize in nothing but the restoration of water-damaged documents, or the cleaning of smoke-damaged fabrics.

You can't be all things to all people, and you don't want to be. You need to understand what you're good at and what's practical to keep in-house. Then subcontract the rest to people that you can count on to give you consistent, professional results.

Put together a team of good subcontractors *before* you need them. Then, when you're asked to handle someone's 200-year-old water-damaged dresser, you can say with confidence that you work with a company that does excellent antique restoration work.

Ask the adjusters you work with for recommendations. Their only goal is to get jobs done quickly and professionally, keep their clients happy, and have as few hassles

REAL STORIES:
Some Things are Just Worth Saving

A very nice older gentleman had a fire in his home. It wasn't a terribly bad fire from a structural standpoint, confined primarily the kitchen and living room. But there was also smoke and water damage in a couple of the adjacent rooms, so a lot of the contents were affected.

With the help of his daughter, he went through everything without too much trouble. We finished our inventory and were starting to pack things out for off-site cleaning, when we got down to the area rug under the living room coffee table. The blue patterned rug was still wet from the firefighting efforts; it was almost completely black with soot and ash, and there was an obvious burn hole in it from a large ember.

We knew from experience that the rug could be sent out to a specialty shop and repaired, but that it would be an expensive undertaking. Usually this type of repair is only done on very valuable Oriental rugs. This rug appeared to be of average quality and we doubted he'd want to invest that much money in it, so we asked him if he wanted to list it on his loss inventory.

It was a long moment before he could answer. "That rug was the very last thing my wife and I bought together, just before she died of cancer last year," he told us quietly. "Please do whatever you can with it. I don't even care if it still has a hole in it. Just make sure it gets cleaned up and brought back."

I doubt there was ever a single contents item that received more attention or more loving care in our shop than that simple blue rug. After the house was repaired, we returned it to the living room with the rest of the contents. We made sure it was arranged so the couch and the table covered the burn mark, and it looked as good as new. His daughter told us later that when he moved back in and saw the rug back in the living room, she could see the tears in his eyes.

Some things are just worth saving.

for themselves as possible, so they're usually a very good source of reliable companies. If you have a good working relationship with one of your competitors, ask them who they use for particular cleaning or repair jobs.

Your suppliers and subs are another good source of names. The restoration industry is surprisingly small, and it doesn't take long for names of good companies to get around — so your carpet cleaner probably knows of a good company that moves pool tables. Once you get started in business, you'll be surprised how quickly *your* name starts circulating.

Once you have some names, do your homework. Interview the people, and ask for a complete rundown on the exact services they provide. Take a tour of their facility, especially if you'll be sending clients there. Ask for references from past clients, and look at some examples of their work. Verify that they have all of the necessary licenses and insurance.

When it comes to contents, there are a number of subcontractors you'll want to use. Let's look at some of them.

Fabric Cleaners

This is one of your most important cleaning subcontractors, and they'll typically handle all your soft goods cleaning. That includes laundry and dry cleaning, linens

Figure 15-5 All of this antique furniture came out of a house that had a large water loss. It's in temporary storage in one of the few dry rooms, awaiting transport to an antique refinishing and repair company. A good antique restoration company is a very valuable subcontractor to have.

and bedding, and draperies and window coverings. Be sure the company has *specific experience* with smoke-damaged fabrics, and with the removal of smoke and other odors. Not all dry cleaners can handle this type of work.

In addition to the cleaning services themselves, look for a company that can assist you on-site, handling the entire soft goods portion of the job under your supervision. That includes inventorying and removing the items from the site; cleaning everything; making any necessary repairs that have been authorized by the owners; storing the items until the structural restoration work is complete; and then redelivering everything.

Drapery Cleaning

Some fabric cleaners can handle clothing, linens, and other soft goods, but aren't set up for draperies. In that case, you'll need to also locate a sub that can clean all types of window coverings, from draperies to blinds. Once again, they need specific smoke restoration experience.

Furniture Repair

A good furniture repair shop should be able to handle everything from simple disconnected joints and loose pieces of trim to veneering and antique restoration. See Figure 15-5. You need a shop that will give you accurate estimates, as this can affect the homeowner's decision regarding whether or not to salvage a piece.

Figure 15-6 This restoration company has the ability to handle area rugs and carpets. The large lifting mechanism uses cables to raise the poles overhead and lift the carpets up for drying.

Rug and Tapestry Cleaning and Repair

The cleaning and repair of area rugs and wall tapestries is a very specialized field. Older hand-knotted rugs contain very sensitive threads and dyes that can easily be damaged if handled incorrectly. Even newer wool rugs have to be cleaned using the proper solutions or they risk damage. The last thing you want to do is be on the hook for damages to a $10,000 area rug because you cleaned it incorrectly, so leave this to the experts. See Figure 15-6.

Rug repair is also very specialized. Rug experts can weave and blend matching threads to repair burn holes and other damage in even the most intricate rugs, and the repair is literally invisible. Repairs such as these are costly, but obviously worth it when you're dealing with highly-valued area rugs or tapestries.

Carpet Cleaning

Carpet cleaning is a good thing to have in-house at some point. It's profitable, and it also simplifies your scheduling when you have your own equipment and crews available whenever you need them.

However, truck-mounted carpet cleaning equipment is a sizeable investment, so until you're ready for that, you need a reliable subcontractor. There are a lot of them to choose from, so you really need to do your homework. Look for someone with fire and water experience, the ability to deal with all types of stains, excellent people skills, and IICRC certification if possible. They also need to do upholstery cleaning, specifically in smoke situations.

If you work with property managers or real estate people, they're the ones to ask. They have to deal with a lot of dirty and stained carpets, and they typically know who the good carpet cleaning companies are.

On-Site Cleaning

If content cleaning is necessary, you have to decide if the cleaning will be done at the site, or back at your facility. That's usually a matter of how many contents you have to clean, how dirty they are, and what type of cleaning facilities you have available at the site.

On-site cleaning is usually reserved for small, simple projects where it doesn't make economic sense to transport the contents. It's also good in situations where extensive deodorization work isn't required, and sometimes as a pre-cleaning step for certain jobs. For example, you may be dealing with a kitchen fire where the cabinets need to be replaced, but first all the smoky dishes have to be taken out and cleaned. If the sink and the rest of the kitchen are still functioning, you could clean the dishes immediately, then pack and store them on-site, rather than risk transporting them.

Processing Smoke-Damaged Contents

In most cases, especially with large fires, you'll be transporting the contents back to your facility for off-site cleaning. Once the smoke-damaged contents arrive at your facility, they need to be processed in an orderly manner. Exactly how you choose to set it up depends on your space, layout, and particular preferences, but the procedure should follow the same basic steps, in the same order, for each job. It's usually a four-step process, which includes: dirty storage, unpacking and cleaning, deodorizing, and repacking and storage or redelivery.

Step One — Dirty Storage

When the contents first arrive, they smell of smoke, so they have the potential to pass that odor on to other contents you already have in storage. Since you've already spent a lot of time and money cleaning and deodorizing other people's contents, the last thing you want to do is contaminate them with new odors. To prevent that, it's imperative that you keep incoming dirty contents separated from clean contents.

Set up a separate dirty contents receiving area specifically for contents that have odors. You can receive other dirty contents there as well, but remember that anything you receive and store in that part of your facility will pick up odors from other smoky contents stored there.

The dirty contents receiving area should open directly onto the unpacking and cleaning areas. These can be separate areas connected by doors, or they can be all one area. The important thing is that the dirty receiving area be separated from the clean repacking and storage area, so that odors don't enter clean areas.

All of the contents should have been inventoried at the jobsite before being packed onto the truck, so when they arrive at your facility, they're already separated into contents belonging only to a particular household. If you're receiving contents from more

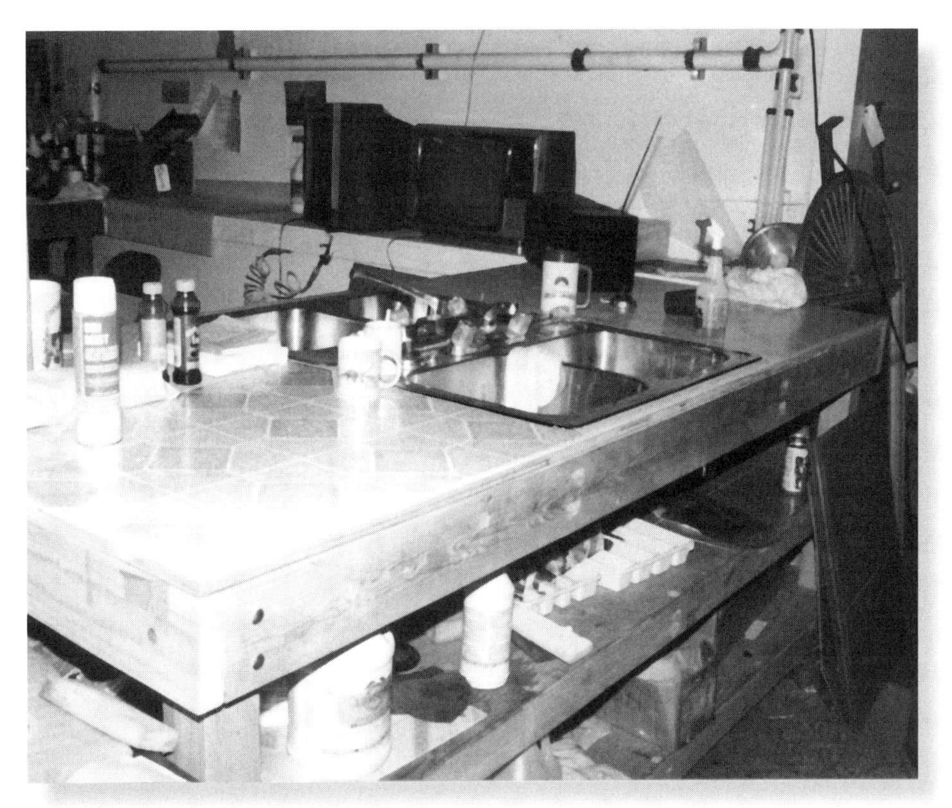

Figure 15-7 Cleaning station at a large restoration company's storage facility. Incoming contents are unboxed at the left end of the counter, then cleaned at one of the two double sinks. Notice the waterproof vinyl covering over the plywood table. This cleaning station is one of several at this restoration company.

than one job at a time, take great care to keep them separated. Create wide aisles on the floor of your warehouse to allow workers and equipment to maneuver safely. Use marking tape to temporarily mark the divisions between different households.

Step Two — Unpacking and Cleaning

After the contents have been unloaded from the truck, they need to be unboxed and cleaned. The best way to do this is to set up unpacking and cleaning stations like the one shown in Figure 15-7.

The process works like this: A box is brought to the station and set on a low table or shelf, where a worker opens it, unpacks the contents, double-checks everything for damage, then sets the contents on a counter at the station, next to a sink. The contents are washed and set aside to dry. The dried contents are then placed on rolling carts, ready to be moved into the ozone chamber for deodorization.

Boxes are broken down as they are emptied, and stored in the dirty storage receiving area if they're to be reused for future demo work. Otherwise, they're taken away for recycling or disposal, along with the packing materials.

(Photo courtesy of Omegasonics)

Figure 15-8 A small ultrasonic tank. The black box sitting to the right is the transducer. The transducer isn't currently attached to the tank, and is shown in this position only for illustration.

(Photo courtesy of Omegasonics)

Figure 15-9 A larger ultrasonic tank, complete with basket, filters (left), casters, lid, and other accessories.

Large, unboxed contents such as furniture are processed next. Each piece needs to be cleaned and restored using the appropriate method required. They are then set on pallets and moved into the ozone chamber.

Ultrasonic Cleaning

As mentioned earlier, there are a number of different combinations of contents cleaning solutions and cleaning methods, like the wet washing method just described. One process that works particularly well for restoration work is ultrasonic cleaning.

Ultrasonics use sound waves transmitted through a liquid to clean. It's a very involved science, but the basic concept is that a transducer is used to generate sound waves. These waves are outside the range of our hearing, typically about 20,000 to 100,000 cycles per second. The sound waves pass through a liquid solution inside a stainless steel tank. See Figures 15-8 and 15-9. The solution may be plain water, another liquid, or combination of liquids.

The sound waves generate millions of micron-sized bubbles in the water that begin to grow, and as they grow, they increase in both pressure and temperature. Eventually, the pressure from the sound waves causes the bubbles to implode, like a balloon implodes when punctured with a pin. When it does, a tiny void, or cavity, is left behind. Water rushes into this cavity with a lot of force, in a process called *cavitation*.

If you place a dirty object into the ultrasonic tank, the cavitation process forces the water or other cleaning solution into the tiniest of crevices in the object. That force, combined with the heat generated by the cavitation, cleans objects remarkably fast and effectively. See Figure 15-10. Depending on the tank size and the type

Before

After

Figure 15-10 Here you see a smoke-covered glass figurine prior to cleaning, and the same figurine after ultrasonic cleaning. Ultrasonic cleaning takes a very short time, and cleans even the tiniest crevices in the most delicate materials.

of solution, you can clean items with smoke, grease, rust, blood, oil, soot, dirt, and a variety of other contaminants. Ultrasonic cleaning also allows you to clean items with multiple parts, such as tools, without disassembling them.

Step Three — Deodorizing

Because contents absorb a lot of smoke odor, cleaning alone won't necessarily eliminate the smell. So, as you clean the contents, you'll typically move them through a deodorization step as well.

As described in Chapter 9, deodorizers are often added to the cleaning solutions used in restoration work. These are water-soluble materials, used in the wash water

REAL STORIES:

That Was Just Nuts!

Late one afternoon, at a big commercial garage that specialized in heavy equipment repairs, a mechanic finished up a welding job on a massive old logging tractor. He closed up the shop for the night, unaware that a welding spark had started a small fire in some old caked-on grease below the engine compartment.

Throughout the night, the fire smoldered inside the old tractor. It had lots of grease to feed on, but not a lot of oxygen, so the flames stayed low and the greasy smoke stayed thick and plentiful. By the time it was discovered the next morning, and the fire department was called, the old logging tractor was a total loss.

Inside the big garage, thick greasy smoke coated just about everything. The worst smoke damage was the area right next to the tractor, where an entire wall of shelves housed scores of small cardboard boxes and bins filled with every conceivable nut, bolt, screw, pin, washer, rivet, fastener or other piece of hardware that you could ever imagine. The collection spanned the many decades that the shop had been in business. Some of the hardware, designed for machines long since out of production, was literally irreplaceable.

The insurance company was in a bind. The parts and hardware were too smoky and dirty to use and the storage bins and boxes were soaked and falling apart from the fire department's efforts. It didn't make sense to discard thousands and thousands of pieces of hardware that were fine except for the greasy smoke. But hand-cleaning each one would have been prohibitively expensive due to the labor involved.

We proposed a way to save everything economically, through a mini-assembly-line cleaning using our ultrasonic cleaner. First, we ordered several cases of replacement cardboard bins in a couple of different sizes to fit their shelving. In the meantime, the mechanic's crew hauled off the old tractor and cleaned up part of the building. Using their own steam-cleaning equipment, they cleaned off part of the building's shell and slab floor and make it ready for us to use as an on-site restoration area.

We set up a series of folding plastic tables, with the ultrasonic tank in the middle. Starting at one end of the shelving, a worker would take a soggy box full of hardware and dump it into a five-gallon bucket with small holes drilled in it. Another worker would process the bucket through the ultrasonic tank for however long it took to get the contents clean — sometimes less than a minute. He removed the perforated bucket from the tank, allowed the solution to drain out, then dumped the contents onto one of several towel-covered tables where air movers quickly dried everything.

The clean, dry parts went into a new cardboard bin, and one of the garage employees — who knew what all the stuff was — labeled the bin and put it back on one of the shelves, which were being cleaned as fast as we could empty them. Even the tiniest and most intricate of parts was cleaned in that way. Within a couple of days, every part and piece of hardware was cleaned and back on the shelf ready to use.

Even the mechanics were amazed — since some of their stuff hadn't looked that good in years!

or in spray cleaners. You can also thermal fog contents in the closed environment of a warehouse, or even inside a truck, van, or storage container.

Ozone

Once in the controlled environment of your facility, perhaps the most effective way of deodorizing fire-damaged contents is through the use of ozone. Ozone works especially well on natural smoke residues, but is less effective on protein and synthetic residues. It also works well on mold and mildew odors.

An oxygen molecule contains two atoms, O_2. When oxygen is split into two individual atoms, and one of those atoms links with another oxygen molecule, it forms

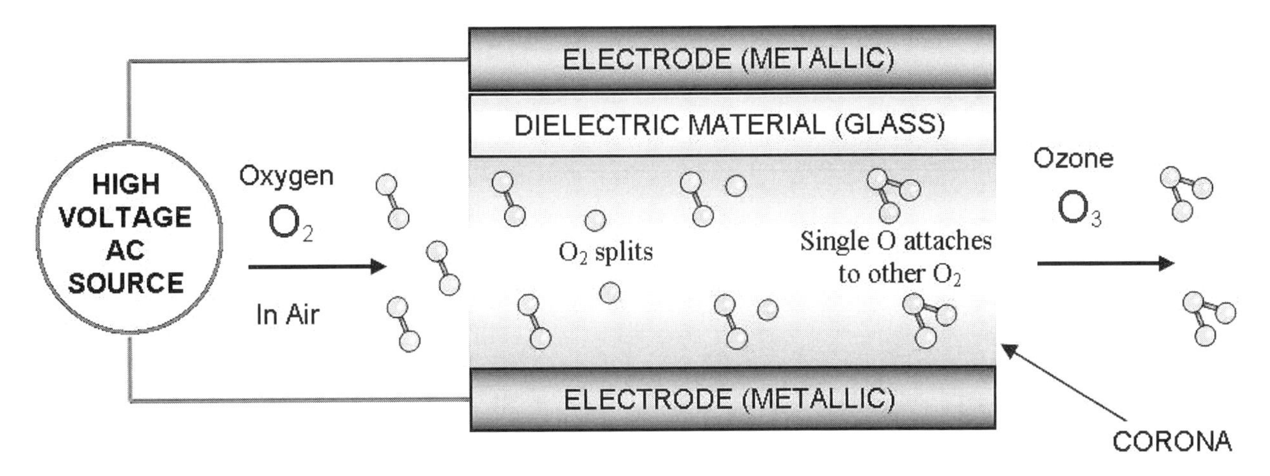

(Photo courtesy of Sonozaire Odor Neutralizer, a division of CB&I)

Figure 15-11 How corona discharge works: Oxygen (O_2) in the air enters the tube on the left. Electricity breaks the bond of some of the atoms, and the free atoms bond with some of the O_2 molecules, forming ozone (O_3).

(Photo courtesy of Sonozaire Odor Neutralizer, a division of CB&I)

Figure 15-12 Several sizes of ozone generators.

ozone, O_3, which has three oxygen atoms. The bond that the third atom has with the ozone molecule is very weak. So, when an ozone molecule comes into contact with an odor molecule, the third atom easily splits off. The odor molecule is completely destroyed by oxidation, which also destroys the third atom. So everything gets used. No odor or extra atom remains, just the original oxygen molecule.

Ozone is produced in ozone generators. The most common way to generate ozone is through a method called *corona discharge*. Air passes through a tube surrounded by an electrical field. The electricity causes a small percentage of the O_2 molecules to split; the free atoms bond with the remaining O_2 molecules, and produce O_3 ozone gas. See Figure 15-11. Ozone generators come in several sizes (Figure 15-12). There's even a small portable generator for use in cars, motor homes and other small spaces (Figure 15-13).

Ozone Chamber

For restoration work, the best way to harness the power of ozone is to construct an ozone chamber. Ozone is a heavier-than-air gas, and since it's unstable, it has a very short life span. It takes anywhere from 24 to 72 hours to work, depending on

(Photo courtesy of Dri-Eaz Products)

Figure 15-13 A small, portable ozone generator, for use in cars, motor homes, hotel rooms, and other smaller spaces.

the type of contents and the severity of the smoke odor. So by using an ozone chamber, the ozone gas is allowed to collect and work for much longer periods of time.

An ozone chamber is just a big room, like the one in Figure 15-14. You can build one to almost any dimension you need. The size would depend on the number of contents you intend to handle and your available space.

Build the room out of 2 x 4s with a drywall interior and exterior. The interior height shouldn't be over 8 feet, so that the gas doesn't dissipate. If you intend to use the space above the room for storage, be sure that the ceiling joists are sized correctly for that use. Select an ozone generator with a capacity that's adequate to match the volume of the room you build.

Place the ozone generator at the top of the room, or on a shelf on the side. Pipe the ozone gas coming from the discharge side of the generator through a flexible hose into the chamber. Attach the hose to the center of a length of 3- or 4-inch perforated PVC pipe fastened to the ceiling of the chamber. The PVC pipe will help distribute the gas evenly throughout the chamber. Arrange the inside of the chamber to suit your needs; you can have an open room, fixed shelves, portable shelves, hanging racks, etc.

Since the ozone gas degrades, it needs to be flushed out and replaced with fresh gas periodically. One way to set up your chamber is to install an exhaust fan, vented to the outside of the building, along with two timers set on two different cycles. One timer activates the ozone generator for a two-hour cycle while the fan is off. The other timer activates the fan for 30 minutes to clean out the room while the generator is off. Alternate the two timers in a continuously repeating cycle for as long as you need the chamber to be active.

The ideal configuration is to have a wide door or a set of double doors at each end of the chamber, like the one in Figure 15-15. One end opens into the chamber from the cleaning and dirty receiving side of your facility. At the other end, the door opens into the clean storage side. That way, you can move your contents from receiving and cleaning, to ozoning, then out of the ozone chamber and into storage, without going back through the dirty receiving area.

Your ozone chamber should only be operated when no one will be in the warehouse. Weekends are a good opportunity for ozoning contents. Be sure that you post warning signs that an ozone operation is in progress, and make sure that the exhaust vent from your ozone chamber fan exhausts above the roof of your building for safety.

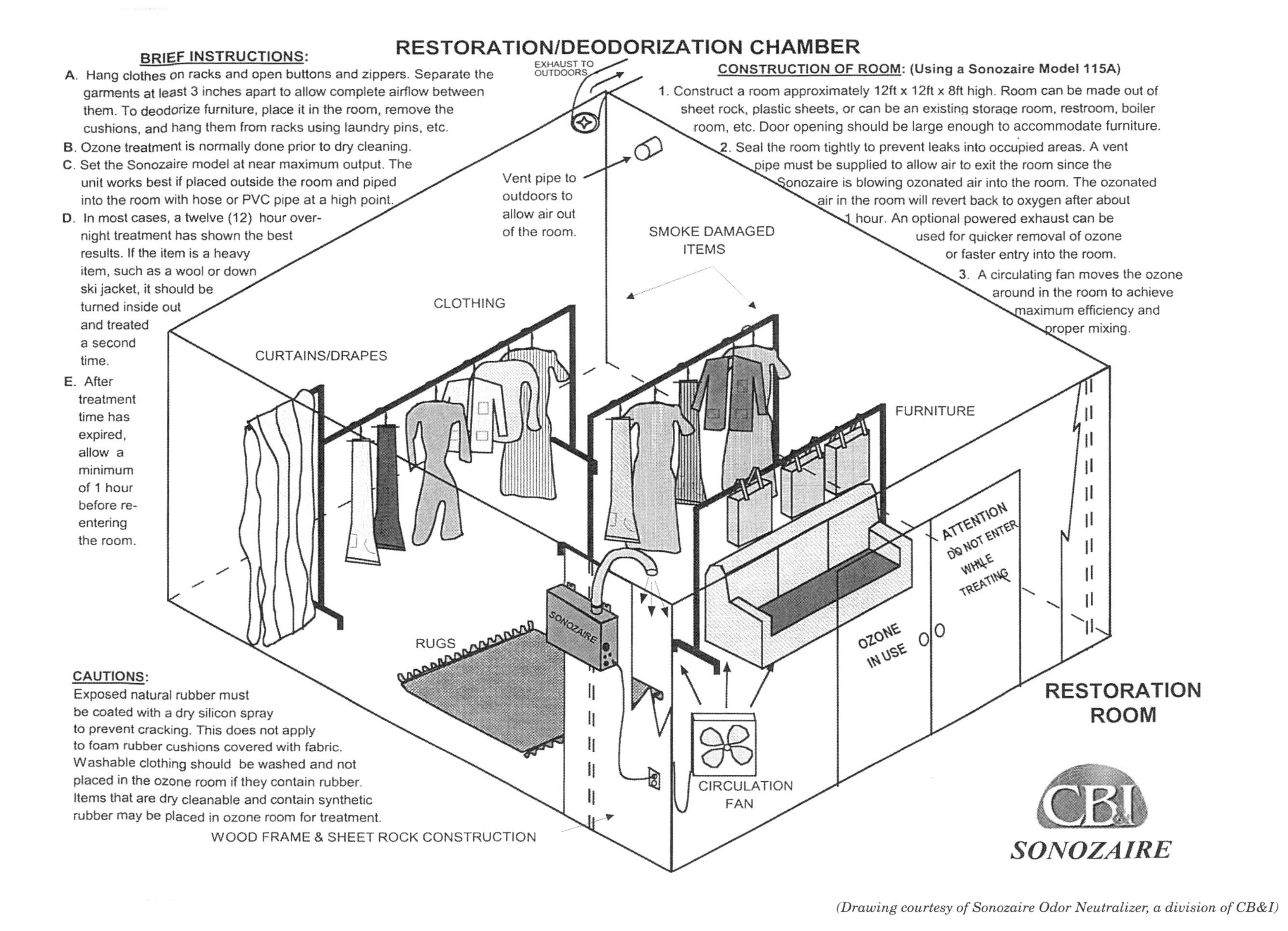

(Drawing courtesy of Sonozaire Odor Neutralizer, a division of CB&I)

Figure 15-14 An ozone chamber.

Figure 15-15 A two-door ozone room at a large restoration company. There are fixed shelves along one side of the room, and rolling carts along the other side. The overhead pipe helps to evenly distribute the ozone gas.

Ozone Disclaimer

Ozone is a very effective deodorizer, but it's also dangerous. You need to understand exactly how and where to use it, and exactly when and where it should *not* be used.

Before purchasing or using an ozone generator, or setting up an ozone chamber, you *must* consult with the manufacturer of the equipment. Obtain an MSDS and complete operating instructions, and *be sure that you read and follow all of the manufacturer's warnings and safety precautions before operating the equipment!*

Talk with your ozone equipment supplier, your insurance carrier, and your local OSHA representative to be sure that your ozone chamber and your method of operation are in complete compliance with all state and local laws and regulations.

REAL STORIES: That's a Sweet Use of the Equipment!

Every year at Christmas, our company purchased several boxes and trays of candy from a local candy store. The company made their own candies at a local facility, and it was some of the best candy to be found. Everyone loved it, and we gave a lot away as gifts each year.

One year, the owner of the candy store called with a dilemma. She had just received a shipment of several hundred small decorative wooden boxes from an overseas supplier that had an odd smell to them. She had been planning to use them as Christmas gift boxes for her candy. She really needed the boxes, but couldn't use them as they were, and there wasn't time to send them back or to try and find replacement boxes. Was there anything we could do?

We went by and picked them up. We looked them over, and we all gave them the "sniff test." Sure enough, they had an odd smell, but none of us knew what it was. They were unfinished wood, so someone suggested that ozone might work. So we opened up all the shipping cartons, and put them in the ozone chamber for 24 hours. The difference was amazing! The odor was completely gone, and the boxes smelled incredibly fresh!

Since we were going to buy candy from her anyway, we worked out a simple trade for our services. She was absolutely delighted with the results, and even threw in an extra gift certificate for my partner and me. It's kinda fun to get a tip once in awhile!

Figure 15-16 Double-decker plywood storage vaults in the warehouse of a restoration company that handles a lot of contents. Notice the stenciled numbers and the sleeves with the inventory lists on the front of each vault. Both are essential for tracking the vault contents and keeping everything organized.

Step Four — Repacking and Storage or Redelivery

As the contents come out of the ozone chamber, repack them for storage or redelivery, depending on where they're going. Remember that everything needs to go into fresh boxes, with fresh packing material. To avoid the possibility of contamination, *never* reuse the original dirty boxes or packing material.

If the contents will be going into long-term storage, follow the guidelines we described earlier in the chapter. Make sure you label the new boxes and the storage vaults. See Figure 15-16. Keep loose furniture in the same area as the storage vaults, ideally on pallets, and tag it with the owner's name or other identification.

If the contents are being redelivered somewhere, use the pack-out inventory or your contents release forms to verify redelivery. Be sure that the homeowners sign for everything.

Items for Immediate Release Back to the Owners

We've talked a lot about cleaning and processing contents and then packing them up for long-term storage while the home or building is renovated. But always keep in mind that fires aren't planned for, the way a remodeling project is; they're life-altering events, and terribly disruptive to your clients and their families. During the course of doing a pack-out after a fire, *always be aware of the effect this has on the people involved.*

The homeowners are going to be overwhelmed by the shear magnitude of what they're going through, and they'll probably not remember that tomorrow they're going to need their cell phone charger or checkbook. Try and work with them, either during the pack-out at the home, or during the unpacking at your facility. Help them sort out some things that they're going to need right away. List them on a separate inventory sheet, and get those items cleaned, deodorized, and returned to them as soon as possible. This will help them out immensely, and will also save you trouble later — because they're almost certain to come looking for some of these things among their stored contents.

Here's a list of items your clients might need right away:

- **Cell phones:** Including cords and chargers.
- **Keys:** Car keys, house and office keys, other vehicle keys, and any other keys they might need.
- **Cash:** Don't pack out cash unsupervised. Ask the client if they have any cash in the house they would like deodorized, and have them get it for you. Note the number and denomination of the bills, and have them sign for it.
- **Medications and prescription records:** This includes medications and records for all family members.
- **Medical records:** Including records for themselves and their children.
- **Glasses:** Including spare glasses, reading glasses, sunglasses, and special prescriptions.
- **Pet medications and records:** This would include all medical records, prescription medications, toys, leashes, cages, and other supplies.
- **Bank documents:** Including their checkbook, savings book, bank statements, and bank account files.
- **Financial records:** This might include insurance papers, car title, mortgage papers, will, financial investment records, current bills, paid bills, etc.
- **Business records:** With so many people working from home, there may be important work or business records that they need. This could include both paper files and computer files.
- **Address books, other records:** Any other names, addresses, phone numbers, or other records that might be needed while getting settled in a temporary home.
- **School records:** Or any other paperwork related to their children.
- **Laptop computer:** This includes cords, cables, disks, case, and anything else that would be needed.

- **Other electronics:** In today's world, this could be a long list. It includes iPods and MP3 players, BlackBerry® smartphones and other devices, PDAs and other organizers, electronic games for the kids, portable DVD players, etc. This list would also include cables, disks, games, and other accessories.

Consider buying some nice plastic file boxes and bins with resealable lids in a couple of different sizes. Have them imprinted with "Courtesy of ..." and your company name on the front. Use these for packing and returning your client's important documents, keys, electronics, and other essential items after you've cleaned them. It's a nice gesture and a good advertising opportunity as well!

16

Getting Set Up for Emergencies

There's no more important aspect to the business of insurance restoration than how well you respond to an emergency. Commonly referred to as ER, your emergency response capability is an essential part of what sets this type of construction work apart from building and remodeling. Rather than thinking about starting a kitchen remodel four weeks from now, with ample time for planning and preparation, you're faced with Mrs. Smith's broken water pipe and flooded kitchen, and she needs you *now!* See Figure 16-1.

Because your ability to respond to emergency calls is a huge part of what will make or break you as a restoration contractor, in this chapter we'll focus on how to get yourself and your crew ready to effectively handle these situations, from the first phone call to the final cleanup.

A Prompt Emergency Response Is Critical

An effective emergency response is your initial chance to get your foot in the door with a prospective client. You've got other restoration contractors that you're competing against for work, and when a homeowner has an emergency, he's going to call one company after another until he gets someone who'll respond *right now.* By being the company that responds promptly, then calmly takes charge of the situation, you become the hero. You'll not only lock up the emergency portion of the work, but you'll also have a very good shot at landing the repair and contents jobs as well.

Think of your response to an emergency call as a job interview. To be honest, being called at 1 AM in a snow storm to respond to a soaking wet house isn't exactly a contractor's vision of the perfect night. *But that first impression is absolutely crucial, and you only get one shot at it.* You may not get the renovation job every time, but the homeowners will remember who responded to their call, and who didn't.

Figure 16-1 The water damage in this house was caused by a broken water line in a second-floor bathroom. It required an immediate response to assist the homeowner and prevent further damage to the building and the contents. Your ability to respond quickly and efficiently to this type of emergency will establish you as a restoration contractor.

Your willingness to respond — and respond well — to an emergency is also important to the client's insurance company. Taking immediate steps to stop further damage saves them money. So your work makes both the insurance company and their clients happy, and that reflects well on you. Insurance adjusters have a tough job, and if you make their life easier, it makes you more valuable to them.

Last, but certainly not least, is the fact that emergency jobs are quite lucrative. You charge a fee to go out to the job, plus an hourly rate for any personnel on-site, often at time-and-a-half overtime rates. You also charge unit-cost rates for everything from water extraction to the removal of wet insulation; rental fees for your drying equipment; and then a profit and overhead percentage on top of that. It's very good work.

Understanding Emergency Response

Suppose a homeowner has a pipe that freezes and breaks, flooding part of the house. Under the terms of his homeowner's insurance policy, he has an obligation to take all necessary steps to mitigate further damage; in other words, to keep the damage from getting any worse.

That doesn't necessarily mean he should crawl under the house and try to do the work of a trained plumber. But he *is* expected to shut the water off, move contents out of the way, and if necessary, seek professional help to begin removing the water and drying out the house as soon as possible.

The insurance company assumes the emergency response work will be limited *only* to loss mitigation, so the cost of that response isn't typically estimated. Instead, the restoration contractor bills for the time and materials required to move and protect contents, remove wet materials, dispose of debris, and dry the structure. The emergency bill is separate from any reconstruction work.

In some instances you may only do the initial emergency, not the repairs. The homeowner may do his own repairs, or he may hire a different contractor. It's important to distinguish between ER work and repair work — and always keep them separate.

Setting Up Crews for Emergencies

To make it easier on yourself and your crew, divide up the duties for emergency responses using an on-call rotation system. How you set this up depends on how many people you have available and their level of training.

Emergency Crew Rotation

It's always best to have as many people as possible on ER rotation. Having lots of people available helps in times of multiple emergencies. It also helps avoid burnout and resentment by allowing the on-call periods to be spaced further apart. No one really likes responding to an emergency in the middle of the night — but there are times when the job will require it. Every crew member, even those in the cleaning division, should be part of the rotation, at least in a backup position. Just so there's no confusion, this should be a condition of employment for each new hire.

The length of an ER rotation can vary, but one week seems to be a good length of time. It's long enough to avoid a lot of crew confusion, but not long enough to seriously interfere with an employee's life. Change the rotation schedule every Monday morning.

Maintaining the Rotation Schedule

Have one person in charge of setting up and maintaining the emergency rotation schedule. It could be you, or it could be your lead emergency technician or your office manager. But to avoid confusion, it should always be the same person.

Sometimes an employee may have to skip a rotation because of illness, vacation, etc. When that happens, *don't let the crew make their own substitutions!* Once you allow that, they'll be making changes all the time and you'll never be sure someone's on call. Make sure that the person who's in charge of the schedule tracks the changes so everyone gets their turn.

Include Yourself in the Rotation

You'll want to be sure to include yourself in the emergency rotation, especially at first when your business is just getting off the ground. Being part of the rotation lets you see first hand what's required on an emergency response, and how homeowners react. You'll also see what is and isn't important to get accomplished during that initial call-out, and be better able to discuss the emergencies with the adjusters.

Going out on emergencies yourself is also great for crew morale. Sopping up water or securing a roof tarp on a stormy night is tough work, and doing it side by side with your crew is a great way to gain their respect. You'll also see what challenges they face, which helps with everything from equipment needs to employee bonuses.

Designate a Lead and a Backup

The lead should be the person with the most training, and the one to receive and evaluate the initial emergency call. When you're starting out, that lead person will probably be you. You can designate an employee to take on that job later, when you have a fully-trained staff.

Once on-site, the lead will determine if he can handle things alone, or if additional help is needed. For example, if there's a lot of wet material that needs to come out right away, or a lot of contents to move, the lead person has the option of calling the backup person in to help.

Train Your Lead People

Since your emergency response personnel are at the site to mitigate damage as quickly as possible, they must be properly trained in evaluating emergency situations. Use the following checklist to help your leads develop the appropriate response techniques. They should be able to:

- **Handle the initial call professionally:** The lead person needs to respond as quickly as possible to calls, and be polite and sympathetic. They need to make the homeowner understand that they're there to help.
- **Evaluate the initial phone call:** Every lead should have a checklist of questions available that they can calmly and patiently go through with the client, gathering all the necessary information to determine whether an immediate response is required.
- **Know what to do on-site:** Every lead must be trained in how to handle different emergencies, including what should be removed and what should be salvaged. They should know what type of equipment to set up, and how much is needed. They must also be able to evaluate whether or not to call out additional help, and what subcontractors they might need to contact right away.
- **Be organized:** It's important that your lead carefully track everything happening on the job, from the initial call-out to the removal of the last bag of debris. Everything must be recorded so that you can correctly bill for it. Emergencies are a very good source of revenue, but how much you're able to bill depends in large part on how organized your lead people are.

Set Standards that Reflect Your Company

Imagine for a moment what it would be like to have a pipe break in your home. You have to call a company that you don't know and ask them to send a crew of strangers to your house — maybe in the middle of the night. You'd certainly want to feel comfortable with the people that show up.

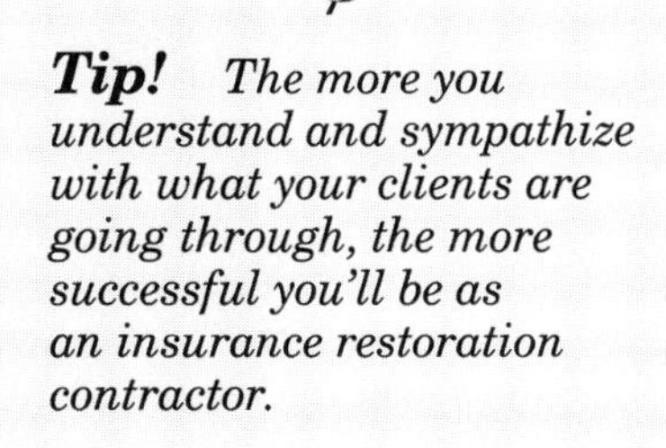
Tip! *The more you understand and sympathize with what your clients are going through, the more successful you'll be as an insurance restoration contractor.*

Always keep that image in mind when setting standards for your own crew's appearance and behavior. Set and enforce policies for your emergency crews that will be a positive reflection on your company and put your clients at ease. You must be able to thoroughly trust every employee entering a stranger's home. Remember, they'll often be required to go through closets and personal contents. Also, talk with your own insurance agent to see if your employees should be bonded.

Determine Wages and Call-Out Fees

It's up to you to determine wages and other compensation for an employee called out after regular business hours. To do that, you'll need to understand what the insurance companies you work with are willing to pay, as well as what laws affect overtime work and other compensation in your state.

Overtime Wages

Typically, workers are paid 1½ times their normal wage for responding to an emergency outside of their normal work hours. Where it gets tricky is determining what *normal work hours* means, both for your company and for the insurance company. You want to understand and clarify as much of this up front as possible, to avoid any confusion and hard feelings with your employees, as well as to avoid hours you can't bill to the insurance companies.

Let's look at some common examples of how this is normally calculated. We'll use two employees, Jeff and Allison, who normally work 8-hour days, from 8 AM to 5 PM with a 1-hour lunch.

Example 1: Jeff and Allison are both called out at midnight and work until 3 AM on an emergency. This is pretty straightforward:

- You can bill the insurance company for a total of 6 hours of after-hours emergency work (2 employees at 3 hours each).
- Three hours of overtime wages would be appropriate for both Jeff and Allison.

Example 2: Jeff and Allison started the day at 8 AM working at Mrs. Smith's house, doing repairs as part of the work you're doing there. Both of them are called off that job at 2 PM to go to Mr. Jones' house on an emergency call. They both stay at Mr. Jones' until 7 PM.

- In this case, you'll be able to bill Mr. Jones' insurance company for a total of 6 hours of emergency work at regular time (2 PM-5 PM for both workers). You'll also bill for 4 hours of overtime work (5 PM-7 PM for both workers).
- Jeff and Allison will each get their normal 8 hours of pay, plus 2 hours of overtime pay (for the 5 PM-7 PM portion of the work).

Example 3: Here's where it gets a little more complicated. You don't have any work for Jeff one day, so he stays home. But at 3 PM, you call him out to help on an emergency. He ends up staying at the emergency job until 7 PM:

- You'll be able to bill the insurance company for 2 hours of Jeff's regular time (3 PM-5 PM), and 2 hours of his overtime (5 PM-7 PM).
- How you pay Jeff is determined by what your agreement is with your employees, and what's required by applicable state laws. He may be entitled to 2 hours of regular pay (3 PM-5 PM), and 2 hours of overtime (5 PM-7 PM), just like you're going to bill the insurance company. Or, he may only be entitled to 4 hours of regular pay for that day (3 PM-7 PM), since technically he wasn't on overtime since didn't work over 8 hours.

Call-Out Fees

A call-out fee is a fee that's paid to each employee who has to go out on an after-hours emergency call — and it's paid in addition to the hourly wages they receive.

Call-out fees are optional, unless you have some type of state law that requires them. They're simply a way of compensating and thanking the crew for the inconvenience of being called out. They also provide a financial incentive that helps motivate the crew to be part of the on-call rotation.

Call-out fees can be included as part of the overall emergency services bill. Some estimating programs have a unit cost category for this as well.

Typically, a call-out fee is paid if an employee is called out after hours, *not* if he's already at work and is called off one job to go to another. Let's go back to the three examples.

- In ***Example 1***, both Jeff and Allison were called away from their homes after hours, so both would get call-out fees.
- In ***Example 2***, both Jeff and Allison were already at work, and left one job to go to an emergency. Even though both of them end up working overtime, neither would be entitled to a call-out fee.
- ***Example 3*** again gets complicated, and it would be up to you to determine policy. You may want to pay Jeff a call-out fee, because

Tip! *Unusual weather activity, such as heavy winds or extremely low temperatures, can result in a large number of emergency calls coming in all at once, forcing your crews to really scramble. When things settle down again, consider buying everyone a pizza lunch, or gift certificates at a local home center or other store. Some form of appreciation to your crew for pulling together in tough times really goes a long way!*

> he was called away from his home and it's the fair thing to do — especially if you're only going to pay him straight time for the 4 hours he worked. But understand that you probably wouldn't be able to bill the insurance company for it, since it wasn't after normal work hours when the call came in.

The call-out fee should be a set amount, such as $50.00, $75.00, or whatever you can afford and feel is fair. It's also easiest to just add it to the employee's paycheck, but clarify this with your accountant or the proper state employment agency to be sure you're doing it correctly. Once the call-out fee policy has been established, make it a written part of your company handbook, and apply it uniformly to everyone.

Handling Emergency Calls 24/7

Some emergency calls come in during regular business hours. But most occur after you've closed for the day, or on the weekend. This is especially true if you live in an area where there are a lot of second homes and vacation rentals. Someone will show up on Friday night to start their relaxing weekend only to discover a wet house from a frozen pipe, or damage from the previous week's wind storm.

Getting Your After-Hours Calls

Obviously, you have to get the call before you can respond to it, so it's critical that you have a reliable and effective method for distressed homeowners to get in touch with you. There are three common methods:

Pager

The caller first dials your business number. A recording gives them a pager number to call, or they can press a button on their phone and have the pager dialed automatically.

Pagers are easy and inexpensive, but the caller doesn't get to talk to a live person right away. They have to leave a message and hope for a return phone call, which probably means that they'll be looking up your competitor's number while waiting for you to call back. For that reason, pagers actually work best when used in combination with an answering service, as described shortly.

Cell Phone

Your regular company phones switch over to an after-hours company cell phone, which is dedicated only for emergency calls. The cell phone is carried by the person who's on call, and is switched between employees when the rotation schedule changes. This can be a low-cost method to implement, and it means the caller gets an immediate, live response.

The substantial drawback is that there's no way to eliminate nonemergency calls. That means you're also going to get calls from existing clients who want to talk about paint colors at 9:00 o'clock in the evening, or a salesman who thinks that Saturday night is a great time to discuss why you need a new table saw.

You can opt to use an answering machine to inform the caller that they can dial the cell number *in the event of an emergency*, which may provide a little bit of screening, but you'll probably still get the paint questions and the table saw sales calls. You also add an extra step with the answering machine, and the panicked homeowner with a genuine emergency hears a machine instead of a live voice.

Answering Service

Although it's more expensive, an answering service offers the best overall option, primarily because it allows the distressed caller to talk to a real person immediately, while allowing your on-call person to avoid unnecessary calls.

The answering service takes the initial call, which gets the ball rolling quickly for the homeowner. They do a limited amount of pre-screening to weed out salespeople and nonemergency calls, then contact the on-call technician using a pager or a cell phone. The service passes along the contact information, and the technician calls the homeowner directly.

For even greater efficiency, have two on-call technicians — one lead and one backup — both with pagers or cell phones. If the lead technician can't be reached for any reason, the answering service calls the backup, which limits the chance of missing the emergency call. More importantly, if the lead technician needs backup help on larger calls, it provides him with quick access to another technician.

The answering service is your company's first after-hours contact with a new client, so it's extremely important that you take the time to find a good one. Interview each one you're considering, in person if possible. See if their services and professionalism are in line with what you're looking for. Here's a checklist of some important questions to ask:

- **Is the service available 24 hours a day, every day of the year?** You obviously want one that operates around the clock, daily, including holidays.
- **Do they charge a flat fee or a fee based on the number of messages they take?** A per-message fee structure can add unnecessarily to your costs, since it encourages the service to take messages even when it isn't necessary.
- **How easy is it to transfer your phones to and from the service?** You want a quick and easy transfer method that allows the service to cover your phones any time you need them to. That way you can use them during lunch hours, or during crucial staff meetings.
- **How will they answer the phone?** You want the service personalized to you and your company.

- **Are they willing to pre-screen calls?** The service should be willing to ask callers a couple of basic questions, which you provide, to help screen out nonemergency calls and keep them from being passed on to your on-call crew.
- **Will they provide you with a time log for all incoming and outgoing calls?** A time log shows when the service receives a call, when they page your on-call person, and when he calls the service back. This is an important tracking device for evaluating the performance of both the service and your crew. Ask the service if they'll fax or e-mail the log to you every morning when the phones are switched back over.
- **Will the service contact more than one person for you?** If the service pages your lead on-call person and doesn't get a response within a set period of time, you want them to contact the backup on-call person.

Once you have an answering service, occasionally test their performance. Call in with a fictitious after-hours problem to see how quickly they answer the call; how they handle themselves on the phone; how well they screen the call using your questions; and how quickly they contact the on-call crew member.

Screening the Emergency Phone Call

You only want the answering service to ask a few simple screening questions. They're not trained in what you do, and you don't want them to unnecessarily aggravate the potential client by delaying the process.

If it's truly an emergency, the homeowner will typically start off the call by saying what's happened to their home: "My living room is flooded and I need someone to come out right away!" Or, "We've had a fire, and there's a big hole in the roof!"

In that case, the answering service might simply take down some basic contact information, and tell the caller that an on-call technician will call them right back. The service should let the homeowner know that they can expect a return call within a certain number of minutes.

It's important that the homeowner have an expected timeframe for the callback. This provides psychological reassurance, and keeps him linked to you for a short period of time. It'll also stop him from calling another company. However, it's very important that your crew member get back to the homeowner within that given timeframe, or you risk losing the job and damaging your reputation. The timeframe for the callback shouldn't be too long, but it also needs to be realistic for your answering service and your crew, such as 15 minutes.

If the caller *doesn't* start off with an a description of what's happened or say that it's an emergency, then the service should politely ask what the call is in regard to. It might be an existing client with a simple question that can wait until the next day, or a client who's really upset about some problems and needs a little handholding as soon as possible. A good answering service will also be able to spot and deflect sales calls.

Talking with the Homeowner

When you or your on-call lead returns the client's call, *make a note of the time.* This starts the clock rolling for billing purposes. Every phone call will be different. Some people are very calm, and will be very grateful for your prompt call. Others will be upset to the point of being angry, argumentative, rude, or nearly incoherent. This is your first contact with them, so it's your job to:

- ➤ Calm them down if necessary.
- ➤ Project your professionalism and sincerity.
- ➤ Assess the nature of their emergency.
- ➤ Accurately get all of the necessary information (Figure 16-2).
- ➤ Tell them clearly when you'll arrive.
- ➤ Tell them clearly what to expect when you get there.

Emergency Evaluation Checklist

To speed the process along and get the most information in the shortest amount of time, have a checklist of questions prepared in advance. It'll help ensure that you arrive at the job with some idea of what you'll be facing, and that you'll have the equipment and supplies you'll need to handle the job.

Initially, address the client as Mr. or Ms. X — never call them by their first name. Identify yourself and your company. Start the conversation with what you already know: "I understand you've had a pipe break," or "I understand that a tree limb fell on your roof."

Next, inform the client that you need to take down some information. Some people get impatient with questions when their living room is flooded or there's rain coming in through a hole in the roof, so it's best to give them a little gentle warning that you have several questions, but it'll just take a few moments. Let them know that by taking a little time now, you'll have what you need to handle the problems at their home much more effectively. A sample checklist of items you need to clarify before you go out to the site is shown in Figure 16-3.

Determine Who You're Talking To

This may seem like an unnecessary step, but it's definitely not. As soon as possible, determine if you're talking to either the owner of the property or an agent of the owner that's authorized to act on his behalf — for example, a property manager. *This is very important!* It's not uncommon to get a call from a tenant, wanting you to come out and work on the house. But tenants don't have any authority over an owner's house, and can't sign anything on their behalf.

If you're talking with a tenant, you'll need to get the name and phone number of the owner or property manager, and then call them first for authorization. *If you neglect this step, you could easily find yourself doing the work and then having the owner refuse to pay for it because he never authorized it.*

Emergency Evaluation Contact Sheet

Date: 6-23-12 Time of call: 11:46 AM (PM)

Client name: Jill Baker

Address: 1222 Green Meadow Ct.

Address: Anytown

Contact person (if different from client): ______

Relationship of contact person to client: ______

Contact phone numbers (circle client's preferred contact number):

Home: 555-1111 Work: 555-9999 Cell: (555-9191) Other: ______

Directions to site: Highway 12 east, take Blue Mt. Rd. exit, L on 12th Ave., R on Green Meadow Ln., R on Green Meadow Ct. Blue house.

Type of loss (circle): (Water) Fire Wind Tree Vehicle Vandalism

Describe loss: Supply line under master bathroom sink came off, flooded bathroom and adjacent closet. Water now off. Wet vinyl and carpet, minor contents.

Insurance Company: All-Country Deductible: $ 500

Agent: Adam Lawrence Adjuster: Not known

Notes: Helper not needed.

Figure 16-2 Emergency Evaluation Contact Sheet. Use this form when talking on the phone with a potential client about their emergency. It'll help you gather the basic information you need to respond to their loss.

Emergency Response Call Checklist

- [X] Is the caller the building owner or an authorized agent? *Owner*
- [X] Will someone at the jobsite be authorized to sign a Work Authorization form? *Yes - owner*
- [X] Is the water, gas, electricity, etc. turned off (if necessary)? *Water's off*
- [X] Is there any sewage or other hazardous materials that you need to be aware of? *None*
- [ ] Are there any openings in the roof or other parts of the home that need to be covered for weather or security purposes?
- [ ] What type of roof will you be dealing with (if applicable)? How steep? What roofing material? Any other information?
- [X] Are there a lot of contents that need to be moved or covered? *Minor*
- [X] Does this require an (immediate) response, or can you respond the next day?
- [X] Based on the site location, how long will it take for you to get there? Give clients a *realistic* timeframe. *20-30 min.*

Figure 16-3 Emergency Response Call Checklist. This is a detailed checklist of items to ask about or clarify during the initial emergency response phone call. It will help prepare you for your response.

Another situation that occurs is that you get a call from a minor child in the house whose parents aren't there. Again, even if the child is old enough and sensible enough to call you, children aren't authorized to sign contracts, and you could be placing yourself in a situation where the owner of the house could refuse to pay you after the work is done. And, unfortunately, in this ugly era of lawyers and lawsuits, you could also be putting yourself and your crew in an uncomfortable position by being in the home alone with a child.

Get the Necessary Information about the Loss

Ask whatever questions are pertinent to the situation, and record the information in the file.

Here are some examples of necessary questions for a water loss:

- Has the water been turned off?
- Do you know where the leak is coming from?
- Is there anything electrical in or near the water?
- Is it clean water or does it contain sewage (for example, from an overflowing toilet or a broken sewer pipe)?
- Have you done anything to remedy the situation, such as calling a plumber?
- Do you have a lot of furniture or contents that will need to be moved?

> ***Tip!*** *Before you respond to any emergency, always verify that the person you're talking to has the authority to authorize the work!*

In the case of a fire or damage from a storm or a vehicle, your questions might include:

- Does the home still have power?
- Are there holes in the roof that require covering?
- Do you know what type of roofing your home has? (Some types of roofing are harder to walk on, and others, like tile, are harder to cover with tarps).
- How many floors does your house have? (This is to help you determine the type of ladder to take. Also, there may be OSHA fall-protection requirements.)
- Do you know the approximate pitch of the roof? Is it fairly steep or relatively flat? (Some states have OSHA requirements for fall protection for steep roofs.)
- Do windows, doors, or other areas of the home require boarding up to make the house secure or weathertight? Do you know approximately how many?

Get Complete Contact Information

Make sure you get all the necessary information that you need to contact the owners and find the house. This should include:

- Owner's full name.
- Complete address of the property.
- Home phone number, as well as other contact phone numbers, such as a work number or a cell phone number.
- Directions to the home.

As you take down the information, read it back to the caller. Verify the spelling of their name and the street name; the house and phone numbers; and the directions. It's all too easy to transpose letters or numbers, or for a stressed-out caller to tell you: "Turn *left* at Maple Street," when they meant turn *right*.

Get Their Insurance Company Information

Get as much information as possible about their insurance. If they don't have that information readily available, or they seem to be getting impatient, simply request that they have the information available for you when you arrive. The information you need is:

- Who is their insurance company?
- Who is their insurance agent?
- Have they contacted the company yet?
- If so, have they been assigned an adjuster or other contact person?
- Do they know what their deductible is?

Work Authorizations

Explain to the caller that you'll be bringing a *Work Authorization* for the owner or the authorized agent to sign. *It's very important that you get this signed!* At this stage of the job, your Work Authorization is the only document you have proving that you informed the owner of what you were going to do, what the hourly rate would be, and that the owner acknowledged and agreed to it.

Explain in as much detail as necessary what the Work Authorization covers. Most people understand why you want them to sign it, and don't have any problems with it. But you'll occasionally get a person who refuses to sign it for any number of reasons. They want their insurance company to see it first, or their attorney, or they simply refuse to sign anything in advance of the work being done.

If this is the case, you need to politely explain that you can't do the work without the authorization, and then suggest that they call someone else. Otherwise you're putting your company at risk of not being paid — or worse — and it's simply not worth it.

Your Work Authorization shows your standard and overtime hourly rates, but you can certainly explain that to the person over the phone if they ask. You may also be asked to estimate how much the total cost will be, but that's virtually impossible to do. Instead, just explain that all work is done on a time and materials basis, based on your company's hourly rates.

Finally, answer any other questions that the client might have. Explain that you may be setting up drying equipment, moving contents, removing wet materials, and any other steps that you think you may be taking when you arrive. Conclude by giving them an *accurate* estimate of how long it will be until you arrive, and then thank them for calling.

After the Call

Take a moment to make sure you've got everything handled. It's better to use up a couple of extra minutes now and go over everything, rather than get to the job and realize that you've forgotten something important.

To help you with your post-call evaluation, you can refer to the checklist in Figure 16-4. This is just a suggested set of questions for you to go over, so be sure and make additions to the list as needed.

Emergency Response Call Preparation Checklist

- [x] Did you log the time of the call?
- [ ] Do you need to call for backup?
- [x] Do you have an emergency binder and other paperwork?
- [x] Do you have complete directions to the job, and a map if needed?
- [x] Do you have a Work Authorization form?
- [x] Do you have your MSDS binder?
- [ ] Do you need any other tools or equipment that are not currently in the van?
- [ ] Are there any special safety precautions you should prepare for?
- [ ] Do you have the appropriate warning signs (if needed)?
- [x] Is your uniform clean?
- [x] Do you have your cell phone and pager?

Figure 16-4 Emergency Response Call Preparation Checklist. This checklist will help to ensure that you haven't forgotten anything before heading out on the call.

Figure 16-5 A van with a sliding side door.

Figure 16-6 A cargo van with swinging side doors. Notice the roof rack. It can be used for carrying ladders, lumber, and other supplies.

An Emergency Response Van

One of the best ways to be ready to respond to an emergency call is to have a van set up with the all the tools, equipment, and supplies you need. It not only improves your response time and ensures that you have what you need when you arrive, but it also greatly simplifies the equipment tracking and documentation that you'll need for accurate billing.

An emergency response van also makes a positive impression at the jobsite. It shows the client you're a professional, and helps assure them of your ability to handle the work. It's also an ideal piece of company advertising sitting in front of someone's house, or while on the road between jobs.

Selecting the Van

Select a late model cargo van, which has fewer windows than a passenger van, with a long wheelbase that can hold all your equipment. Vans are available with one large side sliding door (Figure 16-5), or two swinging doors (Figure 16-6). The single sliding door is generally easier for loading and unloading drying equipment. Both styles also have two swinging doors in the rear.

Another option is a small cube van or a box truck. These vehicles give you a little more cargo room, as well as more interior head room. On the down side, most only have one large rear door, so access is less convenient.

A used van is fine, as long as it's in good condition. Remember that your vehicles both represent and promote your company and its name recognition. They need to be clean and professional, with a uniform paint scheme and bright, clear lettering. A dirty, beat-up van tells people you don't take your company very seriously, so why should they?

(Photo courtesy of American Van Equipment)

Figure 16-7 An interior bulkhead partition divides the cab of the van from the storage area for safety.

Setting Up the Inside of the Van

The interior of the van should be set up with storage racks and bins, which you can build yourself or buy ready-made from one of the van accessory manufacturers. Ready-made equipment is more expensive, but it's generally preferable because it's sturdier and offers more layout and storage flexibility. Most importantly, it's designed and built to take up less interior space than anything you can build.

If you want to build your own storage, and you're not a welder, you can use perforated angle iron for constructing the frames. It's readily available at home centers and hardware stores, and bolts together with standard hardware. Finish it off with ½-inch or ¾-inch plywood shelving. You can customize it and change it as needed to suit your work. Let's look at elements that work well in an emergency response van.

Bulkhead Partition

A bulkhead partition, like the one shown in Figure 16-7, divides the cab from the cargo area. This is an important feature because it prevents equipment in the rear of the van from flying forward and injuring the occupants in the event of an accident. Bulkhead partitions are required by OSHA in most states, but even if they're not, they're a critical safety feature. Install the bulkhead partition first, and then plan the rest of the shelving around the remaining space.

The partition is typically an open metal mesh with a solid frame that bolts to the van. They're sized for specific vans, so they're easy to install. Most have a door that allows access from the cab to the cargo area, but you'll probably never use it; instead, use the space for more storage shelves or bins in both the cab and cargo area.

Bulkhead partitions are available from van accessory suppliers and by special order through most auto parts stores and car dealerships.

Floor Drawers

Install floor drawers next, since everything else is built on top of them. Floor drawers are basically a false floor in the van, usually 3 to 6 inches high, into which drawers are installed that can slide out through the side and/or rear doors. They're great for extension cords, nails, tools, and small parts. You can install removable bins to keep small stuff organized, or you can build in drawer dividers.

You can make your own drawers out of wood by installing 2 x 3s or 2 x 4s on edge on the van floor to create a framework. Design the frame so you have a combination of both long and short drawer openings to accommodate a variety of items, and so that they're situated for unobstructed access to the doors of the van. Build the drawers using ¾-inch plywood for the sides and ¼-inch plywood for the bottoms, sized to fit smoothly between the 2 bys. Finish by installing ¾-inch plywood over the framing to create a new floor for the van.

Pre-made steel drawers, ordered from van equipment suppliers, are more expensive but have the advantage of being lower and taking up less room. They come in different sizes and configurations and have pre-slotted, adjustable interior dividers. On top of the drawers, install a rubber cargo mat, as in Figure 16-8. The mat cushions the equipment that's on the floor, and helps keep it from sliding around.

Shelving

You'll need a combination of shelf sizes:

- **Deep Side Shelves** — Install deep shelves along the side wall, opposite the side door. Use this area for storing air movers and other bulky items. Placing them here allows you to access them from both the side and the back. You can usually get one row of air movers on the van floor, and a second row on the shelf above, with another shelf above that for short storage.

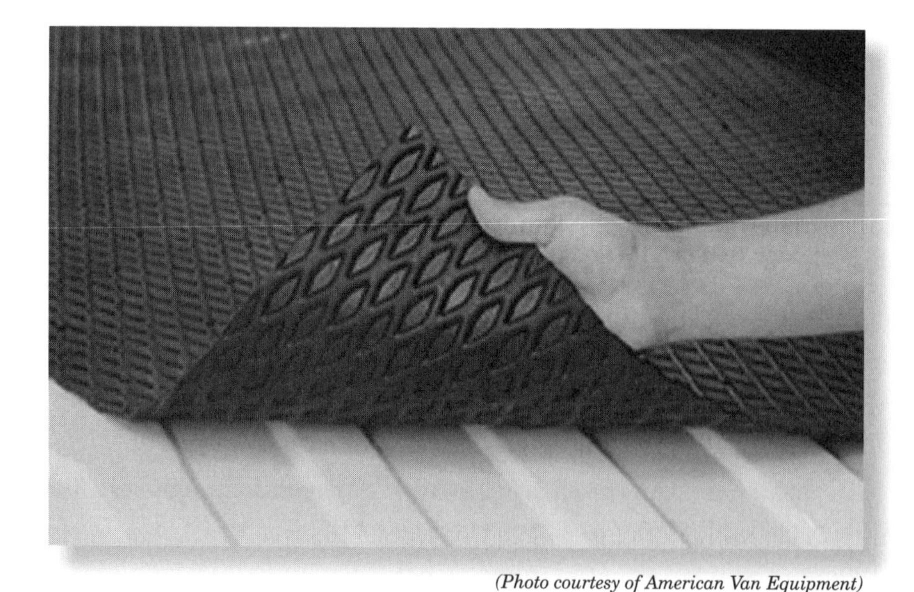

(Photo courtesy of American Van Equipment)

Figure 16-8 Heavy-duty floor mats like this one are a big help in protecting your equipment. They keep the equipment from sliding around, and also help prevent scratches and dings.

- **Shallow Side Shelves** — On the wall opposite the deep shelving, install shallow shelving from the end of the side door all the way to the back. You can use this to hold a wide range of miscellaneous items, from tool boxes to trash bags.
- **Door Shelves** — On the two rear swinging doors, install a couple of door shelves or bins to hold spray bottles, rags, MSDS or job binders, and other smaller items. If your van has swinging side doors, you can install door shelves on them as well, as long as they don't cut down on access to the equipment inside.

Hooks, Clips, Bins, and Other Organizers

Once all the big shelving's been installed, look at the remaining space for other storage opportunities, such as hooks for extension cords, or spring clips for brooms or the wand for the water extractor. A van accessory catalog or home center will give you other ideas. Plastic bins with lids in different sizes are also very helpful. Use them for anything from organizing and storing small parts to holding rolled up extractor hose.

Floor Space

The floor of the van is where you'll be storing larger equipment, such as dehumidifiers. Incorporate some tie-downs, so that you can strap the equipment down to prevent shifting and damage while in transit. Ring-style tie-downs that drop down flat and out of the way when not in use work best.

Roof Racks

Roof storage racks should have enough cross dividers so that you can safely and easily carry both long extension ladders and short step ladders. You'll also want to be able to carry a small amount of lumber for certain types of emergencies. Roof racks need to have convenient tie-downs for securing the load safely. You can equip them with permanently installed ratcheting straps, or carry bungee cords or ropes.

Keep Your Van Stocked and Organized

To be effective, you need to keep the van clean, organized, and well stocked. Figure 16-9 shows a basic inventory list. Use this list as a sample and adjust your list as needed to suit the size of your van and your particular needs. (Remember that water emergencies often require more equipment than what's listed here. Always carry as much drying equipment as your van will hold.)

Once you have an inventory you're satisfied with, print it out and put copies in plastic sleeves. Keep a copy in the cab, one in the cargo area, and one on the wall in the warehouse, near where you store your drying equipment. Use these inventory sheets as a checklist for stocking the van, and as a way to double check that your employees are keeping up on things.

Emergency Response Van Inventory Checklist

General Supplies:

- ☐ (4) 25-foot 12-gauge extension cords
- ☐ (4) 3-tap cord ends (These can be separate adapters that plug into the extension cords, or they can be cords that already have 3-way female ends on them.)
- ☐ Cordless drill/driver with charger and extra battery (If possible, carry two complete sets, including two drills, four batteries, and two chargers.)
- ☐ Tool box with selection of basic hand tools
- ☐ Small socket set with metric and SAE sockets
- ☐ Step ladder
- ☐ Extension ladder (An adjustable "multi-ladder" can be used in place of the step and extension ladders)
- ☐ Broom, dust pan
- ☐ Painter's plastic
- ☐ Trash bags, heavy duty
- ☐ Assortment of screws and nails
- ☐ Lockboxes (For temporary keys)
- ☐ Assortment of drill bits (This should include a set of standard twist bits and a set of larger spade bits up to 1½ inches for boring holes in drywall for wall boots.)
- ☐ Caution tape
- ☐ Ice melter or sand
- ☐ Flashlight
- ☐ Work lights
- ☐ Tape – duct, masking, and blue painter's tape
- ☐ Small shop vacuum (optional)

(Page 1 of 4)

Figure 16-9 Emergency Response Van Inventory Checklist. A sample inventory list of items of emergency supplies and equipment.

Supplies for Water Emergencies:

- [ ] (6-10) Air movers
- [ ] (2) Dehumidifiers
- [ ] (2) 48-inch boots
- [ ] (2) Sets of snout adapters
- [ ] (2) Sets of mini boots
- [ ] 4-inch hose with hose clamps
- [ ] Water extractor with hose and wand
- [ ] Thermo-hygrometer
- [ ] Non-penetrating moisture meter
- [ ] Penetrating moisture meter
- [ ] Foam blocks
- [ ] Carpet kicker
- [ ] Pump sprayer
- [ ] Antimicrobial
- [ ] Disposable gloves
- [ ] Disposable suits and booties
- [ ] Airsled (optional)
- [ ] Carpet probe moisture meter (optional)
- [ ] Electrical spider box (optional)
- [ ] Ramps for loading/unloading equipment (optional)

(Page 2 of 4)

Figure 16-9 Emergency Response Van Inventory Checklist *(continued)*

Supplies for Temporary Roof Covering:

- ☐ Lath
- ☐ 6-mil black plastic
- ☐ Roofing nails
- ☐ Tarps (Some roof covering situations are better handled with tarps rather than plastic sheeting.)
- ☐ Lumber, as needed (During the initial call, or after the initial visit, you may identify a weak roof or a large hole that requires shoring up prior to covering.)
- ☐ OSB or plywood, as needed (Sheet goods are used to cover and stabilize large holes prior to covering them with plastic or tarps.)
- ☐ Safety harnesses (Safety harnesses may be required to comply with OSHA regulations on roofs of certain heights or pitches.)

Vehicle Supplies:

- ☐ Fire extinguisher
- ☐ Tire chains
- ☐ Flares and road emergency gear
- ☐ First aid kit
- ☐ Rope and bungee cords
- ☐ Flashlight
- ☐ Jumper cables
- ☐ Tire pressure gauge
- ☐ Maps
- ☐ Portable GPS (optional)

(Page 3 of 4)

Figure 16-9 Emergency Response Van Inventory Checklist *(continued)*

Technician-Specific Supplies:

- ☐ Respirator
- ☐ Other individual PPE
- ☐ Cell phone
- ☐ Water
- ☐ Other personal gear

Paperwork:

- ☐ MSDS binder
- ☐ Extra copies of MSDS (Have copies of MSDS for antimicrobial and other products you anticipate using on an emergency that you might want to leave with the homeowner.)
- ☐ Emergency binders
- ☐ Work authorizations (These should be printed on NCR paper so you can leave a signed copy with the homeowner.)
- ☐ Warning signs, pre-printed
- ☐ Blank warning sign paper and pens, for making additional signs on-site
- ☐ Sleeves for hanging warning signs
- ☐ Job-specific maps

(Page 4 of 4)

Figure 16-9 Emergency Response Van Inventory Checklist *(continued)*

It's critical that the van be kept stocked and ready to roll at a moment's notice, or you defeat much of the purpose of having it. There are two ways to do this:

1. The best method is to restock the van as soon as you return from your emergency call. This can be considered part of the same emergency response, and the restocking time can be charged to the emergency call.
2. The alternative is to restock the van when you get the next emergency call, and then charge the time to the *new* emergency. *This isn't nearly as effective.* For one thing, it delays your response time to the new call. Also, it sometimes happens that when you arrive at the site, the homeowner refuses to sign the Work Authorization, or has hired someone else that he called earlier and didn't tell you about. In that case, you run the risk of not having a job to charge the restocking time to.

In addition to the basic restocking done after each run, plan on doing a complete cleaning and checking periodically that includes washing the van and cleaning out the entire interior. Every bin should be gone through to make sure it's clean, organized, and stocked according to the inventory list. Any items requiring maintenance should also be taken care of at that time. Assign a specific employee and a specific day of the week or month to this task, so that you're sure it doesn't get overlooked.

It's One of the Best Parts of Restoration!

Emergency work is challenging, lucrative, and, yes, even fun sometimes. It's a crucial part of this industry, and it's something you'll want to pay a lot of attention to as you get your company set up. With the proper training, equipment, and systems in place, you'll find this to be a *very* rewarding part of being a restoration contractor!

17

Responding to Emergency Calls

In the last chapter, we looked at how to get everything ready to roll when those emergency calls start coming in. From the van to the paperwork, you're ready to go! Now what happens?

Figure 17-1 Some emergency calls require moving a lot of contents in order to deal with the problem. Here, the owners had already begun moving contents out of the way of the water in this bedroom. This many contents on an after-hours call often justifies a two-person response.

When that emergency call comes in, one of the first things you need to decide is whether or not you'll need the help of the secondary on-call person. The caller may describe a situation where there's a lot of contents that you'll have to deal with (Figure 17-1), or maybe the description makes you think there's a considerable amount of wet materials to be removed. If you're not sure you need help after speaking to the client, then wait until you arrive at the location to make the determination.

If you do call in the secondary person, note the time that you called him. For billing purposes, that's when his time will start. Decide if you'll meet up at the shop and ride together — which is more efficient from a billing standpoint — or if you'll meet at the jobsite.

Your van should have almost everything you'll need already on it. However, if you determine from the phone call that some additional materials or tools are needed, such as plywood for a board-up, or maybe a taller extension ladder than you normally carry, you'll have to load those items onto the van as well. Whatever you need should be there at the shop.

Arriving at the Job

Your phone conversation with the homeowner should have given you an idea of what to expect when you arrive. You'll have some clues as to how calm or stressed the people are, and the extent of the damage. However, until you get a few of these under your belt, you'll probably find that the reality of an emergency call is different from what you had visualized while talking on the phone.

You may arrive at some water-damage calls to find the homeowners working frantically, while at others the people will seem dazed and overwhelmed. One house may have rooms of furniture tossed around haphazardly, even though there's very little water, while others may have rooms that are soaking wet but the contents are still on the floor, sopping up all that moisture. Some people may have already taken a knife to their carpet, or a crowbar to their hardwood floor!

Fire calls can be even worse, especially if you've never been in a fire-damaged home. As mentioned earlier, water-damaged homes look relatively normal, just wet. But with fire-damaged homes you're dealing with structural damage, contaminated air, loss of power, and a number of other hazardous situations.

Your job in *any* emergency situation is to calmly take charge, which is why you were called out in the first place. On jobs with extensive damage or frazzled homeowners, your calm authority is even more important.

Your Initial Steps

You'll find it's helpful if you and your crew establish a logical pattern of doing things when you arrive at the site, and always stick to it. This order calms the homeowners and shows them that you know what you're doing, so they'll be more likely to step back and let you take over — with less interference.

An orderly response means you're not as likely to miss something, and helps prevent potentially-costly mistakes. Being organized is also the key to keeping your paperwork in order, and that will *greatly* simplify your billing.

The following list of four initial steps will help get you started. You can change the order to suit the work and your own methods, but remember to be consistent in your approach to every job, and always stay organized.

1. ***Introduction:*** Introduce yourself and anyone else you've brought with you. This makes the homeowner feel more comfortable with the group of strangers they now have in their home. Give them a couple of business cards and show them on the card who you are and how to get in touch with you in the future. This may seem a little silly, but in stressful situations people process information in different ways. They don't always hear or remember what they're being told, so do what you can to help them out.

2. ***Recap the loss and explain the process:*** Start with a brief recap of what you know: "I understand that a pipe under the bathroom sink broke." That way you're certain you understand the situation, and they have a chance to add more information now that you're actually on-site.

   Follow this with a brief explanation of how the process works. Let them know that you're going to assess the situation and see what needs to be done. Make them feel comfortable, and assure them that you'll keep them informed every step of the way. Don't spend too much time talking at this point, because you need to start stabilizing the house — as yet you don't know much about what's happened.

3. ***Paperwork:*** Get your paperwork taken care of right away. You explained the work authorization over the phone, so they should be expecting it. Give them a chance to read it over and answer any questions they have — then get it signed! *Don't do any work without first having the work authorization signed by an authorized person! If you don't get a signature, you have to walk away from the job!*

   If your state or a particular insurance company requires that any paperwork be given out, do that now as well. Do your homework so that you know in advance what paperwork will be required, and have it ready in the binder. Have a place on the work authorization where the homeowner can initial that they received the paperwork, so there are no questions later on.

4. ***Inspect the home:*** Before you start, as a courtesy, you should explain to the owners about the inspection process. Verify that it's okay for you to enter different rooms, closets, storage areas, etc. Ask if there are any pets in the house that you need to be aware of. Ask if there are any areas they don't want you entering, or any personal contents they don't want you to handle. If there are, and it turns out that these are areas or contents that have to be addressed as part of the damage, you can discuss it further with them later.

Mitigating the Loss

There are literally thousands of possible damage situations with just as many combinations of building materials to deal with. One job will have wet carpeting, the next may have wet hardwood flooring, and the next have saturated ceramic tile. A falling tree limb may cause a minor hole in one house, but the next tree-limb call you get could involve cracked rafters, massive roofing repairs and damage to the fireplace.

Despite that, your basic approach to the loss should be the same. Do your initial inspection in a slow and orderly manner, beginning from the spot where the loss initially occurred and working out from there. Remember that you're *not* there to do an estimate, and you're also *not* there to perform extensive repairs. *The sole purpose of your first visit is to assess the extent of the damage, and assist the property owners in mitigating the loss.*

In the following sections we'll cover the basic steps you'll usually want to take for loss mitigation.

Remove the Source of the Damage

In most cases, this is as simple as making sure that the water is shut off. But in some cases, it might mean removing a tree from a roof or performing some other type of work.

Stop Any Further Damage to the Structure

One of the key things you'll be expected to do is prevent any additional damage from occurring. Once you've determined what caused the loss, you want to do whatever you can to protect the property.

In the case of a water loss, this usually means removing wet materials such as carpet and pad, and setting up drying equipment (Figure 17-2). In a storm situation, it might mean putting a tarp over a hole in the roof (Figure 17-3). In the case of a fire or damage from a vehicle, it might mean covering damaged windows and doors (Figure 17-4), or building a temporary wall to support damaged structural members to keep them from giving way.

Protect the Contents

As soon as you know that the structure is safe to work in, your next priority is protecting the contents. Often, dealing with the contents is something that you do at the same time you're protecting the house from additional damage. It could be as simple as moving a piece of furniture from a wet area to a dry area, or putting foam blocks under furniture (Figure 17-5). At other times it can involve moving all the furniture from several rooms out of harm's way.

Figure 17-2 Setting drying equipment in place is a common part of an emergency response. Here, an air mover and a set of small boots is being used to push air into a wall cavity behind a set of kitchen cabinets. This type of prompt loss mitigation is crucial to minimizing damage.

Figure 17-3 Covering a roof to limit further damage to the structure is another task common to an emergency response. This roof is covered with 6-mil plastic sheeting. Wood lath was rolled up into the top and bottom edges of the plastic to give it weight and a surface to nail through.

Figure 17-4 An emergency board-up in the aftermath of a structure fire.

Figure 17-5 This piece of antique furniture was set on foam blocks to protect it from the wet subfloor. This simple action alone can save hundreds or even thousands of dollars in furniture repair and refinishing costs.

Figure 17-6 Hasty actions like this in the midst of an emergency can do a lot of damage. If you accidentally scratch the shower pan by putting the toilet in it, you'll be liable for the repairs. Remember to put some type of protection down first. A scrap piece of carpet pad often works well.

Don't Do Any Further Damage

Not doing any further damage seems like an obvious piece of advice, but it happens quite often in the rush to deal with an emergency call to a building.

Think things through, even simple tasks. Don't haphazardly stack things on top of each other. You can do some serious damage that you may find yourself on the hook for later. Don't come into a soaking wet house and immediately crank up the heat — that can result in secondary damage, especially to wood.

It's also easy to cause unnecessary damage with tools, fixtures, and other things when you're moving them around, especially if you're tired and in a hurry. Be careful about setting anything down on countertops, hardwood floors, inside bathtubs and showers (Figure 17-6), on or in contents, or on any other areas that could be damaged. If you're working in an area where setting things down on finished surfaces can't be avoided, place them on a towel, mover's pad, or even a piece of cardboard. When you're moving toilets, specialized toilet dollies like the one shown in Figure 17-7 can be a real help.

Tip! *Working in an emergency situation is no different from remodeling someone's kitchen:* You can still be held liable for any damage you do to the property, so you need to be careful!

Figure 17-7 Toilet dollies are a great way to handle toilets on an emergency job. There are two threaded studs on the dolly so the toilet can be secured in place, keeping it from tipping over. The hollow plastic housing traps any water that leaks out of the toilet, keeping the floor dry. And the four casters allow you to easily move the toilet. The number painted on the front of the dolly allows it to be tracked on a company equipment log.

Careful Demolition

Don't tear out any carpet, drywall, or other materials unless you're *sure* they can't be salvaged. See Figure 17-8. Talk to the adjuster first if you have any doubts, even if that means setting drying equipment on wet carpeting (that you're pretty sure isn't salvageable), and then waiting a day until you can get hold of him. It's better to err on the side of caution than to tear something out that you shouldn't have. If you go ahead and tear materials out, in most cases the insurance company will want you to prove that whatever you tore out really needed to come out.

Never let the homeowner talk you into removing something. There will be instances where the homeowners really want new carpet, and they'll try and convince you to remove the old carpet against your better judgment. *Wait for the adjuster.* If the homeowners have already started their own tear-out, then document what's been done and go from there.

Secure the Structure

Always make sure that the structure is secured before you leave. Check that windows and exterior doors are locked. If they're damaged, then be sure that they're either boarded up or screwed shut so that no one can get in.

When damage occurs to a house from a fire or a storm, newspapers and other media often print or broadcast the address. Unfortunately, that's an open invitation for anyone, from thieves to lookie-loos, to drop by. Be sure the structure is completely secure to protect both your client's contents and your own liability.

In addition to boarding-up, if you consider a house or any portion of it unsafe to enter, cordon it off with caution tape. This will tend to keep most people out, and will also help document that you did everything possible to secure the structure.

Carry an assortment of warning signs in your van, such as *Do Not Enter*, *Caution — Wet Floors*, *Caution — Tripping Hazard*, and others that fit the situations you're going to encounter.

Figure 17-8 The backing on this carpet is delaminating due to water, and probably isn't salvageable. However, you'd still want the adjuster's approval before tearing it out and discarding it.

> ***Tip!*** *Print your warning signs on red 8½- x 11-inch card stock, then slip them into clear plastic protective sleeves. This protects the sign, and the holes in the sleeve allow you to hang it easily.*

You can print your own signs on your computer. See Figure 17-9. That allows you to customize the wording to fit the situation, and also gives you the opportunity to add your company's name and contact information. However, even though you'll want to print your signs out on a computer so they look professional, sometimes you may find something unexpected. So, keep a supply of blank paper and some large, black and red marker pens in the van so you can hand-write a specific sign when necessary to protect a jobsite.

Do No More and No Less Than Necessary

How much to do on the initial visit is one area of loss mitigation where you really walk a fine line. On the one hand, the insurance company expects you to do everything possible to prevent further damage; but on the other hand, they're typically paying you emergency overtime rates, so they don't want you out there any longer than necessary. After-hours emergencies are also physically tough on your crew, so you don't want them there any longer than they need to be.

CAUTION!

WATER DAMAGE
RESTORATION IN
PROGRESS

PLEASE DO NOT
ENTER
CONTAINMENT
AREA

Pro-Done Restoration
123 Builder's Lane
Anytown, CA 99988
(555) 123-4567

Figure 17-9 A typical warning sign for posting on a jobsite. Signs like these can be made up on your computer as needed.

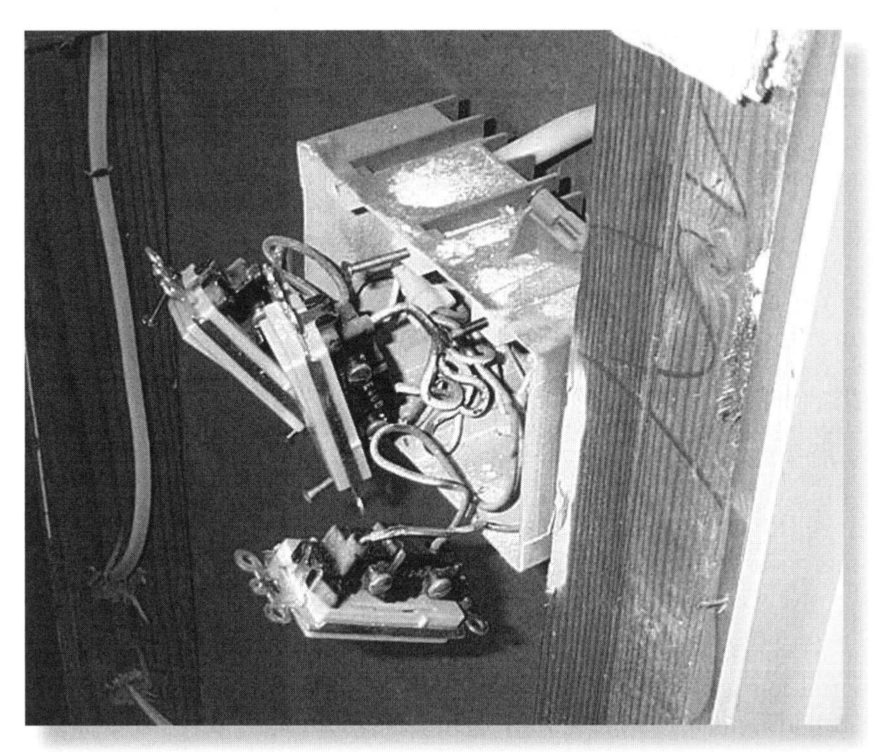

Figure 17-10 Once the drywall removal is complete on an emergency job, be sure that switches and outlets are resecured to their boxes for safety. Don't leave any potentially hazardous situations like this one behind.

Part of your initial plan is to determine what you need to do right away, and what can be put off until the following day. That generally means that all the contents are moved out of danger; any unsalvageable wet material is removed; drying equipment is set as needed; and the house is secure and weathertight.

Guidelines for Cleanup

Another gray area of your emergency response is exactly how much cleanup you should do. You need to make a good impression on the homeowner and, more importantly, make the house safe. But the insurance company isn't going to want to pay you overtime rates to clean up every last speck of dust.

Here are some basic guidelines for cleaning up a jobsite at the end of an emergency response:

- **Remove anything that's dangerous:** This includes protruding nails and staples, loose boards, etc. If you've torn out any carpeting, be sure and remove any exposed tack strip. Don't try and salvage tack strip — it isn't worth it. If you've torn out any drywall, be sure that all electrical wiring is safe. See Figure 17-10.
- **Cover contents with plastic:** Plastic protects the contents from dust and possible damage. It also makes a very good impression on the homeowner, and is something that's easy to bill to the

insurance company. However, follow the precautions outlined in Chapter 13 when covering contents that are wet.

- **Do some basic cleanup:** Do a decent general sweeping. Leave the site in broom-clean condition.
- **Remove debris:** If possible, take the debris with you. Smaller amounts of debris can be placed in garbage bags and hauled off, which makes the homeowner happy and eliminates one more area of liability for you. If there's too much to deal with, stack it up as neatly as possible for removal later. Don't, however, leave the debris in a walkway or driveway where it's a nuisance or a tripping hazard for the homeowner. Check with the homeowner to find the best place for temporary debris storage until it can be hauled off. Loose debris should be covered with a tarp to reduce liability, and to prevent it from possibly blowing around.

Be Careful

Never lose sight of the fact that damaged buildings are dangerous, and the emergency working conditions that you and your crew will be operating in can be very hazardous. During storms, you'll be experiencing wet roofs, high winds, and heavy ice, rain, and snow. You'll encounter floors, walls, and roofs that are structurally damaged and highly unstable after a fire or impact damage.

Don't take chances, and never try to be a hero and single-handedly tackle some job that requires extra help. You can cause even more structural damage to the home, not to mention injury to yourself, trying to take on something big — like a tree toppled against the side of a house — rather than waiting for help.

Take your time, move slowly and deliberately, and be smart in the choices you make.

Wrapping Up Your Visit

At the end of your emergency call, take a moment to wrap things up with the homeowners. Go back over what you've done, and why you did it. Explain what equipment you set up, what it does, and explain anything they need to know about it. Tell them what happens next, when you'll be back, and what you'll be doing when you return.

Different homeowners will process the information you're giving them in different ways, so take your clues from their reactions. If they appear tired and not very focused, then just give them the basics. If they seem very interested in what's going on and have lots of questions, then explain things in a little more detail. Finally, make sure they have all the necessary job-related paperwork, and give them a copy of the Material Safety Data Sheets (MSDS) for any products you used. Finally, give them a couple of extra business cards and remind them how to get in touch with you if needed.

Establish What Your Future Access Will Be

In most cases, you'll need access to the house the following day, and perhaps for several days after that, to check and move equipment and possibly complete your estimate. Explain this to the homeowners, and ask them what future access arrangements for you or your crew they'll feel comfortable with.

What you'll probably be told is something like: "Oh don't worry about that, we're always home." Or, "I work right around the corner. Just give me a call and I'll come right over." Or, "My neighbor in the blue house across the street has a key and can let you in anytime you want."

The problem with any of these scenarios is that they never work as planned. When you stop by, they'll have stepped out for some reason; if you call at work, they're in the middle of a meeting and can't be disturbed; if you call the neighbor, she's taking a nap or out grocery shopping. These are all wasted hours, and you can't bill for them.

> ***Tip!*** *Set your own access code for your lockboxes, and make them all the same to avoid confusion. Select a set of numbers or letters that are easy to remember — but not too obvious. For example, if your company name is Smith Restoration Company, don't use SRC as a code. That's too easy to guess. At least every six months, collect all your lockboxes and change the code to something new. If you don't, employees, ex-employees, subcontractors, past clients, and others will all have your lockbox access code.*

Instead, ask the homeowner if you can set up a lockbox with a key to the house inside. Portable lockboxes are available from home centers, hardware stores, and locksmiths, and are very easy to use. They have a long, looped handle that's secured over a doorknob, porch railing, pipe, or other secure spot outside the house. The lockbox is accessed like a combination lock by turning the knob to specific letters or numbers.

Discuss the lockbox option with the homeowner. Explain the importance of having access — that you don't know exactly what time you'll be by, and that this option is more convenient not only for you, but for them. Let them know you'll call first before coming out if they prefer, and you'll always knock before you enter to make sure no one is home. Once you explain all that, most people are fine with the lockbox idea.

When you set up the lockbox, let the homeowners know what the combination is, and also check to make sure the key they give you works correctly. It's not unusual for a homeowner to provide the wrong key, which obviously defeats the whole purpose!

Emergencies in Unoccupied and Vacant Homes

Quite often, you'll get an emergency call to respond to a home that's unoccupied but furnished. The owners may be away from home on vacation and a neighbor's watching the house, or it could be a person's second home, or a vacation rental.

You'll also get calls to vacant, unfurnished homes and buildings. These include homes that are for sale or rent, or houses that have been vacated by the owners as the result of foreclosure and are now owned by a bank or other lending institution.

Your response to an emergency at an unoccupied or vacant home will be very much the same as to one in an occupied home. When you get the initial call, you'll still need to determine what type of damage has occurred, the extent of it, and all the other details. As before, use your Emergency Evaluation Checklist as a guide to asking all the necessary questions and writing down the information during the emergency call.

Work Authorizations

What's going to be more challenging is getting the Work Authorization signed, but it's still critical that you do. When you take the call, you'll need to determine who the caller is, and what their relationship is to the owner of the property. It could be a property manager or an apartment manager, both of whom are typically considered legal agents of the owner. It could be a person representing the lending institution that owns the property, or it could be a real estate agent who's selling the property and is acting on the owner's behalf.

In some cases, the caller will be a friend, relative, or neighbor who's watching the house in the owner's absence. The laws regarding who is and isn't the legal agent of a piece of property can vary widely from state to state. Your best bet is to sit down with your attorney before this situation comes up, and discuss some of the various scenarios that you might find yourself in with regard to Work Authorizations. Get his advice on how to handle each one, and go from there — do the best you can, but be sure and get a Work Authorization signed by somebody!

REAL STORIES:

Good Thing We Had a Signature!

We once responded to a small water loss in a vacant apartment. It was in a large apartment complex; the manager discovered the damage and made the call to our company. He was also the one who signed the Work Authorization. We did the emergency work, and sent a bill to the apartment complex in care of the manager.

A couple of weeks later, we were quite shocked to receive a very nasty letter from the owner of the complex. He told us in no uncertain terms that his manager had no right and no authority to call us, and therefore he was refusing to pay the bill.

We immediately checked with our attorney. He told us that state law clearly showed that the manager of an apartment building was an agent of the owner, and was therefore able to sign on his behalf. We ended up having to file in Small Claims Court, but we eventually got our money!

Good thing we had that signed Work Authorization.

Obtaining Access

Once you have the information and have taken care of the Work Authorization, your next challenge with a vacant or unoccupied house is access.

Access is an important element of the work. You need to be able to monitor the job, and to not waste a lot of time doing it. As before, your best option is a lockbox. However, you may have a hard time getting that responsible neighbor to part with a key. You may also run into real estate companies and property managers who have set rules about releasing keys to properties they're responsible for.

Be prepared to be flexible. You may have to make a compromise or two in these situations. But don't allow unreasonable constraints to be placed on you. If you have a bad feeling about how the job is going to flow with absentee owners, don't be afraid to turn the job down and walk away.

Once you have access, proceed with the emergency work and the estimating in the same manner as you would with an occupied building. But remember, you have a greater amount of potential liability in an unoccupied house, especially one that still has the owner's belongings in it. Be especially aware of your paperwork and your log notes. Always note the date and exact arrival and departure times of each visit you make to the home. Take lots of photos, and carefully document everything that you do.

Multiple Losses

When it rains, it pours — sometimes quite literally. There'll be times of extreme weather when you'll be hit with multiple emergencies at the same time. Wind storms, hail storms, extended periods of below-freezing weather, wildfires, even hurricanes — all can create a tremendous volume of calls for insurance restoration services.

Multiple-loss situations are tough to plan for. You can't invest in a large amount of equipment or hire a huge crew, and then have both sit idle until they're needed. So instead, you do the best you can with what you have.

When calls start coming in on multiple losses, let your crew know what's ahead. Tell them that you'll need everyone to pitch in wherever possible. Have them start working on getting equipment and supplies ready. See which of your current in-progress jobs can be temporarily put on hold.

Next, organize the calls. Look at which ones are real emergencies, such as a hole in the roof with water pouring in, and which ones can wait, like the house with just a few shingles blown off. You want to get to the worst emergencies first; your overriding goal is to help mitigate damage.

Also look at which insurance companies insure the various callers. While you're trying to deal with the worst emergencies first, you also want to give some priority to the companies that are your primary sources of work. If you get two simultaneous calls for storm-damaged roofs, and you can only get to one of them, go with the company you work with all the time. Your loyalty will be noted and appreciated.

REAL STORIES:

It's Gonna be One Tough Monday!

When you're in construction, keeping one eye on the weather is a fact of life, and nowhere is that more true than in the restoration field. For the new home builder, an approaching storm might signal a few days with no work. But for the restoration contractor, that approaching storm can mean things are about to get hectic!

One fall weekend, the weather was mild and calm all day Saturday and well into Sunday. Then Sunday afternoon, the wind started to pick up. By Sunday night, it was really blowing, and showing no signs of letting up anytime soon. I can remember watching the trees in my yard whipping back and forth and thinking to myself "It's gonna be one long, tough Monday!"

Sure enough, when we arrived at the office at 6:00 AM Monday morning, the phone was already ringing. Storm damage calls, mostly roofing-related, were coming in from all over town. We put our existing jobs on hold for the day, and told the crew to hang on while we got things sorted out. Within an hour, we had taken 24 calls. By noon, the number was approaching 40.

We ended up with a huge piece of masking paper pinned to the wall in our break room. With each incoming call, we were frantically writing down and organizing names and addresses, along with the severity of the damage and the names of the insurance companies.

Crews spent the day racing around, temporarily patching holes. We then spent the rest of the week writing estimates. Within a few days, we had booked several months' worth of new work.

It definitely was, as predicted, a long, tough Monday! But it didn't do our bottom line any harm.

When your crews go out on an emergency and the calls are backed up, tell them to do only what's *absolutely necessary* and get on to the next job as soon as possible. For example, they may not have time to do a good roof patch, but they can get some plastic over the hole for the time being, and return later to do a proper repair.

Use your drying equipment sparingly during multiple loss situations. For example, if you encounter a wet house where you'd normally put six air movers and three dehumidifiers, you may want to consider setting four and two respectively, as long as you feel you're not running the risk of mold or secondary damage. This is a judgment call, but, within reason, it'll let you help more clients and more insurance companies, and build more revenue for your company.

Eventually, no matter how hard you try, you're going to reach your saturation point. Your crews will be stretched to the limit, and you'll be completely out of equipment. When that happens, don't be afraid to turn down jobs. The insurance companies would much prefer that you be honest with them rather than make promises that you simply can't keep.

Price Gouging

During times of severe weather and multiple losses, you can charge pretty much whatever you want, and in order to get the jobs done for their clients, most insurance companies are stuck paying it. Here's the best way to handle that:

Don't do it!

Overcharging insurance companies will make you some extra cash in the short term, but in the long run, it'll hurt your business. Never lose sight of the fact that the emergency will be over soon, and things will return to normal. When they do, you'll be facing the same insurance companies that you gouged when things were tough, and they'll remember just what you did. If you stick with your standard hourly and overtime rates, and your standard material markup, even when demand is high, you'll make much more money over the long run.

Paperwork

As we've discussed, insurance restoration work requires a lot of paperwork. You need to establish and maintain a clear paper trail on every job, and it begins the moment you get an emergency call.

REAL STORIES:

Making a Name for Yourself!

We hadn't been in business very long when our area was hit with a series of severe snow storms, one right after another. All that snow, combined with day after day of subfreezing temperatures, created some horrendous ice damming.

We were getting call after call. But until the weather warmed up, there really wasn't anything we could do except help people get some of the snow off their roofs. The insurance companies realized that paying for snow removal was a good investment, since it helped mitigate further damage. So they were willing to pay to have it done.

Some of the companies in town started doing snow removal, and were charging four and five times the normal hourly rate. When the weather warmed enough to start doing repairs, they were again charging far in excess of the going rate for roof repairs. The insurance companies grudgingly paid their prices because they needed the work done. My partner and I had made the decision to only take on what we could, and to only charge our regular rates, plus overtime when it was due.

The snow finally melted, and things returned to normal, as they always do. And guess what? Our new company started getting call after call for repairs. And the insurance companies told us over and over again how grateful they were for our honesty and cooperation during that tough time. By ignoring the temptation for some short-term gains and staying focused on our long-term growth, we had enhanced our reputation and really put our company on the map!

As you begin to work on emergency jobs, consider what method will work best for you from the standpoint of keeping all the paper effectively corralled. One good method is a three-ring binder. This can be an inexpensive ½-inch binder that you buy in bulk at a discount office supply retailer or warehouse store. You'll also need a three-hole punch to punch your papers so you can put them into the binders.

Make the binders up in advance with all the paperwork you know you're going to need. Keep a couple in your emergency van, and make sure that your leads and backup people have a couple at home whenever they're on call. Each binder should have a copy of the Emergency Evaluation Checklist in it, so whoever's on call has a list of all the questions to ask on the initial call.

Select a specific color for your emergency binders, such as red. If you like the binder idea and want to use them for non-emergency client job files as well, select a different color for those, such as white. Buy the type of binders that have a clear plastic sleeve on the spine so you can place the client's name and job or ID number in the sleeve for fast identification.

The binder should include any other forms that are needed on the job. As additional paperwork gets generated, such as sketches, notes, equipment logs, receipts, etc., they can easily be added to the binder. This keeps all the paperwork together in one place, and simplifies the entire process of tracking information and preparing invoices.

When the emergency's done, take all the paperwork out of the binder and add it to the client's regular job folder, so you have it all in one, organized location. You can then restock and reuse the emergency binder.

Document, Document, Document

I've said it before, and it really can't be overemphasized: *For anything from billing to liability issues, you have to keep track of everything that you're doing on the job!*

Good documentation includes four basic categories:

- Forms
- Notes
- Photos
- Samples

> ***Tip!*** *As you obtain each piece of equipment, put your company name on it and then assign it a number in sequence. The combination of a numbering system and an equipment log will help you keep from losing track of your equipment.*

Forms

We've talked about the different forms that you'll be using on an emergency job, such as the Emergency Evaluation Checklist and the Work Authorization form. Copies of both of these forms should be part of your emergency binder. Any other forms required by your state or local contractor's boards or other agencies need to be collected and filed away as well.

On a typical water loss, you'll be setting a lot of drying and dehumidification equipment, and it's extremely important that you keep track of all of it. This equipment is very expensive, so you obviously want to know where it is at all times. Equally important is the fact that you're renting it to insurance companies, so you need to know exactly what to charge for, and how long to charge for it.

The best way to do this is with a simple equipment log (Figure 17-11). This form basically lists the piece of equipment, the date it was set, and the date it was removed. As with all your other forms, make three hole punches in it and keep several blank forms in each emergency binder.

Equipment Log

Client name: Jill Baker

Address: 1222 Green Meadow Ct.

Type of equipment	Equipment number	Location set	Date set	Date removed
Air mover	123	Master bath	6-23	6-25
Air mover	144	Master bath	6-23	6-25
Mini boot set	6	Master bath	6-23	6-25
Air mover	103	Master bed. Clst.	6-23	6-24
Dehumidifier	37	Master bed.	6-23	6-25

Figure 17-11 A simple equipment log sheet used for tracking equipment at various jobsites.

REAL STORIES:
Do You Know Where Your Equipment Is?

One of our competitors was hired to dry a house after a water loss emergency. Once the other company was done with the drying, the homeowner hired our company to do the reconstruction.

In the course of doing the repairs, one of our crew went down into the crawlspace. Still sitting there, under the house, were two of the other company's air movers!

We placed a call to the other company, and one of their rather embarrassed employees came by and collected their lost equipment! They clearly didn't have a good tracking system!

Notes

Notes are an important part of any emergency job. You need to train yourself and your crew to take as many as possible. Keep them in the emergency binder for future reference.

Good notes make it easier to explain your actions to the adjusters and the clients. They simplify billing, and they make it faster and more accurate to pass a job off from one crew member to another. And while you hope that it never happens to you, your notes will be vitally important if you're accused of any wrongdoing or end up in court for any reason.

(Photo courtesy of Ryobi)

Figure 17-12 This digital camera was developed specifically for jobsite use. It's impact-resistant, dustproof and waterproof, and operates off a rechargeable battery. It's ideal for a restoration environment. Digital cameras have made it easy and inexpensive to photo-document your jobs.

Photos

As the old saying goes, a picture is worth a thousand words. So don't hesitate to take lots of them, especially in this age of digital photography! Use your camera to document *everything* you do. From the moment you arrive at the jobsite, start taking pictures. Photograph the cause of the damage, the location and extent of the damage, the contents — everything. Also, photograph the trim, flooring, cabinets, and any other existing materials that you want to document for future matching.

You don't have to be a great photographer to put a good collection of jobsite photos together, and you don't need to have a bag full of expensive equipment. On the other hand, you don't want to waste time and money taking pictures that are too poor to be useful. Invest in a medium-quality, point-and-shoot digital camera with good resolution and a built-in zoom lens, like the one in Figure 17-12. The store

Figure 17-13 Document pre-existing damage to furniture and other contents. The back of this cabinet was coming loose, and there's a scratch on one side. This is the type of thing you want to be sure to document with both pictures and notes in your file.

where you buy the camera can show you a few good ones to choose from, and they may also offer an introductory class. If not, you can usually find a good basic digital photography class at your local community college. A class will not only help you with the quality of your pictures, but you'll also learn how to size them, store them, print them, and transmit them electronically.

A good collection of job photos will help you immensely in estimating and billing. When you're working on your estimate, the pictures will remind you of details you may have left out of your notes. During reconstruction, pictures can help your crew with some of the details of how things go back together, and they'll also be a reminder of where all the contents go when you move the owners back in.

In order to protect yourself from damage claims, document preexisting damage to the structure or the contents with photos. See Figure 17-13. It's pretty hard for someone to accuse you of scratching a piece of furniture or breaking a light fixture when you have photos documenting that they were in that condition when you arrived!

Many insurance companies now require that you document the loss with digital photos, so talk with the insurance company in advance to determine how many photos they need, and what they want photographed. Knowing what the insurance companies want and providing it is one more way to make their job easier, and that makes you all the more valuable to them!

Samples

One of the big challenges with insurance restoration is matching materials, and also maintaining comparable quality to existing materials. Keeping a good collection of samples helps considerably with this.

As you do your demolition work, retain small pieces of materials. Create a sample collection of baseboards, casing, trim, tile, and hardwood flooring. Cut out a square foot or so of carpet and carpet pad. Keep anything you might need to refer to at a later date. Have some zip-lock plastic bags available in your truck to store smaller items. Use a permanent marker and note on each sample the client's name, the date, the insurance company, the room the sample came from, and any other information you might need to refer to later.

You'll need to hold onto the samples until the job is finished. And, if you have any problems on the job, or if the insurance company requests it, you might want to keep them even longer. To simplify storage, buy some inexpensive stackable cardboard boxes with lids — file storage boxes or Banker's Boxes. You can buy them in bulk at many office supply stores and some big warehouse stores. Tape a piece of paper on the outside of the box with the client's name for easy reference. When you're sure you won't need the samples again, take the paper off the outside, discard the contents, and reuse the box.

As with photographs, some insurance companies will require you to collect samples, especially for floor coverings, and keep them for a specific time period.

18

Your Business Plan

As we discussed in the beginning of the book, some contractors don't put as much thought as they should into getting their business up and running. They acquire knowledge and training in the trades working for another contractor, get their contracting license, and then start their own business. This often involves little more than having some cards printed, sticking a sign on the side of their truck, and taking out an ad in the local newspaper.

There's nothing wrong with starting small; in fact, not taking on a lot of debt or growing too quickly is definitely a good thing. But making the move into the field of insurance restoration requires more of a commitment than many other types of construction work. Besides the training and the equipment, homeowners and adjusters expect a level of professionalism. It can be difficult to earn their trust without it.

The difference between a small, one- or two-man remodeling or homebuilding company and a full-blown insurance restoration company is similar to the difference between a sandwich shop and a restaurant. Both are fine endeavors. Both will make you money, and both serve a need in the community.

However, the restaurant takes more commitment and a higher level of expertise. It requires more staffing, more equipment, more training, more flexibility, and a greater financial investment. But the payoff is greater income and steadier work. A restaurant generally has a higher standing in the community, and as a business, it's better positioned for eventual sale.

A contractor once said to me, after making the move up from a small remodeling company operating out of his house to an insurance restoration company with an office and a warehouse, "Now I feel like I have a *real* business!"

Making a Business Plan

Talk with any banker, business consultant, or other professional who's in the field of helping businesses grow and succeed, and you'll hear the term *business plan*. Many small contractors — and other small business owners, for that matter — think a business plan is only for "big" companies. If that's your mindset as well, you need to reconsider.

A business plan is a document that helps you do three things:

1. Define goals for your business.
2. Set a plan for achieving those goals.
3. Set short- and long-range timeframes for reaching those goals.

That sounds pretty simple, and in many ways it is. But you have to be honest with yourself, and you have to do your homework. If you do, your chances of success with your business will be greatly improved.

Let's look at the three items on the list in more detail.

Establish Business Goals

Your first step is to ask yourself some questions about your business. Here are a few examples to get you started:

- What is your goal for your insurance restoration company?
- How big would you like to grow?
- How many locations would you like to have?
- What type of salary would you, as the owner, like to take home?
- Would you like to own your own building?
- At what age do you want to retire?
- How will your business make retirement possible?
- Are there other goals you'd like to set for yourself and your business?

There aren't any right or wrong answers. Your goals and your definition of success are personal. For you, success may be money, early retirement, a greater sense of community, or having a company that helps others. Or it may be as simple as having a career that puts a smile on your face when you head out to work each morning.

Establish a Specific Plan for Attaining Those Goals

Now that you've decided what you'd like to do with your business, the next part of your business plan is to decide *how* you're going to do it.

Let's take one specific example. Suppose one of your goals is to own a building that will house your business. There are different ways you can achieve that goal.

- You can buy a brand new building that's move-in ready. That way you can set it up and get on with running your company.
- You can buy an older building and put some time and effort into remodeling it. That gives you the financial advantage of some sweat-equity, plus you can remodel the space to suit your business. But that would take time away from running your company while the work gets done, so you'd have to balance the needs of both your business and the building remodel.
- You can buy a piece of land and build your own building. That would allow you construct exactly what you want and need for your business. This option may offer the greatest potential return on your investment, but it's a huge undertaking, involving everything from meetings with city planners to picking out paint colors. Again, this is time away from your business.

In order to make an informed choice, you first need to consider how much each of these options will cost. You'll probably need the help of a real estate agent or other professional.

You also need to consider how each of these options would meet your current and future needs. This will require some educated guesses about where you see your company going in the coming years. Does your plan include more contents storage that'll require a larger warehouse space? Does your plan require more vehicle or material storage? And, how about the need for additional office space for more staff?

From an investment standpoint, which of these options will have the best resale value in the future? This is a tough question, but it's also the type of projection that should influence your decision-making process. It will have an impact on how banks view your business plan for possible financing.

Establish Specific Timelines

A business plan typically includes some short-range timelines for goals to be reached within six to 12 months. It also has long-range timelines for goals to be reached five, or sometimes even 10 years, down the road. Timelines are important for how you view your company. They help you set a path toward achieving your goals, and become an important part of managing your growth and your finances.

Let's set some timelines for your goal of owning your own building. If you're just starting out and you're working out of your home, the written timelines in your business plan might look like this:

- **Within 3 Months:** Move into rented office/warehouse space. Ideal location: Eastside Industrial Park.
- **Within 1 Year:** Invest in bare land. Can't build right away, but should invest in the land as soon as possible. Eastside

Industrial Park location is centrally located; land prices are still reasonable; the area is growing and expanding; has good investment potential!

- **3 Years:** Work on design for building. Make it a work in progress. Refine to meet work flow and projected growth needs.
- **Within 4 Years:** Break ground for new building.
- **5 Years:** In my own building! My own space — no more rent!

Getting Some Help

Asking yourself the questions that go into creating your business plan will make a huge difference in how you view and operate your business. They're essential to setting yourself up to *succeed*, instead of setting yourself up to fail! Also, business plans are often required by lenders if you need to borrow money for your business.

REAL STORIES:
Can I Get Pepperoni on That Business Plan?

When my partner and I first started, we were doing remodeling work, mostly kitchens and additions, as subcontractors for an established kitchen and bath store in town. One day over a pizza lunch, we started talking about opening our own kitchen shop. However, there were already several in town and it wasn't *exactly* what we wanted to do.

Then my partner started talking about insurance restoration. Both of us had done some small insurance jobs in the past, and we both had an interest in the field. The ideas started flowing, and by the time the last of the pepperoni pizza and soft drinks were gone, we had a basic idea of where we wanted our business to go. It wasn't a business plan by any stretch of the imagination, but it was a start. It was certainly a lot more than we'd had when we arrived at the pizza parlor!

More lunches followed over the next few weeks. We tossed around goals over hamburgers, equipment needs over Chinese food, and Mexican lunches yielded some basic timeframes. We piled up pages of notes, and pretty soon we were off and running with a plan for our new business.

Over the years, our business plan evolved along with our business. Every month, my partner and I, along with our General Manager, would have a management meeting at a local restaurant and discuss everything from company picnics to major equipment purchases.

And once a year, just after New Years, the three of us would have an all-day "retreat" at one of our homes to just talk. We'd look at the problems and successes of the prior year. We'd discuss goals for the future, and what we could do to improve the company. Donuts and chocolate cake were always involved. Maybe it was the sugar, but those were always very productive meetings, and we'd generate lots of notes. At the end of the day we'd summarize everything, each take a copy, and that summary would become our guiding force for the coming year.

Yes, we also had a business plan, along with lots of other formal written documents, from financial statements and balance sheets to marketing plans and projected timelines. But our informal monthly lunchtime meeting notes and annual summaries were still the heart of things, because it was a method that worked for us — and we were *consistent* about following up. That definitely contributed to our success. And it probably contributed to the success of several restaurants in town as well!

Figure 18-1 shows an outline of what should be included in a good business plan. There are lots of places to go to find help in preparing your business plan. There are a number of very good books on the subject available at libraries and in bookstores. Many local colleges, as well as Chambers of Commerce, offer classes. And, through the federal government, you can take advantage of the Small Business Administration (SBA), which offers help for people in starting, financing, and managing small businesses. See the Appendix for more information on the SBA.

Writing a Business Plan

1. Cover sheet
2. Statement of purpose
3. Table of contents

I. The Business:
   - A. Description of business
   - B. Marketing
   - C. Competition
   - D. Operating procedures
   - E. Personnel
   - F. Business insurance

II. Financial Data
   - A. Loan applications
   - B. Capital equipment and supply list
   - C. Balance sheet
   - D. Breakeven analysis
   - E. Profit & loss statements
   - F. Three-year summary
   - G. Detail by month, first year
   - H. Detail by quarters, second and third years
   - I. Assumptions upon which projections were based
   - J. Pro-forma cash flow

III. Supporting Documents
   - A. Tax returns of principals for last three years. Personal financial statements (all banks have these forms)
   - B. For franchised businesses, a copy of franchise contract and all supporting documents provided by the franchisor
   - C. Copy of proposed lease or purchase agreement for building space
   - D. Copy of licenses and other legal documents
   - E. Copy of resumes of all principals
   - F. Copies of letters of intent from suppliers, etc.

Figure 18-1 This is the SBA's outline of a good business plan. Although there's no single formula for developing a business plan, there are some common elements. The Small Business Administration (SBA) recommends that the plan include an executive summary, supporting documents, and financial projections.

Gathering Your Team of Professionals

Even though you're great at putting people's homes back together, you can't be expected to also be great at understanding *all* the complicated aspects of running a small business. For that, you need a group of key professionals, from an attorney to an accountant.

Your professional team will help you along the way, from the initial setup of your business to its eventual sale when you retire. I can't stress enough how important these people will be to you and the long-term success of your company. Select them with care.

Here are a few tips:

- **Look for specific small business experience:** Small business operations have very specific needs that are different from those of large companies. Keep that in mind, and look for professionals that *specialize* in small business operations.
- **Conduct interviews:** Remember, you're hiring this person and his company just like any other employee, so interview him first. Be sure his professional skills meet your needs, that your personalities mesh, and that his business philosophy is compatible with yours.
- **Think long-term:** Once you have your team in place, try and keep them for the long run. Working with the same people over the years builds and strengthens relationships. It helps them understand you and your company, which puts them in a better position to advise you as time goes on.

Insurance Agents

It's an unfortunate fact of life when you're in business for yourself — bad things happen, some that may have catastrophic consequences; even financial ruin. The only protection is to have the right type of insurance, in the proper amounts. While it would be nice to be insured against anything that could possibly occur, it would also be prohibitively expensive. That's why you need the help and advice of a good insurance agent.

Insurance is a very complex subject, so good advice is essential. What follows is a very basic overview of some of the different types of insurance that an insurance restoration company might require — but only you and your agent can determine your exact needs.

- **Liability insurance:** Protects you against damage you cause to a person or property, accidentally or through negligence on the part of you, your crew, or your subs. Minimum liability insurance amounts for contractors are often set by the state you're licensed in.

- **Worker's compensation insurance:** Protects you in the case of injuries to your employees. The minimum level for worker's compensation coverage is usually set by the state.
- **Bonds:** Most, if not all, states require that contractors be bonded, which essentially protects consumers if you fail to complete work on a project. The minimum bond amount is typically set by the state.
- **Automobile insurance:** Insures your vehicles against loss from theft, accidents, vandalism, and other problems. It also offers protection to other vehicles, their occupants and damaged property in the event of an accident.
- **Business renter's insurance:** If you're renting space for your business and something happens to the building, such as a fire, the building owner's insurance typically won't cover you. Renter's insurance covers your property in the office and warehouse in the event of a loss. It also extends coverage for liability issues if a person visiting your business is injured. Renter's insurance will cover your costs if the building you're renting becomes damaged and you have to move to another location. *If you rent, you really need renter's insurance!*
- **Property insurance:** If you own your own building, then you'll need property owner's insurance instead of renter's insurance. This covers the building itself, as well as most of the contents in the building. It also offers liability protection if someone visiting your building is injured. Property insurance will cover your expenses if you have to relocate due to damage to the building.
- **Business disruption insurance:** If the building you're in is damaged, you may be temporarily unable to operate your business. This coverage will compensate you for any money you lose while your business is disrupted.
- **Loss of rent insurance:** Suppose you own the building where you operate your business, and you also rent part of the building out to someone else. If the building becomes damaged and your renter needs to move out, loss of rent coverage will cover the rental income you lose.
- **Care and custody insurance:** If your company will be dealing with a homeowner's contents — packing, transporting, cleaning, and repairing — then you'll need specific insurance coverage for those contents. Care and custody insurance covers you if you break something, lose something, or damage something while you're handling it. It *may* also extend coverage to cover something stolen by one of your employees. Care and custody insurance typically has very specific clauses and policy limitations, so be sure you understand them.

- **Content storage insurance:** If you'll be storing client contents in your own building or at a storage facility that you rent, you'll need to insure them against loss, theft, or damage. The price you pay for this insurance is based on the total value of the contents you have in storage, so look for a policy that can be adjusted up or down depending on what you're currently storing.
- **Health care insurance:** If you choose to provide health care for your employees — and you should if you can — there are policies available. The prices vary widely, depending on coverage. Many employers pay for part of the coverage and have the employee pay for the rest, which might be the best way to make it affordable for everyone.
- **Employee discrimination and sexual harassment insurance:** This insurance covers you if someone files suit against you for sexual harassment, or for discrimination based on race, religion, age, or other factors. It also provides coverage if you're sued by any job applicants who feel you discriminated against them during the interview and hiring process.
- **Employee theft insurance:** This covers you if an employee steals from you. This typically includes jobsite theft, as well as embezzlement.
- **Umbrella policies:** An umbrella policy is basically additional coverage that can be applied anywhere. For example, your umbrella policy might kick in additional coverage for liability issues, or in the event of an accident involving a company vehicle.

Looking for an Insurance Agent

There are basically two types of insurance agents: company agents and independent agents. A company agent represents a specific insurance company only, such as State Farm or Allstate. An independent agent offers policies from several different companies. Because they have more policies and a wider range of prices to choose from, an independent agent may be your best place to start.

Given its complexities, and the fact that having the wrong insurance can be detrimental to your company, you need an insurance agent who can explain the policies in a way that you understand. He also needs to understand the insurance restoration business, especially the handling, transporting, and storage of contents, and tailor your coverage accordingly.

Both of you should plan on meeting *every year* to review all of your policies, and adjust them as the needs of your company change.

Attorneys

Attorneys are important to modern business life, and the one you select should be able to assist you with all of the following legal needs:

- **Your legal business structure:** This is how you initially set your company up for tax and legal purposes; for example, a sole proprietorship, partnership, corporation, etc. The structure is crucial to protecting you from liability and for maximizing tax benefits.
- **Drafting legal forms:** You'll need a contract form that meets your specific needs, as well as the legal requirements of your state. You can draft your own contracts using a software program called *Construction Contract Writer*. The program complies with all the requirements of your state and allows you to customize your contracts to your business. A free 30-day trial version is available for download at www.constructioncontractwriter.com. If you decide to create your own contract, it should still be checked by your attorney. You'll also need other forms, such as a Work Authorization, drawn up by an attorney.
- **Reviewing legal documents:** There are any number of legal documents that you'll encounter while doing business, from contracts and lien notices to loan documents and escrow papers, and all these should be reviewed by your attorney.
- **Collections and filing litigation:** There'll be times when you're unable to collect on a debt, and you may want an attorney's help. Also, at some point, you may need to file suit against someone, and an attorney is essential for that.
- **Defending litigation:** If you're in business, you always run the risk of being sued. In today's lawsuit-happy society, you want to have an attorney that can handle your defense needs.
- **Selling your business:** One of the nicest things about building an insurance restoration company is that you're creating a business that can eventually be sold. When that day comes, your attorney will be able to help structure the sale, and handle all the legal paperwork involved.

Looking for an Attorney

Your attorney should be well-versed in business and tax law, as well as contracts. He should be intimately familiar with the state and local laws that affect small businesses. Small law firms are typically a little less expensive, and are usually more responsive to the needs of smaller companies. It's helpful if the attorney has a legal assistant, since you'll generally be charged a lower hourly rate for routine duties taken care of by an assistant.

Accountants and Bookkeepers

The fastest way for a small business to fail is for the owner to lose track of his finances. The need for a good accountant and a good bookkeeper can't be overstated. Your accountant will help your attorney set up the business structure for your new company, with the goal of maximizing your profits and minimizing your taxes. That's always a good thing!

In addition, he'll assist you with your annual tax planning. A good accountant should help you find and take advantage of every legal deduction possible; it's your money, so you want to keep as much of it as you can. Plan on meeting with your accountant at least twice a year, and always with an eye on how to reduce your tax burden.

Your bookkeeper's role is to keep your bills paid on time (Accounts Payable, or A/P), send out your billing statements, and collect and account for the money that comes in (Accounts Receivable, or A/R). The bookkeeper also prepares payroll, and files your quarterly taxes, both to the state and the federal governments.

Looking for an Accountant and Bookkeeper

The accountant you select will depend on how complex your accounting needs are, and how much you can do yourself. Certified Public Accountants (CPAs) have additional schooling and are required to pass a very rigorous certification test, so they're a good choice for complicated tax advice.

Look for a small accounting or CPA firm that specializes in tax preparation and tax consulting for small businesses. Small firms are less expensive and will be more responsive to your needs, and they tend to stay up on all the latest tax codes and regulations that pertain to small businesses.

You may find it best to keep most of your basic bookkeeping work in-house. This saves you money, and keeps you more in touch with the daily financial workings of your company. If you need bookkeeping help with payroll or quarterly reports, then consider a bookkeeping service. Some services are simple single-person home-office operations, specializing in basic A/R, A/P, payroll, and other similar tasks.

There are also larger bookkeeping companies that include some basic accounting services, as well as larger full-service CPA firms that provide some bookkeeping services. Typically, the larger the firm and the more services they provide, the higher the hourly rate.

Bankers

Having a bank you can depend on involves a lot more than opening a checking account. You want a *relationship* with your bank. So, along with your insurance agent, attorney, and accountant, you need a specific banker as the fourth member of your professional team.

Bank services vary significantly, so check several before you settle on one. Here are a few things you'll want from your bank and your banker:

- **Checking account:** You want a business checking account that has a low monthly fee and as few limitations as possible.
- **Savings account:** Plan on opening a money market account if possible, since they usually pay a slightly-higher interest rate than

a standard account. Check the minimum balance requirement, the interest rate, and how many withdrawals and transfers you can make each month without charge.

- **Line of Credit:** This is basically a pot of money that you can borrow from as needed, without applying for a new loan each time. A line of credit is intended for short-term borrowing, for things such as payroll or material purchases that you need to cover until you receive payment for the job. It's there to tide you over in an emergency, and should *never* be used for regular funding of your business's day-to-day operations. A line of credit will have a certain maximum amount that you can borrow, with specific interest and repayment terms. Lines of credit can be very important to your business, and in almost all cases, they're preferable to using a credit card to buy materials. Credit cards typically charge about the highest interest rates you can find.
- **Credit card:** Your banker should be able to arrange a business credit card for you. Credit cards are convenient for taking people out to lunch, for business travel, and for certain small purchases. Be *very careful* how you use credit cards. Many small business owners have gotten into deep financial trouble by using credit cards as a substitute for a positive cash flow. Use restraint and common sense, *and pay the card off in full each month*. A credit card is a convenience, not a business financing tool.
- **Long-term financing:** Talk to the bank about their lending practices. As your business grows, you'll have a variety of financing needs, including vehicle purchases, equipment purchases, and real estate financing. A good bank will say "yes" when the purchase makes sense and your financial stability warrants it, but they'll also know how to say "no" if the purchase could put your company at risk.

Looking for a Banker

If possible, use a local bank that has a stake in the community. Check their fees, and also how liberal or conservative their lending policies are.

When you think you've found a bank that fits your needs, before you open any accounts, set up an appointment with the person who handles their small business accounts. Tell him or her about your business, and explain your business plan. Discuss what your banking and lending needs are going to be, and have an honest discussion about your strengths and weaknesses.

Bankers are only human. If you introduce yourself, and take the time to become a face and a personality with a set of hopes and dreams, you'll be amazed at what a difference that banker can make to your business! If you don't, you'll never be more than just a name and a number on an account — and your banker won't ever be a true member of your team.

Your Business Structure

As you start setting your business up, you'll want to consider how the business should be structured. This will have an impact on how you keep your books, how your payroll is done, and how you're taxed at the end of the year.

Here are four basic types of small-business structures:

1. ***Sole Proprietor:*** This is a single person operating the business alone. Basically, there's no distinction between the person and the business. All profits and losses go directly to the owner and are subject to taxes. All assets are owned directly by the sole proprietor. All debts of the business are his debts, and he must pay them from his personal resources.

   A sole proprietorship is quick, easy, and inexpensive to set up, which is its primary advantage. However, there are very few tax advantages, and the sole proprietor has unlimited liability, which puts him at great personal risk in the event of a lawsuit.

2. ***Partnership:*** A partnership is very similar to a sole proprietorship, except that it's operated by two or more people instead of a single individual. In a legal partnership, the division of ownership must be defined by the partners; for example, each will have a 50 percent share of the business, or it will be divided 60/40 or 70/30, etc.

   As with a sole proprietorship, partnerships are easy and inexpensive to set up, but there are no tax advantages and tremendous personal liability for the owners.

3. ***Limited Liability Company (LLC):*** This is something of a cross between a sole proprietorship and a corporation. Profits and losses pass through to the owner, and it offers only limited tax advantages. However, this type of business entity limits the owner's liability in the event of a lawsuit.

   An LLC is relatively easy to set up, and can often be done without the help of an attorney. It offers better protection for the owner than a sole proprietorship, and more flexibility than a corporation. So it may prove to be a good choice for an individual business owner.

4. ***Subchapter S-Corporation (commonly called an S-Corporation or an S-Corp):*** In an S-Corporation, there are no individual owners. Instead, there are shareholders in the corporation. The corporation itself does not pay taxes on its income, although it may pay a specific corporate tax to the state. Instead, profits and losses are divided among the shareholders based on their percentages of ownership. The shareholders must then report the income or loss on their personal income tax returns.

An S-Corporation is more difficult to set up initially than a sole proprietorship, partnership, or LLC. It also requires some on-going documentation, including shareholder meetings and minutes. However, it offers a number of tax advantages over the other three ownership structures. And since the corporation is the one doing the construction jobs, not the individual shareholders, an S-Corporation also offers a much higher level of liability protection.

The Best Choice for You

The best business structure for your company is the one that you, your accountant, and your attorney decide on jointly. It needs to fit the size of your company, your finances, and the size and complexity of the jobs you do.

Most restoration companies choose to set up as either an LLC or an S-Corporation. We live in a country where lawsuits are far too common, and insurance restoration work has more than its share of liability issues. These two business structures offer considerably more protection than a sole proprietorship or a standard partnership, and that protection is something that you really need to take advantage of.

There's also the tax issue to consider. The odds are stacked against small businesses in many ways, even though they're the backbone of America. Take advantage of anything you can that helps you financially. From a tax standpoint, the S-Corporation is usually the best choice, but you need to discuss that with your accountant.

Buying an Existing Insurance Restoration Company

Insurance restoration companies, like other types of businesses, occasionally come up for sale. The current owners may be looking to retire, move, or simply change careers. Buying an established company offers an alternative to starting and building your own company from the ground up, and it's certainly something worth considering.

The advantages to buying an existing company are:

- You don't need to start from scratch. All of the procedures, techniques, forms, paperwork, and other information and material that you need should already be in place.
- The staff is already there and everyone is trained and knows their jobs.
- You don't need to select and acquire all the equipment. Everything from vehicles to paper clips should be ready to go.
- All of the necessary relationships have been established. You'll have vendors, suppliers, subcontractors, a banker, and

an attorney — everybody you need already in place. Most importantly, the connections with insurance companies, agents, and adjusters have already been established.

But there are some definite downsides to this option as well:

- It'll cost you more initially than if you slowly built your business from the ground up — usually substantially more.
- You'll be incurring a lot of debt all at once, rather than slowly over time. Unless you have a sufficient amount of cash to purchase the business outright, you'll have a large loan payment to deal with each month.
- You'll also acquire the company's problems, including aging equipment, mediocre employees, disgruntled clients, and perhaps even a poor reputation in town.

Considerations When Buying an Existing Company

You can find insurance restoration companies by doing an Internet search, or by contacting companies that specialize in the sale of businesses. You may not be lucky enough to find one right in your own home town, so before starting your search, you might also want to consider whether or not you're willing to relocate.

Once you've located a business you think you may be interested in buying, make a thorough evaluation of the company. This process is known as *due diligence*. The due diligence process can last for weeks or even months. It's your opportunity to learn as much as you can about the company before you decide to buy it.

Since the owners of the company will be giving you a lot of very important and private information, you'll be asked to sign a *nondisclosure* or *confidentiality agreement*. Don't be afraid to sign it, but be sure you understand what your rights and obligations are. You're basically agreeing that you'll keep everything you learn confidential, that you won't use any of the information against the company in any way, and that you won't release any of the information to the public.

During the due diligence period, you'll be inspecting the company from top to bottom. There's a lot to this process, but it all breaks down into four basic areas of concern:

1. ***What are the details of the sale?*** This includes the purchase price of the business, and the terms of the sale. For example, the seller may be willing to finance the purchase for you, or they may want to be cashed out in full, which means you'll need to arrange for your own financing or have the cash available.
2. ***What's included in the sale?*** Basically, you'll get a complete list of everything that's included with the sale, from office chairs and computers to tools and vehicles. You'll also need to inspect everything and make sure it's all in good working order. If the sale includes real estate, you'll need to inspect that as well, and check on taxes, maintenance, etc.

You also need to know if you're buying all of their client contacts, known as their *book of business*. The seller needs to be willing to sign a non-compete agreement, which prevents them from opening up a competing business, or even going to work for one of your competitors.

3. ***What's the financial health of the company?*** This includes a complete review of their financial records. Look at their current income, as well as the income from past years. Check their accounts receivable to see how much money is currently owed to them, and how successful they've been in collecting it. You need to know all their current and long-term debts, and see if there are any legal judgments or tax liens against the business.

   Be aware that the seller will be trying to make the company's financial health appear as strong as possible. Some unscrupulous sellers may even alter the books — slightly or substantially — to make their business appear better than it is. To guard against this, you need the help of a CPA with specific experience in small business sales and auditing. He'll know what to look for in the books, and will have the knowledge to spot potential problems and request additional documents where necessary.

4. ***How solid is the company?*** Finally, determine if the company has a good reputation or not. It could be that the company looks good on the outside, but if it's been badly run or mismanaged, you may have a tough road ahead to regain people's trust. As part of your due diligence, have some candid conversations with as many of the people who deal with the company as possible.

   Talk to subcontractors, vendors and material suppliers. Contact people in the insurance industry who might be sending the company work, and get their impression of the company. Also, talk to key employees about the company, and find out whether they plan to stay on after the business is sold. Losing experienced personnel could undermine your future success.

As you can see, there's a lot to consider when buying an existing company. There are many different ways the sale can be structured, and there are a number of potential problems you may encounter along the way. So, if you're thinking about buying an existing business, no matter how small, be sure you work very closely with both your attorney and your accountant in examining that business and making that decision.

The Franchise Option

Another way you can get into insurance restoration without starting a business from the ground up is to purchase a franchise. And, like buying an existing company, a franchise has advantages and disadvantages.

REAL STORIES:

Wait — Aren't You the Apartment Manager?

Just because a person owns a nationally-recognized franchise, it doesn't make him an expert in the field of insurance restoration. Most franchisors don't require any prior experience, so this can lead to some unusual situations.

One day, we learned that one of the big national franchises had just opened a business in our town. Because of their contacts with some of the larger insurance companies, they immediately got called out on an emergency water job at a local home.

We were quite surprised when, the following day, we got a call from the owner of that very same home. He wanted us to come out and take over the job as soon as possible. When we arrived, we were curious about what had happened. Why had the other contractor — the brand new franchise owner — been asked to leave the job so quickly?

"This is my first home, and I just recently moved in," the owner explained. "Before that, I lived in an apartment. I don't know a lot about houses, so I really want to hire people that do. I had some serious doubts about the other contractor's experience."

"Why's that?" my partner asked.

"Because I recognized the guy right away. Up until a couple of weeks ago, he was the manager of the apartment building I lived in. I definitely think I'd be more comfortable with someone who's had more than two weeks experience in the building trades!"

Franchises are basically "ready-made" businesses, and that's their primary advantage. The franchise company, known as the *franchisor*, has already done all the hard work for you. They've set up office procedures, devised and printed different forms and other paperwork, and even selected all the necessary equipment and supplies you'll need. The franchisor offers training and support, and you get the advantage of their buying power. You also have the use of their trademarks and techniques.

A big advantage to owning a franchise is that you get to take advantage of their reputation and advertising campaigns. Many of the larger franchisors conduct professionally-produced national advertising campaigns, and this creates instant name recognition for you. Some also have national agreements with the larger insurance companies, providing an instant "foot-in-the-door" with these companies that your independent competitors might not have.

When you buy a franchise, there are financing options available through the franchisor. This can be very helpful in tough economic times, when standard bank financing may not be readily available.

For women or minorities there may be additional advantages to buying a franchise. Some franchisors offer special incentives to help them get started, which may include lower prices, additional business consulting, and other programs not available to non-minority males. Depending on your particular situation, this could prove an important feature of buying a franchise.

Of course, none of these benefits comes without a price. When you purchase a franchise, you agree to practice the franchisor's business philosophy. You agree to be bound by their business practices, and to limit yourself to what fits within those parameters. For example, you may have to take on certain types of cleaning jobs that you'd prefer not to do. Or you may not be allowed to do remodeling.

As a franchise holder, you'll need to accept these tradeoffs, and agree to surrender much of your independence. Everything from the color and age of your vehicles to the types of shirts you wear is often regulated. You typically agree to purchase certain products and equipment from the franchisor, and you may also be limited in the products you can purchase on the open market.

And, most importantly, you agree to pay them a certain portion of what you earn. This is usually in the form of an initial royalty fee or business startup fee, plus a percentage of your monthly gross income. Generally, the larger the franchisor, and the more national clout they have, the more you'll have to pay in fees.

The exact agreement that you enter into will vary with the franchisor. The length of the agreement can be from 5 to 30 years. Premature cancellation or termination of the contract can carry serious consequences for the franchisee.

Because franchisors typically don't require prior experience in the field, a franchise can be a great way to enter and learn a new business. And since it's in the best interests of the franchisor to help you make your business as successful as possible, you can usually count on state-of-the-art equipment and extensive and on-going training.

Considerations When Buying a Franchise

As you can readily see, there's a lot you need to be aware of when you buy a franchise. Some of the things to take into consideration include:

- **What Are the Basic Details of Your Agreement?** How long does your agreement with the franchisor last? What happens if you're unable or unwilling to continue with the agreement? What constitutes a breach of the agreement? If you leave the franchisor, under what terms can you break or not complete the agreement? What are the penalties?
- **What Are the Financial Terms?** Exactly what are your initial and on-going fees? If the fees include a percentage of your gross income, exactly how is that percentage calculated? If you're financing the purchase, what are the terms of the loan, the interest rate, any potential penalties, etc.?
- **What Do You Agree To Do and Not Do?** Under the terms of your agreement, you'll be expected to agree to a number of conditions. These might include having a certain number of people available for emergency calls, or having an office of a certain size, or in a particular location. You may be required to do some types of work, such as structural cleaning, but not be allowed to do others, such as remodeling. Be sure you're clear on *all* of these details, no matter how small!

- **What Support Will You Receive?** A big part of what you're paying for is training, advertising, and on-going support, so be clear on all the details. How much initial training will you receive, and what will it cover? For example, will it cover field work, office work, accounting, and business practices? If the training requires traveling, are you responsible for your own expenses? How often will continuing-education be made available? Do you get on-going support on a daily basis? What does the franchisor's advertising campaign look like? Are there national ads? Regional ads? Will there be ads in your specific local market?

- **What Do You Have to Buy?** In order to conduct business to the standards set by the franchisor, you usually agree to purchase a certain amount of equipment, vehicles, supplies, and inventory from them. Get all the details, including your on-going obligations.

While a franchise can offer one possible option for getting into the restoration industry, it's a step you need to take with extreme caution. Remember that the first goal of any franchisor is to make a profit for themselves. The franchise will be set up to their advantage, and in most cases the franchisor will make money on the deal, even if you don't. In fact, some franchise arrangements can be so one-sided that they actually contribute to the failure of the franchisee.

The bottom line is this: *You need to fully understand exactly what you're getting into.* Research the company, read some books about franchise operations and purchases, and talk with other franchise owners. Discuss this thoroughly with your accountant, attorney, and banker. Do your homework, and don't be fooled by a persuasive sales pitch!

19

Setting Up Your Company

There's no single "right" way to set up your business. How you do it will depend on you, your location, your budget, and how your company is structured. In this chapter we'll look at topics you should think about as you get started; some of which you'll probably want to start on right away, and others you can consider doing over time.

Name Recognition

Insurance restoration services are very difficult to advertise and market. Since most people don't think they'll ever need your services, they don't pay much attention to traditional ads. So, one of the key factors in the success of an insurance restoration business is *name recognition*.

Name recognition is what will make people think of your company when they have an emergency, and why they'll dial your number first and not someone else's. Everything from logos to colors helps with name recognition, so you should *never* miss an opportunity to get your name and your logo in front of the community!

What's In a Name?

One of the first things you need is a memorable company name. Your company name should convey something about what your company does, and also your work philosophy or ethic. Ideally, it should be something that's both easy to say and easy to remember, that suggests professionalism, and that sets you apart in people's minds.

Take time developing your company name and identity. Make a list of words and phrases that convey something about what you want your company to stand for. You might try words like these:

- Professional
- Restoration
- Customer service
- Refurbish
- Cleaning
- Rebuild
- Building
- Complete
- We do it all

Just relax and toss around as many thoughts or ideas as you can. Then combine the words, or even parts of the words, in different orders and see what you come up with. Two of the largest insurance restoration franchises, ServiceMaster® and Servpro®, did exactly that. As you can see, these names sound much more professional and focused than *Bernie's Construction.*

Logos and Company Identities

Once you have a name, develop a company identity around it. Design a logo that's easily recognizable, and decide on one or two colors that can be incorporated into your company at every turn. Keep it clean and simple, and if you're not good at design and artwork, have a pro do it. It's not that expensive to work with a sign professional or a graphic designer to come up with a unique design.

A professional designer will provide you with the logo design on a digital computer file, which makes it very easy for you to use your logo whenever and wherever you need it. For example, you can supply the digital images to a shirt company to put on shirts and hats, or you can give it to a printer to incorporate onto your letterhead.

Once you've developed a logo and some colors you like, stick with them. Don't change them for several years; you want people to get used to seeing them and identifying them with your company. Put that logo on everything, from shirts, hats and vehicles to pens, calendars, business cards, and anything else you can think of.

Tag Lines

In addition to your logo, develop a short, descriptive phrase — called a *tag line* — which you can use along with your name. It gives you one more quick and easy way to put your message across to people.

Think of what you want to say about your company, in a few quick words. For example, ServiceMaster uses *Restoring peace of mind*®, and Servpro uses *Like it never even happened*®. Notice that both the company names and their slogans or tag lines are registered trademarks.

Copyrights, Trademarks, and Other Legal Stuff

As you start establishing your new company identity, you want to be as creative as possible. But remember that certain things, from names and written material to pictures and even colors, may be legally protected. Look for copyrighted (©) or registered trademark (®) symbols, which indicate material that's legally protected and can't be used without permission.

Once you have some possibilities for names and tag lines, do some online research. Enter them into Google or another search engine and see if you can find if anything you want to use is already legally protected. If you really have concerns about your proposed name, you can hire an attorney who specializes in intellectual property rights and have him do the necessary searches. He can also copyright or trademark your name, logo, and other company information for you, so you can protect them from illegal use by others.

Tip! *Never borrow or use someone else's work, even in the most innocent way. That's an invitation to a lawsuit.*

Yard Signs

When you have your name, colors, logo, and tag line figured out, have some yard signs made up. This is a very inexpensive form of advertising, and offers you one more way to get your name out there.

There's one caution about yard signs. Before setting one up, you need to get approval, not only from the homeowner, but sometimes the Homeowner's Association as well. Some homeowners may object to having yard signs in front of their homes. And, neighborhood associations may restrict their use, or require that they be a certain color, or that they not exceed a certain size. Yard signs are excellent advertising and you should never miss a chance to use them, but get the necessary permissions first.

A Space to Operate Your Business

Ideally, you should find a small office space for your business in a public location. That gives you a place where you can meet with clients, and where you can store equipment and vehicles. You can start small and build up. Here's what you'll need initially.

Office Space

A 12- by 12-foot space is large enough to accommodate a couple of desks for you and a secretary, or you and a partner. If there are more than two of you in the office, increase the space accordingly.

Figure 19-1 The insurance restoration business requires a lot of expensive equipment. You need to be sure your equipment is kept secure and protected against theft, damage, weather and other hazards.

Eventually, you'll want a private space for the owners and managers, and a shared space for other office staff. You'll also want a conference room to provide an attractive and quiet space for meeting with clients and adjusters. If you design it well and keep it clean, the conference room can double as a coffee area and break room.

Space for Tools, Equipment and Supplies

Restoration equipment is very valuable, and it needs to be kept in a secure and protected space (Figure 19-1). In addition to your specialized equipment, you'll have an assortment of tools and other items that need to be stored, tracked and maintained (Figure 19-2). You'll also need space for the plastic sheeting, lumber, plywood, nails, and other basic supplies you need on hand to respond to emergencies on short notice, often after hours.

Figure 19-2 Warehouse space offers secure storage, where you can keep your tools and other equipment organized.

Vehicle Storage

Your emergency response van will be stocked with additional valuable equipment, so you need to provide a secure place to store it inside when it's not in use. See Figure 19-3. It should be kept where it's easily accessible to crew members. If you have other vehicles, you should look for secure indoor or outdoor storage for them as well.

Location

The best solution when you're first starting out is often a small rental office/warehouse space. Many communities have areas with office/warehouse rental complexes, like the one shown in Figure 19-4, that are geared for start-up companies. They offer decent, secure space at reasonable rents.

Moving your business around is both expensive and disruptive, so try and rent a space that will meet your needs now, as well as for a year or two down the road. A

Figure 19-3 Your emergency response van should always be kept locked up inside, due to the expensive equipment stored in it. Notice the racks of additional equipment, primarily air movers, stored to the left of the van for easy access.

Figure 19-4 Small office/warehouse complexes like this one offer secure, attractive rental units. They're ideal for small- to medium-sized restoration companies, and are usually reasonably priced.

space of about 1,200 to 1,500 square feet, including both office and warehouse, should be adequate. Figure 19-5 shows a simple, but practical layout.

As your business increases and you begin to define your company, you can decide on where to go next. If you get into content restoration and storage, you'll need more square footage of secure warehouse space. If you end up with several vehicles, you may want a place with a large fenced outdoor parking area rather than more warehouse space. This is a perfect example of how your needs will change as time goes on, and how a flexible, written business plan will help you adjust to meet your changing needs.

Company Uniforms

A successful insurance restoration business requires that you establish a level of trust with the homeowners you work with, and the first impression you make when you show up for a job is part of that trust. A clean, well-kept uniform enhances that impression, and gives you another opportunity to promote your company and improve name recognition.

What you choose to provide in the way of uniforms is totally up to you. You may want to supply a certain number of shirts when the employee is first hired, then provide additional shirts as needed (within reason). You may also decide to provide a jacket once a year, or to split the cost of a jacket. Whatever you decide, make the uniform policy part of your Employee Handbook, and apply it equally to all employees.

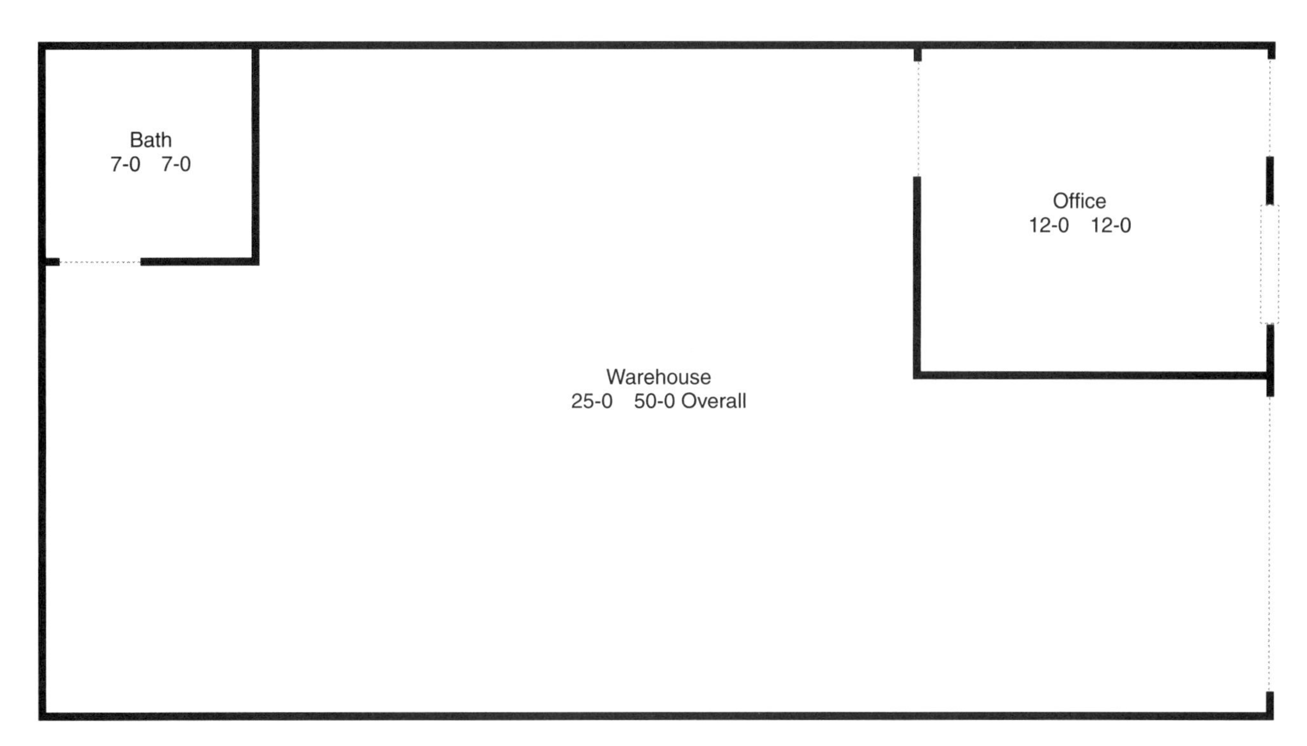

Figure 19-5 A typical office/warehouse floor plan. This unit has 1,250 square feet overall, with a 12 x 12 office, a bathroom, a large roll-up door, and plenty of warehouse space for storage.

Uniform Suggestions

Here are a few uniform suggestions to consider:

- **Polo shirts ($20 – $30):** Polo shirts are ideal for management personnel, office staff, and estimators. They're a nice upgrade from standard tee shirts, but are still fairly inexpensive. Carpenters tend to be a little rough on them, but you might want to consider them for your cleaning staff. Select one or two colors that are in keeping with your company colors, and have your company name and logo embroidered on them.
- **Tee shirts ($8 – $15):** Tee shirts work fine for the rest of the crew. Again, pick one or two colors, and add the company name and logo. Tee shirts are also a nice giveaway item for friends, family, subcontractors, material suppliers, and others, and provide an easy way to promote name recognition.
- **Hats ($10 – $15):** You may want to have some hats in your company colors, with the company name and logo on the front. These are also nice giveaway items.
- **Overalls ($35 – $45):** A lot of the jobs in insurance restoration involve crawling in attics, crawlspaces, and other dirty locations. You might want to consider overalls with your company name and logo on them.

Tip! *Jackets and parkas can be expensive. You might want to consider issuing a company jacket or parka to an employee when they've been with the company for a year. Or, have a program where you pay half the cost and the employee pays for the other half. People tend to take better care of something that they've either earned or had to pay for.*

- **Aprons ($10 – $15):** For your cleaning staff, aprons can be a great addition. They keep clothes clean, offer additional pockets for supplies, and can be printed with your company name and logo.
- **Outerwear:** Jackets ($35 – $55), parkas ($99 – $150), or sweatshirts ($20 – $45) with the company name and logo on them are great for colder weather. You may not be able to exactly match your company colors, but remember to stick with a consistent color scheme to enhance name recognition.

Dress Code

It's an unfortunate reality that employees often push the envelope of what's allowable in the way of dress. And some employees show a surprising lack of common sense that can affect your business. Since a clean, consistent, professional appearance is very important, you're going to need to have a very clear dress code as part of your company policy.

In reality, you may not care if an employee wears a *John Deere* hat or a *Nike* tee-shirt. But if you allow that, then you open the door to all sorts of hats and shirts with political slogans, religious sayings, crude and offensive comments or drawings, and other inappropriate attire. Believe it not, if you hire someone who used to work for one of your competitors, chances are he'll show up for work wearing one of his ex-employer's hats!

So the easiest policy is to not allow anything that doesn't have your company name on it — period. Your dress code should state something like this: *Clothing with other logos, company names, slogans, or other writing or illustrations will not be allowed.*

Be as specific as you can, within reason. Here're some suggestions for your company handbook regarding your dress code:

- **Shirts:** Company-issued shirts only. Clean, no tears. No sleeveless or "muscle" shirts, including company shirts that have been altered. At no time will an employee work without a shirt on.
- **Hats:** Company-issued hats only. (That's obviously assuming that you are issuing hats. If not, be sure and include some language similar to what's listed above, about not wearing hats with other names or slogans on them).

- **Pants:** Clean, no tears. Jeans or cargo pants are acceptable. Shorts are acceptable, as long as they are hemmed. Cutoffs are not allowed. Minimum inseam length for shorts is ___ inches (Note: you'll need to fill in a length that you think is appropriate, such as 7 inches. This prevents people from wearing shorts that are inappropriately short and might be offensive).
- **Outerwear:** Clean, no tears. Long sleeves shall not be loose enough to become tangled in tools or equipment.
- **Footwear:** No sandals or open-toed shoes allowed. Shoes shall be laced at all times. At no time shall an employee work without shoes on. (Note: if your state has specific requirements for footwear, such as steel-toes or non-slip soles, be sure that's listed here).

Company Vehicles

Vehicles are an obvious part of construction work, but they're also a huge expense. So, which vehicles you decide to buy and who gets to use them is a big decision.

Your vehicles are more than just a mode of transportation — they're also a *very* important part of your company's advertising plan. Vehicles provide a "rolling billboard" for your company, and you really need to take advantage of it! It's a great feeling when you mention the name of your company and someone says "Oh yeah, I've heard of you guys. I see your trucks all over town."

Here are five important rules for your company's vehicles:

1. ***Signage:*** Your vans and trucks have lots of blank space on them, so use it. Work with a sign company to develop a signage theme with your logo for all your vehicles, and remember the three C's:
   - **Clean.** The design should be uncluttered and easy to read. Avoid chaotic designs that are hard to decipher rapidly, especially from a distance.
   - **Conspicuous.** Your name and your message should stand out to people passing by on the street or in other vehicles.
   - **Consistent.** The size of your vehicles varies, so the actual sign sizes will vary as well. But with that variation, make sure that the look is consistent. Colors, logo, font, and all the rest needs to be the same from vehicle to vehicle.
2. ***Color:*** Color is an important element in making your vehicles recognizable, so as you acquire them, get them all in the same color. Better yet, have them all painted in a distinctive,

identifiable color that's all your own (a marketing ploy used to great advantage by both ServiceMaster and Servpro). That way, you won't have to stick with one brand of vehicle to ensure a consistent color choice.

A less-expensive option is to just use white vehicles. They're easy to find, and easy to keep clean. Then make them distinctively yours through the use of your colors and logos.

3. ***Condition:*** Impressions are very important, so make sure yours is a good one. Your vans and trucks should all be kept in good condition, not dented or beat up, and not smoking or dripping oil.
4. ***Cleanliness:*** Your vehicles should be spotlessly clean. This makes a statement about your business, and tells people that you'll care as much about their homes as you care about your vehicles. This is especially important as you get into the cleaning aspect of restoration.
5. ***Legal:*** Everything about your vehicles must be legal. If there are state laws requiring something, make sure your vehicles comply with those laws completely. And that extends to your drivers as well. Make it a company policy that they wear their seat belts, observe speed limits, and never have more people in the vehicle than they're legally allowed to have.

What Vehicles Should You Have?

Your company's vehicle requirements will vary with the size of your crew, your location, and the type of work you take on. The following is a list of the types of vehicles a typical restoration company may need.

- **Estimating vehicle:** This can be your personal vehicle, or it could be a van or pickup truck that's also used for construction projects. If it's your personal vehicle, it still needs your company name and logo, and should be equipped with ladders, water-testing meters, and any other gear you need.
- **Emergency van:** As we discussed earlier, a dedicated emergency response van is essential to your company. Ideally, the van can remain at your warehouse where the on-call crews have access to it. Another option is to have the lead emergency crewman take it home at night.
- **Moving van:** As you begin handling more contents, a box truck is a great addition. You can buy good used box trucks from rental companies. These trucks allow you to move contents easily, and you can even temporarily store contents in them, on-site or in your own warehouse.
- **Cleaning van:** This is the van that's dedicated to your Cleaning Division. It's typically outfitted with shelves and bins, and is

used for transporting your cleaners and their supplies to and from jobs. Larger restoration companies often have several designated cleaning vans.

- **Carpet cleaning van:** Cleaning carpets is a big part of restoration work, and as your company grows, you can gain both convenience and extra income by having your own carpet cleaning equipment. Most commercial carpet cleaners are *truck-mount* units, which require their own van or other type of enclosed truck. See Figure 19-6. Truck-mount carpet cleaning equipment also provides very powerful suction, so once you invest in this equipment, you can also use it to extract water on larger water losses.

- **Specialty vans:** If you do a lot of painting, drywall work, or other specialty trades, having a dedicated van might make sense. It allows you to keep all of the equipment and supplies for that trade in one place, ready to go.

- **Debris truck or trailer:** At some point, you might want to consider a trailer or a flatbed truck with sides for hauling debris. If it has a dump bed, you'll save crew time when unloading. The alternative is to use Dumpsters, which are more expensive, but also more convenient.

Tip! *Check with the insurance companies that you work with most often to see what their preference is in regard to debris disposal. Many of them are actually more lenient when it comes to you billing for Dumpsters rather than billing for the use of your own debris truck.*

Figure 19-6 A truck-mount carpet cleaning van. Notice the large hose reel in the center, and all the bins for storing small tools and cleaning chemicals. Remember that every cleaning solution on this van needs to have an MSDS with it.

REAL STORIES:

Hey Man, I Don't Know *Where* It Went!

At one point, we were hauling enough debris that we decided it would be worthwhile to invest in a truck. We found a nice used one with a 12-foot flatbed, sides and a rear gate. We also invested in a hydraulic lift for the bed, to save the time required for hand unloading at the landfill.

The frame consisted of two long C-shaped steel channels, so we had a steel ramp made up that slid right into the channels. When we got to the job, we could slide the ramp out from between the rails, and it was perfect for walking up or pushing wheelbarrows up into the bed of the truck. A steel pin locked it in place when it wasn't in use.

The landfill we used at the time was a large pit. You'd back up to the edge, and unload your debris right into the pit. With the dump-bed truck, our crew could get to the landfill and unload in no time at all.

One day my partner and I were out at a fire job. We were watching our crew load the truck, and they were tossing things into the bed. "Why don't you pull out the ramp and use that to walk up into the truck?" I asked. "It would make loading a whole lot easier."

That question was greeted with lots of blank stares — the kind that you usually only get from your kids. We walked over to the truck, and looked at the frame. Sure enough, the ramp was nowhere to be seen. The obvious "Where's the ramp?" question was greeted with more blank stares and an ominous silence.

Finally, one of the crew members spoke up. "It's at the dump," he said. As it turned out, the crew went out to the landfill one day, and they lifted the dump bed on the truck without first checking to see that the pins holding the ramp in place had been inserted. As the bed came up, the ramp shot out of the frame and down into the pit! And of course, no one thought to rescue it — or to tell us about it.

And who got the blame? Needless to say, it was a kid who'd quit to take another job about a month before. Seems like that was always the way!

Use of Company-Owned Vehicles

As soon as you begin acquiring company-owned vehicles, you need to set policies for the employees that use them. These need to be incorporated into your Employee Handbook, and strictly enforced.

One of the most important considerations is whether or not the employee can take the vehicle home. There are some advantages for you in allowing this:

- The employee can go straight to the jobsite in the morning or straight home from the job at night. This increases productivity, and saves time and gas.
- In the case of the emergency van, it may improve your response time to the job.
- The vehicle may be more secure at the employee's home than at your shop at night.
- It's that much more road time for your rolling billboard to be seen by the public.

On the other hand, you have to really trust your employees. Here's a vehicle with your company name plastered all over the sides, out of your control for the evening. That could be a *tremendous* potential liability for you. Is one of your pickup trucks sitting in front of the local bar? Is your van hauling a bunch of kids around without enough seatbelts for them? Is your flatbed truck hauling furniture or firewood on the weekends?

Here are some policies you'll want to consider for employees who drive company vehicles:

- The vehicle is not for personal use.
- It's the employee's responsibility to check the gas and oil.
- They need to be responsible for keeping the vehicle clean.
- They need to obey all the seatbelt laws and speed limits.
- To avoid potential complaints and conflicts between employees, there should be no smoking in company vehicles.

Use of Non-Company-Owned Vehicles

Unless you've reached a size where you can afford company vehicles for all of your crew, at least some of them will be going to the jobsites in their own vehicles. Here again, you need to set some policies and document them in your Employee Handbook.

Mileage Reimbursements

Obviously, your employees are responsible for getting to work in the morning on their own, and for getting home again afterward, but it's what happens in between that can become an issue.

It can get expensive for your crew to take their own vehicles to jobsites each day, especially if your company services a large geographical area, or if they have to go to multiple jobs. So, you may want to offer a mileage reimbursement.

> ***Tip!*** *If your employees track their* actual *mileage, on their tax returns they can claim the difference between the mileage reimbursement you give them and the amount they're entitled to under the IRS rules.*

One method is to require that the employee track his actual miles, from the time he leaves your shop until the time he finishes his day. You can then agree to reimburse him for those miles, at a set amount per mile. The trouble with this arrangement is that your employees will almost certainly not track the miles accurately. They'll guess at the number of miles because they forgot to record them, or they'll record the miles they drive from the jobsite to the convenience store at lunch time.

A simpler solution for both of you is to consider setting up a mileage reimbursement chart, like the one in Figure 19-7. Using this method, rather than tracking the actual mileage, you simply establish set reimbursements for specific areas. This makes it much easier for them to track and you to verify. They write their mileage on

Mileage Reimbursement Chart

Effective date: January 1, 2011

Employees shall be reimbursed for transportation expense at the rate of $0.20 per mile, based on the following mileage chart:

Distance from shop	***Mileage reimbursement***
Within 5 miles of the shop	No reimbursement
Between 5 and 15 miles from the shop	10 miles
Smithburg	15 miles
Rock River City	20 miles
Johnson Village	35 miles
Adamstown Ranch	50 miles

Employees will be reimbursed with the first pay check of the month for the month previously worked. In order to receive reimbursement, employees must include an asterisk (*) on their time card, indicating that they drove a personal vehicle to a job site outside the local area. The job name must also be included on the time card.

Employees are responsible for their own expenses for transportation between their homes and the shop.

Figure 19-7 A sample mileage reimbursement chart. Charts like this avoid a lot of confusion, and save your employees the trouble of having to track their mileage to and from every job.

their timecards, and then you reimburse them once or twice a month. Figure out a chart and a policy that works well for everyone, and make it a part of your Employee Handbook.

Named Insured

Another issue with having your employees drive their own vehicles is the potential liability. If they're in an accident with their vehicle while driving it on company time, and in the pursuit of company business, then liability may extend to you.

You can help offset that liability by having your company listed as a *named insured* on the employee's automobile insurance policy. This doesn't cost them anything, and it should offer you some additional liability protection. Talk with your insurance agent and your attorney for more details and suggestions about how to set this up.

> ***Tip!*** *This stipulation of having your company listed as a* named insured *should be made a condition of employment for all new hires who'll be using their own vehicles. However, confirm this with your attorney first.*

Company Operations

Because of the emergency calls you'll be getting, insurance restoration is a 24-hour, 365-day operation. However, your office isn't open to the public 24 hours a day. You need to establish *normal* work hours for the non-emergency time that your company will be open, and normal work hours for your crew members. Normal business hours are generally 8:00 AM to 5:00 PM, Monday through Friday.

Scheduling and Crew Call-Ins

Your crew hours may be a little different from your company office hours. You might want to have them arrive at 7:00 AM so you can get them organized and out the door before you officially open to the public. And if you have them work until 4:00 PM with a 1-hour lunch, you have time to wrap things up at the office after the crew has left.

For the most part, you're going to find it easiest to have the crew show up at the office in the morning to get their assignments. But, unless the job is finished, they usually won't come back into the office at the end of the day. Instead, they'll head home directly from the jobsite, which is easier on them and more economical for you.

> ***Tip!*** *A call-in policy is a great tool for keeping yourself informed about what's happening on the jobs. It keeps your schedule up to date, and your crews working efficiently with minimum downtime. If you decide on a mandatory call-in for job updates, be sure that's included in the Company Handbook.*

The downside to this arrangement is that you don't have a chance to talk to them at the end of the day, so you're not always sure exactly what they got accomplished on the job. This can complicate laying out the schedule for the next morning. To compensate for that, consider having a call-in policy. Each of the crew leaders calls into the office at the end of their day, and gives you an update on things. Or, if you and your crew leaders prefer, you can have them text or email their daily updates. But make sure that they clearly communicate what's been accomplished and where they are on the job. I personally prefer talking directly with my employees, as there's less chance of a misunderstanding.

Payroll and Timecards

You and your employees have to maintain accurate hourly work records. This affects everything from the amount of their paycheck to the legality of overtime work and payroll reporting. Accurate time records make it easier to bill the insurance companies correctly, and can also save you money on your worker's compensation insurance.

Since it's not practical to have your employees punch a time clock, the next best thing is an accurate and easy-to-use timecard. You can buy preprinted timecards, but they rarely give you all the necessary information. You're better off designing and printing your own, so you'll have exactly the information you need in a format that makes sense to you.

There are several things you'll want to include on your timecards:

- **Employee name.** Typically at the top of the card.
- **Pay period.** For example, June 1 – June 15, 2012. Also at the top of the card.
- **Date.** This is the date actually worked, usually the first column on the left side of the card.
- **Job name.** This can be the next column on the left.
- **Work performed**. You can have a blank space next to the job name where the employee can list the type of work done that day — framing, painting, cleaning, etc.

  Have a list of work categories printed on the back of the timecard that correspond to your worker's compensation categories, with a code next to each one. Then all the employee has to do is put down the code and the number of hours worked that day in whatever trade or trades he performed.
- **Mileage reimbursement.** If you're using some type of mileage reimbursement chart or other system, provide a column for the employee to note the number of reimbursable miles he drove that day.
- **Totals.** At the bottom of the card, have a place for total hours in each trade, total hours for the pay period, and total miles for reimbursement.

Depending on your company's specific needs, you may want to add other information as well. *Make it a policy that employees fill out their time cards at the end of each day!* If they skip a few days and then try and fill it out later, they'll end up guessing at how many hours were spent on different trades. Be sure to print your timecards on heavy card stock so they'll hold up to jobsite abuse.

Paydays

Paydays can be any day of your choosing, subject to any restrictions in the labor laws of your state. Doing payroll twice a month is typically the most common practice.

Payroll Advances and Loans

You'll no doubt be getting requests from employees for advances on their pay, and possibly even loans. The policy you decide to set is up to you, but by far the best policy is to not do it. Payroll advances create additional paperwork for you, and if the employee leaves your company before the advance is paid back, you have very little chance of ever recovering it.

Once you decide on a policy for advances and loan requests, make it part of the Employee Handbook. It saves a lot of questions and complaints later on.

Employee Garnishments

If you have employees, you'll almost certainly receive a garnishment notice at one point or another. A garnishment is a legal document specifying a certain amount of money that you're required to withhold from an employee's paycheck, along with details about where the withheld money is to be sent.

Garnishments may be for child support, for legal judgments against the employee for back taxes, or for other debts, such as student loans. Once you receive a garnishment and verify it with your attorney, you need to comply exactly with its terms. You may not agree with what the garnishment is for, you may feel sorry for the employee, or you may resent the fact that you have to go through the additional paperwork. That doesn't matter. You have a legal obligation to comply, and you put yourself and your company in serious legal jeopardy if you don't.

Safety Committees

Many states require companies to have a safety committee. These committees usually include one person from management and a certain number of employee members. The committee makes regular inspections of the shop, holds regular safety meetings, and makes specific recommendations for improving any unsafe equipment or working practices.

Safety committees are a good idea no matter what. But if your state requires that you have one and you don't, you can be opening yourself up to some serious fines. Check with OSHA or any other state safety boards to see what the requirements are for your state.

Merchant Accounts

In order to operate your company, you'll need to set up accounts with local merchants and suppliers, such as lumber yards and paint stores. The sooner you can get the ball rolling on these accounts, the better.

If you're a sole proprietor or a partnership, you'll be opening the accounts in your own name, with the business name listed as well. Remember that you're personally responsible for these accounts, so if you close the business, the accounts are still on your credit rating.

If your company is set up as a corporation, you'll probably be asked to sign a *personal guarantee* for the accounts. That means even if the account is in the name of the corporation, you'll still be legally responsible for it as an individual. If you have any questions about what you're signing, discuss them with your attorney.

If you're a brand new business, setting up accounts may be a little difficult. The better your personal credit rating is, the easier it will be. If your credit is shaky, you may initially need to have a cosigner on the account. This is a person with a good credit standing who's guaranteeing to pay the bills if you can't. Another option is to ask for a very low monthly limit on the account.

Once you've established the accounts, pay them on time! It's crucial that you establish and then maintain a good credit rating. A good credit rating is essential for getting loans, opening other accounts, leasing equipment, and other financial dealings for your company — so protect it!

Your Team of Subcontractors

Just about any type of construction requires subcontractors, and insurance restoration is no different. You'll want to establish a team of good, reliable subs that you can count on, especially in an emergency. The following are the primary subs you'll want to have, and the types of services they should provide.

Plumber

You'll be dealing with a lot of broken pipes, fixtures, and fittings. If you live in a cold climate, you'll also be dealing with frozen pipes. You need a plumber who specializes in service work, as opposed to new work. He should be willing to take on small jobs, including repairs of all types. And ideally, he should be available on very short notice, and for emergencies.

You'll also be doing larger fire jobs that require extensive replumbing. If your plumbing contractor can't handle jobs at both ends of the spectrum, you'll want to cultivate two plumbers: one for service work and repairs, and one for large replumbing jobs.

Electrician

The best electrician for insurance restoration work is one who can diagnose and repair a wide variety of electrical problems. You'll be encountering wet electrical panels, partially scorched wiring, strange malfunctions caused by lightning strikes, as well as other more routine repairs.

In the event of a fire, the electric utility company may disconnect the electrical meter until the wiring can be assessed. You may need your electrician to respond quickly so you can have partial power restored to a home in order to keep the heat on or power going to other essential areas. This can sometimes happen on very short notice, so it's nice to have someone on call. He'll also be required to set temporary electrical panels for large repair jobs. On major fire jobs, he'll need to be able to handle a complete rewiring project.

Drywaller

Insurance restoration involves a lot of drywall patchwork. You need a drywall sub who'll do the little jobs, as well as the larger ones that most drywallers like to specialize in. You also want someone who can match different types of textures.

Painter

Just about every restoration job involves painting, ranging from a complete interior and exterior paint job to a single room, or even a single wall. The ideal painting sub should be able to handle jobs of all sizes, and be able to match paint well, especially on new drywall patches.

You also want a painter who can do stain work. Many jobs require cabinet or woodwork refinishing. The painter must be able to match the stains and finishes on existing woodwork.

Flooring Contractor

You'll be doing a lot of floor covering work, ranging from single rooms to entire houses. A good flooring contractor should be able to handle both your carpet and vinyl jobs, and be willing to do just a single room of flooring if needed.

Hardwood Floor Contractor

Water losses often include hardwood flooring. You want a hardwood flooring contractor who can handle it all for you, from patching small damaged areas, to sanding and refinishing or complete new installations.

Legal Paperwork

Finally, let's talk about some of the legal paperwork you'll need. Your attorney can walk you through exactly what's necessary to comply with all of the local, state, and federal requirements. Your specific paperwork needs may vary with where you live and what types of work you do, but there are things you need to be concerned with regardless of your location. These are:

- Contracts and change orders
- Work authorizations
- Material Safety Data Sheets (MSDS)

Contracts and Change Orders

You'll need a contract form that suits your specific business. Have one prepared by your attorney, rather than buying preprinted forms, which rarely include all the disclosures now required in most jurisdictions. Your attorney will be familiar with all of the specific laws in your state, and will draft a strongly-worded contract designed especially for your insurance restoration business. Or, use a computer-generated contract, like Craftsman's *Construction Contract Writer*.

In addition to the contract form, you should have a change order form to go with it. This gives you a legal means of documenting changes that occur on the job, and makes them part of the original contract. If the changes alter the amount of the original estimate, the change order will document that as well. It's also used to change job specifications, or to extend completion dates.

But be sure and have your attorney check all your contracts and forms carefully.

Work Authorizations

Your work authorization is an extremely important document. It's your legal authorization to work on the house. If there's an issue later, this document is proof that you were given permission to enter the home to work or make repairs. It's also proof that the owners were informed of your hourly rate and other pertinent information relating to the job.

Depending on the laws of your state, some type of signed authorization or contract is required if the work on a home will exceed a certain dollar amount. In the case of an emergency call, you obviously won't know how much the final bill will be, so your work authorization is the easiest way to ensure that you've complied with state laws.

Have your attorney assist you with the preparation of a work authorization that suits your specific needs. Once it's been drawn up, you can store it on your computer and print out copies as needed, including a duplicate for the homeowner's records.

Material Safety Data Sheets

As a restoration contractor, you'll constantly be coming into contact with products that could potentially be harmful to you, your crew and your clients. There are warnings and precautions on the labels of most of these products, but they may not provide you with all of the information you need. In the case of common products, like a can of gasoline, you have no information at all.

To provide the necessary safety and medical data, the government developed a mandatory information system of Material Safety Data Sheets, or MSDS. If a product or natural compound is available for public use, an MSDS exists for it, and you're required to maintain a file of MSDS information for every product you use. The MSDS must be organized and readily accessible to anyone who's using or could be exposed to the product.

Material Safety Data Sheets are required to be available wherever a product is sold. This is true whether it's at the retail or wholesale level, or ordered directly from the manufacturer. All you have to do is ask for them. You can also download them from the manufacturer's website.

When it comes to MSDS, you have certain responsibilities, including:

- *You must obtain one for each and every product you use.*
- You need to make copies and organize them into binders or some other type of easily accessible format.
- You need to have a complete binder available in your office, in your warehouse, and in each vehicle that's out in the field. In other words, any and all of your employees and clients who'll be exposed to any of the products you use need to have access to the MSDS.
- You must update the binders periodically as you change or add products.
- You must be sure that all of your employees acknowledge that they've received the binder, and that they understand how to read and use the information the sheets contain.

Your crew and your clients need the MSDS information to keep them safe from harm. And you, as the business owner, are subject to some *very sizeable* fines if you don't comply with the law.

You need to take these requirements very seriously. Be sure you discuss your specific MSDS needs with your attorney or with your local OSHA representative!

Keep it Legal

For your own peace of mind, and for the good of your company and society, *run your business legally!* Pay your taxes in full and on time, file all the paperwork you're supposed to file, keep your insurance in force, keep your workers safe, and don't hire illegal workers or pay employees *under the table* in cash. Comply with all federal and local laws and requirements. They're there for your protection — and it's the right thing to do.

20

Staffing Considerations

Running a thriving insurance restoration company requires good employees. The number of people you hire and their job positions depends on the size of your company and the areas of restoration you decide to specialize in.

In earlier chapters, we looked at the types of projects you may be working on. That information should help you decide on the skills your employees will need for the positions you're filling. Let's look more closely at the staffing positions commonly found in a restoration company, along with some hiring and interviewing suggestions.

Heading the Company

We'll start with your role in the company. How is your company structured? If you're a sole proprietorship, you make all the decisions. If not, your authority is shared. Let's look at how that affects you.

Sole Proprietor

As sole owner, you can't be all things at all times, but you can choose the things you like to do and hire out the rest. For example, you may decide to do the estimating and hire someone to do the carpentry; or maybe it's the other way around — you do the carpentry and you hire an estimator. Or you may find that your role is best suited to being the company's overall manager, and you hire out everything else. There's no right or wrong answer, but it's something you need to think about.

Partnerships

Partnerships are good if handled correctly, but they can also be very tough, and a lot of them don't work out. You need to take special precautions to ensure everyone feels they're doing equal amounts of work and sharing an equal amount of the responsibilities. Here are a few tips to help improve your chances of success.

Make Your Job Roles as Equal as Possible

No matter how hard each of you works, if you're doing completely different things it will inevitably lead to resentment. Suppose you hate public relations and your partner loves it, so he takes on that task while you take on running the jobs. One day, he's going to be out on the golf course wooing an insurance adjuster while you're crawling through a hot, smoky attic. A week later, your partner will be stuck at a boring late night business dinner while you're home relaxing with a beer and watching the ballgame.

Either way, it's human nature to think that you're the one doing the most work and he's the one who has it easy — even if you each chose your role! In reality, you can't always split everything equally, and there'll naturally be things that each of you does better than the other. Just try to always keep your job descriptions and your roles in the company as equal as possible.

Play to Each Other's Strengths

While it's important to do essentially the same jobs, it's inevitable that each of you will have particular strengths and weaknesses. Figure out what those are, and work with them.

Communicate

A partnership is a relationship, and a tough one at that, since you often spend more time with your partner than you do with your spouse. As with any relationship, good communication is essential. Make all major decisions jointly. Don't be afraid to compromise. And most importantly, don't sit on your anger or your resentment — talk about it, work it through, and then move on.

Watch the Money Issues

Agree to a policy that either partner can spend up to a set amount of money on his own, without the other's approval. It could be $100 or $1,000, whatever works for both of you. Any amount over that needs to be jointly discussed and agreed on. That allows each partner the daily freedom to run company business, but prevents either partner from making big financial commitments unilaterally.

Share Equally, Both the Good and the Bad

If an employee's getting a bonus, do it together so you both enjoy it. If someone must be fired or laid off, do it together so that neither of you has to endure the stress alone.

REAL STORIES:
When Your Accountant is Also Your Psychologist

My partner and I had known each other for years, and we had a pretty solid friendship. When we first started our business together, we carefully talked out our strong points, and split up our work duties up accordingly.

We had barely opened our doors when we sat down with our accountant to discuss our accounting needs. "You guys are frickin' nuts!" he exclaimed, after listening to our plans for dividing up duties. "You've just described the perfect recipe for disaster in a partnership. Trust me — I've seen it way too many times! You'll end up resenting each other in no time. Take my advice and work things out so you both do the same jobs."

He was a specialist in small business operations, and clearly, the last word you'd use to describe him would be shy. Also, at 6 feet 4 inches, he was too big to argue with — so we didn't. Good thing too. From that day forward, we set our company up so that we both always did essentially the same thing, and it worked out beautifully.

Just goes to show you — never mess with a big accountant!

Your Staff

When you first start out, you may take all your calls yourself and have an answering service pick up when you can't. But you'll quickly learn that having someone in the office to take calls, set up appointments, and greet clients will take a huge load off your shoulders. So the first position you may want to fill is your front desk person.

Front Desk/Secretary

Never underestimate the value of your front desk person! People only call an insurance restoration company when something traumatic happens to their home. They're stressed out, and in need of help and guidance right away. The first person they'll generally encounter at your company is your front desk person, so the impression he or she makes with the client on that initial call can make or break the job in the first moment!

What to Look for When Hiring

Take your time in interviewing, and look for someone with both a pleasant personality and good general office skills. *And don't be cheap!* The importance of this person's role in your company can't be overstressed. If you want your company to grow and prosper, don't look at this as a low-paid entry-level position. Be prepared to pay more to get a really exceptional person at your front desk!

There are a number of qualities to look for when filling this key position:

- **Exceptional phone skills:** Phone skills are first and foremost. The person you select needs to be able to listen carefully; sort through the information that the caller provides; reassure the caller; and get the essential facts for your crew.

- **Calm efficiency at all times:** You need someone who is efficient, but also friendly, caring, and very patient.
- **Office skills:** In most companies, the front desk person also does secretarial work, such as correspondence, making phone calls, filing paperwork, and general office organization.
- **Willingness to learn:** The desk person in a restoration company will need to have some training in what's involved with typical water and fire emergencies. That'll make it easier for them to answer many of their caller's more common questions. They also need to be aware of the different insurance companies you work with, as well as their programs and preferences, and become familiar with and recognize agents and adjusters when they call.
- **Bookkeeping skills:** In many companies, this position also includes basic accounting or bookkeeping, such as sending out invoices and paying bills. This is a very valuable and helpful addition to this office position.

Office Manager

As your company grows, you may want to hire an office manager. This position oversees the entire office, from the stocking of supplies to the preparation of payroll. The office manager should become very familiar with what your company does; should spend time touring jobsites to see different types of jobs in progress; and should also become familiar with the different pieces of equipment and the different procedures you use.

What to Look for When Hiring

An office manager will need to have a variety of skills in order to be able to run a busy office efficiently:

- **Management ability:** The office manager needs to have the ability to manage people efficiently, as well as being proficient in all aspects of office operations and general office procedures.
- **Accounting skills:** A good understanding of basic accounting principles is essential, since the office manager will be overseeing and reviewing many of your financial records.
- **Organizational skills:** An effective office manager must be organized and able to handle multiple tasks simultaneously, often under stressful conditions.
- **Personality factors:** Your office manager needs to have the right combination of personality and business knowledge to suit your particular company. You need someone who's friendly and cordial with staff and clients, but not afraid to be tough when needed; a person who can handle a variety of situations

and people calmly and efficiently; and, since you'll be working together closely, you want to be sure that your personalities and business styles mesh as well.

Carpenters

As your business grows, you'll eventually need to hire one or more carpenters. They'll need to be able to do a wide variety of tasks, from demolition and complex framing to drywall and painting. You'll probably discover that it's difficult to find carpenters with specific restoration experience, so the next best thing is good general remodeling experience — and the more skills, the better.

What to Look for When Hiring

When interviewing for carpenters, here are a few things to keep in mind:

- **Variety:** Make sure that they like variety. This job requires a lot of bouncing around between jobs and trades, and they have to be okay with that.
- **Ability to work independently:** A typical restoration company generally has more than one reconstruction job going at a time, and in the middle of those jobs, the crew often has to drop everything and respond to an emergency. So it's essential that you have people who don't have to be constantly supervised, and who can think through problems and make sound decisions on their own.
- **Appearance:** You've no doubt met a lot of excellent carpenters over the years, but between the torn tee-shirts, blaring music, and slobbering dogs, you may not be comfortable sending some of them into an occupied home full of antiques. So think about their appearance and professionalism when hiring — it counts!
- **Tools:** The ideal carpenter should have the tools for lots of different tasks. Unless you plan on providing company tools (which is *not* a good idea), make sure that they either have their own tools or are willing to acquire them within a reasonable amount of time.

Specialty Trades

If you find yourself doing a lot of one trade, such as painting or drywall, it might make financial sense to have someone on staff specifically for that trade.

When hiring for a specialty trade position, look for the same type of worker as you would when hiring a carpenter. You want someone with a high level of skill, who enjoys variety, and who already has or is willing to buy all the necessary tools for his trade. And again, you need people with a decent appearance who can work with minimal supervision in occupied homes.

REAL STORIES:
I Can Do This at Home!

We once interviewed a carpenter with great skills. He'd done a lot of framing, including some complicated roofs. He'd built decks, done siding, and performed lots of other carpentry tasks.

Whenever we interviewed a potential employee, we always made it a point to be as honest as possible about everything involved in the work we do. We probably scared a few people off with our stories of sooty fire tearouts and crawls through attics and basements, but it's better that they be prepared up front than to hire them and lose them later.

So we told this carpenter all the different things he'd be involved in. "No problem," was his standard reply. We were delighted with the interview, and hired him immediately.

After about a month on the job, we were very pleased with his performance. He did whatever he was asked, and did it well. So, we were quite shocked when he walked in one morning and quit! "Why?" was the first question out of my mouth.

"This job is one giant 'honey-do' list. It's something different every day, and I can't ever seem to get organized. I never even know what tools to have on the truck. Heck, I do this at home — I don't need to do it at work too. I'm going back to framing!"

We lost a good carpenter, but we gained a new interviewing technique. From that day on, whenever we interviewed a carpenter, we always compared the job to a big honey-do list. Guys can definitely relate to that; and after that we only hired people who said that was fine with them!

Helpers

Depending on the size and complexity of your company, you may at some point choose to hire one or more helpers to free up your higher-paid carpenters or tradesmen for more skilled tasks. Helpers can assist with demolition, cleanup, moving contents, running materials, maintaining the warehouse, and other jobs.

When hiring for helper positions, remember that this can be a tough and dirty job, so look for people who are willing to work hard and get dirty. Helpers should also be people who'd like to learn the trades. A helper position is a great way to provide a career path for someone, while also training them to work well within your particular company. A good helper candidate should have some basic tools, and be willing to buy more over time.

Construction Supervisor

A construction supervisor's role is to oversee and coordinate all of your construction jobs. The supervisor sets schedules, orders materials, coordinates subs, updates clients, and, depending on your company's structure, may also be responsible for hiring and firing within the division.

What to Look for When Hiring

A construction supervisor needs a good, diverse set of skills, including:

- **Strong construction background:** The supervisor needs to understand construction and construction scheduling.

- **Construction management:** The supervisor will have to be organized enough to juggle multiple jobs at the same time; be able to set his own schedules, and work independently without your supervision. This is a fast-paced job with a lot of responsibility, and it's one that's typically fairly well-paid. Ideally, this position should take a lot of weight off your shoulders. But that won't happen if you constantly have to supervise your supervisor!
- **Willingness to be trained:** Your supervisor will need specific fire and water training, as well as training in how to deal with the insurance industry and the specifics of insurance restoration estimates and job flow.
- **Strong people skills:** The construction supervisor will have to interact with a diverse group of people, from insurance adjusters to clients of different economic levels. He should be capable of handling all of the hiring, encouraging, correcting, disciplining, and firing tasks that go along with running a crew.
- **Appearance:** Anyone in a supervisory role sets an example for other employees, and should, therefore, be clean-cut and well-mannered. Remember, your supervisor is a direct reflection of you and your company.

Emergency Technicians

Everyone on your field staff should know how to help out with emergency jobs. But as your business grows, you're going to find it useful to have people who are specifically trained for emergency response work.

Your emergency technicians, first and foremost, need to know everything about responding to an emergency. They have to know what paperwork to take, what notes and other documentation to gather, and how to assist upset homeowners so that everything runs smoothly. They also need to be able to identify what needs to be taken care of immediately and what can wait, what to tear out and what to leave, and what equipment to set. They should know how to set up the van for emergencies, what extra equipment or supplies might be needed for the job they're responding to, and how to clean and service all the equipment when they're done.

Keep in mind that emergencies don't happen every day, so your emergency technicians should be able to help out as carpenters, cleaners, helpers, or even office backup — wherever they're needed in the everyday operations of the company.

Tip! *You'll invest a lot of time and money training a good emergency tech, so do everything possible to keep him employed full time. You don't want him to go looking for work elsewhere!*

What to Look for When Hiring

This is an important position, and in some ways it's a difficult one to interview for. Here are several things to keep in mind:

- **Flexible and organized:** The person you hire needs to be both flexible and organized. The position requires a lot of documentation, and a methodical approach to every loss situation. It also requires a willingness to respond quickly to emergencies at any time of the day or night.
- **People skills:** Emergency techs need to be calm, empathetic, friendly, and able to remain in control when dealing with people who are upset.
- **Physical condition:** This is a physical job, involving a lot of lifting, climbing, crawling, and demolition, so your tech needs to be reasonably physically fit.
- **Carpentry skills:** The tech should have, or be able to learn, basic carpentry skills; and also have or be willing to acquire some basic tools.

Emergency Division Supervisor

As your company grows, you'll want a strong, well-staffed and well-organized Emergency Division, overseen by a full time Emergency Division supervisor. The job of the supervisor is to be responsible for and manage all of the staff in the division, deal directly with the homeowners, and interact with the insurance adjusters.

The supervisor should know the exact status of every emergency job, and know when the job is ready to estimate. He's also responsible for the condition and location of all the division's vehicles and equipment.

What to Look for When Hiring

To be honest, this isn't an easy position to fill. It's pretty specialized, so it's hard to have a sizable selection of applicants coming through the door. You may find a qualified person who used to work for another restoration company, but more than likely you'll want to consider promoting one of your existing emergency technicians.

The Emergency Division supervisor should meet all the requirements for both the emergency technician and construction supervisor positions. But you'll want to put additional emphasis on the people skills, since he'll constantly be dealing with people in emergency situations, and also with insurance adjusters. This requires a good balance of patience, understanding, listening ability, and negotiating skills.

Cleaners

A big part of insurance restoration work, especially fire restoration, involves cleaning — both structural and contents. Cleaning contents also involves inventorying and packing those contents, so good attention to detail should be part of the general job description. As your business grows, so will your staff of cleaners.

Don't make the mistake of thinking that this is a low-paying job that anyone can do. Plan on paying a wage high enough to attract good quality people, with good work ethics, and who are also good with details.

Your cleaning staff needs to be neat and presentable, and you have to be able to trust them in people's homes and around expensive contents. If your own insurance company requires that your employees be bonded, then the ability to be bonded needs to be a condition of employment.

Cleaning Division Supervisor

When you're doing enough jobs, you'll want a Cleaning Division supervisor to oversee cleaning and contents. The supervisor will track all upcoming and in-progress jobs, both in the field and in the shop; be responsible for inventorying and tracking contents; and meet with adjusters and clients.

This can be another position that's tough to fill. The Cleaning Division supervisor needs extensive training in smoke cleaning, deodorization, and fabric and carpet cleaning. If you can't find someone with that specific training, start with someone who's an experienced cleaner and manager, and then provide the necessary training. Getting and maintaining the proper IICRC certifications needs to be a condition of both hiring and ongoing employment.

This job requires a person who's *highly organized!* If someone is overseeing the pack-out of a quarter million dollars worth of contents, you better be confident they can track each and every item. This person also needs to be exceptional with people. Nothing is more personal than your client's belongings, and they'll let you know it. Your supervisor will need to be able to work with a wide variety of people in a wide variety of homes, under very, very difficult circumstances.

Estimator

The estimator's role in the restoration process is a key one, since the quality and completeness of the estimate sets the stage for how profitable a job will be.

The estimator meets with clients and adjusters on-site; examines, measures, and sketches the building; determines what's salvageable and what isn't; suggests creative alternatives for making repairs and salvaging materials; and, along with the adjuster, decides if certain materials can be matched or not.

The estimator needs to carefully document the job as well. He should have a good eye for detail, take numerous pictures, and make careful notes about the extent of the damage. He needs to be able to identify and document pre-existing conditions, and understand how they will or won't affect the claim.

The estimator prepares the actual written estimate and scope of repairs, which can take anywhere from 20 minutes to several days. He needs a thorough understanding of construction techniques, and how long it takes to complete each task.

As the owner of the company, you should do the estimating initially, since you're the one who'll be most affected by an inaccurate estimate. Doing your own estimates lets you better understand your jobs, connect with your clients, and interact with adjusters and insurance companies.

What to Look for When Hiring

If you decide to hire an estimator, you'll probably have to train him in the specifics of fire and water restoration. You'll want to begin with someone who has a solid base of construction knowledge. Since most of today's estimating is done by computer, he'll have to know — or be able to learn — computer estimating software.

Your estimator is also one of the first people to show up on the job, so the impression he makes needs to be a positive reflection on your company. He should be presentable, well-spoken, calm, and patient. He should spend some time getting to know the client and listening to the story of what happened to their home. He may not be particularly interested in the events that led up to the disaster, but if the client wants to tell him, he needs to listen. As with all your staff, it's important that your estimator be good with people.

He needs to be organized, and take a methodical approach to each loss. A restoration estimator also needs to be a good negotiator. He'll often be meeting with adjusters who are seeking to minimize damage claims. He needs to understand when and where he can compromise, while never undermining the quality of the work your company does.

General Manager

Finally we come to the position of General Manager. This is a position you'll establish when your company has grown to the point where you can no longer manage all the operations yourself. The General Manager oversees everything, from finances to jobs to people. Short of estimating and hammering nails, there isn't much that the General Manager won't be involved with.

What to Look for When Hiring

Like the estimator, this is a crucial position. It encompasses everything you did for the company until it grew too large for you to do it alone.

It's difficult to describe exactly what you're looking for in a General Manager, but basically you're looking for another version of you — but maybe even better. A lot depends on how your company is set up, how much responsibility you want to give this person, what you want him or her to assist with, and what you want to keep on your own plate.

Here are some general traits to look for in a General Manager:

- **Financials:** The General Manager needs to have a good background in financials. This includes billing practices, collections, payroll, A/R and A/P, banking, and preparing and reading financial statements.

- **Office practices:** Unless you have a very large office, the General Manager often takes over the duties of the office manager. In fact, you may want to consider promoting your office manager to General Manager. So, this person needs to understand common office procedures, from filing and inventorying to business letters and computers.
- **Human resources:** Human resources are a huge part of the General Manager's job. A General Manager needs a good working knowledge of state and federal employment laws, including payroll laws and payroll reporting practices. The GM will also be identifying staffing needs, advertising for new employees, interviewing, and hiring. Performance reviews, making bonus and reprimand recommendations to the owners, and maintaining employee records are also the responsibility of the General Manager.
- **Company operations:** Your GM will manage daily operations as well as do long-term planning. This could encompass anything from reviewing insurance policies to meeting with the accountant to budget for a future major expense.
- **People skills:** Whether it's sweet-talking a supplier or calming down an irate adjuster, the General Manager needs to possess great people skills. Ideally, you want a person who's calm, even-tempered, and not easily rattled.

Tip! *Due to the private and sensitive nature of the information your General Manager will acquire about your company, it's a good idea to require a specific Confidentiality Agreement as part of the hiring contract. You can talk with your attorney about having one drawn up.*

- **Marketing:** If the General Manager's job will involve marketing, make sure the person you hire has the skill set for it. This could include the ability to design and develop brochures or other marketing materials; write promotional spots for newspapers, radio, and other media; develop a website; or similar creative duties.
- **Community involvement:** Insurance restoration is a difficult business to directly advertise, so a General Manager who's active in the community and can help promote name recognition for your company is a big asset.
- **The "X" factor:** This is the hardest thing to define. A good General Manager almost becomes a partner. This person will be privy to every secret about the company, from finances to employees, so you really want someone you can connect with, feel comfortable around, and trust completely.

REAL STORIES:
"Yeah, This is Pretty Much Me."

My partner and I had decided that we needed a full-fledged General Manager for our company. We were concentrating most of our time and energy on estimating, and we had a couple of people in key supervisory roles, so it was time to take the next big step.

We contacted a local employment agency, and two of their representatives met with us for a long time. We discussed the type of person we were looking for, and then the search began. They sent us several resumes, but only a few looked promising. Then came the dreaded day when we had to start interviewing.

Neither of us really liked interviewing. We were a couple of old-time carpenters who had built up a successful business, and now here we were, in a fancy office at the employment agency, interviewing for an extremely important position. I think my partner and I were even more uncomfortable than the applicants.

With the help of the same two agency representatives, we started slogging through the interviews. A couple of the applicants were okay, but no one stood out. We were down to the second-to-last appointment, and looking at her resumé, she didn't seem like one of the stronger candidates. But as soon as the interview began, you could feel the tension in the room relax. Within a few minutes, it was like we were chatting with an old friend rather than interviewing a potential employee.

Because of the nature of our business, we knew we wanted someone who would be calm under pressure — especially since my partner and I could both be a little volatile at times! As the interview progressed, I was amazed at how calm she was. All the other interviewees had been nervous, just like you'd expect. Finally, I couldn't help but ask, "Are you always this calm?" She just smiled, nodded, and said "Yeah, this is pretty much me."

After that, we knew we'd found the perfect candidate. The last interviewee that day was actually someone we had pretty strong hopes for, based on her resumé, but she never stood a chance. We'd already made our choice.

Never underestimate the power of that "X factor"!

Finding and Hiring Good Employees

Once you've decided on the positions you need to fill, the next problem is where to find the right people. There are several avenues open to you. The ones you choose, and how effective they are, will depend on the particular job opening you want to fill and the condition of the economy. Here are a few possibilities, with their pros and cons:

- **Newspaper ads:** This has always been the traditional route for finding employees, and it usually works well for finding helpers, cleaners, and sometimes even for specialty trades, such as carpenters, painters and drywallers. The drawback is that newspaper ads are expensive, and the more words you use to describe the job, the more the ad costs.

Tip! *If you're hiring for several positions, combine them into one newspaper ad — you'll only have to give the basic information about your company once, and that'll save some money. A multi-job ad also shows the size and strength of your company.*

And, a growing number of businesses are turning away from traditional newspaper ads in favor of the Internet, and that's where many people, especially younger people, will go to look for jobs.

- **Internet:** There are a number of good Internet sources for job postings, both local and national. You pay a single fee for the ad, and as a rule, you're not limited in the number of words you use to describe the job.

  The Internet can be a good source for employees at all levels, but it's especially helpful for supervisory and management positions. The larger and more popular the website, the more people you'll reach — but the more the ad will cost. On the downside, this type of advertising can open you up to unwanted emails, ads, and spam.

- **Employment Agencies:** Agencies do the employee search for you, and they screen out unqualified people. A good employment agency will help you with the interview process if desired, especially for upper-level management positions. An agency can be a big help in filling specialized positions, such as an Estimator or General Manager. However, be aware that all the searching comes at a price, and can set you back hundreds or even thousands of dollars.

Hiring the Right People

Good employees can make or break your company, so when you're hiring, do everything to stack the deck in your favor. Carefully review resumés and applications, and verify references. Make a checklist of questions and desired qualifications, and have it on hand as an interviewing aid — and be sure to follow it. Checklists increase your chances of finding good employees. They're also helpful in documenting the interview process, which in today's society has become increasingly important!

Job Applications and Resumés

Have each potential employee fill out a simple, standardized job application. You can buy preprinted applications, download them off the Internet, or create your own to meet your specific requirements. Standard job applications make it easier to compare potential employees, and it gives you a written document for your file in case you need to refer back to it later.

You can also request a resumé. If you choose to do that, here are a few things to keep in mind:

- Laborers and cleaners may not have a resumé to give you, so don't be too put off by that.
- Carpenters may not be great at resumés either, and there's two ways you can look at that. On one hand, if a person wants a job badly enough, they'll take the time to make an impression with

a well-prepared resumé. On the other hand, carpenters aren't writers, so a written document isn't necessarily a good judge of their skills in the field. You'll see some great carpenters with lousy resumés, and vise versa.

- For office help and supervisory or management positions, a resumé becomes increasingly important. It's certainly not the sole judge of an applicant's abilities, but it *is* a good initial indicator of a person's willingness to make an effort.

Written Job Descriptions

Having written job descriptions for the various positions in your company is a tremendous help. You can provide the job description to the applicant up front, so there's no misunderstanding about what the job entails and what qualifications are required. It's also an important document for your file, especially if you have to fire the person later because they didn't have the skill set they said they did.

You can also use the written job description when placing an ad for an employee. This is especially true when advertising on the Internet. Since you have unlimited words to describe the job, you can simply post your standard job description and that will cover it.

Organizational Chart

Organizational charts aren't just for large companies. As your company grows, a simple organization chart will be very helpful. It defines the different positions in the company, and gets everyone on the same page about who the supervisors and crew leaders are. It's also a valuable tool to use when interviewing, doing reviews, and determining the structure of your company for future growth.

Organizational charts are easy to create, and don't need to be anything fancy. See Figure 20-1. You can identify each person by job title, by name, or both.

Company Policies

Part of the process of finding good employees and creating a company that thrives and prospers is establishing strong company policies for employee conduct. It's up to you decide on those policies, but here are some to consider:

- **Drug testing:** Having a drug testing policy as a condition of employment makes a lot of sense. It sends a strong message about your expectations for employee safety and conduct, and it can also be a benefit for your liability and worker's compensation insurance.
- **Driving record:** A lot of your employees will be driving company vehicles, as well as driving their own vehicles on company business. You need to know if they have a bad driving record, especially accidents, unpaid tickets, or a history of DUIs. Make a good driving record a condition of employment.

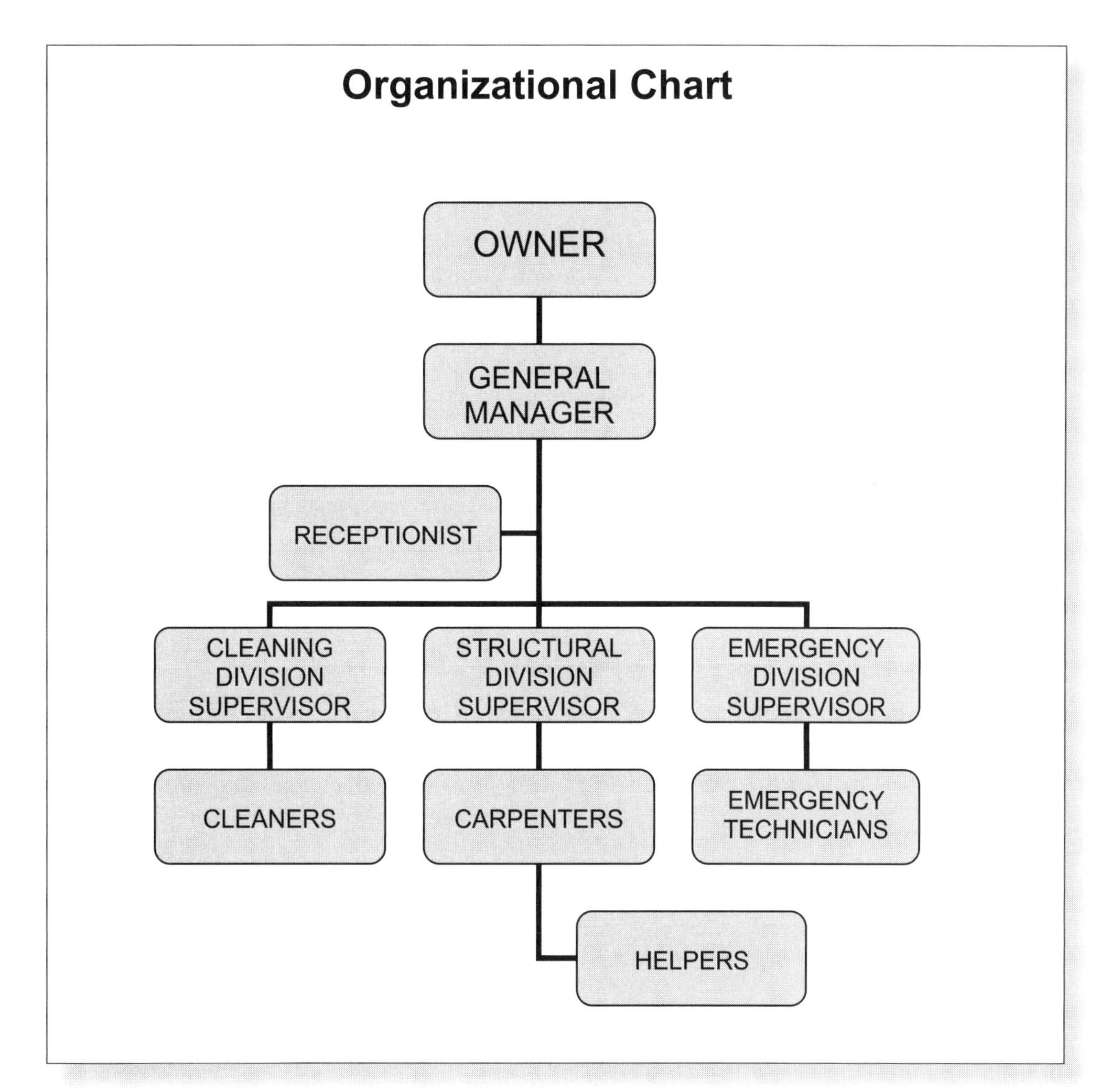

Figure 20-1 A typical organizational or "org" chart. This is a standard, top-to-bottom organizational layout. The lines indicate who reports to whom. For example, the Owner is in overall control of the company. Next in line is the General Manager. Directly under the GM are the Supervisors, who in turn have people under each of them. Also under the GM is the Receptionist, but you can see by the lines that the receptionist does not supervise anyone else.

Organizational charts like this one are very helpful in defining roles and reporting duties. They are often used during the interviewing and hiring process, and are included as part of the employee handbook.

- **Criminal background check:** Your employees will be in a lot of occupied homes, around children, and handling some very valuable contents. You have an obligation to yourself, your company, and your clients to know their history. A criminal background check really should be a condition of employment.
- **Credit checks:** Credit checks on employees are something that's relatively new. Again, you're going to be putting your people where they may be handling valuable contents, including

jewelry. If your employee has serious financial troubles, that might prove to be a serious temptation.

- **Alcohol:** You should have a zero-tolerance policy for alcohol. Anyone caught drinking on the job or showing up for work intoxicated should be terminated immediately. And, when someone is on-call, that means no alcohol!

Make sure these and other company policies are set up correctly. Talk with your attorney first, and make sure that the requirements are legal. Be sure you understand how to administer a pre-employment drug testing program. Also, make sure you understand how to do random drug testing during the course of employment, if needed.

You need to have all of your policies clearly spelled out in your employee handbook so everyone understands exactly what's required, and then apply them to everyone equally. We'll talk more about this a little later in the chapter.

Employee Training

There are a number of very good training classes available to help you learn about insurance restoration. In addition to getting the training for yourself, at some point you'll also want your key employees to take some classes.

However, that training can be expensive. By the time you purchase the class time, pay the employee's travel expenses, and lose his productivity on the job, you'll have invested a couple thousand dollars in that employee. If the employee then leaves your company, even if he's not going to work for one of your competitors, you're still losing all that training that you paid for.

One option is to ask the employee to pay for half of the training himself, with the understanding that he'll be reimbursed after a certain amount of time, say one year. It's in the employee's best interest to get the training, because it makes him more valuable to you. It's in your best interest because he gets the training he needs, and you know you'll probably keep him for at least a year.

If you decide to have this type of arrangement, it should be explained in detail in your employee handbook. That clarifies your policy and ensures that all of your employees are being treated equally.

Keeping Employee Files

It's time for a few hard truths about having employees. We live in a society that's a minefield for employers, especially those with small businesses. In many ways, the laws are heavily stacked in favor of the employee, and there are legions of attorneys out there waiting to sue for any reason whatsoever.

You may fire or even just reprimand an employee for a perfectly sound, commonsense reason, and then find out you're being sued for that a year or two later. You may hardly remember who the person was, and suddenly you have to defend yourself in a frivolous lawsuit.

As an employer, *you need to document and keep track of everything related to your employees*. This is especially true where medical problems, taxes, or other personal issues are concerned.

Believe it or not, the documentation process should begin even *before* the person is hired. It starts with the initial interview. When you're conducting your interviews, it's important that you keep good notes. When the interview is over, keep the notes in a file until you decide whether or not to hire the person.

If you interview a person *but don't hire him*, transfer the interview notes to a "not hired" file, along with their job application, resumé, and any other information related to the interview. There's always the possibility that the person may bring suit against you for some type of hiring discrimination. This could be based on age, gender, race, religion, or other grounds. If you've kept good notes about the interview and why you decided not to hire the person, those notes can be absolutely crucial in defending your actions.

If you *do* hire the person, start an employment file on him *immediately*. There are state and federal requirements on how the file should be set up and maintained, so be sure you understand those requirements. For example, you may be required to keep work-related medical files separate from other files. You may also be required to keep all or part of the file locked up, or otherwise limit access to it.

Once the file is set up, transfer the interview notes, resumé, and employment application into it right away. All that documentation will be very helpful when doing future performance evaluations.

The employee file needs to contain all of the information on that employee, including his basic contact information, reviews, reprimands, raises, work-related medical records, worker's compensation records, vacation and sick days, correspondence to or from the employee — even simple notes. Anything and everything pertaining to that employee should be documented in his file.

When people leave your employ, no matter what the reason, keep their employee files in a separate, inactive file area. Your accountant, attorney, or state labor board can tell you how long to keep a file after someone has left your company. Once the required time has passed, *don't just throw the employee's file away*. You're still responsible for its safe disposal, so make sure it's properly shredded.

The Employee Handbook

An *employee handbook*, also called a *policies and procedures handbook*, has become an incredibly important and useful document in recent years. The employee handbook describes in detail all of the policies relating to your specific company. It includes everything you want your employees to know, from hours of operation and types of uniforms, to discrimination and termination information.

For all the reasons we've discussed, *you need to have an employee handbook*. It gives the employees a standardized document they can refer to so everyone is aware of what the company policies are, and that those policies apply equally to everyone.

You can write your own employee handbook, or you can purchase a prewritten one that has most of what you need in it, and then alter it to fit your specific needs. Your attorney can help you with exactly how to set it up.

If you purchase a franchise or an existing restoration company, they'll probably have an employee handbook already in place. In either case, make sure you review it carefully to be sure that you understand and agree with the policies. This is essentially a contract between you and your employees, so be sure that it's correct!

The employee handbook is a *living document*, meaning it can change as you change. If you change your health care policy, or the length of your lunch breaks, that constitutes a change to the employee handbook. When changes occur, you don't need to rewrite the entire document. Simply issue an addendum to all employees, stating what the old policy was and how the new policy replaces or alters the old one.

What Your Employee Handbook Should Cover

What you include in your employee handbook will vary widely from company to company. But here are some basics it should contain:

- **A welcome statement:** This tells the employee a little bit about the company and its history. It may introduce the owners. It usually describes the company's philosophy, and may also include a mission statement. It may also describe what's expected of new employees, and how they can contribute to the company's success.
- **Orientation procedures:** This gives employees information about the Human Resources (HR) Department, the HR contact person, and what forms new employees will be filling out. If uniforms or other equipment will be issued, that'll be described here. If a drug test is required, that's usually here as well.
- **Pay and benefits:** This section covers how pay and benefits are calculated. This includes vacations, health insurance, and profit-sharing programs. It also tells when employees will be initially eligible for benefits. If the benefits increase periodically — for example, an extra week of vacation after three years — that'll be explained here.
- **Timekeeping:** This section explains the difference between full- and part-time employees. This area also describes timekeeping procedures, such as defining a work day, work week, and daily break periods. If employees need to keep a daily timecard, those procedures will be outlined in this area as well.
- **Conduct and discipline policies:** This covers general conduct expectations, including specific policies such as theft, sexual harassment, and drug and alcohol use. If the company has a smoking policy, that will be covered as well. This section will also include policies for reprimands and termination.

- **Performance reviews:** Included here will be the company policies on how and when performance reviews are given, and who gives them.
- **Dress code:** This section covers the company dress code, if there is one. It includes company-issued uniforms, as well as acceptable shirts, shorts, hats, coats, or other specific hot- or cold-weather clothing.
- **Company equipment:** This section covers anything that's issued by the company, including cell phones, pagers, tools, vehicles, uniforms, etc.
- **Employee vehicles:** If the company has a policy regarding employees using their own vehicles for work, that will be covered here. It includes gas and mileage reimbursements, carpooling, and other policies.
- **Injuries and illness:** This covers procedures for dealing with injuries, including worker's compensation. It also covers sick days, and who employees should report to when calling in sick.
- **Confidentiality:** Most handbooks have a requirement that the employee keep the company's business practices and procedures confidential. If the company requires an employee to sign a specific confidentiality agreement, that'll be covered here as well.
- **Acknowledgement and Acceptance:** The final part of the document is the acknowledgement and acceptance page. This is where your employees acknowledge that *they've read and understood the policies*. They sign this page to show their acceptance of the handbook and the policies it contains.

Tip! *Be sure you keep a copy of the signed Acknowledgement and Acceptance form in each employee's file.*

Some Final Thoughts about Employees

Good employees can greatly decrease your work load and your stress levels. Take your time when hiring, and look for really exceptional people. Offer as good a salary as you can, and provide whatever benefits you're financially comfortable with. Establish reasonable, common-sense policies and apply them equally to everyone. If you make your company a fair and pleasant place to work, you'll be able to retain good employees, which will make both yourself and them happy. And, that will go a long way in aiding your company's ultimate success!

21

Finding Work

Up to this point, we've covered a lot of ground getting you ready to tackle the world of insurance restoration. Now it's time to look at the most important step of all — moving from the theoretical to the practical, and winning some jobs!

One of the challenges of insurance restoration work is that it's difficult to advertise and promote. It's not something you think about needing, like shoes, for example. If you need a pair of shoes, you probably know of at least three places to go and buy them. You frequently see ads in the paper for retailers who carry shoes, and you pay attention to these ads, because you've bought shoes in the past and you'll need shoes again in the future.

Now think about waking up in the middle of the night to a broken water pipe and a flooded living room. You've never encountered this before, and more than likely, neither have any of your friends. You don't know who to call. Most people simply have no frame of reference to help them handle this situation.

That's the difficulty of getting started in insurance restoration work. People know they'll need shoes some day, so they follow the ads. But people rarely pay any attention to ads or promotions for a company that does fire damage repair, cleans smoke, or dries out flooded houses, since they don't expect to ever need one. So you may have to consider some different approaches to attracting clients.

Promotional Materials

Your first step is to get your name out to the public. You need to tell people who you are and what you do. We discussed creating a company name, logo, color scheme and a tag line for name recognition in Chapter 19. You should plan on investing $500 to $1,000 for some professional-grade promotional materials with your name and logo and have them ready on Day 1. If you're a genuinely artistic person, you might be able to do the design work yourself. But most of us aren't. *Professional-grade* means just that — professional — so plan on working with a good graphic designer. It'll be well worth the added cost.

Your promotional materials should do four things:

1. Tell people who you are and what you do.
2. Provide basic contact information for your company.
3. Promote name recognition for your company.
4. Show that you're a professional businessperson who takes his business seriously.

Well-designed, eye-catching promotional materials are *very* important because you're probably going to be competing against national franchise companies and established local companies. If the best you can do is a cheap business card, you'll be lost in the shuffle. Start with some nice-looking, professionally-produced business cards, truck signs and brochures, and expand from there.

Let's look at the basic promotional materials you'll need to begin your business. But remember, these are just the basics. As you start to circulate and get to know the people and companies that you'll want to work with, keep thinking of better and more creative ways to promote your company.

Business Cards

Business cards are the obvious first step. At a minimum, your card must have your name and title, your company name and logo, your contact information, and your license number. Your card can also give a very brief list of the things you do — fire and water repair, cleaning, etc. Consider a color other than white. The card should be eye-catching, but professional — never tacky.

Some states require that any advertising — cards, brochures, and even vehicle signs — include your contractor's license number. Verify this with your state contractor's board *before* you have anything printed.

Brochures

A brochure is an excellent form of advertising to hand out. It should give a brief background of you, your company, and the services you provide.

Be creative in your wording. Instead of saying "we do fire repair," consider descriptions like: structural fire reconstruction, smoke deodorization, structural fire cleaning, content restoration, and so on. Use the words *restoration* and *reconstruction*, instead of just *repair.*

Keep the brochure attractive and interesting, without being too wordy. If a brochure looks like it's too hard or too long to read, people won't take the time. Use easy-to-read fonts and simple colors with lots of open area, called *white space*, between the blocks of type.

Add some photos for visual interest. These can quickly tell people about your company, without the need for a lot of words. Before and after pictures of your work can be very effective, as are pictures of your shop, some of your equipment, or a job in progress. If you don't have those, consider stock photos of a fire- or water-damaged home. You can find free or low-cost pictures on the Internet; just be sure you abide by any copyrights, and don't claim the pictures as your own.

Your brochure doesn't need to be expensive to be attractive. The more colors you add, the more it costs to print. A very nice, two-color design can be fairly reasonable — just don't skimp on the paper quality.

Other Written Materials

Your promotional material will be targeted at two different groups of people: property owners; and agents, adjusters and other professionals. Most everything you develop can be used for both groups, but you may find the need to add specific materials geared toward individual groups.

For example, you may want to have a list of client references to hand out to homeowners. Any interested clients may appreciate a separate sheet that lists all the towns and geographical areas your company services, or a sheet that lists the major pieces of restoration equipment you have, and what each one does. Some of the equipment manufacturers have glossy, full-color spec sheets on their equipment that might make a nice promotional item to add to your brochure. You might consider a single sheet that looks like an invitation, inviting the person to come by for a personal tour of your shop. Adjusters may like to have a handout that describes your estimating software, or some sample forms that you use to document and track moisture levels.

You have an opportunity to provide potential clients with information about your company, so take advantage of it. They may not read all of it at once, but they may certainly want to refer back to it in the future. The main thing is to make sure any materials you hand out are professional and cohesive, and not just a hodgepodge of photocopied sheets that you've thrown together.

Presentation Folder

Now that you have all this material, you can't very well walk into someone's office and drop it all on their desk. So have some presentation folders printed with your company name, logo, and contact information on the front. This gives the person a compact and organized way of storing and referencing your materials, and the nicer it is, the less likely it is to get thrown away.

Simple Promotional Items

You can have pens, small paper tablets, and sticky notes made up with your company name, logo, and contact information on them. These are things everyone needs, so they'll tend to hold onto them, and that keeps your name in front of them. Refrigerator magnets are also good, and they're generally right in front of a person during an emergency.

Tip! *Spend a little extra and get good-quality items. It can be a frustrating reflection on your company when a cheap pen doesn't write well or a sticky note doesn't stick. If it has your company name on it, make sure it's going to work.*

Promotional Gift Items

Since name recognition is so important, promotional gifts can be a great way to spread that recognition. These include common items like coffee mugs and calendars, as well as unusual items such as gym bags and small tool sets. Anything that can get your name in front of someone and keep it there is a good gift. When a disaster strikes a person's home, you want them to remember who you are, and know how to get a hold of you quickly.

Tip! *Don't go crazy with promotional items, just get a couple of different ones to use as the occasion warrants. And be aware that some companies have minimum orders — you may like that gym bag, but you don't want 100 of them!*

Flip through any catalog of promotional items and you'll find thousands of ideas. Almost anything can have a name, phone number, and logo imprinted or even laser-engraved on it. The trick is to select items that you think people will appreciate, keep around, and use. Again, stick with good-quality items that work well and look professional.

You'll be giving these items to people to use or keep in their offices; people such as insurance agents, property managers, real estate agents, and others who you'll want to think of your company first when a loss occurs in a home. A gift that goes home to their kids might be appreciated, but it won't put your phone number in front of them when it counts. Think of promotional items that will stay on or in their desk, like a pocket calculator, a small flashlight, or a nice pocket knife.

We'll talk more about presenting promotional gift items to adjusters later on in the chapter.

Advertising

As mentioned earlier in the chapter, insurance restoration work is difficult to advertise. However, some forms of traditional advertising are still very important, however, so let's look at a few of them.

Phone Book Ads

There's actually one place where advertising pays off for restoration contracting, and that's the phone book. You *absolutely* need a real, honest-to-goodness ad in the Yellow Pages, not just a listing in the White Pages.

Take a moment to think about waking up to that broken pipe and a flooded house. Assuming that the name of a restoration company isn't going to pop right into your head, what's the first thing you think of? You're probably going to grab the phone book and look for a plumber or someone who deals in water damage. Phone book advertising reaches a tremendous number of people. It's at their fingertips immediately. And for those who check the Internet first, you can easily get a link in the online version of the phone book.

A Yellow Pages sales representative will help you with the size and design of your ad, and advise you on the most effective headings to list your company under. The sales reps are very good about making suggestions, and they typically won't push you into something you don't need or can't afford. The ads need to be completed several months in advance of the book's publication, so be sure that you plan ahead!

Your Budget

Advertising in the Yellow Pages is expensive, and will probably represent the single largest expense in your advertising budget. However, it's also the most important.

Ad rates vary considerably, depending on the market that the phone book reaches, and the size and layout of the ad. You can call their toll-free number listed in your phone book to get a rough idea of the average cost of different size ads, and also find out when the ad cutoff dates are.

This will most likely be your primary source of actual advertising. Don't overdo it, but spend enough to be sure that your ads are professional and easy to find. Incidentally, you usually don't have to pay for the ad in full. Instead, you'll be billed monthly as part of your phone bill.

Choosing the Right Headings

There are several possibilities for headings to be listed under, with pros and cons for each. Think about the services your company provides, then put yourself in the shoes of someone looking for those services. Where would you look? Ask your friends and family where they'd look.

Fire and Water Damage Restoration is one heading to consider. Many phone books also have a *Water Damage Restoration* category. To ensure that you reach the most people, you're probably going to want ads under both *Fire and Water Damage* and *Water Damage* categories. *Carpet Cleaning*, *Cleaning*, *Furniture Repair*, *General Contractors*, and *Remodeling Contractors* are additional possibilities as well.

The Size and Design of Your Ad

There are full-page ads, half-page ads, and small boxes, all the way down to just a single-line listing. A full-page ad is the most eye-catching and it gives you the most room to describe your services, but it's also the most expensive, so you have to weigh the pros and cons.

Since more people experience water damage than fire damage, go with the largest ad you can afford under the *Water Damage* category, then compromise with a smaller ad in the *Fire and Water Damage* category. If you can only afford one display ad, put it in the *Water Damage* category, then have them include a line-listing under *Fire and Water Damage*.

Your ad should be eye-catching, but not cluttered. There are unlimited possibilities here, so work with your sales rep to design something that looks good and represents your company in a professional manner. Adding red won't increase the overall cost by much, but it can definitely draw attention to your ad.

Website

A website is essential to any business today. For the restoration contractor, it not only provides a means for people to find you in an emergency, it's also an opportunity to describe your company and what you offer. Websites offer the opportunity for ongoing updates, with fresh pictures and client references.

As with all promotional materials associated with your company, your website should be professionally done. It needs to be easy to navigate, and clearly broken up into sections that quickly provide the users with the information they're looking for.

People expect a visual experience as well as an informational one when they visit your website. This is a great place for before and after photos of your jobs. Pictures offer a dramatic example of your services in a way that written descriptions can't. Consider having an exciting before and after photo group prominently displayed on your home page, and then a section where people can go to look at before and after photos of some of your other jobs. A short written description should accompany the photos. Be sure the text is as exciting and interesting as the photos themselves.

Here are some other things to consider including on your website:

- A complete description of the services you offer.
- Photos and written introductions of yourself and your key personnel.
- Photos of the inside of your facility, along with an invitation to take a tour. Be sure you also include photos of your vehicles.
- Photos and descriptions of key pieces of equipment, including drying equipment, ultrasonic tanks, and your ozone chamber.
- Photos and a written description of your contents storage areas.

- A list of all your certifications and affiliations. Consider links to the certifying organizations as well, so people can learn more about them.
- A list of client references (if you chose to do this, be sure it's kept current, and that you have your clients' permission to use their names).
- Social media connections, such as Facebook®.
- Copies of press releases.
- A Frequently-Asked Questions section.
- Current job openings.
- Include a copyright notice, so no one can use your photos, text, or other parts of your site without permission.

Tip! *Remember to keep your website fresh. Have the site designer incorporate software that makes it easy for you to add new photos and text whenever you want. Set yourself a regular schedule for updating the site.*

Top of Mind Awareness

There's a concept in advertising known as *Top of Mind Awareness*, or TOMA, which is a fancy title for name recognition. Simply stated, if you ask a consumer about a particular product or service, TOMA means that they'll think of your company first. For instance, ask about fast-food hamburgers and McDonalds comes to mind — that's TOMA.

ServiceMaster and Servpro have Top of Mind Awareness because of their national advertising campaigns and their marketing savvy utilizing bright colored trucks. You want to achieve this within your local market, so never miss an opportunity to promote your business. Let's look at some examples of how to do this.

Passive Advertising

Consider some passive advertising in your local market area where people will be exposed to your company name all the time. There are lots of possibilities. Many schools have advertising boards in the outfields of their baseball stadiums, which are inexpensive and provide a big, ongoing spot for your name to be prominently displayed. Fairgrounds and rodeo grounds offer similar opportunities. You might even find some less-expected opportunities, such as the fence on the way into your local landfill or other city or county facilities!

Tip! *Remember, restoration is a difficult business to advertise, so being creative in keeping your name in front of the public is the most effective way to promote Top of Mind Awareness. If they have a loss, you want them thinking of your company first!*

With local schools and municipalities scrambling for ways to generate funds for their operating budgets, this is a great time to be creative. Keep your eyes open for opportunities to get your name up. Don't be afraid to go into places and suggest a win-win opportunity for placing a tasteful advertising board somewhere on their property, but in the public's eye.

Press Releases

Press releases in your local newspapers are free, and they're an excellent way of promoting yourself, your employees, and your company. Did someone in the company complete a training class? Did you just acquire an important new piece of equipment? Was someone promoted? Did your company do something significant in the community? Whatever it is, write it up and send it to the paper.

Each paper will have specific rules and deadlines, so be aware of them. Always comply exactly with what they request, from the method of contact to the requested maximum number of words. Even if the release is about a person, be sure that your company name is included as well.

Here's a typical sample press release:

DATE: January 2, 2012

CONTACT: Genie Lawrence, All-Jobs Restoration, 555-355-1155

FOR IMMEDIATE RELEASE:

> Genie Lawrence of All-Jobs Restoration was recently awarded Journeyman Fire & Smoke Restorer status by the Institute of Inspection, Cleaning and Restoration Certification (IICRC). Journeyman status is only awarded to those people who have successfully completed intensive classroom training and have demonstrated a minimum of three years of practical on-the-job restoration work. All-Jobs Restoration specializes in disaster restoration, reconstruction and cleaning, including fire and smoke restoration, restorative drying for water damage, and 24-hour emergency service for losses of all types.

Home Shows

Home shows are typically put on by local builder's associations — something you should be a member of. Participating in a home show is an excellent way for you to introduce your services to the community.

Don't expect anything immediate to come from the home show. It's not as though the Smiths are going to come to the home show looking for a restoration contractor to repair their fire damage. But having a booth set up with all your cool-looking equipment gives you a chance to show people what you do and how you do it, answer their questions about your type of work, and to pass out your cards and brochures. Mr. and Mrs. Smith may be at the show looking for ideas on a kitchen remodel, but Mr. Smith may find *your* booth more interesting — and perhaps more memorable. This is a good venue to give away some promotional refrigerator magnets.

If you have carpet-cleaning equipment, consider having a drawing to give away a free carpet cleaning, or perhaps some upholstery cleaning. It's a way to attract people to your booth who might not otherwise come over, and while they're there you can introduce yourself and explain what your company does.

Community Involvement

Community involvement is another excellent opportunity for passive advertising and promoting your TOMA. Depending on your particular interests, you can:

- Sponsor a sports team, such as Little League or Youth Soccer. Consider sponsoring both a boy's team and a girl's team, or a co-ed team.
- Sponsor a community blood drive or a canned food drive. You can team up with a local lumberyard to help defray costs. Talk with your local Red Cross or food bank, and also with a local radio station about some free publicity.
- Sponsor a pet adoption day in conjunction with your local humane society.
- Help out with a Boys and Girls Club fundraiser, or sponsor a local spelling bee at the library to help promote literacy.

All of these events help your community and raise your name awareness at the same time. Just beware of doing anything that can possibly alienate potential clients, such as anything that's overly religious or political.

REAL STORIES:
We'll Help with That!

We once worked a small water loss for a member of the local police department. While I was there doing the estimate, we got to talking about a local abandoned house the police had to deal with. Kids had been using it as a party house, neighbors had been complaining, and the court had recently given legal clearance to demolish it.

The police had a guy with a backhoe lined up to do the actual demo, and some teenagers doing community service work to help with the cleanup, but they were looking for a local company to provide some tools and help coordinate the effort. I volunteered immediately!

On the big day, I made sure our moving van showed up with the tools in the back. It was only shovels, rakes, and wheelbarrows, but the moving van had a much bigger sign than any of our pickup trucks! We spent the day out there, helping with the demo and cleanup of that beat-up old house, much to the delight of the neighbors. Our truck got a prominent shot on the local news that evening, our company got a very complimentary write-up in the paper, we received a commendation from the police department, and the community got a clean lot. It was win-win for everyone, but especially for us!

Pretty good deal for one Saturday's effort!

Introducing Your Company

The next step in finding work is to introduce yourself and your company. In the insurance restoration field, there's simply no substitute for meeting people face to face. You need to make them aware of who you are and what you do.

Making Cold-Calls

Making *cold-calls* is simply stopping in to see someone who hasn't specifically requested or agreed to a meeting. To be honest, it isn't easy to go out and introduce yourself to strangers and convince them to take a chance on your company. Unless you're just naturally outgoing, this part of the business is difficult for a lot of people. But much of your work will come directly through referrals from insurance companies and other sources that you've met in this way — there's really no better or more effective means of making contacts.

As the owner of the company, you need to do the in-person presentations. A hired salesperson won't have your passion, or your understanding of the work that your company does. In this industry, a lot depends on your personality and how trustworthy and honest you appear to be. The best way to build those perceptions and nurture relationships is for people to meet you and get to know you.

Cold-Calling Basics

A lot of the do's and don'ts of effective cold-calling are based on simple common sense.

- **Decide who to contact:** Do some research and take the time to come up with a good contact list. Think of every possible contact you can (more on that next). Write down the name of the company, the contact person, and the contact information. You can get a lot of this from the phone book, off the Internet, and from local sources, such as your library and Chamber of Commerce.
- **Prioritize your list:** Organize the list in order of who you want to contact first, second, third, etc.
- **Be professional:** When you do your cold-calls, dress like you take yourself and your company seriously. You should be clean and freshly showered, not coming straight from a jobsite. Wear neat pants and a clean company shirt. A blazer or sport coat is a nice touch. Be outgoing and friendly, but not loud, boisterous or overbearing. *Never do anything that disrupts their office in any way.*
- **Ask for a specific person if possible:** Know the person you want to see. If you don't know how to pronounce their name, find out before you meet them. Ask to speak to him or her personally. Be

assertive, but polite. If the person's not available, ask if you can wait until he's free, or when a good time would be to come back. Never be pushy or rude.

- **Be prepared:** Know how you want to introduce yourself, and what you want to say. Know something about the person you're meeting with, what his position is, and what their company does. Carry your promotional materials in a briefcase or other professional case so they're organized and ready to present.
- **Don't over- or under-stay your welcome:** Most people you'll be meeting with are busy. They may take a moment to talk to you, but don't take up too much of their time. On the other hand, if the person is interested, give them as much time as they want. The last thing you want to do is to appear impatient to be off to another appointment when the person wants to know more about you and your services!
- **Leave the door open for future contact:** Tell them you'd like to check back in a couple of weeks, to see if they have any questions about your materials. Invite them to stop by your office for a tour. And above all, ask them to please give you a call if they have any jobs they'd like you to take a look at.

REAL STORIES:
"I Don't Know — What Do You Think?"

Cold-calling is no fun, but we did a lot of it in the early days when we were getting our business up and running. Once we became established, it always seemed odd to have other business owners make cold-calls on us, introducing themselves and soliciting our business. It was, however, pretty interesting to observe the mistakes other people made.

One afternoon, a guy came into the office and asked to speak with one of the owners. He didn't specifically know who to ask for, so he hadn't done his homework. However, he had a service that our front-desk person thought we'd be interested in, so she came and got me out of my office.

The guy was in jeans and a tee-shirt, which was a bit of a turnoff for me. He introduced himself, and handed me a business card. He was starting a mobile sharpening service, and he said he'd come around to our shop on a regular basis and sharpen whatever blades, bits, and other tools we needed sharpened. Since it was a pain for us to have to run out to the sharpening shop, this actually did seem like a service we'd be interested in trying.

I asked him what his prices were. "I'm not really sure," he said. "How much are you paying now?" I told him that was confidential, and then I asked how often he would come around to do the sharpening. "I guess every couple of weeks or so. What do you think — does that sound about right?"

Then I asked him if he had a brochure or any other information he could leave with me. "To be honest," he replied, "I wasn't expecting to get to talk to anyone. I just wanted to drop off a card —I don't have anything else with me."

Talk about completely unprepared! I cut him off at that point, and told him to come back some other time when he had a price list, a schedule, and other basic information about his business. We never heard from him again, and as far as I know, he never did get his business up and running. Pretty obvious why.

An Introductory Letter or Email

Some people prefer to send an introductory letter before they make their cold-call — to help break the ice. You might want to consider writing a nice letter on company letterhead, and mailing it to the first group of targeted people on your list. Keep it short and to the point. Introduce yourself, tell them you're a new company and what you specialize in, and say that you'll be coming by their office to introduce yourself in person the following week. Include a business card, and invite them to call you in the meantime if they have any questions or have any immediate need for your services.

Another option is to send out targeted emails. Emails allow you to reach a number of potential prospects in a short amount of time. As in the letter, briefly describe your services and let the person know that you'll be calling on them soon.

Emailing is common business practice today, but remember that emails from unknown senders often get deleted immediately, or get caught in spam filters. Don't do mass mailings or send an email with an attachment. Those are often associated with viruses and are much more likely to be deleted.

This mailing or emailing step is optional, but it makes the person you're calling on aware of who you are, which takes a little pressure off you during the initial in-person call. On the downside, it requires additional up-front work and expense.

Here's something very important to keep in mind: don't bite off more than you can chew. If you send out 50 letters or emails promising that you'll stop in to visit in the next week, you won't be able to do it. When you show up three weeks later, they'll either have forgotten who you are, or formed an impression that you're not someone who can be counted on.

Incidentally, if you can't write a decent letter, ask someone else to do it for you, or skip this step. When you're trying to make a good impression, a poorly written letter can do more harm than good.

Make a Friend at the Front Desk

The people who occupy the front desk in an office are often known as *gatekeepers*, and it's a great description. They guard their bosses, and you have to get past them to see the person in charge. So make a friend of the front desk person wherever you can. Be polite and friendly, and involve them in casual conversation. If you see a picture of their child or their dog, ask about them. If you see any indication of a hobby or other interest they might have, use that as an opportunity to start a conversation.

When you leave the office, make a note of the person's name and interests. The next time you come in, review your notes. Ask how the child's school play went, or if they'd had a chance to take the dog down to the river yet this year. You might even show up with some pastries or a special coffee for them sometimes.

Having a friendly relationship with the front desk person will be a benefit to you over the coming years. It makes it much easier to get in to see the boss, and it simplifies receiving and passing along information. It can also result in job referrals, since the front desk person often wields considerable influence in a small office setting.

Your Contact List

There are several businesses that are "musts" for your introductory contact list. We'll cover them in their order of importance to the insurance restoration business, but of course, you can change the order on your contact list to fit your own circumstances.

Insurance Companies

Insurance companies are the most obvious targets for your services, but they're also going to be those most heavily solicited by other contractors. Regardless, it's a market that you need to approach.

The following is a list of the 10 largest homeowner's insurance companies in the country, as of this writing. Each company is followed by their approximate percentage of market share, in terms of actual policies written. The order and percentages don't necessarily reflect how good or bad any given company is to work with.

State Farm Mutual (21.6 percent)

Allstate Corporation (10.0 percent)

Zurich Financial Services Ltd. — includes Farmer's Insurance (7.4 percent)

Liberty Mutual Holding Co. — includes Safeco (5.1 percent)

Travelers Companies (4.5 percent)

USAA Insurance Group (4.4 percent)

Nationwide Mutual Group (4.3 percent)

Chubb Corporation (2.6 percent)

American Family Mutual Insurance Group (2.1 percent)

Hartford Financial Services (1.7 percent)

As you can see, State Farm writes the lion's share of homeowner's policies, and therefore has the majority of losses, generating the most work for restoration contractors. They're also one of the hardest companies to break in with, so you might want to begin with some of the smaller companies. They may not turn as many losses, but they're not as heavily solicited by contractors looking for work either.

Insurance Agents

When a property owner has a loss, the first person they call after the emergency is under control is their insurance agent. Agents are an excellent source of work, and should be at the top of your contact list.

It's easier to start with someone you know, so talk to your own insurance agent first. Bring him your promotional items, talk with him about the types of restoration jobs you're looking to handle initially, and candidly ask him for advice on how best to approach other agents. Listen carefully to his suggestions, especially his opinions about what he does and doesn't like about cold-calls from contractors. Then ask for the opportunity to handle a few of his smaller losses.

From there, expand your personal visits to other insurance agents in your immediate area, and to other outlying areas as desired. Smaller companies have fewer clients and fewer losses than major companies like Allstate, Farmers, and State Farm. But they're often overlooked by other contractors, so they can be a great opportunity to get your foot in the door.

Agents will often tell you that they don't refer specific contractors to their clients. That's usually true, since whoever they recommend reflects on them. Also, they may have restoration contractors as clients that they don't want to offend. Instead, an agent usually keeps a *list* of contractors that he refers, so simply ask if your company can be included on their list.

Independent Insurance Agents

Don't overlook the independent insurance agencies! These agencies represent several insurance companies, both large and small. They also tend to have a little more flexibility in how they refer jobs, so they're an excellent source of work.

Staff Insurance Adjusters

Adjusters are another group of people in the insurance industry that you want to contact. Policy holders may know their agents, but not usually their adjuster. The adjusters don't have much contact with the people their company insures — until there's a loss. But once the loss has been reported to an agent, the agent or the insurance company passes the claim along to the adjuster.

Establishing yourself with adjusters is very important. When they go out to see a loss, the homeowner often doesn't have a contractor yet, so the adjuster is in a great position to make suggestions.

You can begin with the staff adjusters for the major companies. Be forewarned, these guys are solicited on a regular basis by restoration contractors, so don't expect to get a bunch of jobs right off the bat. Instead, your goal is to make sure they know who you are, and what your company has to offer.

Before you ever contact a staff adjuster, do your homework. If you're going to get grilled by anyone, the staff adjusters are the ones most likely to do it. They have a tough-enough job as it is, and they don't have time to deal with contractors who don't know what they're doing, or who treat restoration work as a sideline occupation to dabble in when remodeling work is slow.

Like agents, adjusters often don't recommend specific contractors, at least not until you're very well established. Instead, they'll give the homeowner a list of contractors they know who do the type of work that's needed. Initially, the best you can hope for is to get your name included on their list.

As you get to know the adjusters better, they'll gain more confidence in your work. When they do, you'll find that they may move your name up on their list. They can also make subtle suggestions to a homeowner, such as "Here's a list of contractors that do this type of work. I know that John Doe Restoration does a lot of jobs like this one." Something as simple as that can direct a lot of work in your direction.

Preferred Contractor Programs

Many insurance companies maintain what are known as *preferred contractor programs*, or *preferred vendor programs*. The programs have their pros and cons, but they're definitely something you need to know about.

Under a preferred contractor program, an insurance company will select a group of local restoration contractors, either directly or through a third-party screening company that they contract with. How many contractors they put on the program varies with the size of the geographical area and the number of losses the company has to service. It typically ranges from about three to six.

The selected companies are usually contractors that the insurance company has worked with in the past, and that they trust. The contractors are carefully screened based on a number of criteria established by the insurance company. It generally includes IICRC certification; a certain number of years in business; a credit background check; and other requirements.

If they meet those requirements, they're then placed on a preferred contractor list, and put into the loss rotation. When one of the insurance company's policyholders has a loss, the contractor next on the rotation list receives the assignment. If the policyholder doesn't want to use that company, the assignment is given to the next company on the rotation, or the policyholder can choose any other contractor on the list.

Policyholders don't have to choose from the list or participate in the program. However, the insurance company typically offers them an incentive to select a preferred contractor, such as expedited claim service, or an extended warranty on the contractor's work. Some companies put a little pressure on the policyholders, because they feel it's in everyone's best interest to have a preferred contractor do the work. However, if the homeowner's don't choose to participate, it doesn't change their coverage in any way, and the loss will be handled like any other.

In theory, everyone wins with a preferred contractor arrangement:

- **The insurance company:** They have the benefit of knowing that their policyholder is working with a qualified contractor. The paperwork gets done faster, and the bids are more uniform. Most importantly for them, they're able to maintain some additional control over the entire process, and control translates into profits.
- **The homeowner:** They're relieved of the task of finding a qualified contractor. Their paperwork is also simplified, plus they get whatever incentive the insurance company is offering. For the most part, preferred contractor programs seem to work out well for homeowners, with a lot more positive experiences than negative.

- **The contractor:** For contractors, it's a little more of a mixed bag. On the plus side, they have a steady flow of work from the insurance company, with reduced competition for jobs. Some programs also offer material supplier arrangements with large home centers, which relieves some of the contractor's cash flow burden.

On the down side, there's usually even more paperwork for the contractor. Estimates have to be done within a certain number of days, and within a specific format. You'll have to agree to use only approved estimating software, and an approved price list. Every insurance company has different parameters for their programs; so if you're on more than one program, you'll end up processing your estimates differently for different companies, and you'll have to keep all that straight. Some programs may require that you — and sometimes even your subs — purchase materials only from certain suppliers who are also selected and approved by the insurance company.

If you're invited to participate in a preferred vendor program, get all the details of how the program works, and what your rights and obligations are. If possible, get the names of other contractors who're in the program in other cities. Give them a call, and see how the program's working for them. As long as they know you're not their competitor, you're likely to get some pretty candid opinions.

Tip! *When cold-calling a staff adjuster, you may be told that the only way to break in with them is to get into a preferred vendor program, and that they don't currently have any openings.* Don't be put off. *You can still get work with that company through agents, property managers, and other sources, and the more jobs you do, the more you'll come in contact with that adjuster. Do your jobs well, keep checking back with the adjuster, and you'll soon find yourself in line for the next slot in the program that opens up!*

Conflicts of Interest

In years past, preferred contractor assignments for water losses were given to a single contractor, assuming the company was qualified to handle all phases of the loss. The company would do the emergency mitigation work to dry the structure, then prepare the estimate and do the repair work.

Today, much of that has changed. Some of the larger insurance companies began to see an inherent conflict of interest in that method. They felt that an unscrupulous contractor might do more demolition work during the mitigation phase than was needed, knowing that they'd be the ones bidding to put it all back together. The contractor would win on both ends — more money on demo, and a higher reconstruction bid.

So, many insurance companies now separate the two services within their preferred contractor programs. When a water loss occurs, a mitigation contractor is called in to do the drying and whatever demolition work is needed. When he's ready to certify that the building is dry, his work is done and he turns it over to the general restoration contractor for reconstruction. These insurance companies utilize a single mitigation contractor — usually one of the national franchise firms — and a rotating group of local restoration contractors for the reconstruction.

This change in the way the insurance industry handles work is one you need to be aware of. Whether it continues or not is anyone's guess. Since it seems to lessen the insurance companies' potential liability and expenses, it probably will. So if you're looking to participate in a preferred contractor program, be aware that unless you're a national franchise owner, your best bet in the future will probably be in the reconstruction rotation.

Current Programs

As of this writing, here are the top four insurers with preferred contractor programs in place:

State Farm: Preferred Service Provider (PSP) is one of the longest lived and most comprehensive of all of the programs, and one of the most complicated. They use ServiceMaster for mitigation work. Contact a State Farm Claims Representative (they're not called adjusters at this company) for information on joining the program.

Allstate: Simply called the *preferred contractor program*, it's usually one of the easier ones to get into. Contact an Allstate adjuster or agent for information.

Farmers (Zurich): The Emergency Preferred Vendor Program (EPV). Contact Farmers HelpPoint® Claims Services at 1-800-435-7764 for more information about the program.

Safeco (Liberty Mutual): They use Innovation Managed Property Network as a screening company for their contractor referrals. Refer to their website at http://www.us.innovation-group.com/Property.aspx for information on their screening process. Safeco currently uses Servpro for their mitigation work.

Preferred contractor programs are implemented, changed, dropped, and reinstated at different times by the insurance companies, so do your homework before contacting an agent or adjuster about being including in one.

Tip! *If you want to get in with an adjuster, offer to take on the lousy jobs. Tell him all you want is a chance to prove yourself, and you'll do anything. Adjusters sometimes have trouble finding contractors who will take on a small job, or one that's particularly dirty or not very profitable. Or, they may have a less than pleasant homeowner they're dealing with. If you can make their life a little easier, they'll remember you for it in the future.*

Independent Adjusters

Independent adjusting companies are like independent agents; they represent several different companies. As such, they're often not as limited in who they can recommend, and they can be a great source of work. They also don't tend to get as much attention from other restoration contractors as the big staff adjusters do.

Do some research and see if there are any independent adjusting companies in your area. Some of the smaller insurance companies may not have staff adjusters in a particular area, but instead operate through the independents. It may take some time on the phone or the Internet to find them, but it's all part of putting your contact list together.

Adjusters and Promotional Items

A few decades ago, it wasn't unheard of for large insurance restoration companies to influence adjusters by offering them vacation trips and other expensive gifts — even cars! The insurance companies have since put an end to all this, and rightfully so.

Today, many of the larger insurance companies are extremely restrictive about what their adjusters can accept in the way of gifts. Some companies won't even allow their staff adjusters to accept a calendar or a pen if it has a restoration contractor's name on it. Smaller insurance companies tend to be less restrictive, and most independent adjusters have few, if any, restrictions.

These restrictions don't extend to items such as brochures, so you might try pushing the envelope a little. If you're promoting yourself to an adjuster, bring him a presentation folder. It can have your card, brochure, and any other handouts you have. Then you can slip in a pen and a tablet, or perhaps a small calendar. The adjuster can then keep it if he wants, or give it away (which is still good for name recognition). At worst, he'll tell you that he can't accept certain promotional items, and give them back to you. Just make sure he keeps all the written material!

Property Managers

Property managers oversee the rental and maintenance of all types of properties. They encounter a lot of tenant mishaps, from overflowing washers to kitchen fires and worse, so they need to be high on your list of people to contact.

Property management companies come in all shapes and sizes. Some handle residential properties, some specialize in commercial, others do both. You'll also find companies handling multi-family properties, huge commercial projects, and very high-end properties. Do your research to identify the best companies to approach.

When it comes to property managers, you pretty much have to take the good with the bad. They usually need to work with a restoration contractor who can handle small water jobs, large fires, and anything in between. If you offer to take on some of their smaller, tougher jobs, and you get them done with a minimum of time and effort on their part, you may just get your foot in the door.

Tip! *Most property managers are considered a legal representative of the owner. As such, they can commit to spending money for repairs on a property, up to a certain dollar amount. That amount varies between properties, depending on their agreement with the property owner. Be sure and confirm the amount before you start work.*

Offer a Free Service

Offering some type of free service or free incentive is another way to attract business. For example, you might go around to the insurance agents and adjusters on your list and let them know that you're offering a free board-up service to their clients. In the event of a fire, impact loss, break-in, or other emergency, you'll come out and provide the labor and materials to cover any broken windows and doors — with no obligation.

This doesn't cost you very much, and chances are pretty good that when the homeowner is ready for an estimate of the damage, your company is the one they'll call.

During storms, you might consider offering the same service for emergency roof temps. Again, providing the labor and materials for this is pretty inexpensive, but it could lead to getting your foot in the door on some nice roof and damage repair jobs.

One more option would be to waive all or part of a person's deductible if they sign with your company for the repairs. This is one that you'll have to look at carefully and put a few restrictions on, since you don't want to get in the position of absorbing a $1,000 deductible on a $2,000 job. But worded correctly, this can be a very nice incentive for people who don't have the ready cash to cover large deductibles.

If you know that people want to do some remodeling in addition to their repairs, you can offer a discount there as well. For example, if they sign with your company for the restoration work, you can offer 10 percent or more off of any remodeling work.

Ambulance Chasing

There's the inevitable temptation to simply go up to a homeowner who's had a fire and offer your services. It seems like a pretty innocent combination: You have a service to provide, and they're in need of that service. You'd think it would be a perfect match. However, in the real world, things don't tend to work out that way.

For one thing, you don't know what you're walking into. You could be intruding on a very emotional situation, or even on a crime scene. Also, you know nothing about the person you're approaching — not even if they have insurance, and if they do, in what quantity.

There's a certain perception about contractors who cold-call on homeowners after a disaster. They're called *ambulance chasers*, and they're viewed as opportunists, whose only motivation is greed and profit at the expense of others. While this may be obviously unfair — since every restoration contractor makes a profit when bad things happen to other people's homes — how you approach that work can create a perception of you as either a good guy or a predator.

The story on the next page, *The Wisdom of Getting Work the Old Fashioned Way*, illustrates that point. In my opinion, you're better off sticking with the proven method of contacting insurance companies, adjusters and property managers, and don't intrude on other people's lives.

Success Will Come

If you want to be successful, you need to be prepared. So before you walk out the door, I want to recap some of the really important, make-or-break issues that we've talked about in the prior chapters. I bring these up now because if you're not ready to check the following items off on your mental checklist, then you're not ready to go out and approach the insurance industry for work.

REAL STORIES:

The Wisdom of Getting Work the Old Fashioned Way

As a brand new business, we were still trying to get ourselves known and establish a steady flow of work. We were in the cold-calling stage, and fighting a lot of competition in town. But being new, we were willing to try different things to see how they worked.

One day, we read about a fire that had happened the night before. It was near where I lived, so I drove by the place on my way home. I slowed down, looked it over, then drove on. I went around the block, and approached the home again. This time, I saw the homeowner coming out of the backyard with a shovel in his hand. I pulled over, got out of my truck, and approached him.

I introduced myself, and offered him a business card. He just looked at me with a dazed look, then started sobbing. "My dog died in that fire last night. I just finished burying him. I really don't care if the damn house ever gets fixed!"

Then he turned and walked away without another word, leaving me standing there holding my business card, feeling like the biggest jerk who ever lived.

After that experience, my partner and I both made a commitment to never again approach someone who hadn't asked for our help.

The wisdom of that decision was proven a couple of years later, when we were better established. An insurance company asked us to go out and meet with a homeowner who'd just had a fairly major fire. He didn't have a contractor and was in need of some help and advice.

I called and made an appointment with the man for that afternoon, and pulled up in front of the house right on schedule. We had signs on our trucks, so it was obvious that I was a contractor. The homeowner was standing on his front porch, and I assumed he was there waiting to greet me.

I wasn't halfway up the front walk before he practically growled at me: "What do *you* want?!!" I was momentarily startled by his question and his obvious anger. So I introduced myself — still from a bit of a distance — and reminded him that I had called and made an appointment with him to look at his fire damage.

He looked startled for a moment, then immediately softened and came over and shook my hand. "I'm sorry," he said right away. "Didn't mean to bite your head off. It's just that since the fire, several different contractors have come by and left cards. They're like a bunch of damn vultures, just waiting to pick over the carcass of my house. I wouldn't let them do the work if they offered to do it for free!"

In that moment, I was reminded once again of the man who'd lost his dog and of the commitment my partner and I had made to not do any ambulance chasing. It was definitely the right decision!

- **Commitment:** Insurance restoration isn't something you dabble in — not any more. If you hope to attract the attention of the major insurance companies, this can't be a sideline occupation. It takes a 100 percent commitment of your mental, financial, and physical resources.

- **A Business Plan:** Who is your company? What do you plan to do tomorrow, and five years from now? How will you get there? You need to *plan* for your success if you want to *achieve* your success. It's also the only way that lending institutions will even consider working with you.

- **A Company Identity:** Know who you are, and let the world know it, too. That includes a name, a logo, colors, advertising materials, everything. Be ready from Day 1 to present your company to the world at large. Remember that this is your job interview, and your one and only shot at that first impression. Make it count!

- **A Proper Business Location:** You wouldn't take your wife to an anniversary dinner at the corner hot dog cart, and you wouldn't be comfortable with an attorney who worked out of his car. Insurance restoration companies are no different. They have an office and a warehouse, and that's expected. The days of working out of your truck and your garage are over. If that's what you're hoping to do, then you're better off in another profession.

- **Training and Certification:** You need to be IICRC certified. Insurance agents and adjusters have become increasingly well educated about the science behind proper drying and proper deodorization. And, thanks to the instant availability of information on the Internet, so have homeowners. They're also all very much aware of the liability issues surrounding mold and smoke odor, and the need to disclose property damage to potential buyers. You *have* to be able to demonstrate that you're knowledgeable and properly trained.

- **Equipment:** You'll need to have the necessary testing, monitoring, and drying equipment. You don't need a warehousefull when you're first starting out, but you need enough to immediately handle a couple of jobs when they're given to you.

- **Unit Cost Estimating:** The insurance restoration industry relies solely on unit cost estimating, which we'll cover in later chapters. The minute you're given your first restoration job, you need to be prepared for this kind of estimating.

No new business, whether it's a restaurant, shoe store, or insurance restoration contracting, opens its doors to great success on Day 1. Starting a business and finding work takes effort and perseverance, along with a solid plan, a positive attitude, and a professional appearance.

But the rewards are well worth it, and there's nothing to stop you from achieving anything you want to achieve!

22

Working with Insurance Adjusters

As an insurance restoration contractor, you'll be working very closely with insurance adjusters, so it's important that you understand what they do and how they do it. This is especially true when you run into some of the more difficult situations that we'll be covering toward the end of the chapter.

Many elements of an insurance policy and a damage claim are open to interpretation, and adjusters have a considerable amount of discretionary power when it comes to settling claims (Figure 22-1). Good relationships with adjusters will benefit your business, while a bad relationship can make your life miserable every time you have to deal with that particular insurance company.

You can help the adjusters, and yourself, in a number of ways:

- **Learn how they interact with their policyholders, and what interaction they want from you.** Some adjusters spend a great deal of time explaining things to the homeowner, and prefer that you not interrupt. Others like to get off the jobsite quickly, so the more explaining you can do to the homeowner while the adjuster's working, the better they like it.
- **Find out what their policy is on walk-throughs.** Some adjusters do a thorough walk-through before the estimate's done, especially on larger losses, to get both of you on the same page. Others want to wait until the estimate's complete. Some never bother with a walk-through.
- **Understand how they like their bids prepared.** Some just want to see numbers, others like to see explanations of what's included in that number. Some like to see lots of pictures or drawings, others don't care. Some will insist on line-item notes, others don't want them.

Figure 22-1 Odd construction, such as the mismatched roof slopes found on this home, is one of the many areas where an insurance adjuster has considerable discretion in how he interprets a policy. A good adjuster can make things a lot easier on you when it comes to preparing an estimate.

- **Understand their motivations and their pet peeves.** What's a major issue to one adjuster is often no big deal to another. For example, some adjusters argue that your estimate should only include painting walls that are damaged, while others will allow you to paint the entire room.

> ***Tip!*** *You may have heard the old saying that it's easier to ask forgiveness later rather than to ask permission up front.* Not so with insurance adjusters!

There's one thing that's universally consistent among adjusters: you don't do anything without their prior approval. If you tear out a room full of carpet or toss some cabinets you don't think are salvageable, and you haven't discussed it with them first, you could easily find yourself having to pay for the replacement materials out of your own pocket.

The Role of the Insurance Adjuster

Once a loss occurs to a policyholder's property, the adjuster steps in. The insurance adjuster's job covers several responsibilities during the restoration process, including:

- **Representing the insurance company:** It's up to the adjuster to explain what's going on to the homeowner. He makes all the decisions related to the loss, based on his company's policies.

In the event of a dispute over coverage, he gets the final word, not you and not the homeowner.

- **Determining coverage:** The adjuster has to determine if the damage is a covered loss. He inspects the property, reviews the policy, and decides what is and isn't going to be included in the claim.
- **Initiating and/or overseeing emergency services:** If the loss requires emergency mitigation, such as drying services or a board-up, the adjuster will oversee the process. In some instances, he may be the one to initially set the process in motion. He'll also review the drying logs; oversee the extent of the mitigation and drying work; and review and authorize payment of the emergency service bill.

> ***Tip!*** *Insurance adjusters aren't always treated politely by the people they interact with; so when you have dealings with them, be sure to invoke the* Golden Rule — *"Do unto others as you would have them do unto you." You'll be surprised at what a huge difference this can make ... a little common (or, as it now seems, uncommon) courtesy goes a long way!*

REAL STORIES:

So *This* Is What it Feels Like on the Other End!

There was an independent adjuster in our town, let's call him Dan, who we dealt with on a fairly regular basis. He was a nice enough guy, but definitely a tough negotiator. There wasn't a single job that we estimated for him where he didn't argue over the need for or the cost of something we were proposing.

One day, Dan called us with a new fire job — his own home! He'd had a pretty bad fire the night before, and he wanted us to do the repairs. Now it's a pretty big compliment when an adjuster trusts your company enough to want you to fix his own house. But I have to admit we were a little leery about taking on the project.

My partner and I went out to see the damage and work on the estimate together, as much for moral support as to share the work. While we were there, the staff adjuster for Dan's insurance company showed up to meet with Dan and discuss his claim. They walked through the house together, with us in tow, discussing the damage. We stayed very quiet on this one, and only spoke when spoken to.

So it was with no little amusement that we listened to the conversation Dan was having with the adjuster — now that the shoe was on the other foot. Dan would point out a damaged area and tell the adjuster what he wanted done. Most of the time the adjuster agreed, but there were several times when he said he felt something should be cleaned instead of replaced, or he'd agree to replace part of something, but not all of it. Turns out this adjuster was every bit the tough negotiator that Dan was!

They finally came to an agreement on the scope of the loss, but I don't think Dan was fully satisfied. After the adjuster left, he looked at us and asked "Is this what I put you guys through?" We just grinned and nodded. As he walked off, he just shook his head. "Man, I can't believe how hard it is to be on the other end!"

I'd certainly never want another adjuster to have to go through a fire in his home like Dan did, but I've often wished there was some way we could pass this lesson on to all the other adjusters we had to deal with!

- **Investigating questionable claims:** If the adjuster has questions about the claim, he can take steps to get those questions answered. That might involve hiring a structural engineer, a cause-and-origin expert, an arson investigator, or other experts.
- **Preparing and/or reviewing the estimates:** The adjuster reviews the contractor's estimate, and decides if it's fair and if it includes all of the necessary items. In some cases, adjusters write their own estimates, either as a way to check and compare costs against the contractor's estimate, or as a way of establishing the final payment value of that particular claim.
- **Negotiating with the homeowner and the contractor:** Most jobs have some financial issues that need to be worked through. The homeowner may think something needs to be replaced that wasn't in the original estimate, or there may be a disagreement on how long something is going to take to repair. All of these things have to be negotiated among the homeowner, the adjuster, and you; conducting those negotiations is the adjuster's job.
- **Preparing and/or reviewing supplemental estimates:** If there's additional damage that couldn't initially be seen, the contractor writes a supplemental estimate for it, which the adjuster will review.
- **Working on contents, outside living expenses and other elements of a claim:** Larger losses often involve a number of

REAL STORIES:

Learn From One of the Best

I recently met with a senior claims representative for one of the major insurance companies. He's a man I've worked closely with throughout my career, and whose opinion I respect. I asked him what he looked for in a contractor, and how he thought a contractor could work at building a good working relationship with adjusters and gaining respect that will last well into the future.

This is what he told me, so pay attention and learn from one of the best.

"Here are the three rules that I make a point of telling all my contractors:

"First, have good communication with your homeowners. Tell them what's going on. Have a system in place so that when they call you, or when I call you, you know the status of their job, or you can find out within a few minutes. Don't leave anyone hanging, and don't put them in the position of calling me to ask what's happening on their job. Homeowners want to be kept in the loop, and I expect that from all my contractors.

"Second, I want accurate estimates. That's a given. Write a good scope. Know your unit costs and know your software. Give me good line-item notes to explain what you're doing.

"And third, I want quality workmanship — no cutting corners. For me, that's always expected from any contractor I deal with."

There you go. Some simple, straightforward advice that's well worth following!

complicated elements, and the adjuster is involved in all of them. Depending on the type of loss, these can include the replacement, cleaning, transport and storage of contents; arrangements for outside living expenses; business relocation; tenant issues; replacement of loss of business or rental income; condominium issues; subrogation claims; and any other issues related to the loss.

- **Acting as an arbitrator:** A good adjuster wants to see a job run smoothly, and he wants the policyholders, contractors, and superiors to all be happy. So, he'll step in to arbitrate minor disputes along the way whenever possible.
- **Finalizing and paying the claim:** When the job is done, the adjuster takes care of the final paperwork. This procedure varies between companies, but it essentially comes down to making sure the job's been done for the agreed-upon price and the policyholder is happy. When that's accomplished, the adjuster authorizes the company to issue payments.

Understanding their Pressures and Motivations

Insurance adjusters come in all shapes and sizes. Some are young and fresh out of college; others are older and perhaps nearing retirement. Some know a lot about construction, which makes it easy for you to communicate with them. Others know very little about construction, which gives you a chance to be helpful. And there are a few — the worst kind — who know very little, but don't want to admit it.

Insurance companies generally suffer from a bad reputation when it comes to settling claims — sometimes deserved, and sometimes not. But the adjusters have to take the brunt of the fallout from their company's reputation. A lot of homeowners enter into a loss situation assuming the insurance company isn't going to deal with them fairly, and by extension, that it's the adjuster's job to be the bad guy and nickel and dime the claim wherever possible.

Quite often, homeowners get what they pay for when it comes to insurance coverage. Homeowners insured by small companies offering low-cost policies may find they don't get very good claim services; while the larger, well-known companies offering comprehensive, but more expensive policies, try hard to deal fairly with both their clients and their contractors. Also, the more honest the client, the better treatment they receive. Stretching the truth with claims only makes everyone suspicious, and the process more difficult.

Adjusters may have some leeway when it comes to settling claims, but all adjusters are at the mercy of the companies they work for. They have to follow the company guidelines. When a company's procedures change, the adjuster has to follow along — whether he likes it or not. Every adjuster has to answer to someone above him in rank, and how well he enforces company policy in handling claims usually determines how far he'll go with the company.

REAL STORIES:

Do You Think This is Covered?

I once went out on a storm damage claim that was a little unusual. The storm had done some damage to the home's roof and siding that was pretty straightforward, and I estimated that first. I then moved on to damage that had occurred to the home's decks, patio, and fencing. "Oh I wouldn't bother with that," the homeowner told me. "The insurance adjuster said none of that was covered, and I can't afford to fix it on my own."

That surprised me, since I had seen this type of storm damage before and these items had always been covered. I had only met this adjuster once before, and I knew that he was very young and very new at his job. I also knew that he worked for an insurance company that had a reputation for not training its adjusters very well. So when I got back to the office, I called him.

"Oh," he said after I'd described what I'd seen. "I saw that, but I didn't think it was covered. Do *you* think it's covered?" I told him that I thought it was, and why. He met me at the job the next day, policy in hand. He looked at the damage, read the policy, looked some more and then read some more. Finally, he looked at me with a sheepish grin. "I guess you're right. According to this, I think it *is* covered!"

I added the additional damage to my estimate, and his company paid for the whole thing. The homeowner was delighted to get the additional repairs done. The adjuster was happy because he avoided making a mistake that one of his superiors would definitely have caught. Our company got the benefit of more repair work. And from that day on, that adjuster always used our company when he could, and he rarely questioned our estimates.

Talk about a winning situation for everyone involved!

Helping Adjusters in the Real World

It's not surprising that most insurance adjusters know a lot about insurance policies. But what might surprise you is that they're not always clear on how to interpret those policies in real-world situations.

For example, the adjuster may clearly understand that a water loss is covered, and a mold loss isn't. But if you have a wet, moldy piece of drywall, how is that handled? Or what happens when you have a house with wet carpeting, which *is* covered, but you have to move the tenant's contents to get to the carpet, and moving those contents *isn't* covered? The better you understand some of these issues, the better you'll be able to assist the adjuster in making these determinations.

There are a lot of gray areas in insurance policies, some of which may even work to your advantage. So don't be afraid to suggest a compromise, or another way of looking at things. If you make the adjuster's job easier, and help them look good to their superiors, they'll definitely remember you.

Different Types of Adjusters

There are several types of insurance adjusters, and it helps to understand the differences among them.

Staff Adjusters

A staff adjuster is someone who's employed by a specific company, such as State Farm or Allstate. He deals only with the policies and policyholders for that particular company. In medium- to large-size cities where there are enough claims to justify the cost, insurance companies will typically maintain an office for their staff adjusters, with one or more adjusters working out of that office.

Very large cities may have more than one office, with several adjusters in each one. In smaller areas, staff adjusters may work from their homes, or they may work out of an office in another city and only visit the other locations in their territory as needed.

Cat Teams

Occasionally, a natural disaster such as a hurricane or a massive ice or rain storm may hit an area. This can result in dozens, hundreds, or even thousands of claims coming in at the same time. When this happens, insurance companies typically mobilize catastrophe teams, or *cat teams* as they're commonly known. Depending on the size of the disaster, a cat team may consist of dozens or even hundreds of adjusters and support personnel brought in from all over the country.

Cat adjusters tend to process claims very quickly. Because of the need to expedite claim settlements, they can be more lenient in some ways, and stricter in others. However, that same need for fast settlement means that you have to write your estimates quickly and usually in accordance with a specific set of guidelines.

Also, remember that a natural disaster affects several insurance companies at once. So, you could have cat teams in your city from six or eight different companies, all with different ways of doing things. If you're in that situation, find out as quickly as you can what they want. Contact one of your local adjusters, and simply ask what the cat team policies are going to be. Determine who'll be in town, what the expected turnaround time is for estimates, and any tips he might have on how best to be helpful.

Independent Adjusters

Independent adjusters work for themselves, rather than for a specific company. They contract with different insurance companies to adjust losses for them within a certain geographical area. Independent adjusters may do work for companies that don't have enough claims in the area to justify a staff adjuster; or, they may work for a large company and pick up some of the overflow that adjusters from other companies in their area aren't able to handle.

Independent adjusters do much the same thing that staff adjusters do, but they sometimes do it a little differently. For one thing, they don't have a boss to answer to. They also may have a little more leeway in how they process claims, or in the timeframe for completion.

On the other hand, the independent adjuster is still working for an insurance company. That company is their client, and without their client, they wouldn't be in business.

Tip! *Don't get ahead of yourself. It may be very tempting to bid on an apartment fire with repairs that will run $600,000, or a major flood in a hotel that could easily top three quarters of a million dollars. But it's also a good way to get your company in big trouble if things don't go well. Stick with smaller losses initially, and don't go after the really big jobs until you're well established and financially stable.*

So conduct yourself with an independent adjuster the same way you would with a company adjuster. Your goal is to form a relationship that benefits both of you.

Large-Loss Adjusters

As the name implies, a large-loss adjuster is one who specializes in losses over a certain dollar amount, such as $500,000 or perhaps $1 million. These adjusters primarily deal with very large residential or commercial fires, floods, and other losses.

Large-loss adjusters can be either company adjusters or independent adjusters. They're typically very experienced and very knowledgeable, and are used to dealing with experienced contractors.

Public Adjusters

Public adjusters work directly for the homeowner, as opposed to working for the insurance company as either a company or an independent adjuster.

A public adjuster usually enters the picture when a policyholder is dissatisfied with the amount the insurance company is willing to settle on their loss. Or they may come in if the settlement of the loss has been delayed. Public adjusters typically work on a percentage of the value of the loss, which is an important point to remember; the higher the loss amount, the larger their fee.

All of the costs associated with a public adjuster are borne by the policyholder. So, to make it worthwhile for the policyholder to hire a public adjuster, the claim must be settled for more than the company insurance adjuster was willing to cover. That can make negotiations very stressful for all sides.

Tip! *When working with a cat team or a public adjuster, be very careful not to give in to greed. Never change your own estimating policies, or up your prices for a quick profit. After the cat team has left town and the public adjuster has moved on to another client, you'll be dealing directly with the insurance companies for your future jobs — don't jeopardize your reputation.*

Long-Distance Claims Processing

There's an unfortunate trend in the insurance industry toward *long-distance claims processing*. This is a cost-cutting measure implemented by the insurance companies that greatly complicates your work by eliminating some on-site visits by adjusters.

Figure 22-2 A claims processor's guidelines may specify a certain amount of time per square foot for removing hardwood flooring. But it may take considerably longer than what those guidelines have allowed, especially if the flooring is glued in place over an OSB subfloor and it comes up one small sliver at a time. Situations like these can sometimes lead to misunderstandings.

Let's say you're called out on an emergency water loss. You do the necessary emergency work, and write an estimate for repairs. In addition to the estimate, you also take a number of photos. You then electronically upload your emergency bill, your estimate, and your photos to the insurance company's central claims office, which is usually in another city, or even another state.

A claims processor reviews everything, and then makes a decision on both your emergency bill and your estimate. This obviously isn't easy to do, since he hasn't seen the job. He must rely solely on your loss photos and his company guidelines regarding the average cost of a loss of this type. See Figure 22-2.

The claims processor will often question or even deny parts of your estimate if they don't match his "typical damage" guidelines. He may refuse to pay for painting a wall if your photos don't clearly show a water stain on it. He may refuse to pay for part of your debris removal bill if the amount of debris you hauled off seems "excessive for this type of loss."

The other unfortunate fact is that these claims are handled through very large claims centers, with lots of processors. You may not deal with the same people often enough to get to know them, and in some instances, you may even have to deal with multiple processors on the same claim. It's hard to learn what your claims processor wants under these circumstances.

Figure 22-3 Certain homes defy placement into standard categories. This stately mansion from the early 1900s suffered a water loss. It had a number of intricate, custom moldings that were damaged. Losses in homes such as this one require careful coordination with an on-site adjuster.

This long-distance method of dealing with claims is here to stay, so you need to adjust your thinking — and your patience level — accordingly. You'll need to learn where you can compromise, and where you can put your foot down and stand firm. See Figure 22-3.

Working Out-of-Town Jobs for Adjusters

Adjusters typically have specific geographical areas that they deal with. It may be limited to one part of a large city, or in more rural locations it may cover hundreds of square miles. Ask the adjuster what area he covers. If your company has the capability of covering the same area that he does, be sure to let him know. If you can do his out-of-town jobs in addition to his local ones, you become that much more valuable to him.

When working out of town, you'll need to decide if the location is close enough to bring your crews back home each day, or if they need to remain at the remote location during the week. This decision is usually based on the distance from your shop to the remote job, and which one makes the most economic sense. Either way, clarify this with the adjuster prior to starting the job.

Round-Trip Driving

If you're driving back and forth each day, you'll need to determine how much time your crew will be on the road. For example, if the job is 90 minutes away, then each crew member will be spending three hours per day driving. If you have three guys working on that job, that's 45 hours per week just in driving time!

When you're working out what to charge an insurance company for out-of-town driving time, there are several items to consider:

1. How long will the job take, and how many people will be commuting?
2. How many overtime hours, if any, will you have to charge? If your crew's on the road three hours per day, that only leaves five hours of work time. So you may want to work 10- or 11-hour days so that the crew has adequate time on-site. That means you'll have to pay overtime. If so, you'll need to adjust your billing rate accordingly.
3. Should you consider a discounted rate for driving time? For example, if your normal shop rate is $55 per manhour, you could bill driving time at $35. That allows you to make a small

REAL STORIES:

Fine, Then *You* Come Down Here and Do It!

We had a good-sized water loss in a master bathroom that extended into the master bedroom and a huge walk-in closet. The closet had a complicated shelf and organizer system that had to be completely disassembled in order to remove the damaged carpeting. It took us four hours to take it apart and mark all the various pieces so we would know how to put it back together again. In our estimate, we included six hours of labor to reassemble it after the new carpet was installed.

I got a call from the claims processor, telling me he would only pay two hours labor to disassemble it, and three hours to put it back, because that was all his guidelines for closet organizers would allow him to pay. We went back and forth on the phone, and the conversation became increasingly heated. He accused me of adding hours that I didn't need to pad the bill "in case I had a problem." I asked him how long he'd been a contractor, and how many closet organizers he'd installed. He wouldn't budge, and neither would I.

I finally hung up and called the homeowner. I explained the problem to her, and told her that unfortunately her master closet was going to remain in pieces in her garage for the time being. I told her that the insurance company refused to pay for the necessary time to reassemble it, but that I'd continue trying to get something worked out with them.

About 30 minutes later, I got a call from the homeowner, telling me everything was okay. She had called the insurance company, got a hold of the processor's supervisor, and very sweetly informed him that my bill and my estimate were justified. She then invited the supervisor to send the claims processor out to her house to put the closet organizer back together himself.

"If he can do it in less time than my contractor," she told the supervisor, "then I'll refund the difference personally, right on the spot. If it takes him longer, I intend to have my contractor bill you additional, since it will be obvious at that point that he *under*estimated, and should be entitled to more money!"

She said the supervisor really didn't have a comeback to that one, and agreed to my bill and estimate as written. I wish I could have hired her as an estimator!

profit on the driving, and save the insurance company some money as well. This will also indicate your willingness to cooperate.

4. Figure your mileage compensation for gas plus wear and tear on the vehicles. Calculate the number of round trip miles in your estimate, and agree on a per-mile reimbursement rate with the insurance company. For simplicity, you can use the current IRS reimbursement rate ($0.51 per mile, at the time of this writing).
5. Find out if your subs are going to charge you additional trip charges. Adjusters typically have no problem with subcontractor trip charges, so long as you can document them in writing. A copy of the sub's bill is usually proof enough, but verify that with the adjuster.

Staying Out of Town

Tip! *Insurance adjusters are usually grateful to have a restoration contractor that they trust take on an out-of-town project. But before you commit to the job, be sure that all of the details for mileage, subcontractor trip charges, and per diem rates are agreed upon, so there are no misunderstandings.*

The other option is to stay at the remote location for the duration of the job, or at least during the work week. This eliminates the driving time, and enables your crew to be on the job more hours per day.

Keeping your crew at the remote job site requires you to pay *per diem*, which literally means *per day* or *by the day*, in addition to wages. Per diem is a payment for daily expenses, to cover lodging, meals, and other expenses that are unique to the out-of-town job. The insurance company makes per diem payments to you to cover the cost of having your crew stay out of town, and you in turn make per diem payments to your employees to reimburse them for their expenses.

There are a few different ways you can handle per diem agreements with your employees:

- You can pay them a specific lump sum per day, and they can use it as they wish for lodging and meals. For example, we'll say that you give each employee $150 per day. One may choose to stay in a nice motel for $100 a day, and use the other $50 for meals. Or, you may have a couple of others who agree to share a $75 motel, eat donuts and sandwiches, and pocket whatever money's left.
- You can use an Expense Reimbursement Form. First set an agreed-upon daily expense limit per employee, such as $150. The employee uses his own cash or credit card, and keeps his receipts for motel charges and meal purchases. He then turns in the receipts and the Expense Reimbursement Form at the end of the job, and gets reimbursed for the amount he spent, up to the allowed limit.

REAL STORIES:

It'll be Rough, but I'll Take That Job!

One of the larger insurance companies we dealt with called us one summer day with a possible out-of-town job. They had a fairly substantial fire loss in a large rental cabin at a popular lake resort about 90 minutes away from us, and they wanted an estimate. My partner met with the adjuster and the owners of the resort, and got the contract for the repairs.

The lake was too far away to effectively shuttle crews back and forth each day, so he discussed a per diem arrangement with the adjuster. There were no motels close by, and almost all of the rental cabins at the lake were booked up for the days that we'd be working there.

So, my partner got creative. He proposed renting a travel trailer and putting it on-site for a couple of the crew members to use, plus having use of one of the smaller rental cabins that hadn't been booked for the period. The adjuster and the resort owners readily agreed. It was an economical solution for the insurance company, and the resort owners got reimbursed for rent they would have lost on the cabin. He negotiated a per diem rate for meals at the resort's café, and the resort owner even tossed in the use of a little aluminum fishing boat for the crew on the weekends!

As we did with all our out-of-town jobs, we asked for volunteers among the crew. We explained that it would be pretty tough duty, staying in a nice travel trailer overlooking the lake, eating at the café or fishing for dinner after work, and cruising around in the boat on Saturday afternoon.

It was the first and only time in our history that the entire crew volunteered for an out-of-town job!

- Another method is for you to arrange payment for the lodging in advance, via your company credit card. You can then make per diem payments to the employees for their meals only, again only up to a prearranged limit.

Set your employee per diem policy first, then work out an agreement with the insurance company. That way you'll have a better idea about what motels in the area cost, and approximately what you'll be spending in food expenses.

Once you know that, your per diem arrangement with the insurance company should be based on so much per employee per day. Remember that your company is going to incur additional expenses with the running of an out-of-town job. You'll also have part of your crew unavailable for new or existing jobs, as well as new emergency calls. So it's perfectly reasonable to charge more per day in per diem rates than you're paying your employees.

Insurance and Code Upgrades

The issue of code upgrades can be a sticky and confusing point in insurance policies. It's one that affects the contractor, the homeowner, and the insurance company alike, *and it's a very important issue that you should be aware of* (Figure 22-4). It's also an area where your relationship with the adjuster can help you with solutions to complicated problems.

Figure 22-4 Code upgrades on older homes can range from simple to highly complicated. This older home had structural problems, caused by excessive snow loading, which required complete replacement of the roof structure. All of the rafters and ceiling joists were undersized by today's standards. The attic insulation and much of the wiring didn't meet code either. These homeowners were fortunate; their policy had code-upgrade coverage built into it.

Suppose you've been hired to repair an older home that had a fairly substantial fire. Let's take a look at how these code issues could come into play:

- The house, built in 1965, was totally in compliance with the building codes and accepted building practices in place at that time.
- It has single-pane windows, R-11 attic insulation, and other components that were fine then, but not allowed in construction today.

Under *today's* code, you'll need to install dual-pane windows, and upgrade the attic insulation to, say, R-38. The kitchen requires additional electrical circuits, as well as GFCI outlets. Perhaps your local code also calls for floor insulation. You won't be able to get the necessary permits without all these upgrades, so you include them in your estimate.

The insurance adjuster reviews your estimate and sees your bid for R-38 insulation in the attic, even though the house had R-11 at the time of the fire. He explains that the homeowner can't profit from the loss, and that the insurance company is only obligated to bring the home back to its *pre-loss condition*. The same holds true for the windows, electrical wiring, and other code-related items included in your estimate.

You explain that you can't get permits if you build to 1965 standards. He understands that perfectly, but he refuses to pay for the upgrades and tells you that it's up to the homeowner to foot the bill for the extras. So you remove those items from your insurance company estimate, and prepare a separate estimate for the homeowner, to cover the difference in cost for each of the code upgrades.

However, the homeowner argues that he has a full-replacement policy, and the insurance company is obligated to repair the home so that he can occupy it again, whatever that requires. He also refuses to pay for the upgrades, and tells you that it's up to the insurance company to pay for them.

This isn't an uncommon situation, especially if you live in an area with a lot of older homes. And to make matters worse, both the adjuster and the homeowner have valid points. But valid or not, it leaves you stuck in the middle, with no easy solution. If you find yourself in this situation, you basically have three options:

1. Explain the situation to both parties, and then step aside. At issue is the contract between the insurance company and the homeowner, and it's up to them to rectify it while you remain a neutral third party.
2. If the job is large enough and profitable enough to warrant it, and the code upgrade is relatively minor — like a few hundred dollars worth of attic insulation — you can offer to pay for the upgrade yourself.
3. You can suggest a compromise, such as the idea that you, the insurance company, and the homeowner all share equally in the cost of the insulation upgrade. Or, you might be able to get your electrician to include the additional circuits in his overall wiring estimate. Or, you might have your window supplier provide a letter saying that the cost of special-ordering single-pane windows would be higher than obtaining in-stock dual-pane windows, in which case the insurance company would actually *save* money. And that savings could be put toward the cost of the other upgrades. (See Figure 22-5.)

You all want the house repaired, so see if you can use your imagination and work out a creative compromise (Figure 22-6).

Insurance and Lead Paint

As a restoration contractor, you'll be working on buildings of every age and description. And when dealing with older buildings, one of the things that you're likely to encounter is lead paint. For decades, lead was a common ingredient in

Figure 22-5 In this older home, a compromise was reached on a few of the code upgrades. With the owner's approval, MDF was substituted for the original clear fir moldings and paneling used in this wainscoting. The money saved was applied to code upgrades needed in other areas.

Figure 22-6 This older-model wood stove had to be removed due to structural damage caused by a fire in the attic above it. It wasn't an EPA-approved stove, so by code it couldn't be reinstalled. In this case, the insurance company and the homeowner worked out a compromise. The homeowner purchased a new wood stove, and the insurance company paid for the new flue pipe, all the fittings, and the installation labor.

> ***Tip!*** *The presence of lead-based paint can't be determined just by looking at it. It needs to be verified through testing of paint-dust samples by a professional testing lab. There are test kits available for doing your own lead paint testing, but due to the inherent conflict of interest in doing your own testing, you should have the testing done by an independent testing facility.*

virtually all types of paint sold for residential and commercial applications. Its use in paint for housing was curtailed by the federal government in 1978. But one estimate puts the number of pre-1978 homes and buildings containing lead-based paint at somewhere around 72 million.

Even if that pre-1978 home has been repainted several times, lurking underneath all those layers of latex there's probably a layer of lead paint. When that lead paint begins to peel and chip, or when you begin to work on it during renovation, it can pose a significant health risk — especially for children. It's also a huge liability for you.

In April of 2010, the Environmental Protection Agency (EPA) issued a new rule that's designed to focus the efforts of consumers and contractors to protect against potential lead-based paint health hazards. Called the Lead Paint Renovation, Repair and Painting Rule (RRP), this new rule affects contractors and subcontractors who work on older homes, so it definitely affects restoration contractors.

Under the RRP rule, contractors working in pre-1978 homes, schools, and day-care centers who disrupt more than six square feet of lead paint are required to become EPA Certified in lead-safe work practices. Realistically, as a restoration contractor, you never know when you'll be called out to work on a pre-1978 building with lead paint, *so you need to be properly certified from the start.*

Adjusters are well aware of these new rules. As with IIRC certification, adjusters will expect your company to have the proper EPA lead-safe certification in place if you intend to work with their clients.

Contractors are required to take a one-day training course, and then their company must send in an application to the EPA to complete the process. The EPA certification is good for five years. Doing work without proper certification exposes you and your company to fines from the EPA, as well as potential civil liability from building owners.

If the age of the house causes you to suspect there might be lead-based paint, talk to your adjuster before proceeding any further. Insurance companies are obligated to cover the cost of lead-paint testing, and if necessary, remediation. This is an important issue to clarify up-front with your adjuster, and you'll need to work closely with him throughout the testing and remediation process.

For more information on certification in the RRP program, check with the EPA, listed in Appendix A.

Insurance and Asbestos

Asbestos was once widely-used in the building trades. Due to increasing health concerns, manufacturers began removing it from many building materials and products in the 1980s. In 1989, the Environmental Protection Agency (EPA) banned asbestos and began the process of phasing it out. That ruling was overturned in 1991, and as a result, some consumer products still legally contain trace amounts of asbestos.

Asbestos can be found in a variety of products in older homes and commercial buildings, including some you might not expect:

- Acoustic (popcorn) ceiling texture
- Many sheet and tile hard-surface flooring products
- Some older types of duct insulation
- Some varieties of plumbing and steam pipe wrapping
- Some roofing and felt paper products
- Many older ceiling tiles
- Some varieties of siding materials, especially cement-fiber siding shingles
- Vermiculite insulation

Asbestos becomes a problem when the fibers are crushed and broken, allowing them to become airborne. When that happens, they can be inhaled into your lungs and can create a significant health hazard.

What happens with many insurance losses in older homes is that either the loss itself or the subsequent repair work can disturb the asbestos materials, putting the fibers into the air. Dealing with asbestos in a loss situation is much the same as dealing with lead paint. The insurance company has an obligation under the policy to pay for the costs associated with testing suspected materials to see if they contain asbestos. If they do, the insurance company will pay for the cost of proper remediation.

Again, you need to work very closely with your adjuster. *As soon as you suspect the presence of asbestos, stop work immediately and consult with your adjuster to get authorization for the proper testing.* If asbestos is present, have a certified asbestos remediation contractor estimate the cost of proper removal and disposal, and discuss those steps and their related costs with the adjuster and the homeowner prior to proceeding.

> ***Tip!*** *Asbestos remediation requires a number of very specific steps, from air sampling to proper disposal of the contaminated material. This is something you'll want to sub out to a licensed and certified asbestos abatement contractor.*

Insurance Companies, Lead Paint, and Asbestos

Insurance companies take the stance that lead paint and asbestos are *not* claims in and of themselves, but *are* covered as part of a larger loss. In other words, homeowners can't put in a claim because there's asbestos in the popcorn ceiling texture in their living room, or because lead paint was used to paint their siding. But if that living room ceiling or that siding gets damaged as part of a covered loss, then their coverage will extend to whatever asbestos or lead-paint remediation is necessary to deal with the problem.

Overlapping Claims

You may sometimes be involved in a situation where there are overlapping claims. This can take a number of different forms, and all involve ongoing communication and careful coordination with your adjuster to avoid any problems or misunderstandings.

Condominiums

Condominium work is an area where you're very likely to become involved in an overlapping policy situation. Condominiums have unique and specific insurance policies. The named perils are typically the same as with a standard homeowner's policy, but how the coverage is handled can be much different.

In most condominium purchase agreements, the homeowner owns everything starting from the interior surface of the exterior wall inward. The condominium association owns everything from the outside of the exterior wall outward. In other words, if you can see it from the inside of the unit, it's probably the condo owner's responsibility. If you can see it from the outside of the unit, it's probably the condo association's responsibility. That can make for some tough situations, especially in emergencies.

Suppose you have a fire in a building of four attached condo units. We'll call the units A, B, C, and D. The fire starts in Unit B. There's heavy structural damage in Unit B, and lighter structural damage in Unit C. In both those units, the fire breaks through the roof and the exterior walls. There's smoke damage to both the inside and outside of Unit A. There's light smoke damage to the inside of Unit D.

Should you be hired to make all the repairs, you could potentially be dealing with *five* different insurance companies on the same job. Each homeowner could have a different insurance company, and the condo association could have the fifth. You would have to write five different estimates in five different formats for the five different insurance companies.

On the other hand, it's also possible that you could be one of five different contractors working on the same job. Or, you could be hired by the condo association to do just the exterior repairs, while the owners of Unit B and Unit C each hire their own contractors to do the interior structural repairs on their units. That means having to coordinate structural repairs with two other contractors at the same time, both of whom are essentially working on different parts of the same building you're working on!

Be very careful when dealing with condos. Talk to the all of the adjusters involved in the claims, as well as all the owners. If necessary, attend a meeting of the condo board. Be absolutely sure that you know exactly who you're working for, and make sure you know and understand where there may be gaps in the coverage responsibilities, or where the coverage may overlap.

Subrogation on Overlapping Claims

In another example, a person buys a brand new single-family home. A week after moving in, a plumbing fitting that was incorrectly installed gives way, flooding the house and the owner's contents. The homeowner has a policy with one insurance company, and the plumber who was the cause of the damage has a policy with another insurance company.

What usually happens is that the homeowner's company will step in as the primary insurer, and will do whatever's necessary to get the homeowner — who's their client — taken care of. At the same time, they'll evaluate the loss, determine the cause, and track down the plumber. The plumber will probably turn in a claim to his insurance company, who will in turn reimburse the first insurance company for the cost of the repairs. In most cases, the plumber's insurance will also reimburse the homeowner for the cost of the deductible that they had to pay.

As mentioned earlier, the process where one insurance company looks to another for reimbursement of expenses is called *subrogation*. Most insurance companies have entire subrogation departments dealing with these situations on a daily basis. For the most part, subrogation losses won't affect you directly. Your contract is with the homeowner, who's being paid by his insurance company.

When you encounter a loss where you think there might be a subrogation claim, work closely with your adjuster. He may want you to save affected parts, such as the plumbing fitting in the previous example, or he may want you to document specific things during your repairs.

Not Every Job is Worth Taking

Building a reputation and relationships with adjusters and insurance companies sometimes means taking on jobs you don't want. It may be a tough homeowner that you'd rather not deal with, or a rundown house that you know you're going to have a hard time repairing correctly. See Figure 22-7. Or perhaps you're swamped at the moment, and fitting in one more project would be difficult. But if taking on that nasty homeowner or repairing that ugly house will help the adjuster out of a bind, you may just score some much-needed points.

However, there are times when you simply have to say no. And, once you accept the fact that you really don't need to take every job that comes through the door, your life will get a whole lot simpler.

Figure 22-7 Occasionally, you have to take the bad with the good; you swallow hard and just work with the adjuster to get the job done. In this case, an old chimney needed to be cleaned up and repaired correctly. It was a difficult job, but it got the adjuster out of a bind, and helped build the restoration contractor's reputation. Restoration work is often a careful balancing act. Sometimes you should do these jobs, and sometimes you should just walk away.

Here are a few situations worth remembering:

- **Bad insurance companies:** Not all insurance companies are worth working for. Some of them are downright cheap, and you'll find yourself spending more time than it's worth arguing over nickels and dimes. Or you'll be forced to compromise your quality to save their company a couple of bucks. When you run into one of those companies, you're better off turning down their work offers and concentrating on companies that are easier and more profitable to work with.
- **Bad insurance adjusters:** Like insurance companies, there are good and bad insurance adjusters. You might encounter one who's on a massive ego trip, or who's in a perpetual bad mood. He seems to take genuine delight in arguing with you, or in finding ways to cut your estimate. Adjusters who want to beat you up before the project even gets started, won't be there for you if there are problems later on. If possible, it's better to just not deal with them.

- **Cat team conflicts:** During natural disasters, you may have multiple cat teams in the area, which can offer a good opportunity to strengthen your relationships with your regular companies, and perhaps forge a new relationship or two. But remember that these guys work fast, so guard against stretching yourself too thin!
- **You have too much work:** Natural catastrophes aren't the only times when you'll have too much work. You could be stretched too thin for any number of reasons, and if you can't take on a new job and be able to devote the full resources of your company to it, turn it down. An adjuster would prefer that you're honest rather than take on something new that you can't properly see through to the end.
- **Too far away:** There are some advantages to taking on out-of-town jobs, but sometimes they can put too much of a strain on your company. If that's the case, decline the work rather than get yourself into a bad spot.
- **Too expensive:** You'll usually be able to get a deposit on a job from the homeowner, or even an upfront payment from the insurance company, to cover initial expenses. But occasionally you'll encounter a job that will require a sizeable outlay of money to get things started. If you consider the job a financial risk because you'll be tying up too much of your ready cash, or you'll have to wait too long for payment, then turn it down.
- **Trust your gut:** It's sad to say, but there are some real jerks out there, and you're sure to encounter your share of them, whether they're clients or adjusters. They may be people with unrealistic timeframes. They may have expectations that you know you'll never be able to meet. They may be abusive to you or to your crew. You may consider them a financial risk. Whatever the reason, when your gut instincts tell you that this is someone you'd rather not deal with, pay attention!

Some Thoughts on Working with Adjusters

And finally, here are a few thoughts based on my experiences and interviews with adjusters:

- Most adjusters view restoration contractors and their crews as extensions of the insurance company. As such, they have certain expectations. They want to see presentable, clean-cut crew members, in uniform, who know what they're doing. They don't want to walk onto a jobsite and see a bunch of people standing around smoking, hear loud music blaring, or observe any activity that might make their clients uncomfortable.

- Adjusters expect you to treat every client the same, from the owner of a small manufactured home to the owner of a million dollar mansion. Whether their client is the nicest little old lady or the biggest jerk in town, they should all get equal treatment.
- Always write a fair estimate, charge a fair price, don't nickel and dime, and follow through with your commitments. Do that consistently, and you'll quickly earn a reputation as an honest contractor that adjusters can count on.

An adjuster once told me, "Remember that restoration work is a series of peaks and valleys. We all go through them. My best advice is to always set a goal of ending on a peak."

23

Structural Estimating

In the beginning of the book, I made the analogy that a restoration company was like a three-legged stool: The three legs being the Emergency Division, the Contents/Cleaning Division and the Structural Division. The two supporting legs, the Emergency Division, and the Contents/Cleaning Division, are both handled almost entirely on a time-and-materials basis, so as you keep the work coming in, those divisions should stay profitable.

The third leg, your Structural Division, can sometimes quite honestly feel like it needs a bit of glue to keep it from falling off. Structural jobs take a lot of work, a lot of manpower, and a lot of materials. They chew up the greatest cash flow, and often seem to create the most headaches.

Unlike projects in the other two divisions, structural jobs require an estimate *before* the project gets underway. With emergency or cleaning and contents jobs, the bills are usually itemized on a unit cost or time-and-materials basis either as the job proceeds or at the end. When doing structural repairs, you have to prepare an estimate for the scope of the structural work and have it accepted before you get the job. How well you're able to write those estimates is directly related to how profitable your structural jobs will be.

Structural estimating for insurance restoration is part science and part art form. In order to do the best job for your client, satisfy the insurance company, and also make a profit for yourself, you have to walk a fine line. You have to know what to take apart and what to leave alone; understand all the little tricks of unit costing, so they work to your advantage; and understand how to write estimates within insurance industry guidelines without sacrificing your company's profitability. See Figure 23-1.

Figure 23-1 A structural estimate, such as one for this major fire loss, requires time, patience, organization, and attention to detail. You need to methodically walk through each step, in each room. To provide an accurate and detailed estimate, you also have to have a clear understanding of how your estimating program is structured, and what your unit costs include.

There are a few specific reasons why insurance restoration estimating is different from other types of construction estimating:

1. ***Urgency:*** Your clients usually aren't excited about the estimate the way they would be with a kitchen remodel they've been dreaming of. They've suffered a loss and they're inconvenienced. They're often out of their house and they want back in. Even though the insurance company is paying outside living expenses, the clock's ticking for them. They feel pressured, so you're pressured to estimate quickly, no matter what.

2. ***Unit costs:*** Most contractors estimate by applying a certain number of hours or days to a job. But with insurance restoration, everything is estimated in unit costs — something you may not yet be familiar with. These are highly-detailed estimates, and probably many pages longer than what you're used to providing.

3. ***The insurance industry:*** When you're estimating remodeling or new home construction, your only worries are pleasing your client and beating your competition's price, and making sure you're going to end up with a profit on the job. In insurance restoration work, you also have an adjuster to worry about — and the adjuster's boss, and the adjuster's boss's superior, and on up the corporate ladder. These are people who'll tell you to charge $40 for a particular trade instead of $45; or that you can only charge 2½ hours to assemble the cabinet that took you 3 hours; or that you can't charge for driving to an out-of-town job unless it's two zip

codes away, no matter how many miles that is; and if it's Tuesday and the moon is full, you work for free. In other words, be prepared for far more oversight of your estimates than you've ever had to deal with before.

Structural restoration estimating is a broad and, in many ways, complicated subject. In this chapter and the next, we'll take a look at some of the basics, including books, forms, classes, and computer programs available to help start you off in the right direction.

Changes in Insurance Restoration Estimating

It would be wonderful if insurance restoration estimating could be as simple as turning in a bid that says *Fix house*, but it's not — far from it. Insurance estimates are some of the most detailed that you'll ever prepare.

Oddly enough, there was actually a time when a *Fix house* bid — or at least something close to it — might have been enough to land a restoration job. Before insurance restoration became the exacting field it is today, estimating was a much less detailed undertaking. But three things have had a substantial impact on the field of estimating: Auto repairs, computers, and changes in the insurance restoration field.

REAL STORIES:

If Only Estimating Could be That Easy!

My partner was once called out to do an estimate for a kitchen fire. When he showed up at the house to inspect the damage, the homeowner explained that she'd already had another contractor out to do an estimate, but the insurance adjuster said he wouldn't accept the first contractor's estimate.

Curious, my partner asked if she knew why the adjuster wouldn't accept the first estimate. She said she wasn't exactly sure, and she gave him the bid to look at. It was a single sheet of paper with the contractor's name at the top. Scrawled across the bottom of the paper was:

"Fix house, $10,000."

The Exact Science of Estimating Claims

If there's one thing that the insurance industry loves, it's *exact* numbers. They have entire offices whose only job is to try and transform everything that happens to people and their possessions into numbers.

When it comes to insurance repairs, you really see this in black and white with automobile repair claims following an accident. Let's say you have a 2009 Ford Taurus with a dented right-front quarter panel. There are books galore that tell *exactly* which parts are needed, and *exactly* how long it takes to make the repairs — and from this information, the insurance company knows *exactly* how much they'll pay for repairs.

The insurance industry absolutely loves that, because it provides proof that *any* body shop should be able to make those repairs within that time frame, for the same cost. Over and over, repairing the damage to any 2009 Ford Taurus with a dented right-front quarter panel should take the same amount of time and use the same parts. If the books say it takes 6 hours, that's what they'll pay for. If it takes your body shop 8 hours, that's too bad.

Figure 23-2 While every kitchen fire will have some similarities, they have more differences. Each one is unique, and has to be approached that way. For example, this kitchen has damage to an unusual rough-sawn, tongue-and-groove wood ceiling, as well as damage from smoke that penetrated into the attic around the vent pipe.

Somewhere along the way, they decided to try and apply the same principles to repairing a damaged house. The reasoning was that most houses have similar construction techniques and similar materials. Therefore, a washer hose that breaks in a laundry room, or a fire on someone's stove, should do roughly the same amount of damage to just about any house.

Using that logic, they've attempted to reduce many parts of residential repair into sets of uniform numbers. Even in the face of overwhelming evidence that there are simply too many variables in home construction and the ways damage can affect homes, they continue to try and get most things to fit into a neat little package. You'll never convince them that a fire in the kitchen of a home that was built in 1921, and has been remodeled five times, is different from a fire in the kitchen of a tract house built last year. They still want both houses to be able to be repaired like the 2009 Ford Taurus. See Figure 23-2.

The result of this exact science of repairs in the insurance industry is the creation of centralized claims processing centers and claims processors, to take the place of local insurance adjusting offices with local adjusters. This trend has been growing steadily for many years now, and shows no sign of changing.

Standardized Pricing and Computers

Of course, changes in how claims are processed have also been greatly impacted by computers and the Internet. They have revolutionized insurance restoration estimating. It's now possible to create, distribute, and continually update unit-cost price lists for adjusters, claims processing offices, contractors, subcontractors, and material suppliers.

This fits in nicely with the industry's ongoing attempt to standardize pricing and centralize claim processing. If an insurance company or a restoration pricing specialist can establish that nine different painting contractors in your city charge $0.50 a square foot to paint, they're in essence standardizing the rate for that region. So while your company can charge $0.60, or whatever you want, it's unlikely that any insurance company you work for will pay over the standard $0.50 rate.

Pricing Specialists

To keep up with the demand for current, accurate price lists, software companies, publishers, and even some insurance companies, employ or subcontract with pricing specialists. It's the job of the pricing specialist to continually examine the prices of a wide range of materials, from lumber and drywall to roofing and concrete.

They also look at labor prices for all the trades, from laborers to journeymen. In addition, they stay on top of worker's compensation laws, local and federal worker taxes, fuel costs, and anything else that affects the cost of rebuilding a damaged home.

Pricing specialists are also responsible for making regional adjustments. Cedar may cost less in Washington and Oregon than it does in Iowa. Southern pine lumber may be more cost effective to use in the South than Douglas fir. A drywall installer may get a higher wage in New York than he does in Oklahoma. For those reasons, the pricing specialists must keep up on costs all over the United States, and then modify those costs for each specific region to reflect local pricing.

> ***Tip!*** *As a recognized restoration contractor, you'll periodically be asked to participate in pricing surveys. This is your opportunity to give input on costs and fair price structures. So if you're asked to participate, be sure that you do.*

From a Sideline to a Specialty Trade

Insurance restoration estimating has also been affected by changes in the industry in general. What was once a sideline for many contractors has now become its own widely-recognized and highly-specialized construction trade.

Much of this transformation is the result of technological advances in restoration equipment, chemicals, and techniques that require specialized training. Gone are the days when the only option for a charred piece of wood or wet piece of drywall was to tear it out and replace it. Drying techniques, smoke deodorization, mold remediation, content restoration, and other areas of the restoration field have improved dramatically, and that's really good for everyone.

Unit Costing

The insurance industry really doesn't like time-and-materials estimates used for structural estimating. They feel time-and-materials is too vague and open to interpretation.

Let's say you told an adjuster, "It'll take 4 hours to fix this wall, and 3 hours to fix that floor." If the adjuster agrees to that, he's basically given you a blank check for 7 hours of labor. He really doesn't know what he's getting for that 7 hours of work. You could later say, "Yeah, but when I told you I'd fix the wall and the floor, I didn't know I'd have to screw down the subfloor and also fix those nail pops. None of that was included in the 7 hours, and now I need 5 hours more."

The same problems can happen with materials. You might tell the adjuster, "I need two doors and 70 feet of casing for this job." You give him a material figure of $270 in your estimate. Then later, you send him a bill for $320 in materials. When he asks why, you tell him you found that the door openings were irregular and $270 wasn't enough to cover the special-order doors, as well as extra shims, nails, and putty it took to install everything.

As a result of these types of occurrences, the insurance industry relies almost exclusively on what are known as *unit costs* for estimating. Unit costs factor everything into the mix, and take as much of the guesswork out of estimates as possible. When labor or materials or both are calculated into the unit cost for a particular operation, both the adjuster and the contractor are dealing with the same information and know what's included.

What Is a Unit Cost?

Understanding unit costs is actually pretty straightforward. In fact, unit costs are something you already use all the time.

A *unit* is whatever the construction industry standard is for measuring something. For example, the standard unit for drywall or painting is the square foot. For carpet or vinyl floor covering, the unit is the square yard. For concrete, it's the cubic yard. For items like baseboards and casings, it's the linear foot. Materials ordered and sold individually, like doors and windows, have *each* as the standard unit of measurement.

Once you know the item and its unit measurement, then you can find the cost associated with that unit — hence the term *unit cost*. The unit cost includes everything associated with performing the operation involved with that item, whatever that operation may be.

Finding Unit Costs

When your estimate says you're going to paint a wall, you use a unit cost to describe that operation. That way, both you and the adjuster will understand exactly what's being done. That's the theory, but in order for the theory to work in the real world, both you and the adjustor have to understand *exactly* what's included in that per-square-foot unit-cost price. You both need to agree on where that unit cost comes from, and who selects it.

As the estimator, you're the one with the initial responsibility of establishing the unit cost. You can do that through the long and tedious process of figuring it out for yourself. Or, much more likely, you can do it by purchasing an estimating book or software program that's specifically written for the insurance restoration contractor. These books and programs do the work for you. They contain thousands upon thousands of unit costs for all sorts of construction operations. They're updated throughout the year, so they remain current and accurate.

We'll talk more about selecting the right estimating book or software program for your work in the next chapter.

Know What the Unit Cost Includes

Let's stick with the wall painting example for the moment. Suppose your software program shows that the unit cost for wall painting is $0.45 a square foot, including labor and materials. You have a wall that's 8 feet high and 10 feet long, or 80 square feet. In your estimate, you'll simply need to state: *Paint wall, 80 square feet @ $0.45, $36.* Both you and the adjuster will understand what that means, based on your agreement of what's included in that unit cost for wall painting.

In your estimating program, you'll look up *Painting,* then the particular type of painting you want to do. How it's described will vary with the program, but it'll probably be something like: *Painting, 1 Coat.* If you look closely at the description under that heading, you'll find more information about what's included in the *Painting, 1 Coat* operation.

Here are some of things that may have been included under that heading in order to arrive at the $0.45 per square foot unit cost:

- ➤ **Labor:** This will be based on a well-researched labor rate for painters.
- ➤ **Materials:** This will include the current rate for paint, typically of a good, generally-accepted quality. Sometimes there may be categories to choose from, such as *average quality*, *good quality*, or *premium quality*.
- ➤ **Preparation time:** If any preparation work is included, such as puttying minor holes in the wall prior to painting, this will be noted.
- ➤ **Masking:** As with preparation work, if masking and tarping surfaces are included, this will be noted. With painting unit costs, what's typically included is normal masking, such as laying out tarps, removing switch plates, etc.
- ➤ **Equipment:** Some estimating programs make an allowance for tools and equipment that wear out normally in the course of a job. In the case of painting, that might be roller covers. Other programs don't allow for wear in the unit price, and leave that as an overhead item.

- **Labor burden:** As a contractor with employees, you're aware of the fact that if you're paying an employee $18 an hour, when you add in Social Security, worker's compensation, other taxes, and perhaps health insurance, your actual cost for that hourly wage is well above $18. The difference is known as *labor burden*, and that figure is reflected somewhere in the estimating program's labor numbers. That's done so that the unit cost is adjusted to more accurately reflect the actual labor amount being paid by the employer.
- **State and regional labor adjustments:** Painting costs vary in different regions, so estimating programs can't have blanket rates covering all areas. Painting a wall in California, where labor rates are high, doesn't cost the same as painting a wall in Iowa or Arkansas, where labor rates are lower. Rates even differ in various parts of the same state. So the programs use state and regional adjustments, or modifiers, to bring prices in line with local costs.

  They start with an average unit cost number, then examine labor rate trends in different states and regions, and adjust the rate up or down accordingly. If you're using an estimating book, the modifiers will typically be in a table, and you'll have to manually make the percentage adjustment somewhere in your estimate. If you're using a software program, you can usually enter your city or zip code information, and the program will automatically make the adjustment for you.

Understand Each Unit Cost

Suppose you're estimating an interior paint job on an entire house, and the estimating program you and the adjuster agree on calls for a unit cost of $0.42 per square foot. That seems a little low to you, based on past experience. But then you notice that there's a *prep wall* category for $0.06 per square foot. You figure you can use that to make up the difference. After all, you only agreed to paint the rooms, and nothing was said about prepping them first.

So you complete the job, and you bill the adjuster for the agreed-upon $0.42 per square foot, plus a supplemental charge for $0.06 per square foot for prep. The adjuster refuses to pay the additional charges, because they're already included. "Check the program," he tells you, and he's right. So where did you make your mistake? Actually, you made two of them:

1. *Never* bill an adjuster for a supplemental charge without first getting his approval. That will do nothing but irritate him. It'll also set him up to deny the supplemental, then you'll have a real fight on your hands trying to collect the money from your clients instead.
2. And, perhaps most important, you didn't look at the unit cost operation's description in the estimating program *before* completing your estimate. If you had, you'd probably have seen a

notation along the lines of *"includes spackling small holes, filling minor cracking, and other normal preparation work,"* or similar wording.

If you'd seen that in advance, you could have checked the walls of the house to see if they required more prep work than the $0.42 unit cost would cover. If so, that would have been the time to estimate additional labor.

Five Basic Estimating Operations

There are five basic work operations commonly covered in insurance restoration estimating. On any given part of a job, you may be estimating one, all five, or a combination of two or more. They are:

1. Remove and Replace
2. Remove and Reinstall
3. Reinstall
4. Replace
5. Remove

To explain the difference in these five operations, let's look at a simple example.

Assume that you've been called out on a water loss at Mr. Jacobs' house. The water has affected his living room, which has carpeting with padding. Here are five different estimating situations that you might face.

Remove and Replace

Remove and replace means that the material has been damaged and isn't salvageable, but it's still in place. You're going to tear it out, discard it, and replace it with new material. Often abbreviated *R&R.*

In Mr. Jacobs' living-room water loss, both the carpet and the pad were ruined by the water. So your estimate is to *remove and replace the carpet and pad.* You're going to tear out both the carpet and the pad, then install new pad and new carpet.

Remove and Reinstall

Remove and reinstall means that you're taking something out, but you're going to save it and then reinstall the same item later, which is pretty common in insurance restoration. Often abbreviated *R&I.*

In Mr. Jacobs' living-room water loss, we'll now say that neither the pad nor the carpet was too wet, but the water caused a couple of squeaks to develop in the subfloor — it needs to be screwed down. In order to do that, you'll have to roll the carpet and pad back out of the way to make the repair, but then you're going to reinstall them. So your estimate would be to *remove and reinstall the carpet and pad.*

Reinstall

Reinstall means that you're estimating to reinstall material that's salvageable, but that's already been removed. In other words, you're estimating the labor for the reinstallation, but *not* the labor for the removal. This differs from *remove and reinstall* in that the remove labor has already been done. This is an example of what you might face when coming onto a job after an emergency response has already taken place.

Let's go back to our living-room water loss example. During the emergency response, we'll say that the carpet and pad were rolled back so that they could be floated and dried. In your estimate, you would simply need to list the necessary labor to *reinstall existing pad* and *reinstall existing carpet.*

Replace

Replace means you're estimating new materials and labor to replace those that have already been removed. Quite often, materials get removed as part of emergency operations, so they're obviously already gone when it comes time for you to do your estimate.

For our water loss example, we'll assume that neither the carpet nor the pad were salvageable. Both were removed and discarded during the emergency phase of the job, so now your estimate is to *install new pad and new carpet.*

Remove

Remove means that you're only removing the materials. Maybe you were only hired to do the demolition work, or just the drying. Or you could be removing materials, but the owner intends to put in something different on his own after he's settled with the insurance company.

In our water loss example, we'll say that Mr. Jacobs intends to do some remodeling work on his own. He only wants you to come in and give him an estimate to take out the old floor covering. So your estimate would just be to *remove the old carpet and pad.*

Unit Costs for Different Operations

Each operation you'll be estimating may have a different unit cost. *Be sure that you're charging the correct amount for the operation you're performing. It's very important that you understand exactly what the unit cost operation is.*

Unit Cost to Remove and Replace (R&R)

Let's assume that you have a loss that includes tearing out and replacing some ½-inch drywall. In some estimating books or computer programs, you may find a listing that reads something like this: *Drywall, remove and replace, ½-inch, taped and textured, ready for paint.* Or you may find two separate listings, such as: *Drywall, remove, ½-inch,* and *Drywall, ½-inch, taped and textured, ready for paint.* In this last case, you'll have to add the unit costs to remove and then replace together.

Either way, you'll be performing an entire set of operations. The unit costs cover all the materials and labor for the removal of the old material and the installation of the new material.

- The unit cost would be per square foot of area to be drywalled
- Labor to remove the old drywall, pull the old fasteners and prep the framing
- Labor to install the new drywall, tape and finish it, and texture it
- All the drywall materials, including the drywall, screws, tape, mud, and texture
- Normal delivery and stacking of all materials
- Normal wear and tear on installation and finishing tools
- Labor burden
- Normal cleanup of the work area

The cost of hauling the drywall debris away and disposing of it at a disposal site is usually *not* included in the unit cost.

Unit Cost for Labor and Materials

This unit cost includes both the materials and the installation labor for the drywall, but it assumes that the installation site is already prepared. It's similar to the first category, but doesn't include the additional labor for removal and wall preparation.

Unit Cost to Remove and Reinstall (R&I)

This unit cost covers the removal and reinstallation of materials that are salvageable, such as doors, cabinets, trim, carpet, appliances, and fixtures. These are all examples of materials that are commonly removed, salvaged, and then reinstalled. Within an R&I unit cost, you'll typically find:

- Labor
- Labor burden
- Incidental materials such as fasteners, caulk, adhesive, etc.

Unit Cost to Reinstall Only

When something has already been removed and needs to be reinstalled, this unit cost covers the labor for that operation, and includes:

- Labor
- Labor burden
- Incidental materials

Unit Cost for Labor Only

This unit cost covers *only* labor. You need to know what type of labor is being done, and select the proper category. If the labor is for the installation of a material, this category *does not* include the cost of the material itself. It only includes:

- Labor
- Labor burden

Unit Cost for Material Only

In this unit cost, the cost of the material is included, but not the labor to install it. For the drywall example, it would include:

- Drywall, screws, tape, mud, and texture material
- Normal delivery and stocking

Unit Cost for Remove Labor Only

This unit cost is only for the labor associated with removing a particular material. It doesn't include the material, or any reinstallation labor. For the drywall example, this category might actually be a little more specific than just a labor category. You could select a category such as *Remove drywall and bag for disposal*. In this case, the unit cost would typically include:

- Labor to remove the drywall and fasteners, and prep the framing
- Labor to cut or break the drywall into manageable lengths to fit into trash bags
- Labor burden

Some unit costs will include the trash bags. Others will either prompt you to add a category for that, or they'll include them as overhead. Again, it doesn't include the actual cost of debris removal to an offsite location.

Unit Cost Add-Ons

It would be nearly impossible for estimating books to list every possible unit cost scenario. If they did, the books would be thousands of pages long. Instead, they use what are known as *unit cost add-ons*. These are simply numbers that you can add to a base unit cost to adjust for different situations.

Suppose you have a job with vinyl windows. Some of them are standard windows, but some have Low-E glass, others have obscure glass, and some have grids. Your estimating book would need list after list of windows, with all the sizes, and prices for each of those different combinations of options for each size.

> ***Tip!*** Add-ons are a very important part of estimating. *Be sure you understand what they are, and where to find them in your estimating book or estimating software program.*

Instead, what you usually find is a list of standard windows, in all the available sizes and configurations. After that there'll be a short list of add-ons. It will say something like: *for Low-E glass, add 9 percent per square foot; for obscure glass, add $3.50 per square foot;* etc.

Other add-ons might include extra costs for hanging drywall over a certain wall or ceiling height, or for installing roofing on roofs over a certain pitch. There might be add-ons for OSHA-required safety gear, or working in cramped or confined areas.

Unit Cost Minimums

Suppose you're estimating a loss where you need to replace a lot of flooring, and you need to paint one wall, but that's the only wall that needs painting on the entire job. The wall is 8 x 10 feet, and your unit cost for painting is $0.45. So painting that wall will net only $36. That's obviously not enough money to get your painter involved, and hardly enough even to get one of your own crew out there.

So what you need to consider is the concept of *minimum charges*. Minimum charges need to come into play when what you can charge for a particular operation, based on unit costs, is too low to be profitable.

Each trade typically has a minimum charge. It might be a flat rate, such as $85. Or it might be the equivalent of two or three hours of labor. Check your estimating book or check with the adjustor you're working with on that particular job. See what minimums are being used in your book or by that company. Also, doublecheck with your painter, electrician, plumber or other sub, to be sure that the minimum is sufficient.

Find and Stack the Categories

One of the most important concepts for you to know and understand as an insurance restoration estimator can be summed up in five simple words:

Find and stack the categories.

It doesn't matter whether you're using an estimating book or a software program. Either one will be full of categories, subcategories, and add-ons to those categories. You need to locate *all* of them for every part of the job you're working on. They all contain unit costs, and you need to understand what's included in those unit costs. If they don't include *everything* that you plan to do on your job, then you need to find other unit costs that include those missing items and stack them on. How well you're able to do this will dictate your success and your profitability as an estimator.

Figure 23-3 The door on this commercial building was accidentally struck by a delivery truck. At first, it seemed like a straightforward door replacement. But closer examination revealed damage to the wall framing as well. That led to siding replacement, which in turn led to extensive electrical and telephone work. All of these steps had to be identified, worked through, and included in the estimate.

There are two important steps to finding and stacking the categories:

1. *Work through every step of the job in your mind as you estimate it on the jobsite.*
2. *Find every category and every unit cost that fits those steps.*

Work Through Every Step at the Jobsite

When you walk onto a jobsite to do an estimate, here's what you need to be thinking about:

- How am I going to make repairs and put this building back together?
- What are the steps, one by one, that I want the crew to follow here?
- What categories will I need to use to get *all* those steps into my estimate? (Figure 23-3)

It sounds tedious, but it quickly becomes second nature. Let's work through an example of how the process comes together.

Mrs. Banks had a small water loss at her home. The crew has already done the necessary emergency work. Now the estimator needs to prepare an estimate for repairing the damage.

The first thing the estimator does, either at the office or at the site, is review all of the notes made by the emergency crew. This gets him mentally prepared for the job — he knows what happened, and what areas are affected. He knows what had to be torn out, and what he'll be looking at. He's an experienced estimator; these notes will help him get organized. But he knows *not to jump to any conclusions* about what he's going to see or what he's going to estimate.

His first step is to take accurate measurements. Many estimators use laser distance measuring devices like the one shown in Figures 23-4 and 23-5. These devices make it easy to take accurate measurements on your own. As he goes, he prepares a scale sketch of the room on graph paper, and includes the room dimensions and notes on important items that he needs for his estimate (see Figure 23-6).

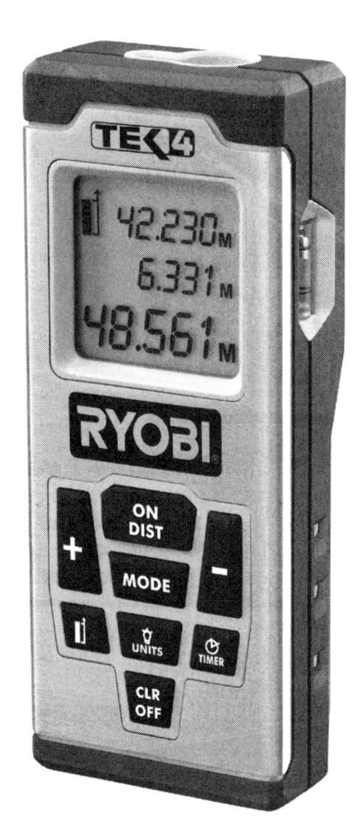

(Photo courtesy of Ryobi)

Figure 23-4 Laser distance measuring devices, such as this one from Ryobi, can be a big help on estimating jobs. It's accurate to 1/16 of an inch, with a distance range of 195 feet.

(Photo courtesy of Ryobi)

Figure 23-5 As an estimator, you'll have a number of large and often awkward situations to measure, usually working alone. While this countertop isn't really one of them, it's a good illustration of how a laser distance measuring device can be used in situations where a conventional tape measure might be difficult to handle on your own.

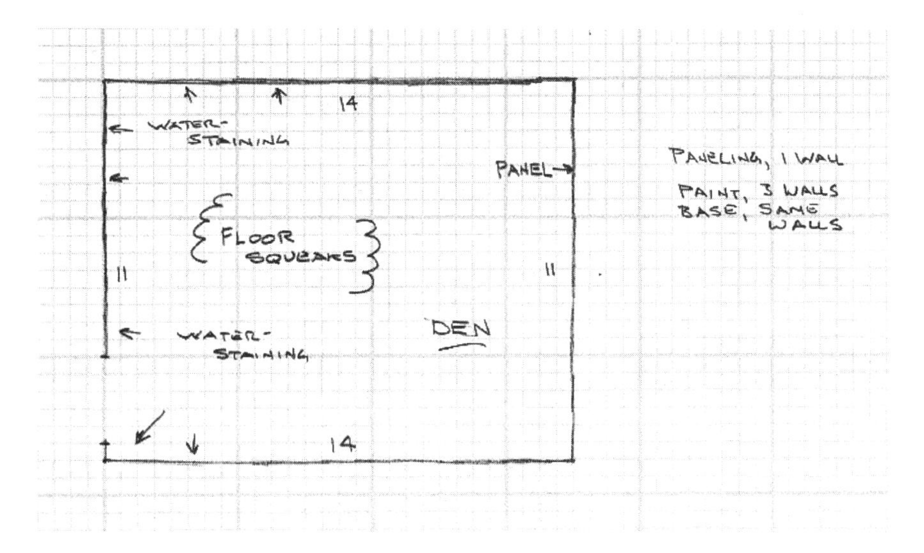

Figure 23-6 An example of an estimator's scaled sketch on graph paper. The sketch is dimensioned, and includes important notes for future reference when writing up the estimate.

He then begins a methodical inspection of the area. As an estimator, you can do this from floor to ceiling, ceiling to floor, around the room, or whatever manner best suits you. The important thing is to be consistent. For this example, we'll say our estimator works around the room, bottom to top, except for floor covering, which he leaves for last.

Here's what he observes about the general condition of the room as he looks around:

- There are water stains on three walls. The fourth wall is paneled, but it's undamaged.
- The paint-grade base has been removed on three walls, but isn't salvageable. The base on the fourth wall is still there.
- The ceiling is painted a different color, and is undamaged.
- The door was removed, but is undamaged and can be reinstalled.
- The carpet and pad were torn out during the emergency, and are not salvageable. About half of the tack strip was removed.
- There's a squeak in the floor. The homeowner reports that the squeak wasn't there before the water loss.
- All of the room's contents were moved out during the emergency, and will have to be moved back.

Now, let's look at how our estimator's jobsite field notes for this loss might look:

1. Seal 3 walls, allow for 50 percent of wall area
2. Prep 3 walls, allow additional 50 percent over normal cost due to indentations from base removal
3. Paint 3 walls, 1 coat
4. Mask 1 wall complete, paneled
5. Additional mask labor, 0.25 hour, due to different color on ceiling
6. Replace 2¼-inch sanitary base, paint-grade wood, 3 walls
7. Seal and paint base, 3 walls
8. Reinstall door, no damage
9. Resecure subfloor, allow for approximately 25 percent of subfloor, as per drawing (no guarantee)
10. Replace tack strip, approximately 50 percent
11. Replace pad
12. Replace carpet, 12 x 14, allowance $18 per square yard

Tip! *Keep in mind that many of these notes would be abbreviated. But for clarity in this example, everything is spelled out. Also, the notes probably wouldn't be numbered. They're numbered here to correspond to the numbered items in next section, which is an explanation of how these items are estimated.*

13. Reset contents, small room
14. Post-construction cleaning, room and contents, 1 hour
15. Debris removal, ½ truck load

Find the Categories to Cover the Steps

After you make your list of repairs, you now need to be sure that you can find all the necessary labor and material unit costs to cover all the steps for those repairs. Let's look at those repairs for our example.

1. ***Seal 3 walls, allow for 50 percent of wall area:*** You can't simply paint the walls and charge a unit cost for that, without sealing them first. If you did, the water stains would show through. But if you sealed the water stains before painting without putting an extra unit cost in your estimate for it, the cost of doing that work wouldn't be covered under the painting category. So you'd lose money.

   The problem is that insurance companies typically won't pay to seal all the walls. They'll only pay you to take care of the water stains at the bottom. You can spot-seal the stains, or you can bid to seal a percentage of the walls. It basically depends on the number of stains, and the amount of area that's affected. In this case, it was easier to estimate to seal 50 percent of the three affected walls.

2. ***Prep 3 walls, allow additional 50 percent over normal due to indentations from base removal:*** This is an area where you, as an estimator, need to understand your unit cost categories and what they include. Normal prep includes spackling minor holes, like those left by picture hangers, and that's usually included in the painting unit cost. If it isn't, additional labor needs to be added to the painting cost.

   In this case, the estimator noted that during the removal of the baseboards, some indentations were left in the walls. That's not unusual in a water loss situation, where tools easily leave marks in the soft drywall. This type of repair is *not* included in the normal prep. However, the estimator caught it and included an additional cost.

   This damage just occurred in a couple of areas. It's easier to use a per-square-foot unit cost for prep, but this damage didn't occur everywhere along the walls. This is something that the adjustor would catch and argue with, so the estimator only bid repairs for 50 percent of the walls. He could have used a straight labor figure instead. This is a judgment call on the part of the estimator.

3. ***Paint 3 walls, 1 coat:*** When the water stains are sealed and the prep work done, you need to paint the walls. Only one coat of paint is estimated for the three affected walls.

4. ***Mask 1 wall complete, paneled:*** Normal masking is usually included in the unit cost for painting, but that's something you need to check. Even if it *is* included, remember that one wall is paneled. That one will need to be masked off completely for protection while the other three are painted.

5. ***Additional mask labor, 0.25 hour, due to different color on ceiling:*** Because the ceiling is a different color, it needs masking around the edges for protection while the three walls are being painted. This wouldn't need to be done if it was the same color as the walls. You might get a little flack from an adjuster on this, but it's a justifiable expense as long as it's not excessive. Unit costs for labor are by the hour, so the 0.25 refers to one-quarter hour.

6. ***Replace 2¼-inch sanitary base, paint-grade wood, 3 walls:*** This is pretty straightforward. The base is already out, so you'll be estimating a *replace* price that includes labor and materials by the linear foot. The estimator has noted that the base is wood, not MDF, since wood has a higher unit cost.

7. ***Seal and paint base, 3 walls:*** This is new base, so it will require a seal coat and a finish coat. The estimate could also be for two paint coats, but typically seal and paint is slightly less expensive, and a good adjustor knows that.

   Here's something else you need to check. Most unit costs include filling nail holes and caulking the base for painting. If it doesn't, you need to catch that and be sure that you include enough additional labor and materials to make up for it.

8. ***Reinstall door, no damage:*** This is a labor-only figure. The estimator made a note to himself indicating there was no damage to the door.

9. ***Resecure subfloor, allow for approximately 25 percent of subfloor, as per drawing (no guarantee):*** This is the labor to screw the subfloor back down to the joists to eliminate the squeaks. Some estimating books and programs have a category for this, based on square footage. If yours does, you would estimate how much of the room needs to be resecured, such as the 25 percent in this example. Otherwise, estimate how long you think it will take, such as *labor, 0.50 hour* for one-half hour.

   Floor squeaks are tricky things. Sometimes you can eliminate them, sometimes you can't. So the estimator made a note to himself to add a disclaimer to his estimate. This will be worded in such a way as to simply let the homeowner and the adjuster know that every effort will be made to try and fix the floor squeak, but that the contractor can't guarantee that it will go away, or that it will stay gone.

10. ***Replace tack strip, approximately 50 percent:*** Here's another example of where you need to know your unit costs. In most of them, tack strip is included with the carpet installation price. But if it isn't, you want to be sure to use an additional line item that allows you to add new tack strip into your costs.

11. ***Replace pad:*** In most books and programs, pad is figured by the square foot. That's because there's no pattern or direction to it, so it can be cut and patched in anywhere while it's being installed. So for this example, you would just figure 154 square feet of pad (11 x 14).

12. ***Replace carpet, 12 x 14, allowance $18 per square yard:*** Carpet, on the other hand, is sold in 12-foot-wide rolls. So when you're estimating, you need to take that into consideration. For this room, the estimator made a note that he'll need a 12 x 14 piece to handle the 11 x 14 room. He's also made a note to allow $18 a square yard for materials in his estimate. That's based on the *like kind and quality* of the materials that were in the home at the time of the loss. (We'll talk more about carpet allowances later in the chapter.)

13. ***Reset contents, small room:*** Since the contents were removed during the emergency, they need to be put back as part of the reconstruction. This can be done as a unit cost if your estimating program has that. Unit costs are usually listed as *small room, medium room, large room,* and *extra-large room.* If that's the way your program has it, be sure you know what that unit cost covers, and select the category that's most appropriate. Otherwise, use a labor figure.

    Also, remember that unit costs are usually either for *remove and reset contents*, meaning you're going to take them out of the room and bring them back; or just for *reset contents*, meaning they're already out and you're just bringing them back in. Our estimate is for *reset contents.*

14. ***Post-construction cleaning, room and contents, 1 hour:*** At the end of a job, you need to return the room to a pre-loss condition. That means it's clean and ready to be reoccupied. You should always allow time for a post-construction cleaning. Some books and programs have this as a unit cost, but it's usually best to estimate this as a labor figure. In our example, the estimator has allowed one hour, and has made a note that it includes cleaning the room and the contents.

15. ***Debris removal, ½ truck load:*** Part of the final cleaning is removing debris. As the estimator, you'll need to make a guess about how much debris will be accumulated during the course of construction. In our example, we'll have masking, carpet and pad scraps, paint cans, and probably a couple of trash bags of miscellaneous debris.

Flooring Allowances

How much you can allow in an estimate for flooring may be tricky. There's a wide range of prices for flooring materials, and many types of flooring look alike, especially to the untrained eye. When determining a material allowance for carpet or vinyl, you have a couple of different options:

- **Homeowner receipts:** If the homeowners had the flooring installed within the last couple of years and still have their receipt, the insurance company will usually be satisfied with that. Remember to adjust what the homeowners paid to reflect current prices.
- **Ask your floor covering shop:** If you have an experienced floor covering shop with knowledgeable salespeople, checking with them is a good tried-and-true method to establish a cost value. They understand the different types of fibers, densities and materials; if you provide them with a sample, they can tell pretty quickly and accurately what a material is worth. They can then give you a written allowance to put in your estimate. That used to be an acceptable means of establishing value — at least until the last few years. It's still acceptable with some companies, but the larger ones are going to a new method of scientific analysis.
- **Scientific analysis:** This method of establishing the price of carpet and vinyl flooring has come into use in the last few years, especially by the larger insurance companies. A piece of flooring, usually about 6 to 12 inches square, is sent to an independent lab for testing. The lab analyzes everything about the sample, and comes up with a price range for it. That pricing is then sent back to both the insurance company and the contractor. The insurance company selects the lab and picks up the cost for testing.

Tip! *Before putting a carpet or vinyl allowance in your estimate, be sure and check with the insurance company to see if they require a sample be sent out for lab analysis.*

Overhead and Profit

Imagine that you walk into a shoe store and see a pair of shoes you like. They have a price tag of $60 on them. You think that's a reasonable price, so you go ahead and buy them. If you're like most people, that's the end of it.

But if you took the time to think about it, you'd know that built into that $60 is a certain amount of money to cover the shoe store's overhead. Without that money, the store owner couldn't pay his rent, or his utility bills. Also included in that $60 is an amount for the store owner's profit, so he can make a living and hopefully have a little left over to help his business grow and prosper.

Now imagine that you go into the same shoe store. But instead of a lump sum price tag on the shoes that says $60, the price tag also lists the owner's overhead and profit, like this:

Shoes, subtotal	$50.00
Overhead, 10 percent	5.00
Profit, 10 percent	5.00
Shoes, total price	$60.00

You're still paying the same $60 for the pair of shoes, but psychologically, there's something about knowing that you just paid $5 toward his power bill and $5 toward his retirement or his next vacation that's a little harder to swallow. That's why you don't see it listed in black and white on every price tag. In fact, you never see those numbers listed anywhere — not in restaurants, or hotels, or at the home center, or on the bill you get from your doctor or your mechanic.

Nowhere, that is, except on the bottom of an insurance estimate.

For whatever reason, it has long been a practice to show overhead and profit figures at the bottom of a restoration estimate. No one seems to know when or why the practice started, but that's the way it is, and it's not going to change. The two numbers together are almost always referred to in the industry as *O&P*.

Overhead

In another one of those longstanding conventions in the insurance restoration field, overhead is calculated at a rate of 10 percent. Why 10 percent is used, or who thinks that's a fair number in today's world, no one knows. Quite honestly, it really doesn't cover your overhead. But to date, it's been impossible to get changed. You can try — and you and others in the industry should definitely continue to do so — but be forewarned, 10 percent is the going rate.

You arrive at your 10 percent overhead figure by simply multiplying the subtotal at the end of the estimate by 0.10. If you have an estimating program, it will make this calculation for you automatically.

Profit

The same thing holds true for the profit figure. It's been fixed at 10 percent forever. There's no particular reason; it just is.

It's important for you to understand how profit is calculated. As with overhead, it's calculated by taking 10 percent of the subtotal. But, that calculation is done *prior* to adding on the overhead figure, not *after* adding on the overhead. So, the overhead and profit figures are always the same.

If you take 10 percent of the subtotal for overhead and 10 percent of the subtotal for profit, you end up with a total O&P of 20 percent:

Estimate subtotal	$25,000
Overhead, 10 percent	2,500
Profit, 10 percent	2,500
Estimate total	$30,000

Stacking the O&P

However, if you take 10 percent of the subtotal for overhead and *add* that to the subtotal, you get a new estimate subtotal. Then, if you take 10 percent of the *new* subtotal for your profit, you'll end up with a higher bottom line figure. The total O&P works out to be 21 percent instead of 20 percent:

Estimate subtotal	$25,000
Overhead, 10 percent	2,500
Estimate subtotal	$27,500
Profit, 10 percent	2,750
Estimate total	$30,250

This practice is known as *stacking the O&P*. It's been tried numerous times in the past with many different insurance companies. You won't get away with it, so just don't bother.

Matching Materials

The issue of matching materials comes up frequently in renovation work, and can cause everyone a lot of stress. (See Figure 23-7.) It has to do with three of the fundamental rules of insurance losses that we've discussed before:

1. A policyholder cannot profit from a loss.
2. Materials should be replaced with items of like kind and quality.
3. The insurance company has an obligation to return the policyholder's home to a pre-loss condition.

Finding Compromises

All three of these rules make sense on their own. But in reality, they often conflict with each other. When that happens, it creates confusion for everyone. Let's look at some common examples of situations you're likely to face.

Figure 23-7 The gable overhang on this older home is made up of individual beaded, interlocking boards. This is the existing material, so this is what should be matched as part of the repair. These material-matching details are the ones that the estimator needs to be aware of, and include as part of the estimate.

Countertops

A homeowner has a kitchen fire that damages one section of a laminate countertop. It's only one short counter, but it's a color that's no longer available. You could replace all of the counters, but then the homeowner would be getting new countertops to replace ones that aren't damaged, and therefore profiting from the loss. You could replace just that one counter with a different material in the same color, but as it wouldn't be the same material as the original — you haven't brought the countertop back to a pre-loss condition. Or you could replace it with a laminate counter in a different color, but again, that's not bringing the home to a pre-loss condition. Before the fire all the countertops were laminate, but they all matched.

There's no easy answer here. You just have to make whatever compromise suggestion you feel is best. Perhaps the homeowner would accept a more expensive counter in that one area, such as a wood top to use as a cutting board, or a piece of marble for candy making. Otherwise, the only reasonable suggestion is to put in all new countertops that match. After that, step aside and let the homeowner and the adjuster work it out.

Paneling

Suppose you have a room with lots of paneling in it. It's outdated, but it's real wood paneling, and now it's water-stained. Only three sections are damaged in the middle of the room. But, you can't get the stains out, and you can't match the paneling.

In this situation, the homeowner may want to get rid of the old paneling. If that's the case, the adjuster may accept a compromise of removing one wall of paneling and installing painted drywall. That could offer a less expensive alternative to replacing an entire wall of real wood paneling.

Paint

Paint always seems to bring on a matching issue. Fortunately, with the quality of matching done at most good paint stores today, it's not all that bad. The decision comes down to one of economics.

For example, let's say the paint on two walls of a room is damaged by water stains. The insurance company may opt to pay for painting only those two walls, since painting all four would let the policyholder profit from the loss.

But in reality, it's often more cost effective to paint the entire room. Otherwise, you have to take the time to try and match the exact paint color on the other two walls, not to mention the time to mask those two walls off to protect them while the two damaged walls are being painted. If it's less labor intensive and thus less expensive to paint the whole room, then that's what the insurance company will usually let you do. However, they usually *won't* include the ceiling.

Painting the exterior usually works a little different. Say there's a fire that damages two walls of siding on a large house. The insurance company will probably tell you to match the paint color, and then to paint the damaged walls from one corner to another, wherever the best break points are.

Wallpaper

It's nearly impossible to match wallpaper. Let's say you have a room with wallpaper on all four walls, and the paper on one wall is damaged. In this situation, if the owner wants the room wallpapered again, your best bet is to simply state that you must estimate the removal of all the old paper and the repapering of the entire room. The adjuster really has no alternative with this one.

Sight Lines

A *sight line* refers to whether or not one room or space can be seen from another room or space, even when they're physically connected to one another.

For example, suppose that instead of one room with wallpaper on all four walls, you have wallpaper on the walls of an entryway. That entryway is open to a hallway, which in turn opens onto a dining room and then the kitchen. All of the walls in the

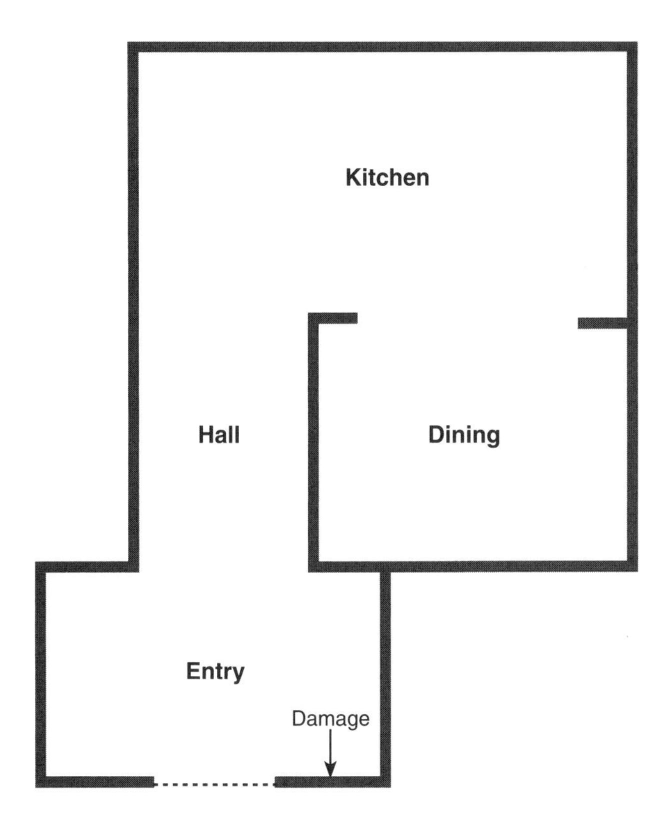

Figure 23-8 This drawing shows a sight-line situation. The entry, hall, dining room, and kitchen all have the same wallpaper. The only damage is to the wall in the entry, to the right of the front door. Since this area can't be seen from the kitchen or the dining room, the adjuster will probably discuss a sight line with the estimator and the homeowner, and limit the amount of wallpaper to be replaced.

hallway, dining room, and kitchen have the same wallpaper as the entryway. But only one wall of wallpaper in the entryway is damaged.

The wallpaper in the entry can't be matched or repaired. But the insurance company isn't going to want to replace the wallpaper in all of those rooms just because there's damage to the paper in the entry.

Instead, they'll take the sight lines into consideration. For example, if the entry is a relatively isolated space, as shown in the drawing in Figure 23-8, and it can't be seen when standing in the kitchen or dining room, then a *break point* will be established. Even though the wallpaper is continuous, the insurance company will want to see it broken at one corner of the entry, where there's no sight line to the break from any of the other rooms.

Sight lines are especially important with floor coverings, such as vinyl, carpet, and ceramic tile. For example, you may have a two-story house with the same carpet on both the upper and lower floors. If the carpet is damaged in the living room, the stairs are usually considered a sight line to the second floor. As such, the living room carpet will be replaced, but it won't be extended up the stairs to the second floor.

Room Breaks

With floor coverings, you also want to be aware of *room breaks*. Insurance companies will almost always want you to break your floor coverings at the doorways to rooms.

Take the example of a house with a master bedroom and two other bedrooms. See Figure 23-9. All three bedrooms are connected by a hallway, and all the rooms have the same carpet. A water loss damages the carpet in the master bedroom and the hallway, but the other two bedrooms aren't affected.

Even though it's the same carpet throughout, the insurance company will only replace the carpet in the master bedroom and hallway. They'll expect you to break the carpet at the doorway of the other two bedrooms. Where they won't expect you to break the carpeting is at closets. They'll assume you'll replace the carpet in both the master bedroom closet and the hall closet.

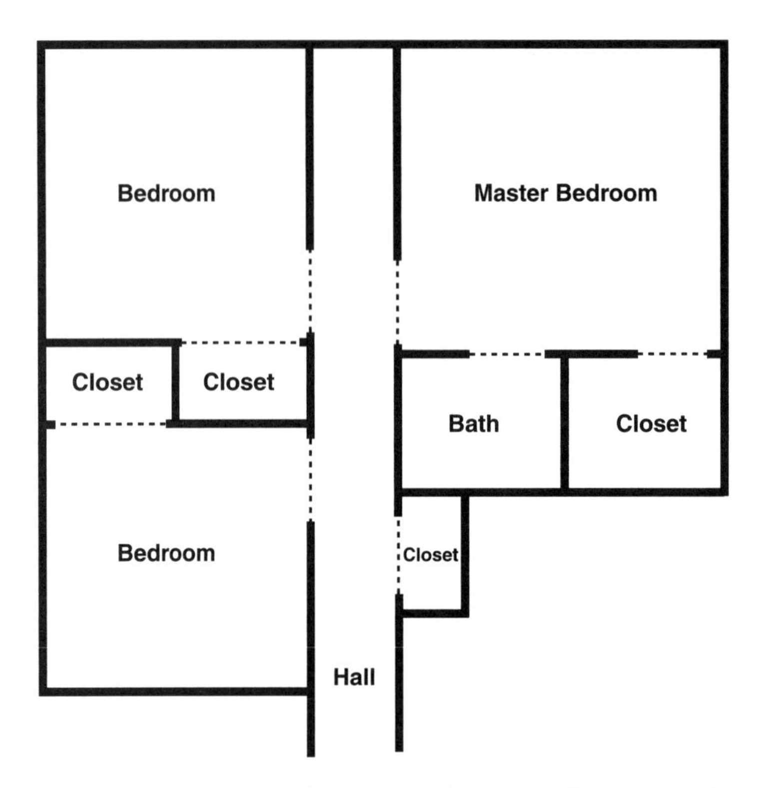

Figure 23-9 In this situation, a water loss occurred in the master bathroom. The carpet in the master bedroom and the hall was damaged. Even though the carpet is the same in all three bedrooms and the hall, the insurance company will usually only replace the carpet in the master bedroom and the hall. They'll stop the replacement at the doorways to the other two bedrooms. However, they'll generally replace the carpet in the hall closet and master bedroom closet even though there was no damage to the carpet in those areas.

Five Rules for Successful Structural Estimating

Estimating a structural restoration job can be tough. And, the bigger the job, the tougher the estimate. But good estimators know that there are certain things you can do to help ensure a successful estimate. Let's sum up this chapter with five rules you should always keep in mind when doing structural estimates.

Rule #1: Only Bid What You Can See

One of the most important things to understand about estimating for insurance restoration work is to only estimate what you can actually see. The insurance company isn't interested in what you *think* might be going on inside the wall. They're not interested in what *might* be happening behind the cabinets, or under the bathtub. They only want to know what you can physically see and document at the moment you do the estimate.

Let's say you're estimating a fire loss, and you see that there's drywall on the inside of a wall that's burned, and siding on the outside of the same wall that's burned.

The drywall has crumbled in a couple of places, and you can see that the insulation and wall framing inside the wall is black. You now have enough visual evidence to know what's going on inside the wall, so you can estimate replacing the insulation, and cleaning and sealing the studs.

But what if the fire's not quite that bad? The drywall is black with smoke, but it's not crumbled through. The siding on the outside is smoky, but not burned. There's enough smoke on the inside and outside of the house for you to *assume* the insulation will have to be replaced, *but you don't know for sure.*

In that case, you could cut a couple of test holes in the drywall and look inside to assess the condition of the insulation. That's a process known as *destructive testing.* If the insulation is black, then you have enough proof to warrant including it in your estimate. Or, you can simply wait until the drywall is torn down and then write a supplemental estimate for replacing the insulation.

Rule #2: Estimate to a Pre-Loss Condition

Sometimes this is a hard rule for a contractor who's used to improving the homes he works on. You see a room that's only three-quarters painted, and you figure you'll estimate painting the whole thing. It's easier, and it will look better when you're done. But you can't. It wasn't that way to start with, and you have to just bid what was there:

- ➤ If it was there at the time of the loss and it got damaged as a result of the loss, then it needs to go into your estimate.
- ➤ If it wasn't there at the time of the loss, or if the loss didn't damage it, then it can't be in your estimate.

Rule #3: Always Estimate Materials of Like Kind and Quality

If the kitchen cabinets were photo-finished particleboard, then photo-finished particleboard cabinets is what you're going to bid. It doesn't matter if the homeowners hated them and want to change them to something else. They can certainly pay you to make an upgrade later. However, the adjuster doesn't care about that, and doesn't want to see it in your estimate. *Always bid what was there at the time of the loss.*

Sometimes, it works out the other way, as a gain for the homeowners. The owners of a 1950s ranch house may hate the dark paneling that covered every wall of their now-flooded finished basement. They're happy to get rid of it and replace it with drywall. But that dark paneling was sheets of ¾-inch furniture-grade stained birch, and quite expensive at today's prices. So, whether they replace it with drywall or not, they're going to get paid for that expensive paneling.

Figure 23-10 During firefighting efforts at the home, firemen came into the back yard through this vinyl side gate. When estimating the loss afterward, the estimator made a very careful inspection of every affected area. By doing so, he noted that the cross member had been knocked out of the gate, and some of the slats had been broken at the bottom. He included a new gate in the estimate.

Rule #4: Bid the Steps, Bid the Categories

Once again, remember the two basic stages of putting your estimate together. When you arrive at the jobsite, walk through the job slowly and methodically. If possible, try and do it alone, without the homeowner or any of your crew in tow.

In your mind, put the job together, step by step. Don't leave anything out, not even the smallest of details. (See Figure 23-10.) The more detail-oriented you are, the better your estimates will be. Write everything down as you go. Make detailed drawings, and take good notes.

Then, using your estimating book or software, find every unit cost and category necessary to cover every one of the items in every step you've written down. Know what's in those unit costs. If the steps in one unit cost don't cover all of the labor or materials you need, then you'll have to find another category and stack it onto the first one. Keep doing that until you've covered all of the materials and labor that you need. If you don't, you're leaving money on the table, and that's no way to make a profit.

Rule #5: Be Honest

Be honest with your adjusters, your clients, and yourself. It's amazing how much easier that makes your life, and how many dividends that pays in the long run.

As soon as you start bending rules, or stretching the truth a little here or there to try and help this adjuster or that client, the problems start to multiply. If you see something on a job, and you're asked about it, just give your honest opinion. If it favors the homeowner over the insurance company, or vice versa, that's just the way it is.

Remember that there are lawyers everywhere. You could be called into court to defend your estimate or your actions on a jobsite at any time. That's another reason to act with complete honesty at all times.

But most of all, it's simply the best way to conduct yourself as a businessperson and as an estimator, and the best way to build your reputation in the community.

24

Estimating Books & Software

In the last chapter, we talked about the steps involved in doing structural estimates. In order to do those structural estimates you'll need a good estimating book or software package. And because of the nature of the insurance restoration industry, it can't be a general-purpose construction or remodeling estimating guide or program. It has to be one written specifically for insurance restoration work.

There are several good estimating books and software packages currently available for insurance restoration contractors. The one or ones you choose will depend on a variety of factors: your budget, the learning curve involved, the requirements of your company, and the requirements of the insurance companies you work with.

In this chapter we'll look at two examples in detail, starting with the *National Renovation & Insurance Repair Estimator* from Craftsman Book Company (currently $74.50). This is one of the best examples of a renovation estimating book currently on the market. The unit cost data contained in the book is also included on the accompanying CD, along with an integrated estimating program. A download of just the database and program is available for under $50.

Next, we'll get an introduction to *Xactimate®* from Xactware Solutions, Inc. ($125 to $250 per month to lease). This is one of the top examples of a dedicated insurance restoration estimating program. It includes a CAD program, reporting software, and quite a bit more.

Once you have an overview of both these references, you'll have an understanding of their differences and the pros and cons of each. This will help you in your decision-making process, as the cost differential is significant.

A Word About Estimating Books and Programs

In general, books have the advantage of being less expensive than dedicated estimating software programs. They don't require a computer, so they're portable enough to carry with you. They're quick to learn, with virtually no learning curve, and you don't have to take any specialized classes to use them.

In order to remain timely, estimating books are typically published annually. If they aren't, don't buy them. The better estimating books are linked to websites that allow you access to an on-line database. The database is updated regularly throughout the year, so the unit costs stay as up-to-date as possible.

Tip! *Never rely on an outdated estimating book, no matter how cheaply you can purchase it at a garage sale or a used book store!!*

On the downside, books can be slow and cumbersome from an estimating standpoint. They often require you to look up, and cross-reference numbers and figures. You also need some sort of manual or computerized spreadsheet form for entering and totaling your numbers, so that you can turn it into a completed estimate.

If you're thinking of using an estimating book, your best bet is to look for one that has an integrated software package included, which works hand-in-hand with the book. That allows you to quickly pull numbers from the book and plug them into your estimate. The software will usually do the math for you as well. That saves considerable time, and also greatly increases your accuracy. Craftsman's *National Renovation & Insurance Repair Estimator* does exactly this.

A dedicated estimating software package is one that's designed strictly for insurance restoration estimating. It will have an extensive database of prices, and cover an extremely wide range of loss situations. The software is continually updated with current pricing information and regional modifiers, all of which are available on-line to purchasers or subscribers of the software package.

The typical estimating software package contains more than just the ability to generate an estimate. You can usually create drawings, reports, material lists, and a number of other documents from the same software.

The advantage to estimating with this type of software system is its speed and accuracy. Once you've learned the system and become proficient at using it, you can prepare an estimate very quickly. You can also modify your estimates to meet changing job conditions, or to write supplementals. This is especially important as your company grows. Once you begin handling a large volume of business, including jobs such as major structure fires, having this type of estimating capability becomes increasingly important.

However, this type of dedicated estimating software package has a couple of drawbacks. First of all, they're fairly expensive. While good estimating books are available from $50 to $100, estimating software packages can run up into the thousand-dollar range, and higher. Some of them aren't even available for purchase, but can only be leased and require ongoing updates. *Xactimate* is one of these.

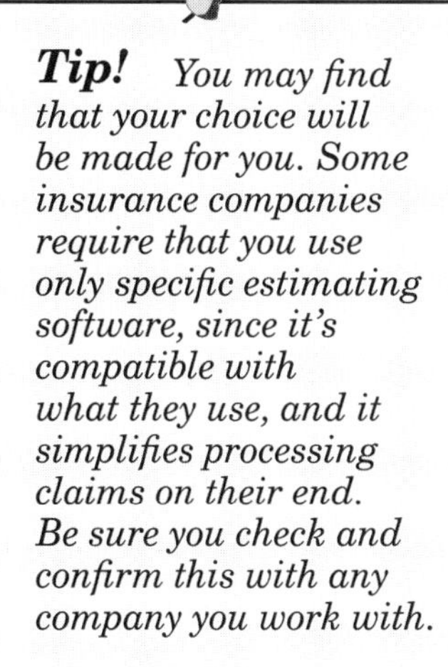

Tip! *You may find that your choice will be made for you. Some insurance companies require that you use only specific estimating software, since it's compatible with what they use, and it simplifies processing claims on their end. Be sure you check and confirm this with any company you work with.*

Their other drawback is the learning curve. Due to their high degree of sophistication, they take time to learn. There are usually tutorials included with the package, but realistically, the best way to learn the software is to attend a training class.

National Renovation & Insurance Repair Estimator

The *National Renovation & Insurance Repair Estimator*, published by Craftsman Book Company, is one of the best and most complete unit cost guides on the market for insurance restoration estimating. It's used and recognized by a number of different insurance companies, and the accuracy of its data is widely accepted.

The *National Renovation & Insurance Repair Estimator* contains thousands of unit costs for all phases of insurance restoration work. The book is published annually, and the prices are revised and updated quarterly. Once you purchase the book, the revised prices are available to you at no charge over the Internet for the entire year that the book's in print.

The book is arranged in categories based on typical construction activities, such as *Doors*, *Drywall*, *Masonry*, *Painting*, etc. Each category shows costs broken down by crew hours, unit, material, labor, and then a total. Many of the categories have drawings associated with them to help you identify the correct type or material. See Figure 24-1.

Most of the categories are further broken down into subcategories. For example, under *Painting*, you'll find *Acoustic ceilings*, *Awnings and carports*, *Drywall*, etc. Underneath those subcategories, you'll have a variety of tasks, such as, *Paint acoustic ceiling texture: prime*, *1 coat*, *2 coats*, *3 coats*, and the breakdown of costs for each of those tasks. See Figure 24-2.

Emergency Services

As we've been discussing, structural estimating is only one leg of the three-legged stool that makes up your restoration company. Another one of those legs is emergency work, and that's an important one. You need to know how to bill accurately and completely for your emergency services.

The *National Renovation & Insurance Repair Estimator* has several sections on different emergency services and equipment. It covers everything from water extraction (Figure 24-3) and thermal fogging to drying equipment.

Doors

	Craft@Hrs	Unit	Material	Labor	Total

Using Door Items. When replacing a door, select the door type, then the jamb type, and add the two together. Door prices are for the door slab only and include routing and drilling for standard hardware and hinges. Door tear-out is for the door slab only. Use this tear-out price when removing the slab and leaving the jamb and casing intact. Jamb and casing prices include the cost to prehang the door in the jamb and the cost of hinges (but not the door lockset). Tear-out includes removal of the entire door. Use this tear-out price when removing the door, jamb and casing.

Minimum charge.

	Craft@Hrs	Unit	Material	Labor	Total
for door work	1C@2.50	ea	38.50	132.00	170.50

Folding Door. Folding doors usually used in closets. Includes folding door section with hardware (including track, pins, hinges and a basic pull as needed). Priced per section of folding door. Does not include jamb or casing. Folding doors are often called bi-fold doors.

Folding door (per section)	Craft@Hrs	Unit	Material	Labor	Total
replace, hardboard smooth or wood-textured	1C@.394	ea	31.90	20.70	52.60
replace, hardboard wood-textured and embossed	1C@.394	ea	36.50	20.70	57.20
replace, mahogany (lauan) or birch veneer	1C@.394	ea	34.10	20.70	54.80
replace, ash or oak veneer	1C@.394	ea	37.20	20.70	57.90
replace, walnut or cherry veneer	1C@.394	ea	42.60	20.70	63.30
replace, paint-grade pine panel	1C@.394	ea	55.50	20.70	76.20
replace, stain-grade pine panel	1C@.394	ea	68.40	20.70	89.10
replace, paint-grade pine full-louvered	1C@.394	ea	65.50	20.70	86.20
replace, stain-grade pine full-louvered	1C@.394	ea	76.00	20.70	96.70
replace, paint-grade pine half-louvered	1C@.394	ea	62.30	20.70	83.00
replace, stain-grade pine half-louvered	1C@.394	ea	72.20	20.70	92.90
replace, red oak folding panel	1C@.394	ea	83.50	20.70	104.20
replace, folding mirrored	1C@.394	ea	95.70	20.70	116.40
remove, folding door	1D@.114	ea	—	4.05	4.05
remove, folding doors for work, reinstall	1C@.148	ea	—	7.78	7.78

standard *panel* *full-louver* *half-louver*

(Courtesy of Craftsman Book Company)

Figure 24-1 This page, from the *National Renovation & Insurance Repair Estimator*, shows some of the thousands of unit costs included in the book. Notice the description at the top that tells you what the overall unit cost includes. The columns under the individual door heading include (left to right), the type of work being performed; the type of material the door is made out of; a picture of the door; a crew code (1C) for the wage that the labor rate is based on; the amount of time allowed for the task (0.394 hours); the unit (ea for each); the material price; the labor price; and the total price. Also notice that there's a Minimum Charge for door work.

Understanding Unit Costs

In the last chapter, we talked about the importance of understanding what goes into a unit cost. You need to know what is and isn't included within a particular unit cost, and when you need to stack additional unit costs together to get everything you need into your estimate. In different parts of the *National Renovation & Insurance Repair Estimator*, you'll find unit cost descriptions that help clarify what's included. It's important that you read these through, and see if there's anything else that you're being told to stack on to the costs listed.

	Craft@Hrs	Unit	Material	Labor	Total

Painting

Painting coats. Prices for one coat, two coats and three coats do **not** include: the primer coat, sealer coat or stain. To estimate painting, add the cost of the primer coat to the number of paint coats. For example, to estimate the cost to prime a wall then paint with two coats, add the prime price to the two coats price. To estimate the cost to stain an item then cover with two coats of varnish, add the stain cost to the two coats price (unless otherwise noted).

	Craft@Hrs	Unit	Material	Labor	Total
Minimum charge.					
for painting work	5F@3.00	ea	40.40	151.00	191.40
Acoustic ceilings.					
Paint acoustic ceiling texture					
prime	5F@.011	sf	.32	.55	.87
1 coat	5F@.011	sf	.35	.55	.90
2 coats	5F@.017	sf	.49	.85	1.34
3 coats	5F@.023	sf	.66	1.15	1.81
Paint acoustical ceiling tile					
prime	5F@.011	sf	.32	.55	.87
1 coat	5F@.011	sf	.35	.55	.90
2 coats	5F@.017	sf	.49	.85	1.34
3 coats	5F@.023	sf	.66	1.15	1.81
Awnings and carports.					
Paint aluminum carport or awning					
prime	5F@.006	sf	.23	.30	.53
1 coat	5F@.006	sf	.26	.30	.56
2 coats	5F@.010	sf	.35	.50	.85
3 coats	5F@.013	sf	.47	.65	1.12
Drywall, plaster and stucco.					
Paint plaster or drywall					
prime	5F@.006	sf	.15	.30	.45
1 coat	5F@.006	sf	.16	.30	.46
2 coats	5F@.010	sf	.23	.50	.73
3 coats	5F@.013	sf	.30	.65	.95
Paint stucco					
prime	5F@.014	sf	.26	.70	.96
1 coat	5F@.015	sf	.28	.75	1.03
2 coats	5F@.023	sf	.38	1.15	1.53
3 coats	5F@.030	sf	.51	1.51	2.02

(Courtesy of Craftsman Book Company)

Figure 24-2 This page of painting costs shows unit costs based on the item to be painted and the number of coats to be applied.

	Craft@Hrs	Unit	Material	Labor	Equip.	Total
Water Extraction						
Emergency Service. All prices are for work done between 7:00 a.m. and 6:00 p.m. Monday through Friday. Add **25%** for emergency work between 6:00 p.m. and 10:00 p.m. Monday through Friday. Add **65%** for emergency work between 10 p.m. and 7:00 a.m. Monday through Friday, on weekends, or on holidays.						
Minimum charge.						
for water extraction work	9S@3.50	ea	—	151.00	111.00	262.00
Equipment delivery, setup, and take-home charge. Per loss charge to deliver water extraction equipment, set the equipment up, and take the equipment home once the water is extracted and drying is complete.						
minimum	—	ea	—	—	34.50	34.50
Extract water from carpet. Lightly soaked: carpet and pad are wet but water does not rise around feet as carpet is stepped on. Typically wet: water in carpet and pad rises around feet as carpet is stepped on. Heavily soaked: water is visible over the surface of the carpet mat. Very heavily soaked: standing water over the surface of the carpet.						
lightly soaked	9S@.005	sf	—	.22	.07	.29
typically wet	9S@.007	sf	—	.30	.08	.38
heavily soaked	9S@.010	sf	—	.43	.09	.52
very heavily soaked	9S@.014	sf	—	.60	.13	.73

(Courtesy of Craftsman Book Company)

Figure 24-3 This shows some of the different unit cost entries for emergency services. The description at the top tells you how to apply the emergency service charges, including the typical hours and suggested add-ons for overtime hours. There are also charges for water extraction, based on how wet the carpet is, and a description of how you determine the wetness of the carpet.

Throughout the book, at the end of each category, you'll find a description that explains the wage and labor burden for the crews performing the tasks listed in that category (see Figure 24-4). Be sure and read those descriptions. If they don't fit with what you're currently paying your employees, you could be losing money by using that particular unit cost. In that case, you'll need to simply go into the program and change the wage rates accordingly. If you need help, Craftsman Book Company offers free technical support for all their costbook programs.

Estimate Add-Ons and Regional Modifiers

It would be cumbersome for any book to list every possible combination of materials. So instead, as mentioned in the last chapter, there are lists of add-ons used to modify prices for those variations in material costs.

For example, when writing a roofing estimate, the *National Renovation & Insurance Repair Estimator* shows you a percentage to add on if the roof is steep or complex. For window estimates, the book shows you how to add on additional costs for window grids, Low-E glass coatings, and other types of glazing.

Regional modifiers are used to adjust material and labor prices up or down to reflect average costs for each state, and then costs within areas of that state. For example, the average costs in Alaska are 19 percent higher than the costs listed in

Mold Remediation Labor					
Laborer	base wage	paid leave	true wage	taxes & ins.	total
Mildew remediation specialist	$18.80	1.47	$20.27	15.63	$35.90
Mildew remediation assistant	$10.50	.82	$11.32	10.38	$21.70

Paid leave is calculated based on two weeks paid vacation, one week sick leave, and seven paid holidays. Employer's matching portion of **FICA** is 7.65 percent. **FUTA** (Federal Unemployment) is .8 percent. **Worker's compensation** for the mold remediation trade was calculated using a national average of 21.43 percent. **Unemployment insurance** was calculated using a national average of 8 percent. **Health insurance** was calculated based on a projected national average for 2011 of $738 per employee (and family when applicable) per month. Employer pays 80 percent for a per month cost of $590 per employee. **Retirement** is based on a 401(k) retirement program with employer matching of 50 percent. Employee contributions to the 401(k) plan are an average of 6 percent of the true wage. **Liability insurance** is based on a national average of 10.0 percent.

(Courtesy of Craftsman Book Company)

Figure 24-4 This shows a breakdown of the crew rates for mold remediation work and an explanation of the labor burden built into those rates.

the book. However, in the Ketchikan region of Alaska, they're only 2 percent higher. In Alabama, the average costs for the state are 7 percent lower than the book, but in the Mobile area, the costs are 4 percent higher. The *National Renovation & Insurance Repair Estimator* also has modifiers for Canadian provinces and regions within those provinces. As with the unit costs, the modifiers are updated on a regular basis to ensure that they remain current.

The National Estimator *Program*

One of the nice features of the *National Renovation & Insurance Repair Estimator* is that it includes *National Estimator*, a very useful estimating program that contains all of the unit costs from the book. The program allows you to quickly and easily pull in line items to prepare an attractive, well-organized estimate. See Figure 24-5. The *National Estimator* program is very easy to learn and use, and is based on many common *Windows*™ commands. It includes a tutorial that teaches you the basics pretty quickly. And, if you need it, there's the free phone-in technical support.

The program has a search feature that allows you to find whatever you're looking for, similar to the index in the book. Once you locate the item, you can copy and paste costs or line items into the estimate you're working on. The unit costs are clearly indicated, as are the units themselves (square foot, linear foot, etc.). It's simply a matter of entering how many square feet or linear feet you want.

While the program is preset with all the unit prices, they're also easy to change. If you don't think that the underlying wage that the unit prices are based on is high enough, you can go in and change the wage. When you do, any unit costs based on that wage will change as well. Or, you can change the number of manhours that have been allowed for a certain job, which will also alter the unit cost. All in all, it allows you enough flexibility to be able to adjust the program and the prices to meet your needs.

Construction Estimate With Unit Costs
File Name: Baker Bathroom Repair Estimate1.est

Qty	Craft@Hours	Unit	Material	Labor	Equipment	Total
Plywood underlayment remove						
Per Unit:	1D@.0700	SY	0.00	3.18	0.00	3.18
4.00	1D@.2800	SY	0.00	12.72	0.00	12.72
Plywood underlayment replace, 1/2"						
Per Unit:	5I@.1680	SY	8.72	9.33	0.00	18.05
4.00	5I@.6720	SY	34.90	37.32	0.00	72.22
Vinyl floor replace, high grade						
Per Unit:	5I@.1830	SY	30.39	10.17	0.00	40.56
5.00	5I@.9150	SY	151.93	50.85	0.00	202.78
Baseboard remove for work, then reinstall						
Per Unit:	6C@.0650	LF	0.00	3.81	0.00	3.81
24.00	6C@1.560	LF	0.00	91.44	0.00	91.44
Paint baseboard, 2 coats						
Per Unit:	5F@.0170	LF	0.24	0.85	0.00	1.09
24.00	5F@.4080	LF	5.69	20.40	0.00	26.09
Paint walls, one coat plaster or drywall						
Per Unit:	5F@.0060	SF	0.18	0.30	0.00	0.48
192.00	5F@1.152	SF	33.62	57.60	0.00	91.22
Toilet remove for work, then reinstall						
Per Unit:	7P@2.160	Ea	2.38	160.40	0.00	162.78
1.00	7P@2.160	Ea	2.38	160.40	0.00	162.78
Debris hauling with pick-up truck per pick-up truck load						
Per Unit:	--@.0000	Ea	0.00	0.00	24.85	24.85
0.50	--@.0000	Ea	0.00	0.00	12.42	12.42
Minimum charge for cleaning work cleaning work						
Per Unit:	1B@1.800	Ea	2.28	68.65	0.00	70.93
1.00	1B@1.800	Ea	2.28	68.65	0.00	70.93

Total Manhours, Material, Labor, and Equipment:					
	8.9	230.78	499.38	12.42	742.58
			Subtotal:		742.58
10% Profit of $74.26 & 10% Overhead of $74.26, shown as:					
			20.00% Overhead:		148.52
			Estimate Total:		891.10
			7.25% Tax on Materials:		16.73
			Grand Total:		907.83

(Courtesy of Craftsman Book Company)

Figure 24-5 This is a sample estimate using costs from the *National Renovation & Insurance Repair Estimator* and written using the *National Estimator* program included with the book. This estimate is for a small water loss in a 5- x 7-foot bathroom. It involves replacing the underlayment and vinyl flooring; removing and reinstalling the base and the toilet; painting the walls and the base; and cleanup and debris removal. The program stacks the O&P, which isn't allowed by most insurance companies. However, you can add a line of text above as shown, and put 20 percent into the Markup dialog box for O&P combined.

You can use the area modification table to adjust the prices to reflect where you live as well. All you need to do is enter your zip code, and the price will automatically be adjusted accordingly. If you're working on an out-of-town job and want the current pricing for the town you're working in, you can switch the pricing in the program to reflect that zip code.

The program also has the ability to add in sales tax on your materials. You simply enter the tax rate for your area when the dialog box for sales tax appears. It will total the sales tax and enter it as a separate line on your estimate.

When you finish your estimate, you can save it into **My Documents**, as you would any other file, so you always have it accessible for future reference on other similar jobs. You can also export the estimate into another program, such as *Word* or *Excel*, or directly into *QuickBooks* for your billing and accounting uses.

Another nice feature of the program is that you can call up an old estimate, and take information from it to paste into a new estimate. For example, say you recently wrote an estimate for the Adams job that included a lot of drywall and painting. Now you're working on an estimate for the Jones job, and it includes much of the same work. You can go into the Adams estimate, and copy all of the line items you need. Then go into the Jones estimate, and paste all the line items there. Simply change the square footages as needed, and you're all set. Keep in mind, however, that the prices won't be updated. You only want to copy lines and prices from a recent job.

At the end, you can set your overhead and profit to whatever you want, but remember, for insurance estimates they both need to be 10 percent. This program automatically stacks the O&P, so you're probably going to have to make a manual adjustment for your overhead and profit. Calculate the 10 percent manually and add a text line showing the breakdown for Overhead and Profit, as shown in Figure 24-5. Then enter 20 percent in the Markup dialog box under Overhead, to get your correct total.

Another benefit of this program is *Job Cost Wizard*, which allows you to produce a professional estimate to give to your client. See Figure 24-6. You have the option here of spreading your overhead and profit throughout the costs, rather than showing them as separate entries at the end of the estimate. You can see the slight difference in the prices between the estimate you produce for yourself and the insurance company and the estimate you give your client. The client's copy has your markup built into the costs, resulting in a difference of $1.32 in the overall price.

Xactimate

When it comes to estimating software for the insurance restoration industry, *Xactimate* is the most well known. *Xactimate* is used by most of the national insurance companies, so it's a name you'll be hearing a lot as you get into the industry. If you participate in any of the Preferred Contractor Programs, *Xactimate* will almost certainly be a requirement. It may be a requirement for many companies even if you're not a preferred contractor.

Estimate

A to Z Renovations

111 Beacon St.
Hillside, OR 92000
999-555-0101
#555777

Date	Estimate #
5/4/2011	1
Customer	**Job**
John J. Baker	Baker Bath Repair

Customer Information
John J. Baker 222 Pine St. Hillside, OR 92000

Description	Qty	Rate	Amount
Plywood underlayment, remove			
Labor, per SY	4	3.82	15.26
Plywood underlayment, replace, 1/2 in.			
Material, per SY	4	10.46	41.86
Labor, per SY	4	11.20	44.78
Vinyl floor, replace, high grade			
Material, per SY	5	36.47	182.34
Labor, per SY	5	12.20	61.02
Baseboard, remove for work, then reinstall			
Labor, per LF	24	4.57	109.73
Paint baseboard, 2 coats			
Material, per LF	24	0.29	6.91
Labor, per LF	24	1.02	24.48
Paint walls, one coat, plaster or drywall			
Material, per SF	192	0.22	41.47
Labor, per SF	192	0.36	69.12
Toilet, remove for work, then reinstall			
Material, per Ea	1	2.86	2.86
Labor, per Ea	1	192.48	192.48
Debris hauling with pick-up truck, per pick-up truck load			
Equipment, per Ea	0.50	29.82	14.91
Minimum charge for cleaning work, cleaning work			
Material, per Ea	1	2.74	2.74
Labor, per Ea	1	82.38	82.38
*Project Subtotal			892.34
Project Total			892.34
7.25% Tax on Materials	1	16.81	16.81
*Tax Charges			16.81
		Total	909.15

Figure 24-6 This estimate produced through *Job Cost Wizard*, part of the *National Estimator* program, is the same as the one in Figure 24-5. It's designed to give to the client. The option of showing overhead and profit as part of each cost rather than a total at the end has been applied.

Xactimate is an extremely powerful software package. It contains a sophisticated 2D and 3D drawing program called Sketch®; an extensive library of unit costs for structural, emergency, cleaning, and content estimating (see the sample in Figure 24-7 on the following page); powerful estimate-writing software; and a variety of sophisticated and useful reports. Built-in wizards help you with everything from estimating plumbing and HVAC systems to designing roofs and pricing insulation.

> ***Tip!*** *If you're thinking of using a dedicated estimating software program,* Xactimate *is the most widely-recognized. And, as your company grows, you'll almost certainly need to invest in it.*

You can use *Xactimate* with a desktop or laptop computer, tablet PC, convertible, or netbook. It works on stand-alone computers, or it can be loaded onto a server and set up as part of a network. As you might expect, the CAD (drawing) program it contains uses up a lot of computing power, so the bigger your computer is, the better the program will operate.

Xactware, like Craftsman, is constantly monitoring and updating their prices. They've created price lists for over 450 separate economic areas in the United States and Canada. All updates are done automatically, via a download over the Internet.

> ***Tip!*** *Before investing in* Xactimate, *make sure your computer meets or exceeds their system requirements. If it doesn't, you'll have to upgrade your system or the program won't work correctly.*

Fast and Interactive

Xactimate is also designed as an interactive link between the insurance companies and the restoration contractors using the program. When an insurance company sets up a file for a new loss claim, the information can be downloaded through *Xactimate* directly to the contractor. The information will include the name, address, and other contact information for the policyholder; the policy number; the claim number; and a description of the loss along with relevant details.

Once the contractor has prepared the estimate, he can upload it directly to the insurance company, again through *Xactimate.* The complete file, including digital photos of the loss, goes directly to the insurance adjuster, the claims processer, or whomever the company designates to receive it. If modifications or supplementals are required on the loss, they can be requested, completed and sent in just as quickly.

The speed of this interactive process is one of *Xactimate's* advantages. It's fast for all the parties involved — the contractor, the insurance company, and the homeowner. Estimates can be viewed almost instantly, and claims can be quickly processed and approved.

Second Level

Second Level

DESCRIPTION	QNTY	UNIT COST	TOTAL
10. 2" x 4" x 10' #2 & better Fir / Larch (material only)	28.00 EA @	3.70 =	103.60
11. 2" x 4" x 8' #2 & better Fir / Larch (material only)	513.00 EA @	2.96 =	1,518.48
12. 2" x 4" x 12' #2 & better Fir / Larch (material only)	18.00 EA @	4.46 =	80.28
13. 2" x 4" x 20' #2 & better Fir / Larch (material only)	7.00 EA @	8.11 =	56.77
14. 2" x 4" x 18' #2 & better Fir / Larch (material only)	2.00 EA @	7.33 =	14.66
15. 2" x 4" x 16' #2 & better Fir / Larch (material only)	4.00 EA @	5.87 =	23.48
16. Labor to frame 2" x 4" non-bearing wall - 16" oc	2,834.54 SF @	0.79 =	2,239.29
17. 2" x 10" x 12' #2 & better Fir / Larch (material only)	2.00 EA @	13.27 =	26.54
18. 2" x 10" x 10' #2 & better Fir / Larch (material only)	1.00 EA @	11.01 =	11.01
19. 2" x 8" x 8' #2 & better Fir / Larch (material only)	21.00 EA @	6.41 =	134.61
20. 2" x 8" x 16' #2 & better Fir / Larch (material only)	20.00 EA @	12.71 =	254.20
21. 2" x 8" x 12' #2 & better Fir / Larch (material only)	5.00 EA @	9.65 =	48.25
23. 2" x 6" x 12' #2 & better Fir / Larch (material only)	4.00 EA @	7.25 =	29.00
24. 2" x 6" x 20' #2 & better Fir / Larch (material only)	2.00 EA @	11.80 =	23.60
26. 2" x 6" x 16' #2 & better Fir / Larch (material only)	5.00 EA @	9.56 =	47.80
27. Sheathing - waferboard - 1/2"	1,280.00 SF @	0.94 =	1,203.20
28. 2" x 10" x 8' #2 & better Fir / Larch (material only)	4.00 EA @	8.81 =	35.24
29. 2" x 6" x 8' #2 & better Fir / Larch (material only)	28.00 EA @	4.82 =	134.96
30. 2" x 8" x 10' #2 & better Fir / Larch (material only)	4.00 EA @	8.01 =	32.04
31. 2" x 6" x 10' #2 & better Fir / Larch (material only)	5.00 EA @	6.02 =	30.10
108. 2" x 4" x 14' #2 & better Fir / Larch (material only)	4.00 EA @	4.00 =	16.00
109. 2" x 4" x 92 5/8" pre-cut stud (for 8' wall, mat only)	89.00 EA @	2.20 =	195.80
110. R&R Sheathing - waferboard - 1/2"	1,600.00 SF @	1.20 =	1,920.00
111. R&R Rafters - 2x6 - Labor only - (using rafter length)	297.02 LF @	2.03 =	602.95
112. R&R Rafters - 2x8 - Labor only - (using rafter length)	482.67 LF @	2.43 =	1,172.89
113. R&R Rafters - 2x10 - Labor only - (using rafter length)	53.72 LF @	2.69 =	144.51

AKEBY_HOUSE

(Courtesy of Xactware Solutions, Inc.)

Figure 24-7 Here you see a sample of some of the unit cost line items in an *Xactimate* estimate. Most of these are for framing lumber, materials only. The last four entries at the bottom are unit costs for labor and materials to remove and replace (R&R) sheathing and rafters.

How Xactimate *Works*

There's a lot to the *Xactimate* software package, and some of it you may only use rarely, or not at all. Here's a very basic idea of how the program works:

- You start an estimate by opening a brand new file on your own, or by opening a new assignment that's been sent to you by an insurance company. As part of this process, you can select a price or cost list from your area to associate with that particular estimate.
- Using your field notes, you create a drawing in Sketch, the *Xactimate* drawing program. Sketch is a fairly sophisticated

CAD program, allowing you to draw in both 2D and 3D. With Sketch, you can draw single- or multi-level homes and connect the levels with stairs. You can also design a roof structure for the house. Figure 24-8 shows a sample 2D drawing.

You can also make duplicate floor plan drawings, or duplicate mirror-image floor plan drawings, such as you'd have in an apartment complex with multiple units. Custom designs, such as angled walls, bays, curves, or other features, are all possible.

- As you draw, Sketch automatically creates a set of rooms for you in the estimating program. The dimensions of the rooms are based on the information you enter into the drawing.
- Working from the *Xactimate* unit cost price lists in the estimating program, and from your own field notes, you then enter the work that needs to be done in each of the rooms. Some categories in the program also offer drawings or photographs of different materials, like kitchen cabinet styles and materials, to help ensure that you're selecting the right type and quality of material.
- As you enter the work to be performed, you also enter a formula or a formula code that will allow *Xactimate* and Sketch to automatically do all the math for you. For example, suppose you wanted to paint the ceiling and all the walls in Room 1. After entering the proper unit cost for the painting you want to do, in the appropriate box you would then click on *WC* for *Walls and Ceiling*. The program will instantly calculate the square footage of the walls and the ceiling in Room 1, and enter that number. You could also just click on *C* for the ceiling or *W* for just the walls, if those were the only areas to be painted. For painting trim, like baseboards, there's *PF* entry which will give you the perimeter distance around the room at the floor level. There are many different options that you can select. For individual items, such as light fixtures, simply enter the quantity needed. See Figure 24-9.
- At the end of the estimate, *Xactimate* totals everything up, and gives you a final grand total page with the total price for the estimate. You'll also get a page headed Grand Total Areas, shown in Figure 24-10, with the total of all the areas you entered into Sketch.

All this computing power comes at a fairly high price. As of this writing, *Xactimate* can only be leased. There are currently two lease plans available for the program: month-to-month, at a cost of $250 per month per user; or annually for $1,500 per user, which works out to be $125 a month per user. If you're on a network with a number of users, it can be quite expensive. The advantage of having the software on a server is that all the users on that network have access to all the estimates stored on the server.

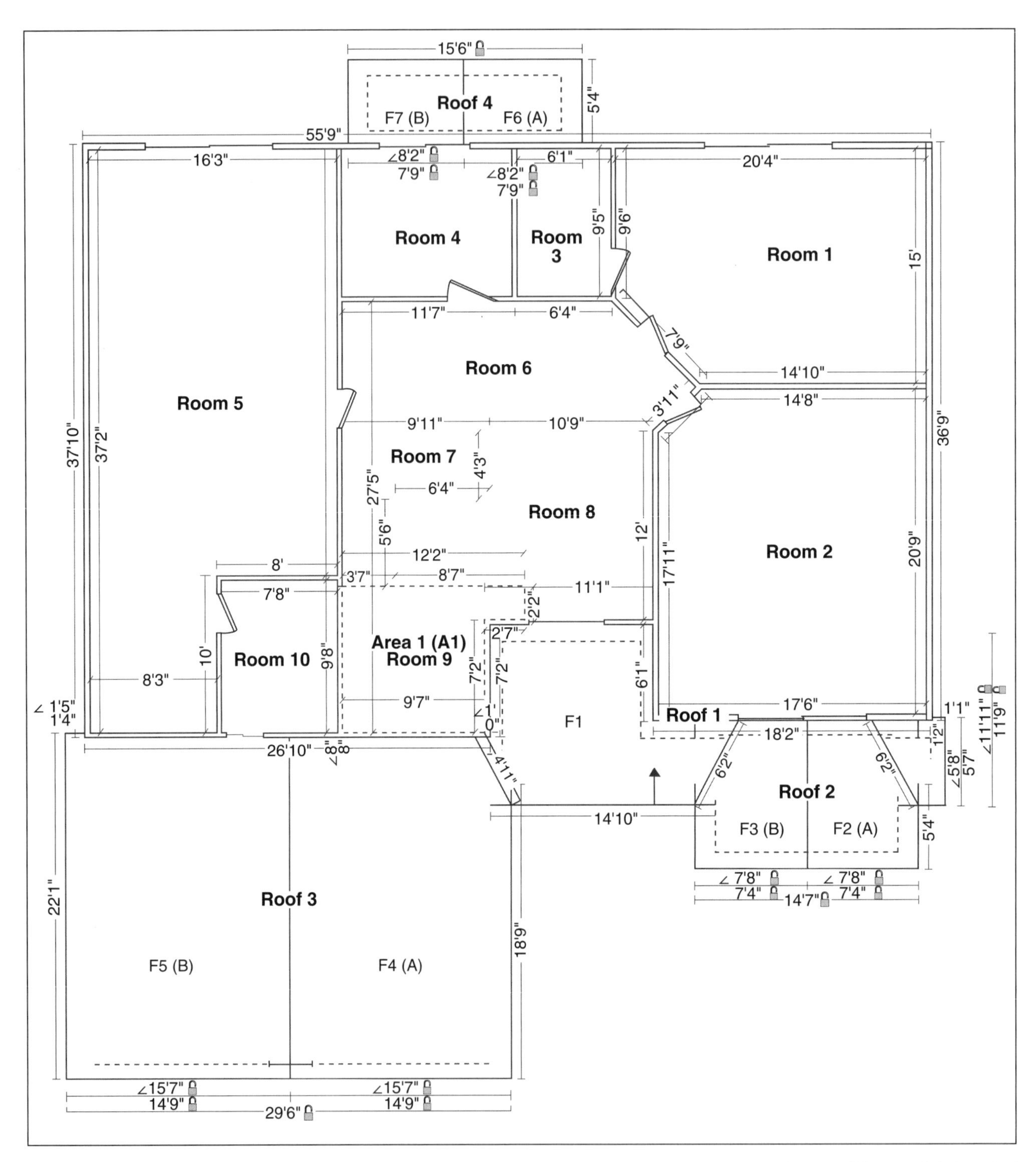

(Courtesy of Xactware Solutions, Inc.)

Figure 24-8 This is an example of a typical 2D floor plan drawing created in Sketch. Each of the rooms created in the drawing are automatically entered into the estimating program as well. The dimensions entered in the drawing are also used by the program to create formulas for calculating the square footages, linear footages, and other measurements used for pricing.

DESCRIPTION	QNTY	UNIT COST	TOTAL
77. R&R Light fixture	2.00 EA @	53.60 =	107.20
78. R&R 110 volt copper wiring run and box - rough in only	2.00 EA @	40.71 =	81.42
79. R&R 110 volt copper wiring run, box and switch	2.00 EA @	50.36 =	100.72
80. R&R 110 volt copper wiring run, box and outlet	8.00 EA @	50.37 =	402.96

Room3 **Height: 8'**

DESCRIPTION	QNTY	UNIT COST	TOTAL
73. R&R Light fixture	1.00 EA @	53.60 =	53.60
74. R&R 110 volt copper wiring run and box - rough in only	1.00 EA @	40.71 =	40.71
75. R&R 110 volt copper wiring run, box and switch	2.00 EA @	50.36 =	100.72
76. R&R 110 volt copper wiring run, box and outlet	4.00 EA @	50.37 =	201.48

Room4 **Height: 8'**

DESCRIPTION	QNTY	UNIT COST	TOTAL
85. R&R Light fixture	2.00 EA @	53.60 =	107.20
86. R&R 110 volt copper wiring run and box - rough in only	2.00 EA @	40.71 =	81.42
87. R&R 110 volt copper wiring run, box and switch	2.00 EA @	50.36 =	100.72
88. R&R 110 volt copper wiring run, box and outlet	8.00 EA @	50.37 =	402.96

Room6 **Height: 8'**

Missing Wall: **1 - 12' X 8'** **Opens into ROOM1** **Goes to Floor/Ceiling**
Missing Wall: **1 - 11' 11 1/2" X 8'** **Opens into BAY1** **Goes to Floor/Ceiling**

DESCRIPTION	QNTY	UNIT COST	TOTAL
57. R&R Light fixture	2.00 EA @	53.60 =	107.20
58. R&R 110 volt copper wiring run and box - rough in only	2.00 EA @	40.71 =	81.42
59. R&R 110 volt copper wiring run, box and switch	2.00 EA @	50.36 =	100.72
60. R&R 110 volt copper wiring run, box and outlet	4.00 EA @	50.37 =	201.48

Room7 **Height: 8'**

Missing Wall: **1 - 16' 5" X 8'** **Opens into ROOM2** **Goes to Floor/Ceiling**

DESCRIPTION	QNTY	UNIT COST	TOTAL

AKEBY_HOUSE

(Courtesy of Xactware Solutions, Inc.)

Figure 24-9 This page shows some of the individual line item entries in an estimate.

Grand Total Areas:

8,036.30	SF Walls	4,655.11	SF Ceiling	12,691.41	SF Walls and Ceiling
4,584.09	SF Floor	509.34	SY Flooring	1,006.97	LF Floor Perimeter
0.00	SF Long Wall	0.00	SF Short Wall	1,023.55	LF Ceil. Perimeter
4,584.09	Floor Area	4,752.27	Total Area	7,715.44	Interior Wall Area
3,801.94	Exterior Wall Area	441.42	Exterior Perimeter of Walls		
3,491.32	Surface Area	34.91	Number of Squares	380.30	Total Perimeter Length
138.24	Total Ridge Length	254.52	Total Hip Length		

(Courtesy of Xactware Solutions, Inc.)

Figure 24-10 The Grand Total Area page at the end of the *Xactimate* estimate provides you with a total of all the dimensions that were entered into Sketch.

If you're working with one of the major insurance companies, they may require that you download assignments and upload completed estimates through *Xactimate*. If that's the case, there's typically a per-transaction or per-month fee associated with that as well.

As you'd imagine, this isn't a program that you can just sit down and use. While it's well-designed and reasonably intuitive for anyone with prior computer and *Windows* experience, there's still a definite learning curve. You should plan on either taking a training class or spending some time working with someone who knows how to use the program and can train you on the job. After that, it's a matter of working with it and gaining experience. However, once you learn it, it's a very powerful tool that will definitely streamline your estimating.

The Need for Estimating Books and Programs

No matter what type of book or software you end up selecting, you can be assured that the days of handwriting insurance restoration estimates based on time and materials are long gone. Highly-detailed, well-organized estimates with unit costs that conform to industry standards are now a requirement, so plan accordingly.

On the bright side, once you get used to it, this type of estimating is faster, more consistent, and more accurate than the types of estimating you've probably done in the past. And that translates into higher profits for you and your company!

25

Dealing with Finances, Remodeling, & Clients

Insurance restoration has the distinct advantage of providing very steady, year-round work, barely affected by the state of the economy. But it's not without its financial ups and downs. You have to be very careful about taking on too much before you're financially ready. There are also some oddities about dealing with the financial side of insurance companies and their clients that are different from what you deal with in other construction fields. In this chapter we'll be looking at those differences.

Managing Finances and Collections

As with everything else the insurance companies do, they have rules and they have paperwork for processing payments, and those rules and that paperwork don't always work in your favor. Even though your client is the homeowner or building owner, how the insurance company processes payment to your clients will have a direct bearing on how and when you get paid.

Mortgage Holders and Multi-Party Drafts

When someone purchases a home, they usually take out a mortgage through a bank or other lending institution. The homeowner owns the house, but the lender retains an interest in the home as security for the loan. In return for lending the money, the homeowner has an obligation to protect the bank or lender's investment until the loan is paid off.

In order to do that, the homeowner is required to purchase a homeowner's insurance policy from whatever insurance company they choose. The lender usually insists that their name be included as a beneficiary on the insurance policy. So the lender becomes what's known as a *named insured*.

When a loss occurs at the house, the insurance company issues payment to the policyholder in the form of a *draft*, which is similar to a money order. When the insurance company issues the draft, the bank immediately verifies that the funds are available in the insurance company's account to cover the amount of the draft. The bank then takes those funds out of the account and holds them, so that they are guaranteed to be available when the draft is cashed or deposited by the homeowner. Drafts are a very reliable form of payment, since they won't bounce the way that a regular check might.

For a small loss, the payment will probably be for the entire amount of the claim, minus the homeowner's deductible. For a larger loss, part of the payment may be held back until the work is completed. We'll cover that in more detail shortly.

A draft can be simple or it can be confusing, depending on the property ownership and the named insured. If there's more than one legal owner of the property, the draft will have all the names of the property owners listed; it will have the contractor's name listed; and it will usually have the mortgage holder's name listed. If the owners have a second mortgage on the property, the second mortgage holder's name will also get tossed into the mix. And so on.

Drafts in a Perfect World

In a perfect world, a draft is no big deal. Suppose Mr. and Mrs. Smith own the house together. They have a first mortgage at a local bank, and no second mortgage. You meet them at the bank, they both sign the draft, you sign it, and their friendly bank officer signs it. They deposit the draft into their account, and then they use the funds to pay your draws as they come due, in accordance with your contract. It's very simple.

Drafts in a Not-So-Perfect World

There are, of course, a number of things that can — and do — happen in the not-so-perfect *real* world. Here are a few examples:

- The local bank that held the mortgage on the Smith house sold it to a bank three states away, who sold it to a holding company on the other side of the country. The Smiths have been sending their mortgage payments to a Post Office box for the last three years. The entire signing process is delayed for several weeks while the correct people in the correct offices are tracked down and the draft relayed to them for signatures.
- The local bank is able to put the Smiths in contact with the new mortgage holder, which is a conglomerate of multinational companies in London. They require that the draft be sent by courier for signature. Their turnaround time for the entire process is eight weeks.

- The local bank still holds the mortgage, but their mortgage department is now in another state. They have a strict policy that they won't sign off on insurance drafts without an inspection of the loss by a bank official. Inspections will also be required before any in-progress draws will be issued. All draw requests must be made two weeks in advance, on bank draw request forms.
- Mr. Smith has been having an affair, and Mrs. Smith threw him out of the house two months before the fire. The house is the subject of a divorce proceeding, and Mr. Smith refuses to sign the draft because he's not living there anymore, and he couldn't care less if Mrs. Smith gets the house fixed or not.
- Mr. and Mrs. Smith are fine, and so is the bank. However, Mr. Smith's grandfather once co-signed a loan so the Smiths could buy a boat, and somehow the house got tied up in the deal as part of the collateral. Through a glitch, the grandfather is named on the draft, but he passed away three years ago. It will take at least a month before the insurance company's main office can process the death certificate and issue a replacement draft.
- Mr. and Mrs. Smith are fine, and so is the bank. Everyone signs the draft, and the money gets deposited. Then the Smiths spend most of it on a trip to Hawaii.

These examples may seem just a little far-fetched, but they're not. They're composites of things that have really happened, and it doesn't take long to get into some serious financial trouble if you're strung out on several large jobs at once. So work closely with the insurance company, the bank, and the homeowners. Stay on top of your draw schedules, and never let your clients get behind in what they owe you. If they do, stop the job until they get caught up.

Tip! *Insurance restoration work requires enough cash flow to be able to sustain yourself and your company when there are delays in the processing of insurance company drafts. Never take on more work than you can afford to handle.*

Hold Backs

On large losses, insurance companies will often do what's called a *hold back*. That means that they'll retain a certain portion of the settlement until the repairs have been completed on the home.

Suppose the Lawrences have a $60,000 fire in their home. The insurance company typically won't just issue them a draft for $60,000. Remember that the purpose of the insurance settlement is to repair the house, and everyone involved has a stake in seeing that process through to completion.

The insurance company will therefore issue the Lawrences an initial draft for the repairs, but they'll hold back part of the total amount, usually about 20 to 25 percent. This is to ensure that the work actually gets done.

As the work nears completion, the contractor notifies the insurance company, and the amount held back will be released. It's important that you know that this money is out there, *but it needs to be requested.* Otherwise, there'll be a delay in getting it when the job wraps up.

Getting Paid

Despite your best efforts, there are times when you're not going to get paid. That's a fact of life in any type of construction. The homeowner may have cashed the insurance draft and spent the money, he may have filed for bankruptcy, or he may not like the job you did, and is holding something back. How you handle the situation is up to you, but here are some possible remedies:

- **Escrow:** If the clients feel you've done something wrong and they're holding back final payment, suggest that the money be put into an escrow account. It protects the money, and it's a great incentive to work things out and avoid further legal action. Issue joint instructions to the escrow company that the money can only be released on the signature of both parties. Typically, escrow charges are split equally between both parties, and the payment of those costs needs to be part of the escrow agreement. In some cases, you may want to pay the fees yourself in order to resolve the dispute quickly.

- **Arbitration:** If you have a local arbitration service, give that a try. Your state contractor's board may offer that service as well.

- **Small Claims Court:** Many cities have cut back on the availability of Small Claims Courts due to budget restraints. And if you win, you still have to collect. But it's often better than ignoring the debt and leaving the money on the table.

- **Collection Agencies:** Don't rule out collection agencies. If you're in the right, especially if you've won a legal judgment, this is one way to collect it. You give up a certain percentage of the money to the agency, but getting part of your hard-earned money is definitely better than nothing at all.

Tip! *You should include an arbitration clause in your contract. Check with your attorney on the proper wording to use. You might want to consider including an escrow clause in your contract as well.*

Figure 25-1 This house had a major structure fire, which affected the garage, part of the living room, and the kitchen and dining area. The owners took the opportunity to enlarge the garage, extend the cramped loft area, and do extensive interior remodeling downstairs.

Doing Additional Remodeling Work

Once the initial shock of the loss has passed, people naturally start thinking about the rebuilding process. For some, that means simply putting everything back exactly as it was. Those are the easy ones — and also the least common. For most people, their insurance loss is an opportunity to make some long-desired upgrades or other remodeling changes. See Figure 25-1.

Remodeling, upgrades, and material changes can run the gamut from minor paint and trim changes to major additions and structural alterations. They present the opportunity for the homeowner to make the house into what they've always wanted it to be, and for you, as a contractor, to generate some additional revenue.

Doing a little remodeling while you're there to repair the house anyway probably seems like no big deal. But keeping everything balanced between the homeowner and the insurance company can get *very* confusing. You run the risk of losing money — sometimes a *lot* of money — if you're not very careful.

Can Policyholders Make any Changes They Want?

In a word, yes. The insurance company is paying for the damage that occurred to the house as a result of the covered loss. After all parties have agreed to the value of the loss, the homeowners can appropriate the money however they wish, so long as the house gets repaired. If there's no mortgage on the house, the homeowners can settle with the insurance and do as little or as much in the way of repairs as they wish.

In some cases, a homeowner may wish to make a change that saves him money. For example, if a fire damages an existing wood shake roof, the insurance company will pay the homeowner for new wood shakes. However, the homeowner decides he wants a new laminated composition roof instead of wood shakes, which is not only fire-resistant, but it's also less expensive. The homeowner can opt for the composition roof, and keep the remaining money or use it towards other improvements.

In other cases, the change may be more expensive. Suppose that same homeowner wants an upgraded tile roof instead of wood shakes. He can pay for the cost difference out of his own pocket, or perhaps he might do his own debris removal and site cleanup to save some money, which you can credit him against the upgrade.

The Difference Between Restoration and Remodeling

The insurance company is obligated to pay to replace whatever was damaged in the loss, with materials as close as possible to what was there when the loss occurred. No more, no less.

In the case of large remodeling projects associated with a loss, the situation is pretty obvious. If a person has an 1,800-square-foot house and they have a fire, they can't rebuild 2,000 square feet and expect the insurance company to pay for the difference.

But what about a small, simple upgrade, like replacing 2¼-inch base with 3½-inch base? What about painting a wall red rather than white like it was before? These little changes seem like no big deal to the homeowner, but they add up, and *the insurance company won't pay for them*. And they shouldn't be expected to.

So unless you want to absorb the extra cost yourself, the homeowners will have to pay for any and all changes to the original estimated repairs. But before they'll open their checkbook, you've got some work to do:

- You're going to have to track and bill the labor and material costs of those changes.
- You're going to have to convince the homeowner that *they* have to pay for it, *not* the insurance company.
- You're going to have to ensure that there's no overlap between what the homeowner is paying for and what the insurance company's paying for.
- And, you're going to have to do all this without interrupting the overall flow of the job.

So, as you can see, it isn't quite as easy it seems.

Figure 25-2 Several of the major beams in the house were damaged, but still salvageable after cleaning and sealing. Others had to be replaced as the result of the fire. Some could have remained if the owners rebuilt the house as it was, but due to the remodeling, had to be changed. This type of project requires a careful balancing act of charges and credits.

Your Initial Estimate and Contract

Let's say you estimate the repairs on a loss at $25,000. The insurance company and the homeowners agree that you've included everything that was damaged, and your estimate allows a fair and reasonable price for the repairs. The insurance company agrees to pay the homeowners $25,000, and the homeowners agree to accept that amount to settle the claim.

Unless additional damage is found, your estimate has set the value for the loss at $25,000. From the insurance company's perspective, everything is done. They'll keep the file open until all the work is complete and the final paperwork is turned in, but for all intents and purposes, they're now out of the picture.

As the insurance company steps away from the picture, everything switches to your relationship with your clients, the homeowners. If they don't want to make any changes, you can write a contract for $25,000 and get on with the work.

If they *do* want to make changes, you then begin a new estimating process. The goal now is to take your initial estimate for repairing the house to a *pre-loss condition* and change it to reflect what the clients *actually* want you to do.

Why Changes Get So Confusing

A simple change in the size of baseboards is pretty easy to manage, but most of the changes you'll be faced with aren't that easy. The big problem is that the insurance company is already paying for part of the work. When the homeowner wants to make a change, it falls to you to calculate how much of the overall work needs to be charged against the amount provided by the insurance company, and how much needs to be charged to the homeowner. See Figure 25-2.

REAL STORIES:

It's Only Another Foot!

I was at my very first construction job, working as a helper for a small remodeling contractor. We were rebuilding an old detached garage that had a lot of dryrot damage. The project included rebuilding two of the walls and installing a section of new foundation. Before we got past the demolition stage, the homeowner asked the contractor, my boss, if he could extend the garage an additional foot.

I watched as my boss took measurements, and spent the time to do the calculations. It took a while, and I started getting impatient with the delay. So, with all the arrogance of youth and inexperience, I finally asked him what the big deal was, "It's only an extra foot, does it really matter all that much?"

Thankfully, he was an ex-Navy officer, and he'd learned to spot a teaching opportunity when it came up. So he took me on a short tour of the garage.

"Just an extra foot, huh? That extra foot requires additional excavation," he explained. "It means more time setting forms, and more concrete for the foundation. It means additional pieces of lumber. It means more sheathing, and more stucco. The rafters have to be extended, which means more roof sheathing, and more shingles. It takes extra manhours. It requires some additional cleanup. It might even mean an alteration to our building permit, which will chew up even more time. And the extra time we spend on this job will delay other jobs down the line.

"I know you hope to be a contractor yourself one day, so keep this one thing in mind: *Every change that a homeowner makes has a price tag!* Even if they end up not doing that extra work, it costs you time to estimate it. The sooner you understand that, the better. Otherwise, you'll constantly be reaching into your wallet and paying for all of it yourself. And that will eat you up in no time at all."

It was a good lesson, and one I've always been grateful to have learned early enough in my career for it to have done me some good!

Most of what you do in relation to a change will be questioned by people on both sides. The homeowner typically feels that since so much work is already being done as part of the insurance loss, anything additional should be pretty cheap — if not free. The insurance company is going to be watching to be sure that nothing that's not related to the original loss is being slipped through on their nickel.

Some of this may seem very trivial and unimportant to you. You may be thinking: "What's an extra few square feet of paint, since I'm painting anyway?" Or, "My framers are already on the job. What difference does it make if I frame the wall here, like it was before, or over there, where they want it now?"

The fact is, it matters a lot! Once you get established, your company will be doing hundreds of different jobs every year. If you leave money on the table at each of those jobs because you don't want to take the time to calculate a fair price for the additional work created by homeowner changes, you'll end up throwing away thousands, even tens of thousands of dollars!

Here are some examples of homeowner changes, and how they can affect you:

- The insurance company is paying to paint two walls in Dan's dining room, including masking, tarping, and moving enough contents to access those two walls. Dan wants the other two walls painted as well. You'll have to calculate how much additional paint is required, as well as the labor and materials for the extra masking, and the labor for the extra content moving.

Be aware that Dan will probably feel he should only be charged for the extra painting labor, since the insurance company's already paying for the paint, the masking, and the content moving.

- Jill has a red bedroom that was damaged by a water loss, and you've estimated to repaint the room. Based on pre-loss conditions, and the like-kind and quality of paint, you've estimated one coat of red paint. However, Jill wants the color changed from red to white. It seems like no big deal to her, since you're painting the room anyway, and the insurance company is paying for it. But to cover all that red is going to require a coat of primer first, so you have additional materials and labor that you're going to have to charge her for.
- Here's another commonly-overlooked painting change, in reverse. Tim has a white bedroom, and you've bid to repaint the walls. However, he wants to change from white walls to dark blue, but leave the ceiling white. Tim assumes there won't be any charge for this, since you're painting the walls anyway. However, the walls are going to now require two coats of blue instead of one coat of white, to get the blue paint to cover evenly. In addition to the extra coat, there's additional labor to cut in the second color at the ceiling.
- Here's one that's a little more involved. A fire at Genie's house damages a wall between the kitchen and the living room. The insurance company's paying to rebuild the wall, which includes repairing two electrical outlets, doing some drywall work, and painting the wall. Genie wants the wall moved out about a foot toward the living room, to make the kitchen larger. It involves the same amount of framing, drywall, wiring, and painting, so she thinks the change should be done at no additional charge.

  However, moving the wall requires new drywall work to be done on the ceiling, and the kitchen ceiling will have to be painted, which wasn't in the original estimate. Painting the ceiling requires masking off the entire kitchen. The one-foot wall move requires one additional foot of new flooring. And what was originally a simple repair of two electrical outlets now requires a junction box and rerouting wires. All in all, it's far from a "no charge" alteration.

Completion Dates

Completion dates are always a sticky part of doing additional remodeling work. If the homeowners aren't able to remain in their home because of the loss, the insurance company will pay for outside living expenses. But if the repair work takes longer

Figure 25-3 The fire on the upper floor of this house destroyed the roof of the garage. That was covered as part of the insurance loss. But because the homeowners decided to add a third stall to the garage, the roof line was changed, which added time to the reconstruction. Changes in completion time, as well as outside living expenses, are factors that need to be taken into consideration when doing remodeling work on an insurance loss.

than expected due to homeowner changes, the insurance company isn't going to be happy about it. See Figure 25-3.

Suppose a fire occurs in a kitchen, and the damage is bad enough that the homeowners need to move out temporarily. You estimate that the repairs are going to take three weeks, so the insurance company agrees to pay for a hotel with a kitchenette for that length of time.

But before the repairs begin, the owners decide to take this opportunity to remodel the kitchen. You work through all the numbers, give them credits for this and charges for that, and come up with an entirely new estimate, which they agree to. The three-week repair job is now an eight-week remodeling job, leaving the homeowners with no kitchen, no hotel, and no additional outside living expenses for an additional five weeks.

That's their problem, not yours. But they'll quickly *make* it your problem. They're going to want you to speed up the job and get them back into their home, or set up a temporary kitchen in their home, which wasn't in your bid. And you'll probably get a phone call from the adjuster as well. It's not an insurmountable problem, but definitely one you need to think about!

Four Primary Categories of Changes

The remodeling changes that homeowners request usually fall into one of four broad categories. Understanding these categories will make it a little easier for you to estimate the changes in a way that's fair to everyone.

One: Changes that Affect Both Material and Labor

This is the most common type of change, and can sometimes be relatively simple. For example, suppose a homeowner had wallpaper in the bathroom which was valued at $25.00 a roll, installed, and you've estimated that you'd need 10 rolls to replace it. The homeowner now wants to switch from wallpaper to paint. It's going to require $300.00 to repair a couple of seams and texture the previously smooth walls, and an additional $150.00 to mask and paint everything. That brings your new estimate total to $450.00. Then you need to credit them back the $250.00 you had in your original estimate for the wallpaper, resulting in a net additional charge to the homeowner of $200.00.

That's a fairly straightforward example, but others are much more involved. While rebuilding a kitchen after a fire, you may be asked to relocate appliances, radically change the cabinet layout, or move walls. While repairing a bathroom after a water loss, the owners may want to change fixture locations, and maybe do a small bump-out to create a big new shower.

These changes require a lot of recalculating and adjustments to the original charges. In addition to the new charges, they require credits for work that was in the original estimate, that won't be done now. *Be very careful not to take these changes too lightly.* Never assume that the cost of what you're *not* doing in one area will be enough to cover what you *are* doing in another. Take the time to work through the additions and subtractions, and come up with an accurate number.

Changes to Building Permits

If a building permit is required as part of a loss, the insurance company will pay for the cost of the permit, as well as your time to go and get it. They'll also cover any related costs, such as drawings or engineering reports. However, if the homeowners make any type of change that alters the permit, the insurance company won't pay for the additional permit fees, or for any additional drawings or other alterations.

Two: Changes that Affect Material Costs

A change in materials is a fairly common request as well. Let's say that the homeowner had wallpaper in the bathroom which was valued at $20.00 a roll, and they pick out new paper priced at $22.00 a roll. The preparation and installation labor is the same, so the only cost difference is the additional $2.00 a roll. If you had estimated 10 rolls, the material change will result in an additional charge of $20.00 to the homeowner.

Material changes can work in the opposite direction as well. Let's say that the homeowners had oak baseboards in the living room, but they want to change to paint-grade MDF. Once again, the labor to cut and install the baseboards is the same. Your original estimate was for 64 linear feet of oak base at $1.90 per foot, for a total of $121.60 in material costs. The new material will be 64 linear feet of MDF base at $0.90 per foot, for a total of $57.60. That results in a credit to the homeowner of $64.00.

Three: Changes that Affect Labor Costs

Some changes will affect the cost of labor without necessarily affecting the cost of materials. For example, let's say a small kitchen fire ruined the countertops, cabinets, and cooktop, and you've estimated to replace them.

When it comes time to do the reconstruction, the owners want a slightly-different layout, which puts the cooktop and range hood in a new location. Your electrician is going to use the same amount of wire and other supplies, so there's no change in his material cost. However, he's going to have to route the wires through a different area, at an additional charge of $65 for one extra hour of labor. The HVAC contractor is going to charge two additional hours of labor, at $55 per hour, because the hood ducting will now be more difficult to install.

So the result of this change is going to be $175 in labor, even though no additional materials have been used.

Four: No-Cost Changes

Some changes don't have any cost attached to them. It could be a simple change in paint color, or it could be a change in the style of baseboard. You only need to document these changes in writing so there's no question later on. The easiest way to do this is with a simple Change Order form (see Figure 25-4) that shows what you've agreed to do:

Change paint color in master bedroom from Acme Polar White to Acme Subtle Beige. *No charge*

or

Substitute MDF colonial base for MDF bullnose base in living room. *No charge*

This gives both you and the homeowner a written record of the changes, so that there's no confusion or disagreement in the future.

Changes in Overhead and Profit

As we discussed previously, insurance companies require you to show overhead and profit as individual line items on your estimate. If the homeowners have seen the same version of your estimate that you provided to the insurance company, then they're well aware of those O&P numbers. They don't usually have a problem with O&P when it's the insurance company that's footing the bill, but when it comes to changes that they're paying for, those numbers can get a little confusing, and even be a source of dispute.

There may be times when you want to use different O&P figures than those you've used with the insurance company. For example, the complexity of the changes may be such that you want to increase the amount of O&P you're going to charge. Or, the changes may result in an estimate that's lower than the initial one, but you don't want to drop the amount of your overall O&P for the job.

Craftsman Restoration
123 Main Street
Alltown, CO

PH (555) 888-1234
FAX (555) 888-4321

CHANGE ORDER #1

DATE: **9/5/11**

CLIENT: **BOB SMITH**

JOB LOCATION: **1212 APPLE LANE**

The following change(s) are hereby authorized and agreed to as specified below:

Description of changes	Amount
Delete shake roofing in original insurance estimate, dated 8/26/11	$ (4,362.48)
Substitute Pine Willow Industries laminated composition shingles, color Aged Barnwood	$ 2,791.10
Total Amount Due:	**$ (1,571.38)**

Acceptance: *The prices and specifications shown above are hereby agreed to in full, and are now a part of the original contract and specifications. It is understood that all terms and conditions of the original contract are hereby extended to include these changes. All change orders are to be documented in writing, signed by all parties, and fully funded prior to commencement of work on that change order. All change orders, implemented or not, are subject to a base fee of $50.00 per change order.*

____________________ **Craftsman Restoration**

____________________ **Client**

____________________ **Date of acceptance**

Figure 25-4 Use a Change Order form like this one for remodeling modifications after the contract with the homeowner has been signed. Make sure it spells out all the details of the proposed changes. The change order should reference the original contract, and make it clear that all of the terms of the original contract are still in force.

Remember that while showing the O&P numbers was required for the insurance estimate, when you're writing up change orders for the homeowner, showing those figures is entirely up to you — and so is the amount you charge.

How you want to handle charging for or crediting back O&P on changes is also up to you. There are two basic methods for estimating and writing up your changes. One method shows the O&P in the same manner as the original estimate, and the other doesn't. There are instances when factoring the O&P into the individual costs of the changes or into the bottom line number will avoid a lot of arguments and requests for credits or discounts. Always remember that O&P is how you keep your business going and how you make your living — you're entitled to it.

Estimating Changes in the Original Format

If you estimate changes using the same format as the original insurance estimate, it's very easy for the homeowners to compare the original estimate and the revised estimate side by side. Using *Xactimate®*, for example, you can take your original estimate, copy it, and then add and subtract from the copy as needed to come up with your new estimate.

This method can also save you a considerable amount of time, especially on large changes that involve many alterations to the unit costs. For instance, you may be changing the type of baseboard in every room, as well as changing from carpet to hardwood in some rooms, and substituting paint for wallpaper in others. Since you've already written the original estimate with all of the linear footage and square footage numbers figured in, altering the original estimate won't take nearly as much time as starting from scratch.

You'll certainly want to include additional O&P charges when a change results in an increase in the price of the job. If you use your original estimating software to make the changes, the O&P changes will automatically be generated as well.

Let's look at a situation in which an owner wants to make a change, and you've estimated the original insurance restoration job using *Xactimate* or another computerized estimating software.

Let's say the Bakers had an icemaker line break in their kitchen, flooding several rooms. You did the emergency work, and wrote up an estimate for the repairs. The house originally had painted MDF baseboards. As part of the demolition during the drying process, you disposed of the baseboards in the living room, dining room, kitchen, master bedroom, and master bathroom. You've estimated to replace and paint the same type and quality of baseboards in each of those rooms.

Mrs. Baker would like to upgrade to stained oak baseboards instead of the MDF in those rooms. She'd also like to have you remove the baseboards in the remaining rooms, and replace them with the stained oak as well.

At this point, you can open up your original estimate, make a copy of it on your computer, and then make your new estimate based on the requested changes. Go back into the rooms where you've calculated replacement baseboard, and change each entry from *replace MDF base* to *replace oak base* using the appropriate categories. Next, change each entry from *paint base* to *stain and finish base*. That takes care of the changes to the rooms you've already bid.

Now, you need to calculate the base in the other rooms where Mrs. Baker wants to replace the baseboards. If you're lucky, she discussed these changes with you the first time you were out measuring and estimating the job, so you went ahead and measured the additional rooms at that time. If not, you need to make another trip out to the site and get the additional room measurements.

You can then add the other rooms to your original drawing, or you can simply create a new, fictitious room called *changes*, and enter the total length of the additional base into that room. Remember that this additional baseboard is still in place, so you want to use the *remove and replace base* category, not just *replace base*. Add in the stain and finish for the new baseboards, as well as a small additional charge for debris removal, and you're done. The computer program will calculate the new numbers for you.

Using this method, in a very short time you've created a revised estimate that shows:

- the credits for replacing the original base in all of the rooms where you had originally listed it;
- the credits for painting the original base in those rooms;
- the charges for removing and disposing of the base in the rooms that weren't part of the original loss;
- the charges for material and labor to install the new oak base in the entire house;
- the charges for material and labor to stain and finish the new oak base in the entire house;
- a revised subtotal, revised O&P figure, and revised total for the entire job.

> ***Tip!*** *When revising an estimate on your computer, use* **Copy & paste** *or your estimating software's* **Duplicate estimate** *feature to make a copy of the original. Label the copy with the name of the client and mark it as a revision. For example, call the revised estimate "Baker 2," or "Baker changes," so you know it isn't the original. Always do your revisions on the copy, not on the original. That way, if something happens during the revision process, you still have a permanent copy of the original.*

Estimating Changes in a Lump-Sum Format

As you can see from the example we just looked at, using your computerized estimating program is a quick way to make revisions to an estimate. However, there are also a couple of drawbacks to that method.

For one thing, when you make the type of changes described, you alter the entire estimate. Suppose that in addition to the changes in the baseboard, Mrs. Baker wanted you to estimate changing all the casings in the house, as well as all the doors. Then, after you plug all those changes into the computer and come up with a new figure, Mrs. Baker realizes that the cost is more than she wants to spend. She asks if you can redo the estimate with the base and casing, but not the doors. You now have to go back into the computer, create "Baker 3," and do another estimate. It's not difficult, but it *is* time consuming.

Another drawback to using your original insurance estimate is that it shows your O&P as two separate line items. That means that if the changes result in a credit, it will reduce your profit and overhead amounts as well, something that you may not want to do. Finally, reusing the original insurance estimating format also sets the overhead and profit percentages at 10 percent each. If you want to adjust those percentages up or down for the new changes, it's difficult to do.

So, you may want to consider an alternate method of writing estimates that includes homeowner changes. Rather than breaking out the O&P at the end of the estimate, you can simply factor it into the individual charges at whatever percentage you want, as we did with the *National Estimator* estimate completed in *Job Cost Wizard* in the last chapter (refer back to Figure 24-6).

However, the easiest way to document changes is to put together a comprehensive list of all the changes the owner wants to make, along with the charge or credit for each one. Then write each of the changes on a Change Order form (Figure 25-4) as a simple bulleted list, along with the bottom-line price, rather than in the format created by your original estimating software. For the Baker example, it might look something like this:

Proposed changes to original estimate for Mr. and Mrs. Baker, 123 Main St., dated June 1, 2011

- **Base, $720.00.** Delete cost of MDF Colonial base in living room, kitchen, dining room, master bedroom, and master bathroom. Delete cost of painting base, same rooms. Supply and install stain-grade oak Colonial base, same rooms. Remove and dispose of existing MDF Colonial base in remainder of house, and replace with stain-grade oak Colonial base. Stain and finish all new base. Note: net price includes credits from original estimate.
- **Casing, $1,290.00.** Remove and dispose of all existing MDF door casings for entire house. Supply and install stain-grade oak Colonial casings. Stain and finish all new casings.

You can then continue in the same manner for the doors and any other changes that Mrs. Baker has asked you to estimate for her.

This method takes a little more work to write up, since you aren't pulling descriptions directly from your estimating software. It's also not quite as detailed, when compared side-by-side with the original estimate. However, it allows the client to see at a glance how much each of the proposed changes will cost, which makes it easier for them to decide what fits their budget. It also gives a clear description of what work you're proposing, as well as what you're crediting them back from the original estimate. And finally, it keeps your O&P hidden. For those who prefer having everything on your computer, you can keep a simple Change Order form on your computer, make a copy of it, fill it out and print it for your client.

Changes Before the Job Begins

Ideally, the homeowners know they want to make remodeling changes before you begin your work. That way you can discuss the changes, estimate them, and incorporate them into the initial contract. This eliminates a lot of confusion, as well as having to write change orders later on.

There are two ways to do this. If you've completely redone the estimate to incorporate the changes, then simply reference the new estimate in the contract. For example, on your contract, under the *description of work* section, insert something such as:

> *All work to be completed as per revised estimate dated September 19, 2011. This revised estimate replaces all prior estimates.*

That clearly states which estimate you're basing your contract on, and it also shows that any estimate you wrote and presented prior to that date isn't being used.

Perhaps you've prepared an estimate for the changes using the change order format. In that case the wording should reference both the original estimate and the written changes, like this:

> *All work to be completed as per insurance estimate dated June 15, 2011, with the exception of those changes described in Change Order Estimate #1, dated June 22, 2011.*

Again, this wording is to clarify which estimate has priority. It clearly describes the date of the insurance estimate being used, as well as the date and number of the change order.

Changes After the Job Begins

Quite often, you'll sign a contract and get the job underway, then encounter changes at a later date. In that case, you'll need to write one or more change orders. Since a change order is basically an extension of the original contract, it has to document whatever change is being made. Change orders include:

- changes in the specifications for the job
- changes in price
- an extension of the completion date
- an alteration in your original contract agreement

Your contract will usually include a clause that says the contract contains the complete agreement between you and the homeowner. It also usually states that any changes to the contract need to be in writing, and need to be agreed to and signed for by all parties.

That's an important point to remember. If you make a change based on a verbal agreement with the homeowner, and then run into a dispute over payment later, most attorneys will tell you that in a dispute between what's written and what's verbal, the written agreement will win out. So *always* get it in writing!

Here's a typical contract clause that addresses exactly that. If you don't already have this or something like it in your contract, you should add it:

> *Any alteration or deviation from the specifications set forth in this contract involving increased or reduced cost of materials or labor will be executed only upon written orders for same, signed by both the homeowner and the contractor, and will become an additional charge or a credit to the sums set forth in this agreement.*

A change order can be handwritten on a blank form, or it can be something that's printed up on your computer. What matters is that it contains the following information:

- **Change order number.** Number the change orders in sequence for each job. On larger jobs, where you end up with multiple change orders, this makes it a lot easier to keep the changes organized. It's also very helpful if you have to make changes to the changes (hopefully you don't have too many of those!). For example: *Change Order #3. Delete oak cabinets as specified in Change Order #1. Substitute maple cabinets, same design and layout.*
- **Job reference.** This can be the client's name, the job address, the job number, or whatever is used in the original contract to identify the job.
- **Date.** This is the date of the change order.
- **Exact description of the changes.** For simple changes, you can put them right into the change order itself. For example: *Add one 4-0 4-0 double-pane sliding vinyl window in dining room, style to match existing window.* For multiple or more complicated changes, refer to the change estimate that you wrote, such as: *Changes to be as per Change Order Estimate #2, dated May 15, 2011.*
- **Changes in the price.** If the change results in a charge or a credit to the original contract amount, clearly state how much that charge or credit is, even if the proposed change doesn't increase or decrease the price. For example: *Change kitchen cabinet handles from Acme #12 in polished brass to Acme #56 in bronze. No change in price.*
- **Changes in the payment schedule.** For larger alterations to the contract, you may need to change the payment schedule. You may want to ask for a deposit against the cost of the changes, or you may add in another payment draw, or even ask that the changes be paid in full before the work begins. Whatever your agreement, document it in the change order.

- **Changes in the completion date.** This is very important, and it's something that's often overlooked. You want to make sure the homeowner understands that the changes you're going to make at his request will delay the completion of the job. Even a simple change can cause problems — like giving a homeowner a price for the new doors he wants, but neglecting to mention that they're special order items with a delivery that's six weeks out.
- **Reference to the original contract.** Be sure that the change order references the original contract. It should state that any parts of the original contract that are not affected by the change order will remain as originally written.
- **Signature and copies.** Sign and date the change order, and make sure the homeowner does the same. Give him an exact copy for his files, and keep one for yours.

Get Your Lawyer Involved

Change orders are going to become a big part of your life as a restoration contractor, especially on large jobs. You want to be absolutely certain that whatever change order format you use is legal and binding. Consult with your attorney and design a change order form that meets all specific legal requirements for your locale, as well as those of the state where you live. That includes making sure that your change orders meet state and federal guidelines for the three-day right of rescission laws, also known as the *cooling-off period.*

To Charge or Not To Charge

Change orders will inevitably become a big part of your life. They can be simple, or they can be complicated. There may only be one on a job, or they may come at you in what seems like an unending stream.

You can't blame the homeowner for asking. They didn't intend to have any work done on their house, then suddenly there's a fire and they have an insurance draft for $50,000 on the way. That would make anyone think it's a great time to make some of the improvements to the house that they've always wanted. They don't know how much these things will cost, so they're going to ask you.

But, you need to make the decision about how often you want to sit down and figure out "How much would it cost if … ?", or "How much would I save if we don't do … ?"

One way to control the number of times they ask that question is to charge an estimating fee or a change order fee. Your initial estimate is always going to be free. Occasionally you'll be hired and paid by an insurance company to write an estimate for a job that they'll tell you up front you're not going to get. But other than that, you're going to be expected to estimate for free. That's just the way it is.

But that's not the case with changes. It's up to you to decide whether or not you want to charge for estimating changes and other additional work. Here are a couple of options to consider:

- **Always do free estimates.** Just assume estimates are part of being a contractor, and provide whatever estimates you're asked for at no charge.
- **Provide a certain number of estimates at no charge.** You may choose to provide the first one or two change order estimates for free. After that, charge an hourly estimating fee, or a fee on a per-estimate basis. If you choose to charge an estimating fee, be sure you specify those charges in your initial contract. Otherwise, you run the risk of angering the client when you hit them with unexpected charges.

  When a client is aware that they'll be charged for the estimates at some point, they tend to be more focused. They usually try and get all of their estimating questions answered at one time, rather than hitting you with a lot of "what ifs" as the job progresses. Also, once you've established that you charge a fee, you have the option of waiving that fee if you want, which becomes a selling feature.
- **Charge a refundable fee.** A refundable fee is especially good with large change estimates, such as room additions or major renovations. You can charge a set fee or a per-hour fee for the estimate, with the understanding that all or part of the fee will be refunded if the client chooses to have the work done. It's a good incentive for them to go ahead with the remodeling project, while still covering you for your estimating time if they don't. Again, be sure that this is spelled out in your contract.

REAL STORIES:
It Was the "Perfect Storm" of Changes!

A couple was doing some exterior work on their home, getting it ready to put on the market to sell. They built a new cedar fence and applied an oil finish to the new wood. Then they put fresh bark mulch around some of the planters next to the fence. Finished with their work, they spread the oily rags out in the yard to dry, and left to run some errands.

While they were gone, a strong wind kicked up. The oily rags were blown into a pile against the fence, where they spontaneously combusted. The burning rags caught the freshly-oiled wood on fire, fed by the fresh, dry bark. The gusting winds pushed the flames against the wood siding where it met the fence, and the flames quickly spread up the wall and into the attic. The result was severe structural damage to the attic and upper floor, and water damage to the lower floor — the result of the firefighting efforts.

Over a period of several days, I inspected and measured the house, and wrote an extensive estimate for the fire and water damage. We eventually

(continued on next page)

REAL STORIES:

It Was the "Perfect Storm" of Changes! *(continued)*

signed a contract, and began the work. As the demolition work got underway, we discovered quite a bit of additional damage inside the walls and in previously-concealed parts of the attic, resulting in a fairly lengthy supplemental estimate. This was followed by an additional set of change orders to the original contract.

When the demo was complete, the owners approached me about some extensive changes they wanted to make. They were still planning on selling the house, and thought the changes would make the home more appealing. They had a big house, and their ideas involved several structural changes, as well as changes to the wiring, plumbing, and heating systems. Since part of the work was covered by the insurance estimate and part of it wasn't, the changes involved a lot of number juggling. Credits had to be issued for work that was no longer being done, while additional charges were necessary for new work that *was* being done.

The complexity of the changes — and then the changes to the changes — made this another multi-day process. More change orders were issued, and reconstruction work finally got underway again.

In the meantime, the adjuster contacted me. He was aware of the changes that the owners were making, and wanted to know how it would affect the completion date and the couple's outside living expenses — which the insurance company was paying for. I was stuck trying to figure out what the time difference would be, taking into account what we were going to do initially versus what we were doing now. I came up with an approximation of the time difference, and then had to meet with the adjuster and the homeowners, who learned that part of the time they'd be out of the house *wouldn't* be covered by the insurance company. Even though the time delay was due to *their* changes, it put us under pressure to get the work done more quickly.

We had just completed all of the rough framing, wiring, and plumbing, when I got a phone call from the homeowners. Through a fluke encounter with a friend of a friend, they'd found a buyer for their still-unfinished house. At first I didn't think too much of it, since our contract was with the current owners, and the new owners wouldn't be taking over until after the job was done.

I couldn't have been more wrong.

The buyers were coming in from another state, and due to some tax considerations they needed to take ownership of the house as soon as possible. The sellers — my current clients — already had their eye on another house in town, so they were perfectly willing to make an immediate sale.

And on top of all that — yep, you guessed it — the buyers wanted to make some changes!!

Just when you think you've seen it all, something new comes along, and this was definitely something new. We had a contract with the current owner, who was selling the house to a new owner, while the contract was still in effect and uncompleted. After meetings with our attorney and some head scratching with my partner and our general manager, we agreed that we would undertake the project. Turns out we had the upper hand here, since they couldn't walk away from the contract without generating a lien, which would have messed up the entire sale.

We insisted that the seller remain responsible for the original contract, and we requested a fairly steep estimating fee before I would begin, once again, with the incredibly tedious process of estimating the next round of changes. In the meantime, the seller had the buyer put an additional amount of money into escrow to cover the changes the buyer was making to what was essentially still the seller's house. Talk about convoluted!

So now I had to meet with the buyers as well as the current owners. Brand new walls that had just been framed had to come back out. Brand new electrical wiring, plumbing, and ductwork had to be moved — again. On top of all that, a couple of the changes required alteration of the existing building permits!

It was a massive paperwork balancing act. Is this stud being charged to the new owner, or the old one? Is this outlet part of what the insurance company's paying for, or did the sellers request that we install it? Who ordered those plumbing changes — the seller or the buyer?

In the end, the house turned out beautifully, but I'm not sure anyone's nerves were ever the same again. Estimating fee or no estimating fee, when the "perfect storm" of estimates, supplementals, change orders, and changes to the changes to the changes start piling up over your head, run for shelter somewhere safe!

Disclosure Laws

While we're on the subject of selling, let's talk about disclosure laws. Most, if not all, states have disclosure laws regarding the sale of real estate. That means that if a seller knows something about the piece of property that he's selling, and what he knows could affect the condition or value of the property, then he has a legal obligation to inform the buyer. That knowledge typically extends to insurance-related losses to the property, including fires, water losses, and especially mold.

As part of this disclosure process, you may occasionally be contacted by a past client or a real estate agent with a request for information about repairs you made on a particular house. The best advice is to tread lightly and carefully; you don't want to provide confidential information to someone who's not entitled to it, but you don't want to deny someone information who has a legal right to it. Here's some advice:

1. Get the information request from the buyer and the seller, or their agent or title company, in writing.
2. Consult with your attorney, and get his advice in writing before complying with the request. Be forewarned that you're probably going to have to pay the attorney fees out of your own pocket — but that's the cost of doing business. You can skip this step, but you risk the liability of granting or withholding information, and that could get you into trouble and cost you even more legal fees.
3. Keep a copy of all documents, notes, responses, and other correspondence.

The Emotional Side of Insurance Restoration

For your clients, an insurance loss can range from a mild inconvenience to a horribly traumatic or even life-altering event. As a restoration contractor, you'll be entering that environment, so it's important to understand how people react to a loss. Your success in this field will greatly depend on how you handle yourself and your clients during the initial contact.

You already know your role as a good contractor. But now you're about to take on some additional roles and become part psychologist, part guidance counselor, and part new-best-friend as well. These are all qualities you'll need in order to deal with your clients after a loss. And just as no two losses are the same, no two people are the same either. You're going to encounter people at their best — and at their very worst.

Having sudden damage occur in their home brings out all kinds of emotions in people. Some people become angry with you, as if it's all somehow your fault. Others treat you like a hero — you're there to help them through a tough spot. Some of your clients will be incredibly selfish, or even outright dishonest. They'll ask you to lie for them and help them try to falsify their claim. Others will be very careful not to take advantage of you or their insurance company. And some will bring you cookies.

REAL STORIES:

Now, About that Bathroom Door

I was once called out on a fairly large fire loss. I knew the adjuster very well, and the people who owned the home didn't have a contractor, so he recommended our company.

I met the adjuster and the clients on the job, and we all toured the damage together. Then we sat on a couple of undamaged chairs in the backyard. The adjuster explained a number of coverage issues to them. I could see they were listening, but beginning to get a little confused.

Then I talked to them a little about what would be involved in the repair process. Both the adjuster and I could tell that all this information was beginning to overwhelm them, so we kept our meeting as short as possible. They asked a couple of questions, but I sensed they were starting to drift.

At one point, the gentleman stared down at the ground for a long moment. He clearly had something important that he wanted to ask, so I paused and waited for his question.

"The bathroom door was painted kind of an ugly blue color," he said at last. "I never really liked it. Will we be able to paint it a different color?"

Here was this man and his wife, in the midst of a big fire-damaged mess. They were facing a couple of months of repair work; they'd lost a lot of irreplaceable contents; they were going to have to move into a temporary home; and they were discussing these matters with two people they'd met only an hour before. And he wanted to know about the color of the bathroom door! That's when I knew anything else we told them simply wasn't going to sink in. Not on that day.

So I just assured him that he could have his bathroom door any color he wanted, and we wrapped up the meeting.

Take it all in stride. When someone's rude or angry with you, put it in the context of the situation, and try not to take it personally. And when someone's genuinely grateful for all you've done to help, be sure and take that in and remember the moment. Rebuilding people's lives is what you do, so whenever you're acknowledged for it, pat yourself on the back.

What People Do — and Don't — Hear

As you do more and more jobs, you'll begin to read people surprisingly well. In time you'll hone your ability to understand and empathize with them, and that goes a long way in this business.

For most people, any loss involving their home, no matter the size or the type, is overwhelming. They have to deal with the physical damage, as well as the insurance agents, adjusters, contractors, cleaners, and a whole bunch of other strangers. Add to that the forms, contracts, authorizations, notices and other paperwork, and their lives quickly become a jumble of information going in one ear and out the other.

People have a limit to what they can take in. The more extensive the loss, the faster that limit's reached, and once it's reached, they no longer absorb what you tell them. They'll listen to you as you speak, and they may even nod as if they understand, but they're not actually *hearing* what you say.

This is an important fact to recognize. You may notice that the person's eyes seem a little glazed or their attention starts wandering. They may ask you to repeat what you just said, or they may repeat a comment that they made just a moment before. Those are your clues that they've reached their saturation point. And, at that point, you need to wrap up your meeting. Call or stop in the next day, and pick it up again where you left off. You'll probably have to go back a few steps, but you'll avoid a lot of misunderstandings later.

In Conclusion

Way back at the beginning of this book, I promised to introduce you to the world of insurance restoration. I hope I've done a worthy job of that, and you've had a pretty comprehensive look behind the curtain, at both the good and the bad.

In conclusion, I offer you my personal best wishes. I hope your career in insurance restoration is every bit as rewarding as mine has been. This really is a blending of the very best of what construction has to offer: Rebuilding homes, and rebuilding lives. Good luck!

Appendix A

Information Sources

Products and Equipment

Absorene Manufacturing Company, Inc.
(Cleaning products)
2141 Cass Avenue
St. Louis, MO 63106
Phone: 314-231-6355 or 1-800-662-7399
Fax: 314-231-4028
Web: www.absorene.com

Airsled, Inc. *(Appliance moving equipment)*
70-A Aleph Drive
Newark, DE 19702
Phone: 800-247-7533
Fax: 302-292-8921
Web: www.airsled.com

American Van Equipment
(Van and truck accessories)
149 Lehigh Avenue
Lakewood, NJ 08701
Phone: 800-526-4743
Fax: 800-833-8266
Web: www.americanvan.com

Aramark *(Uniforms)*
Phone: 800-388-3300
Web: www.aramark-uniform.com

Associated Bag Company
(Boxes and warehouse supplies)
400 West Boden Street
Milwaukee, WI 53207
Phone: 800-926-6100
Fax: 800-926-4610
Web: www.associatedbag.com

Bosch Tools *(Tools)*
Robert Bosch Tool Corp.
1800 W. Central Road
Mount Prospect, IL 60056
Phone: 224-232-2000 or 1-877-267-2499
Fax: 224-232-3169
Web: www.boschtools.com

Branders.com *(Promotional items)*
2551 Casey Avenue
Mountain View, CA 94043
Phone: 650-292-2752
Fax: 650-292-2775
Web: www.branders.com

CDS Moving Equipment, Inc.
(Plywood storage vaults)
Web: www.cds-usa.com
(See website for various locations and phone numbers)

Construction Contract Writer
(Contract-writing software)
Craftsman Book Company
6058 Corte del Cedro
Carlsbad, CA 92011
Phone: 1-760-438-7828 or 1-800-829-8123
Fax: 1-760-438-0398
Web: www.constructioncontractwriter.com

Craftsman Book Company
(Estimating books and software)
6058 Corte del Cedro
Carlsbad, CA 92011
Phone: 1-760-438-7828 or 1-800-829-8123
Fax: 1-760-438-0398
Web: www.craftsman-book.com

Curtis Dyna-Fog, Ltd. *(Thermal-fogging equipment)*
17335 U.S. Highway 31 North
P.O. Box 297
Westfield, IN 46074
Phone: 1-317-896-2561
Fax: 1-317-896-3788
Web: www.dynafog.com

Dri-Eaz Products, Inc.
(Restoration equipment and supplies)
15180 Josh Wilson Road
Burlington, WA 98233
Phone: 360-757-7776 or 1-800-932-3030
Fax: 360-757-7950
Web: www.drieaz.com

Ebac Industrial Products, Inc.
(Restoration equipment and supplies)
700 Thimble Shoals Boulevard, Suite 109
Newport News, VA 23606
Phone: 757-873-6800
Fax: 757-873-3632
Web: www.ebacusa.com

Empire Promotional Products *(Promotional items)*
231 West 29th St., Suite 205
New York, NY 10001
Phone: 212-268-9910 or 877-477-6667
Fax: 212-268-9913
Web: www.empirepromos.com

Harbor Freight Tools *(Tools and equipment)*
3491 Mission Oaks Boulevard
Camarillo, CA 93011
Phone: 1-800-444-3353
Fax: 1-800-905-5215
Web: www.harborfreight.com

Mobile Home Depot
(Manufactured home parts and supplies)
672 N. Milford Road, Suite 104
Highland, MI 48357
Phone: 248-887-3187 or 877-887-3187
Fax: 248-889-4556
Web: www.mobilehomedepotmi.com

Omegasonics Corporation *(Ultrasonic equipment)*
330 E. Easy Street, Suite A
Simi Valley, CA 93065
Phone: 1-805-583-0875 or 1-800-669-8227
Fax: 1-805-583-0561
Web: www.omegasonics.com

On-Site Storage Solutions
(Temporary on-site storage containers)
3637 Glendon Avenue, Suite 101
Los Angeles, CA 90034
Phone: 1-888-667-4834
Web: www.onsitestorage.com

ProRestore Products
(Cleaning and restoration supplies and equipment)
4660 Elizabeth Street
Coraopolis, PA 15108
Phone: 412-264-8340 or 800-332-6037
Fax: 412-262-7150
Web: www.prorestoreproducts.com

Protective Products International, Inc.
(Lead paint abatement products)
140 Kerry Lane
Wauconda, IL 60084
Phone: 847-526-3377
Fax: 847-526-1380
Web: www.protectiveproducts.com

pulsFOG *(Thermal-fogging equipment)*
Abigstr.8
88662 Überlingen
Germany
E-mail: info@pulsfog.com
Web: www.pulsfog.de
(See website for U.S. distributors)

Rotobrush® International LLC
(Duct cleaning equipment)
801 Hanover Drive, Suite 700
Grapevine, TX 76051
Phone: 1-800-535-3878
Fax: 1-877-535-3878
Web: www.rotobrush.com

Ryobi Tools *(Tools)*
Phone: 1-800-525-2579
Web: www.ryobitools.com

Sonozaire *(Ozone equipment)*
A Division of CB&I
3102 E. Fifth Street
Tyler, TX 75701
Phone: 903-510-5337 or 1-800-323-2115
Fax: 903-581-6178
Web: www.sonozaire.com

Uline *(Boxes and warehouse supplies)*
12575 Uline Drive
Pleasant Prairie, WI 53158
Phone: 800-958-5463
Fax: 800-295-5571
Web: www.uline.com

U.S. Products
(Carpet cleaning and water extraction equipment)
11015 47th Avenue, West
Mukilteo, WA 98275
Phone: 425-322-0133 or 800-257-7982
Fax: 425-322-0136
Web: www.usproducts.com

Wm. W. Meyer & Sons, Inc.
(Insulation vacuums and duct cleaning equipment)
1700 Franklin Boulevard
Libertyville, IL 60048
Phone: 847-918-0111 or 800-797-8227
Fax: 847-918-8183
Web: www.meyervacuums.com

Xactware Solutions, Inc. *(Estimating software)*
One, Xactware Plaza
Orem, UT 84097
Phone: 801-764-5900 or 800-758-9228
Fax: 801-932-8013
Web: www.xactware.com

Zefon International
(Mold and other testing equipment)
5350 S.W. 1st Lane
Ocala, FL 34474
Phone: 1-352-854-8080 or 1-800-282-0073
Fax: 352-854-7480
Web: www.zefon.com

Zinsser *(Pigmented shellac and other sealers)*
Rust-Oleum Corporation
11 Hawthorn Parkway
Vernon Hills, IL 60061
Phone: 847-367-7700 or 877-385-8155
Fax: 847-816-2330
Web: www.zinsser.com

ZipWall, LLC *(Portable containment systems)*
37 Broadway
Arlington, MA 02474
Phone: 800-718-2255
Fax: 781-648-8806
Web: www.zipwall.com

Businesses

Certified Restoration Drycleaning Network, LLC
(Textile restoration)
2060 Coolidge Highway
Berkley, MI 48072
Phone: 800-520-2736
Fax: 248-246-7868
Web: www.crdn.com

Doctor Fungus *(Information about mold and fungi)*
Web: www.doctorfungus.com

SCORE *(Non-profit small business counseling)*
Phone: 1-800-634-0245
Web: www.score.org

ServiceMaster *(Restoration franchisor)*
860 Ridge Lake Boulevard
Memphis, TN 38120
Phone: 1-866-348-7672
Web: www.servicemaster.com

Servpro Industries, Inc. *(Restoration franchisor)*
801 Industrial Boulevard
Gallatin, TN 37066
Phone: 615-451-0200 or 1-800-SERVPRO
Fax: 615-451-0291
Web: www.servpro.com

Wise Steps, Inc. *(Industrial hygienist)*
P.O. Box 3895
Salem, OR 97302
Phone: 503-585-4002
Fax: 503-585-3317
Web: www.wisestepsinc.com

Insurance Companies

Allstate Corporation
2775 Sanders Road
Northbrook, IL 60062
Phone: 1-800-255-7828
Web: www.allstate.com

American Family Mutual Insurance Group
6000 American Parkway
Madison, WI 53783
Phone: 1-800-692-6326
Web: www.amfam.com

Chubb Group of Insurance Companies
15 Mountain View Road
Warren, NJ 07059
Phone: 1-800-252-4670
Web: www.chubb.com

Farmer's Insurance
4680 Wilshire Boulevard
Los Angeles, CA 90010
Phone: 1-800-435-7764
Web: www.farmers.com

Foremost Insurance Group
P.O. Box 0915
Carol Stream, IL 60132
Phone: 1-800-527-3905
Fax: 1-800-325-1507
Web: www.foremost.com

Hartford Financial Services Group, Inc.
One, Hartford Plaza
Hartford, CT 06155
Phone: 1-860-547-5000
Web: www.thehartford.com

Liberty Mutual Group, Inc.
175 Berkeley Street
Boston, MA 02116
Phone: 1-888-398-8924
Web: www.libertymutual.com

Nationwide Mutual Group
1 Nationwide Plaza
Columbus, OH 43215
Phone: 1-866-688-9143
Web: www.nationwide.com

State Farm Mutual
One, State Farm Plaza
Bloomington, IL 61791
Phone: 1-800-732-5246
Web: www.statefarm.com

Travelers Companies
One, Tower Square
Hartford, CT 06183
Phone: 1-800-328-2189
Web: www.travelers.com

USAA Insurance Group
9800 Fredericksburg Road
San Antonio, TX 78288
Phone: 1-800-531-8722
Web: www.usaa.com

Zurich Financial Services Ltd
Mythenquai 2
8002 Zurich, Switzerland
Phone: +41 (0)44 625 24 04
Web: www.zurich.com

Trade Associations

American Board of Industrial Hygiene
6015 West St. Joseph, Suite 102
Lansing, MI 48917
Phone: 517-321-2638
Fax: 517-321-4624
Web: www.abih.org

American Industrial Hygiene Association
2700 Prosperity Avenue, Suite 250
Fairfax, VA 22031
Phone: 703-849-8888
Web: www.aiha.org

Independent Insurance Agents & Brokers of America, Inc.
127 South Peyton Street
Alexandria, VA 22314
Phone: 800-221-7917
Fax: 703-683-7556
Web: www.iiaba.net

Institute of Inspection, Cleaning and Restoration Certification (IICRC)
2715 E. Mill Plain Boulevard
Vancouver, WA 98661
Phone: 360-693-5675
Fax: 360-693-4858
Web: www.iicrc.org

National Association of Independent Insurance Adjusters
P.O. Box 807
Geneva, IL 60134
Phone: 630-208-5002
Fax: 630-208-5020
Web: www.naiia.com

National Association of Public Insurance Adjusters
21165 Whitfield Place, #105
Potomac Falls, VA 20165
Phone: 703-433-9217
Fax: 703-433-0369
Web: www.napia.com

National Institute of Restoration
P. O. Box 6790
Charlottesville, VA 22906
Phone: 434-973-4200
Web: www.nir-inc.com

Restoration Industry Association
12339 Carroll Avenue
Rockville, MD 20852
Phone: 301-231-6505
Fax: 301-231-6569
Web: www.restorationindustry.org

Government Agencies

Centers for Disease Control and Prevention (CDC)
1600 Clifton Road
Atlanta, GA 30333
Phone: 1-800-CDC-INFO (1-800-232-4636)
Web: www.cdc.gov

Department of Housing & Urban Development (HUD)
451 7th Street S.W.
Washington, DC 20410
Phone: 202-708-1112 or 1-800-333-4636
Web: www.hud.gov

Environmental Protection Agency (EPA)
Ariel Rios Building
1200 Pennsylvania Avenue, N.W.
Washington, DC 20460
Phone: (202) 272-0167
Web: www.epa.gov

National Institute for Occupational Safety and Health (NIOSH)
(Division of the Centers for Disease Control and Prevention)
1600 Clifton Rd.
Atlanta, GA 30333
Phone: 1-800-CDC-INFO (1-800-232-4636)
Web: www.cdc.gov/NIOSH/

Small Business Administration (SBA)
409 3rd Street, S.W.
Washington, DC 20416
Phone: 1-800-U-ASK-SBA (1-800-827-5722)
Web: www.sba.gov

Appendix B

Institute of Inspection, Cleaning and Restoration Certification (IICRC) Course Descriptions

(Courtesy of IICRC)

IICRC Courses with Descriptions

CCT — Carpet Cleaning Technician (2-day course):

The Carpet Cleaning Technician course teaches the fundamentals of carpet cleaning. Topics include fiber types and characteristics, fiber identification, carpet construction and styles, dyeing at the mill, soil characteristics and the chemistry of cleaning. This course will also introduce the technician to the five methods commonly used in carpet cleaning. The differences between the CCT and CMT courses are, basically, a focus on either general or residential versus commercial applications.

CDS — Commercial Drying Specialist (4.5-day course):

The Commercial Drying Specialist course teaches damage inspection and evaluation, work flow management, process administration and technical methods of effective and timely drying of commercial, industrial, institutional and complex residential water-damaged structures, systems, and furniture, fixtures and equipment (FFE). Resources for this course include: the current ANSI/IICRC S500 Standard and Reference Guide for Professional Water Damage Restoration, reference media, scientific, technical and industry resources.

CMT — Commercial Carpet Maintenance Technician (2-day course):

The Commercial Carpet Maintenance Technician course teaches entry-level commercial carpet cleaning techniques to individuals engaged in the maintenance of commercial facilities. In addition to the basics of carpet construction, emphasis will be placed on teaching IICRC S100 carpet cleaning methodologies, safety procedures, and proactive techniques to individuals who will perform these procedures in the field. Course graduates will have a basic understanding of the importance of preventative, interim, restorative and salvage cleaning and how they contribute to the overall success of the commercial carpet maintenance program.

UFT — Upholstery & Fabric Cleaning Technician (2-day course):

The Upholstery & Fabric Cleaning Technician course covers upholstery fiber categories, fiber identification and testing, manufacturing of the fiber and fabric, chemistry of cleaning, upholstery cleaning methods, protections, spotting and potential problems. A student will have a specific knowledge about fabric and fiber content, as well as furniture construction. This enables students to identify limitations and potential cleaning-related problems on a given piece of upholstery.

OCT — Odor Control Technician (1-day course):

The Odor Control Technician course covers olfaction and odor, odor sources, detection process, theory of odor control, equipment, chemical options and applications. The student will learn how to address odors caused from biological sources such as decomposition, urine contamination, and mold; combustion sources such as fire and smoke damage; and chemical sources such as fuel oil spills or volatile organic chemicals.

CRT — Color Repair Technician (2-day course):

The Color Repair Technician course addresses the history of color, color theory, natural and synthetic dyes, dye methods, types of dyes, types of fibers, carpet styles and dye procedures. Other topics covered include color-related cleaning issues such as fading, color loss due to contamination or bleaching, and the use of cleaning agents that may affect or remove color from carpet.

RRT — Carpet Repair & Reinstallation Technician (2-day course):

The Carpet Repair & Reinstallation Technician course teaches techniques and safety issues related to carpet repair and reinstallation, carpet construction, inspection process prior to cleaning, tools of the trade, floor preparation, adhesives, carpet cushion installation, tackless strip and moldings, seaming, and proper stretching. This class also teaches various repairs that the carpet cleaner/restorer may come up against while dealing with an installed textile. This knowledge will enable the technician to recognize and avoid installation problems that may arise while performing carpet cleaning, as well as how to correct many issues encountered.

WRT — Water Damage Restoration Technician (3-day course):

The Water Damage Restoration Technician course is designed to teach restoration personnel performing remediation work how to deal with water losses and to give them a better concept of water damage, its effects, and the techniques for drying structures. This course will give residential and commercial maintenance personnel the background to understand the procedures necessary to deal with water losses, sewer backflows, and contamination such as mold.

ASD — Applied Structural Drying (3-day course):

The Applied Structural Drying courses covers the effective, efficient and timely drying of water-damaged structures and contents, using comprehensive classroom and hands-on training, in order to facilitate appropriate decision-making within a restorative drying environment. Students will experience live hands-on use of instruments, extraction systems, drying equipment, and chemistry use in an actual flooded building situation.

WRT/ASD Combo Course (5-day course):

This course must be attended in its entirety and completion of both WRT and ASD exams during that same five (5) day time period. If any portion of the course and exam are not completed during the five (5) day period the student would be required to reattend the entire five (5) course to receive both certifications. No portion of the five (5) day course could be applied to either a WRT or ASD course being taught separately. See course description of WRT and ASD for further details of subject coverage.

AMRT — Applied Microbial Remediation Technician (4-day course):

The Applied Microbial Remediation Technician course covers mold and sewage remediation techniques for individuals engaged in property management, property restoration, IEQ investigations or other related professions. Emphasis will be placed on teaching mold and sewage remediation techniques to individuals who will perform these procedures in the field. Course graduates will be adequately equipped to perform remediation services, while protecting the health and safety of workers and occupants.

LCT — Leather Cleaning Technician (2-day course):

The Leather Cleaning Technician course addresses leather identification and cleaning techniques for professional on-location cleaners, restoration and inspection service providers, as well as other related industries. Emphasis shall be placed on theory, practical application, proper identification of leather types, soiling conditions and proper professional solutions to the cleaning challenges faced by the individuals performing the work in the field.

RCT — Rug Cleaning Technician (2-day course):

The Rug Cleaning Technician course covers area rug identification and appropriate cleaning techniques and methods for professional cleaning, restoration and inspection service providers, as well as others in related industries. Emphasis will be placed on teaching cleaning theory, practical application, hands-on techniques, and understanding cleaning limitations.

SMT — Stone, Masonry, & Tile Care Technician (2-day course):

The Stone, Masonry, & Tile Care Technician course provides precise information for the proper maintenance and cleaning of natural and man-made (cultured) stone, ceramic tile, masonry, and grout surfaces. Students will learn the theory behind and reasons for proper tools, chemicals and equipment. Emphasis will be placed on protecting surfaces, technicians, and building occupants. Students completing the course will be adequately equipped to perform maintenance-related tasks involving stone, masonry and ceramic tile surfaces.

SRT — Fire & Smoke Restoration Technician (2-day course):

The Fire & Smoke Restoration Technician course concentrates on technical procedures for successfully completing the restoration of a fire- and smoke-damaged environment. Students will learn how to combine technical procedures with a practical approach to managing the jobsite and how that relates to pricing the job.

HST — Health & Safety Technician (2-day course):

The Health & Safety Technician class is a must for anyone performing restoration or mold remediation! This course is provided to increase safety in your workplace and reduce your company's risk of costly fines and penalties. It is also designed to put you ahead in the restoration industry and to show insurance companies you and your staff are ahead of the competition regarding health and safety. Topics to be covered are: OSHA standards, Inspections, Citations and Penalties, Recordkeeping, Personal Protective Equipment (PPE), Hazard Communication, Hazardous Materials, Confined Spaces, and Bloodborne Pathogens.

FCT — Floor Care Technician (2-day course):

The Floor Care Technician course is an introductory level course in hard surface floor care. Topics covered include Industry Overview, Health, Safety and Liability, Floor Covering Materials, Chemicals, Tools and Equipment, Floor Care Principles and Procedures, Specialized Floor Care Procedures, Problem Solving and Trouble Shooting, Managing the Floor Care Function, and Industry Resources.

RFMT — Resilient Flooring Maintenance Technician (2-day course):

The Resilient Flooring Maintenance Technician course provides training to expand the knowledge and understanding of the professional technician in the maintenance of resilient flooring. Provides students with technical information that will aid in the identification of classifications within the resilient flooring category. Participants will learn about the basic raw materials used in the manufacturing process of each of these classifications. Classification identification will include common and individual characteristics and properties of the individual classifications. Students will learn the life cycle of the classifications of flooring in the resilient category, to include the manufacture, installation, maintenance and demolition of the products. Individual programs for each classification will discuss the impact of environmental conditions on productivity and frequency of cleaning.

RFI — Resilient Flooring Inspector (4-day course):

The Resilient Flooring Inspector course is intended to teach a basic understanding of resilient floor covering inspection. This course was developed for individuals engaged in inspecting, installing, creating and/or interpreting specifications, and for other related professions. Emphasis is placed on teaching participants an overview of resilient floor covering.

MSI — Marble & Stone Inspector (3-day course):

The Marble & Stone Inspector course is intended to teach a basic understanding of marble and stone floor covering inspection. This course was developed for individuals engaged in inspecting, installing, creating and/or interpreting specifications, and for other related professions. Emphasis is placed on teaching participants an overview of marble and stone floor covering.

ISSI — Introduction to Substrate & Subfloor Inspection (3-day course):

The Introduction to Substrate & Subfloor Inspection course is an introductory, prerequisite course created to teach a basic understanding of substrates and subfloors as they relate to finish floor coverings. This course was developed for individuals engaged in inspecting, installing, creating and/or interpreting specifications, and for other related professions. Emphasis is placed on teaching participants an overview of substrates and subfloors.

CTI — Ceramic Tile Inspection (4-day course):

This course teaches an understanding of ceramic tile, identification, and ceramic tile concerns. The course covers the following: types of materials, manufacturing processes, handling, transportation, distribution network, specifications, substrates/subfloors, installation methods, maintenance, cleaning, inspection procedures, warranties, report writing, risk management, expert witness, and accepted industry practices and standards as each relates to inspections.

WLFI — Wood Laminate Floor Inspection (4-day course):

This course teaches an understanding of woods and laminates, identification and wood/laminate concerns. The course covers the following: types of materials, manufacturing processes, handling, transportation, distribution network, specifications, substrates/subfloors, installation methods, maintenance, cleaning, inspection procedures, warranties, report writing, risk management, expert witness, and accepted industry practices and standards as each relates to inspections.

SCI — Carpet Inspector (5-day course):

The Carpet Inspector course focuses on preparing carpet inspectors for certification. It also serves as an introductory floor covering inspection course explaining basic investigation procedures that apply to all floor coverings. Some basic knowledge of fibers, yarn construction, carpet construction and installation is assumed and the emphasis is on field testing, lab testing, inspection techniques, advanced construction, installation and maintenance issues as well as report writing, warranty interpretation, color issues, photography, sample collection, practice and tours of related facilities.

IICRC Advanced Designations

As technicians accumulate various certifications, they become eligible for advanced designations: IICRC Journeyman Status and the cleaning industry's highest technical designation, IICRC Master Status. Requirements to obtain these prestigious designations are outlined below.

JTC — Journeyman Textile Cleaner

Twelve (12) months active service in the industry after original certification date, plus attainment of specific designations as listed below. Designation will automatically be awarded upon attainment of the proper credits:

Certification required in CCT or CCMT, and UFT, and either OCT, CRT or RRT.

JSR — Journeyman Fire & Smoke Restorer

Twelve (12) months active service in the industry after original certification date, plus attainment of specific categories as listed below:

Certification in UFT, OCT and FSRT.

JWR — Journeyman Water Restorer

Twelve (12) months active service in the industry after original certification date, plus attainment of specific categories as listed below:

Certification in CCT or CMT, WRT and RRT.

MTC — Master Textile Cleaner

A minimum of three (3) years after original certification date, plus attainment of specific certifications as listed below:

Certification in CCT or CCMT, UFT, OCT, RRT or BRT, and CRT.

MSR — Master Fire & Smoke Restorer

A minimum of three (3) years after original certification date, plus attainment of specific certifications as listed below:

Certification in CCT or CCMT, UFT, OCT, FSRT and HST or equivalent.

MWR — Master Water Restorer

A minimum of three (3) years after original certification date, plus attainment of specific certifications as listed below:

Certification in CCT or CCMT, RRT, WRT, ASD, AMRT or AMRS, and HST or equivalent.

AMRS — Applied Microbial Remediation Specialist:

One year of active IICRC certification as an AMRT; plus IICRC HST Health & Safety (HST) certification or approved equivalent, and 10 verifiable microbial remediation jobs or 1000 hours verification microbial experience. Contact IICRC Headquarters for further details.

Appendix C

Sample Mold Remediation Protocol

Prepared by an experienced industrial hygienist to establish specific mold remediation worker procedures for a water-damaged basement.

(Courtesy of Wise Steps, Inc.)

Remediation Safety and Health Protocols

These protocols are *minimal* steps. If more mold growth is found on materials or areas during deconstruction, a change in the amount of material to be cleaned or removed will need to be assessed. These protocols have been developed using the EPA document 402-K-01-001 *Mold Remediation in Schools and Commercial Buildings* and the New York City Department of Health, *Guidelines on Assessment and Remediation of Fungi in Indoor Environments* and IICRC S520 *Standard and Reference Guide for Professional Mold Remediation* and IICRC S500 *Standard and Reference Guide for Professional Water Damage Restoration.*

During the remediation, I am available to assist the remediation workers in determining the extent of the mold growth contamination and additional remediation needs beyond what has already been found.

1. Preparation

Make sure there is no air supplied or returned to the central ventilation system from the damaged rooms.

Remove all furnishings from the offices and rooms that had been affected by the flood. Clean these items by damp wiping with a disinfectant or HEPA vacuuming as they are moved. This equipment and furniture should be stored off site in a clean and dry storage facility.

2. Remediation Worker Safety

The remediation contractor shall comply with all applicable OSHA regulations, such as having a Written Respiratory Protection Program and Hazard Communication Program.

All workers at all times inside the containment must be fully equipped with protective equipment. This should include at least:

- Tyvek coveralls
- Gloves
- ½-face respirators with N95 or HEPA cartridges
- Eye protection

The respirators must be worn during the time of demolition and initial cleaning.

Protective equipment should be donned and doffed outside the containment.

3. Containment/Isolation

The remediation contractor must build and maintain structurally sound containment barriers in order to isolate remediation work areas from the non-damaged areas of the building.

Rooms that were not affected by the flood waters should be sealed by using 6-mil flame-retardant plastic sheeting across the door entry.

The base of the stairs to the main floor should also be sealed by using 6-mil flame-retardant plastic sheeting to protect the rest of the building.

Establish negative pressurization in the basement containment work areas. This is usually done by using HEPA air moving equipment which can eventually be used in a recirculation scrub mode to clean the air in the affected rooms after removal of the sheetrock and carpet. Make-up air into the containment should be filtered.

4. Extent of Remediation

Flooring

Remove the water-damaged carpeting.

Walls

Remove the baseboard and coving in all rooms that were flooded. Where there is visible mold growth remove at least 2 feet of sheetrock. (It is preferable to remove the bottom 2 feet of sheetrock on all walls that the flood came in contact with, as well as all insulation where the walls have been opened.)

In the kitchen, the dishwasher should be moved so the back wall can be inspected for mold growth. If growth is found on this wall behind the dishwasher, then all lower cabinets in the kitchen should be removed so the bottom of the walls can be taken out. The underside of the cabinets must then be disinfected and cleaned.

5. Sanitizing and Cleaning

- After the sheetrock and insulation have been removed, apply a broad spectrum sanitizer/disinfectant on all surfaces of the opened wall and floor.
- Dry all the opened walls and floors.
- Final cleaning after all the deconstruction has been completed, disinfected, and dried — must HEPA vacuum all surfaces.
- No dust or construction debris should be left in the rooms before the clearance tests.
- Use a HEPA air scrubber *(in recirculation mode)* in the containment after final cleaning for at least 24 hours before clearance tests are done. Turn the air cleaner off at least 2 hours prior to clearance tests.
- *DO NOT APPLY A SEALANT PRIOR TO POST REMEDIATION ASSESSMENT.*

6. Post Remediation Inspection

I suggest post remediation inspection and clearance sampling be done after the walls and carpeting have been removed and the cleaning job inside the contained areas is complete. This would include assessing the following Remediation Performance Indicators:

- If the containment followed the recommended protocols:
  Leave the containment in place until post remediation testing is completed and the results have been received.
- Removal of all visible mold growth.
- No surface debris remains inside the contained rooms on the floors.
- Air and possibly surface samples in the containment and outside the contained areas should be found at levels that are similar to those found on visibly clean surfaces in non-water-damaged buildings.

Lab Report

Sample Identification	East end of basement				West end of basement				Carpet				Outdoors			
Date Analyzed	12/01/2011				12/01/2011				12/01/2011				12/01/2011			
Volume(M³)	0.0750				0.0750				0.0750				0.0750			
Percent Of Trace Analyzed	100% of Trace at 600x Magnification				100% of Trace at 600x Magnification				100% of Trace at 600x Magnification				100% of Trace at 600x Magnification			
Debris Rating	2				2				4				2			
Analyte	**Total Count**	**Count/M³**		**%**	**Total Count**	**Count/M³**		**%**	**Total Count**	**Count/M³**		**%**	**Total Count**	**Count/M³**		**%**
		Result	**DL**			**Result**	**DL**			**Result**	**DL**			**Result**	**DL**	
Mycelial Fragments	2	27	13	n/a	2	27	13	n/a	45	600	13	n/a				
Pollen	6	253	42	n/a	2	84	42	n/a	43	573	13	n/a				
Total Fungal Spores	144	6050	42	100	1966	82800	42	100	341	14100	42	100	9	321	42	100
	Fungal Spore Identification				**Fungal Spore Identification**				**Fungal Spore Identification**				**Fungal Spore Identification**			
Alternaria													1	13	13	4
Arthrinium																
Ascospores	1	42	42	0.7					4	169	42	1.2				
Aspergillus/Penicillium	135	5700	42	94.2	1950	82300	42	99.4	236	9960	42	70.6	7	295	42	91.9
Basidiospores	6	253	42	4.2	7	295	42	0.4	37	1560	42	11.1				
Bipolaris/Drechslera									1	13	13	0.1				
Botrytis																
Chaetomium					1	13	13	0								
Cladosporium	1	42	42	0.7	2	84	42	0.1	51	2150	42	15.2				
Curvularia									1	13	13	0.1				
Epicoccum									3	40	13	0.3				
Ganoderma									3	127	42	0.9	1	13	13	4
Myxomycetes	1	13	13	0.2	1	13	13	0								
Oidium/Peronospora																
Pithomyces																
Rusts																
Smuts/ Periconia																
Stachybotrys																
Scopulariopsis					3	40	13	0								
Torula																
Ulocladium					2	27	13	0	4	53	13	0.4				
Unclassified Conidia									1	13	13	0.1				

Glossary

A

AAHX: See *air-to-air heat exchanger.*

accelerant: A material, such as gasoline, that causes a fire to burn faster and hotter than it otherwise would.

accounts payable (A/P): The money that a company owes to its vendors for products and services that have been purchased on credit.

accounts receivable (A/R): The money that's owed to a company by its customers for products and services that have been provided, and for which the customer has been sent an invoice.

Act of God: The legal term for events that are outside of human control. These usually include floods, earthquakes, volcanic eruptions, and other natural disasters that no one can be held responsible for.

activated charcoal: Charcoal that's undergone a treatment process that expands the millions of tiny pores between the carbon atoms. This process increases the surface area of the charcoal, making it more effective at trapping and holding impurities. Also called *activated carbon.*

ADA: See *Americans with Disabilities Act.*

add-ons: An additional amount of money that's added to a base unit cost to adjust for particular materials or conditions. Also called *unit cost add-ons.*

adjuster: See *insurance adjuster.*

adsorbent: A material, such as silica, that will take in and give off moisture without a physical or chemical change.

agent: See *insurance agent.*

AHAM: The Association of Home Appliance Manufacturers, a trade association for the home appliance industry. They evaluate and set standards for many home appliances, including dehumidifiers.

air changes per hour: A rating for ventilation equipment that describes the number of times the entire volume of air within a given space can be removed and replaced with new air within one hour. For air cleaning equipment, it's the number of times the entire volume of air within a given space can be cleaned and reintroduced to the space within one hour.

air dolly: See *Airsled®.*

air exchanger: See *air-to-air heat exchanger.*

air mover: A piece of restoration equipment with electric-motor-driven fan blades that create a high-speed stream of air. The air is used to evaporate moisture from various materials for collection by a dehumidifier. Also called a *blower.*

air sampling: Collecting air from within a space, typically for a set amount of time at a set air volume, for the purposes of testing for airborne contaminants.

air scrubber: A type of portable fan unit that draws contaminated air across a HEPA filter. The filter removes the microscopic contaminants from the air, and then discharges the cleaned air back out.

air-to-air heat exchanger (AAHX): A ventilation device that takes in air inside a structure, exhausts it outside, then takes in outside air and brings it inside. The two air streams cross each other, so the heat from the inside air is given off to the outside air as it comes into the building. Also called an *air exchanger.*

airborne: Particles or substances that float in the air.

Airsled®: A tool for moving appliances and other heavy equipment using a stream of air. It consists of an electric motor that generates an air stream, passes the air through hoses and then forces the air out through perforated plastic plates in the unit's base. This results in a strong stream of air underneath the plates that provides the lifting and moving capabilities. The Airsled is placed under the heavy object to easily lift and move it. Sometimes referred to as an *air dolly.*

alpha-numeric: Capable of displaying both letters (alpha) and numbers (numeric). For example, the message displays on most of today's pagers and cell phones are alpha-numeric.

American National Standards Institute (ANSI): A private, not-for-profit organization that promotes and facilitates voluntary standards for U.S. businesses.

American Society of Heating, Refrigeration and Air Conditioning Engineers (ASHRAE): An international organization whose mission is to advance heating, ventilation, air conditioning and refrigeration through research, written standards, publishing and continuing education.

Americans with Disabilities Act (ADA): Federal legislation that prohibits discrimination on the basis of disability in private companies and commercial facilities. The act contains a number of complex requirements pertaining to the construction, alteration, and renovation of commercial buildings.

ANSI: See *American National Standards Institute.*

antimicrobial: A product which is specifically formulated to control, inhibit, or destroy microorganisms.

architectural review committee: A group of paid employees and/or volunteer committee members who establish and/or enforce CC&Rs related to the exterior appearance of the homes within a specific community. This typically includes the size and overall exterior design of the home, exterior building materials and color palettes, fencing and landscaping, and additions and other types of exterior remodeling.

ASHRAE: See *American Society of Heating, Refrigeration and Air Conditioning Engineers.*

ASTM: *ASTM International* (formerly known as the American Society for Testing and Materials) is an international standards organization that develops and publishes voluntary technical standards for a wide range of materials, products, systems, and services.

awl: A hand tool with a wooden or plastic handle and a long slender shaft ending in a sharp point used to mark surfaces or to pierce holes in materials.

axial fan: A type of air mover with vertically-mounted fan blades used to create a high-volume air stream that's directed out of the air mover through a round opening on one end.

B

backing: The back layer of a carpet, to which the face or visible fibers of the carpet are attached. There are typically two backing layers, one that the fibers are actually attached to, and a second one for additional strength.

bacteria: Any of a group of microscopic, single-celled organisms that inhabit virtually all environments.

balanced drying system: A drying system that's been correctly sized and set up for a particular job. When a balanced system is in place, more water vapor is removed from the air by the dehumidifiers than is created with the air movers. That results in a gradual drop in humidity levels in the building.

bank draft: See *draft.*

biocide: A type of antimicrobial product that's specifically formulated to destroy microorganisms rather than just inhibit or control their growth.

black water: Nonpotable water that's contaminated with both liquid and solid waste materials. Also called *Category 3 water.*

blocking: Raising furniture and other contents up onto foam blocks as protection against water damage.

bloodborne pathogens: Microorganisms present in human blood that can cause disease in humans. Examples of these types of pathogens include hepatitis B virus (HBV) and human immunodeficiency virus (HIV). Bloodborne pathogens are a particular hazard when dealing with trauma scenes and contaminated water losses.

blower: See *air mover.*

board-up: Covering windows, doors, and other areas with a temporary covering of plywood or other solid material to provide weather and security protection after a loss.

boot: A wide plastic fitting used on the end of an air mover to spread the air flow over a larger area, or to push air up into a wall cavity. Sometimes called a *wall boot.*

booting a wall: Using an air mover and one or more boots to push air up into wall cavities. This is done to speed the drying of materials inside the wall, such as insulation, framing, and the back side of the drywall.

borescope: An inspection instrument that consists of a tube with a reflecting mirror and an eyepiece used for inspecting inaccessible areas.

bottom barrier or ***board:*** See *road barrier.*

bound water: Water that's trapped within the cells of a piece of wood, or inside relatively dense materials, such as mortar. Bound water is generally more difficult to remove than free water.

boundary layer: A layer of moisture-laden air that's directly in contact with a wet surface. The boundary layer will slow the evaporation of moisture from the wet surface.

broom-clean condition: A term used to indicate that a building has been swept up during or at the end of a construction project, and that all large debris has been removed and disposed of.

BTU: British thermal unit, a unit of heat equal to the amount of heat required to raise one pound of water one degree Fahrenheit at a specified temperature.

building code upgrades: See *code upgrades.*

business plan: A plan that defines the business, sets specific goals for the business, sets forth a plan of action to meet those goals, and sets short- and long-term timelines for reaching those goals. It helps business owners better understand and strategize for the success of their business.

C

C-channel: The bottom plate in a metal stud wall, into which the studs are inserted. The C-channel sits on the floor with the bottom and two sides forming a trough, which has the potential to capture and hold water in a water loss situation.

CAD: Computer-aided design.

call sheet: During an initial phone call to report a loss, a call sheet is used to record contact information and other details about the loss.

capillary action: The process by which liquid pulls itself up or out against the force of gravity, as a result of attraction between molecules.

capillary tube: A small tube that's part of a refrigerant dehumidifier system. The tube restricts the flow of refrigerant between the condenser and the evaporator.

carcinogenic: Anything that either directly causes or contributes to the cause of cancer.

carpet clamp: A clamp used on the end of an air mover to secure carpeting while drying.

carpet dryer: An air mover with a short, wide snout that moves a concentrated stream of air low to the ground. Commonly used for drying carpets and other types of flooring.

carton stand: A metal rack used to sort and store flattened boxes prior to use.

cash value policy: A homeowner's insurance policy that pays the policy holder at the current value of the building or the contents, minus depreciation. Also called a *fair market value policy.*

cat team: See *catastrophe team.*

catastrophe loss: A large number of losses that all occur within a short period of time and within a confined geographic region, such as in the aftermath of a tornado, hurricane, ice storm, or explosion. When this occurs, an insurance company may make additional personnel and financial resources available to handle claims quickly. The claims collectively become known as catastrophe losses. Often referred to by the abbreviated term *cat losses.*

catastrophe team: A group of adjusters and support personnel that's mobilized by an insurance company from within a state, region, or even from throughout the country. Catastrophe teams are formed in the aftermath of a large natural or man-caused disaster. Their role is to quickly process the hundreds or even thousands of insurance claims that are filed over a very short period of time. Commonly known as a *cat team.*

Category 1 water: Water that doesn't pose a significant health risk. Also called *clean water.*

Category 2 water: Water that has a significant amount of contamination, from either chemical, biological, and/or physical contamination. Also called *gray water.*

Category 3 water: Water that's come from a grossly contaminated source, such as a sewer or a septic tank. Also called *black water.*

cause and origin inspector: A person who's trained to inspect a loss, especially a fire loss, to determine where and how the loss occurred.

cavitation: The sudden formation and collapse of bubbles in a liquid. Also, the pocket or cavity that's formed in a liquid by the collapse of the bubbles.

CC&Rs: See *covenants, conditions, and restrictions.*

CDC: Centers for Disease Control, a federal agency that's part of the Department of Health and Human Services, whose primary purpose is to investigate, diagnose, prevent, and control diseases.

CFM: See *cubic feet per minute.*

change order: A written addendum to the original contract, documenting agreed-upon changes to the original project specifications. Change orders include the specific details of the change, the cost or credit associated with that change, and any alteration in the completion date. A change order must be signed by the parties to the contract.

change order fee: An optional fee charged to cover the expense of estimating changes to the original project specifications.

claims representative: See *insurance adjuster.*

Class 1 water loss: A water loss that primarily involves relatively dense materials that do not absorb large amounts of water. The water loss is confined to one room, or just part of a room.

Class 2 water loss: A moderate water loss, involving an entire room or rooms with wet carpet, wet pad, and other porous materials. The drywall has wicked moisture to a height of 24 inches or less.

Class 3 water loss: A severe water loss, involving several rooms with wet carpet, wet pad, and other porous materials. The drywall has wicked moisture to a height in excess of 24 inches.

Class 4 water loss: A water loss where water is concentrated in deep pockets of the materials and is relatively difficult to dry out with standard air movers.

clean room: A small structure or enclosure, used primarily in mold and bacterial remediation jobs, to provide a place for workers to put on and take off protective clothing when entering and leaving a containment area.

clean water: Water that isn't contaminated, and does not pose a health risk. Also called *Category 1 water.*

clearance test: A test or set of tests performed by an industrial hygienist or other qualified professional at the end of a mold remediation, asbestos abatement, or other site cleanup project. The testing is done to determine the quality of the air inside the building, and to ensure the building is safe to occupy following the completion of the work.

closed drying plan: The process of drying a building while keeping it closed off from outside air. A closed drying plan provides more control over temperature and humidity conditions inside the building during the drying process.

code upgrade insurance: Insurance that covers all or part of the costs associated with bringing a building into compliance with current building codes.

code upgrades: Upgrades to a home or other building that are required in order to meet current building codes. Building code upgrades can become necessary during reconstruction even if the building was in compliance with the codes that were in place at the time the building was originally built.

cold call: Telephone call or in-person visit made to someone who doesn't know you and who's not expecting the call or the visit. Cold calling is done primarily to introduce yourself and your company, and to solicit future work.

cold coil: A name for the evaporator coil inside a refrigerant dehumidifier.

combination drying system: Using a combination of drying equipment and outside air to speed the drying process in a building.

completion date: The date when a restoration project is estimated to be substantially completed. The proposed date and the specific standards required for substantial completion will be contained within the contract.

confidentiality agreement: See *nondisclosure agreement.*

containment: Temporary walls or other dividers, usually made of plastic sheeting, used to isolate contaminated sections of a building from those areas that aren't contaminated.

content restoration: The process of returning damaged contents to, or as close as possible to, their pre-loss condition. Content restoration may include cleaning, dry cleaning, ultrasonic cleaning, deodorization, repair, refinishing, reupholstery, and other work.

contents: The generic term used by renovation contractors for all of the personal belongings within a home or business. Contents are generally accepted to include anything owned by the building owner or tenant that can be readily moved out of the building, as opposed to items that are permanently attached to the building.

contract: A legal written document between a contractor and a building owner, providing the details of the work to be performed, the estimated cost, beginning and completion dates, draw schedules, and other important details. The contract must be signed and dated by the parties involved.

corona discharge: A common method for generating ozone gas. Air is passed through a tube surrounded by an electrical field. The electricity causes a small percentage of the oxygen molecules (O_2) in the air to split. The free atoms bond with the remaining O_2 molecules, producing ozone gas (O_3).

covenants, conditions, and restrictions (CC&Rs): Specific regulations and limitations placed on a group of homes, townhomes, or a condominium complex, by a builder, developer, and/or a homeowner's association. The regulations generally cover all common use areas, such as parking spaces and community swimming pools, as well as landscaping, fencing, exterior construction or exterior paint colors of the buildings.

covered loss: Damage to a building or its contents that's specifically covered by an insurance policy.

crossover duct: A large diameter flexible heating duct that connects the rigid ductwork located in each half of a double-wide manufactured home. In multi-unit manufactured homes, such as a triple-wide, there are multiple crossover ducts.

cubic feet per minute (CFM): A rating of air flow.

cubic foot: A three-dimensional unit of measurement equal to 1 foot in length by 1 foot in width by 1 foot in height. Used for measuring volume.

cubic inch: A three-dimensional unit of measurement equal to 1 inch in length by 1 inch in width by 1 inch in height. Used for measuring volume.

cubic yard: A three-dimensional unit of measurement equal to 3 feet in length by 3 feet in width by 3 feet in height. Used for measuring volume.

D

daylight basement: A basement that has an exterior access on at least one side of the home.

decibel (dB): A rating of sound intensity. Normal conversation is approximately 60 dB. With extended exposure, decibel levels of 85 and above can lead to hearing loss.

deductible: That portion of an insured loss to be paid by the policyholder. The deductible may either be a set dollar amount, such as $1,000, or it may be a percentage of the insured value of the building, such as 20 percent.

dehumidifier, desiccant: A type of equipment used to remove excess water vapor from the air. Desiccant dehumidifiers use a rotating silica gel wheel to extract the water vapor.

dehumidifier, refrigerant: A type of equipment used to remove excess water vapor from the air. Refrigerant dehumidifiers use a compressor, a condenser, an evaporator, a fan, and a refrigerant material to extract the water vapor.

delamination: The separation of the two backing layers on a piece of carpet or other laminated material.

Department of Environmental Quality (DEQ): A state agency whose job it is to work with the EPA in overseeing and protecting that state's environment. Each state has a DEQ.

Department of Housing and Urban Development (HUD): A federal agency that assists with homeownership, community development, and access to affordable housing. HUD is also the agency that oversees manufactured home standards.

Department of Transportation (DOT): The federal agency that oversees federal highway, air, railroad, maritime and other transportation administration functions. Each state also has its own DOT, which regulates that state's highway building projects, maintenance, and transportation administration.

depreciation: The reduction in value of building materials or contents due to age, wear, or other factors.

depreciation holdback: A certain percentage of the value of a covered loss that's held back by the insurance company until repairs to the structure are substantially completed, or until all nonsalvageable contents items are replaced.

destructive testing: The process of removing material, cutting holes, or otherwise creating an access to allow for additional visual inspection of a structure.

dew point: The point at which the air becomes saturated with moisture and condensation forms.

dish pack: A special cardboard carton with corrugated inserts used for packing, storing, and transporting dishes.

double-wide: A manufactured or mobile home, built as a two separate units, that's designed to be towed to a site and then assembled into one home, as opposed to a single-wide, which is built as one completely assembled unit and towed to the home site.

draft: A written order issued by a bank against a particular account, very similar to a money order. The bank withdraws the money in advance from the account of the person or company who's distributing the draft, and in return issues a *bank draft* for that amount. Drafts are a very reliable form of payment, since the funds are always in place in advance of issuance.

draw: A payment for work completed up to a predefined point in the project.

draw schedule: A list of dates or completion stages during the course of the project when payments, or draws, are due and payable. A draw schedule will be included in the contract.

dry rot: The decay of timber or other wood surfaces in buildings and other wooden structures caused by certain fungi.

due diligence: The process of investigation that a potential investor performs when considering the purchase of a business, real estate, or when making any other type of financial investment. The due diligence process allows the investor to examine and verify important financial, management, and other factors involved in the investment.

E

EIFS: See *Exterior Insulation and Finish Systems.*

employee handbook: A booklet or manual given an employee when hired that explains the policies, procedures, guidelines, and expectations for employment with that company. Also called a *policies and procedures handbook.*

encapsulant: A liquid additive material mixed with water used in sealing surfaces affected by smoke, mold, and other conditions requiring restoration work.

encapsulate: To apply an encapsulant to a material in order to encase the surface and prevent bacteria, mold spores, smoke odors, or other undesirable odors, particles, or microorganisms from becoming airborne.

engineered lumber: Structural lumber or sheet goods that are made up by bonding wood chips, fibers, veneers, or strips with adhesives under heat and pressure.

Environmental Protection Agency (EPA): The federal agency responsible for reducing air and water pollution, as well as regulating pesticides, toxic chemicals, and hazardous waste.

equilibrium moisture content (EMC): The moisture content at which wood is neither gaining nor losing moisture.

evaporation: The change in water molecules from a liquid state to a vapor state.

Exterior Insulation and Finish Systems (EIFS): An EIFS installation involves three layers: An inner foam insulation board; a polymer and cement base coat reinforced with glass fiber mesh; and a textured top coat. Many EIFS installations have had problems with moisture seeping in and getting trapped behind the layers, causing mold and wood rot in structural members. These moisture-related EIFS problems led to a number of individual and class action lawsuits. More-current installations have a drainage arrangement installed to keep moisture from being trapped behind the layers. Also called *synthetic stucco.*

extraction: The removal of water by mechanical means, such as through the use of a vacuum.

F

fair market value policy: See *cash value policy.*

fall protection: Any of a variety of safety devices designed to prevent workers from falling when working on a roof or other high or steep surface. Specific fall protection requirements are set by OSHA.

Federal Emergency Management Agency (FEMA): A federal agency that's part of the Department of Homeland Security. FEMA helps emergency managers and communities prepare for, respond to, and recover from emergency situations.

Federal Housing Administration (FHA): A division of the Department of Housing and Urban Development, the FHA provides mortgage insurance on loans made by FHA-approved lenders throughout the United States. The FHA insures mortgages on single-family and multifamily homes, including manufactured homes. Loans must meet certain requirements established by the FHA to qualify for insurance.

final cleaning: See *post-construction cleaning.*

fire triangle: The fire triangle consists of the three elements needed for a fire to start and sustain itself: oxygen, fuel, and heat.

float: The lift resulting when a strong stream of air from an air mover is directed beneath a carpet. Floating a carpet is a common method for drying both the carpet and the subfloor below at the same time.

free water: Water that's contained in the open pores of wood and other materials. The water is absorbed by the material relatively easily, and is also easier to remove than bound water.

G

GFCI: See *ground-fault circuit interrupter.*

glass pack: A special type of cardboard carton with corrugated inserts that are divided into squares. Used for packing, storing, and transporting tableware, such as cups, mugs and glasses.

gray water: Nonpotable water that contains some type of liquid contamination, such as urine, bacteria, or chemicals. Also called *Category 2 water.*

ground-fault circuit interrupter (GFCI): A safety device designed to prevent electric shock by detecting faults in the grounding system and instantly disrupting power to the electrical appliance or fixture plugged into the GFCI-protected receptacle.

gusset: To reinforce and secure the joint between two or more structural members by attaching a wooden or metal plate on top of and overlapping the joint. Also, the plate itself.

gusset plate: The wood or metal plate used for gusseting.

H

hammer-probe moisture meter: A moisture-detecting instrument with two long surface-probing pins and a sliding weight. The weight is used to hammer the pins into the surface being tested to determine the depth of the moisture intrusion.

hard goods: Contents items such as furniture, appliances, electronics, tools, dishes, etc., as opposed to soft goods, which are clothing, linens, etc.

hardscape: Those parts of the exterior landscaping that are not alive, including such items as decks, fences, stone walls, concrete or brick patios, gravel and paver paths, water features, etc.

H.A.T.: An acronym for humidity, airflow, and temperature. These are the three important factors involved in drying a structure.

Heating, Ventilation, and Air Conditioning (HVAC): The central function of an HVAC system is to provide thermal comfort and acceptable indoor air quality, within reasonable installation, operation, and maintenance costs.

heightened awareness: A situation in which a building owner is more aware of the condition of their home or building than they might otherwise be, and will spend more time than normal looking for and documenting flaws and problems. Heightened awareness also refers to a person who has experienced a fire loss, and has therefore become more sensitive to smoke odors, real or imagined.

high-efficiency particulate air filter (HEPA): A highly-effective filtering system that is the standard level of filtration for mold remediation. A HEPA filter, by definition, is one that's capable of removing 99.97 percent of airborne particles 0.3 microns or larger.

high-speed air mover: A specific type of fan, used by restoration contractors, which provides a high-speed stream of nonheated air for evaporating moisture.

hold-back: A portion of money from the settlement of a loss that's retained by the insurance company until the repairs on the home have been completed.

Homeowner's Association (HOA): A legally recognized association of the owners of homes and lots within a specified community. The Homeowners Association is responsible for enforcing the rules and regulations of the community, and also for the maintenance and upkeep of any common areas.

hot coil: Another name for the condenser coil inside a refrigerant dehumidifier.

HUD: See Department of Housing and Urban Development.

human resources (HR): The person or department in a company that deals with personnel. Human resources responsibilities typically include recruitment, administration, management, and training, and sometimes hiring and firing.

humidistat: A humidity sensor control, which is used to turn a dehumidifier on or off at a preset humidity level.

humidity: Moisture vapor in the air.

HVAC: See *Heating, Ventilation, and Air Conditioning.*

hygrometer: An instrument for measuring the relative humidity in the air.

I

IAQ: See *indoor air quality.*

ignition point: The minimum temperature at which a material will continue to burn on its own, without applying any additional external heat. Also called the *ignition temperature.*

IICRC: See *Institute of Inspection, Cleaning and Restoration Certification.*

implode: To collapse suddenly and violently inward.

in-house: An operation handled by the company's own staff or crew, as opposed to subcontracting it out to another company.

inches of water lift: The number of inches that a vacuum can pull water up in a tube under controlled conditions. The greater number of inches of water lift, the more powerful the vacuum is. A rating of wet/cry vacuum suction power.

indoor air quality (IAQ): The general condition and quality of the atmosphere inside a building. There are a number of factors that affect the quality of a building's interior air, including ventilation, occupancy, type of air handling equipment, moisture levels, the presence of chemicals and other pollutants, and more.

industrial hygienist: A person with the specific education, training, and experience to anticipate, recognize, evaluate, and control health hazards in working and living environments.

infiltration: Air that enters a building through small cracks or gaps in the structure.

inspection: A site visit and examination done by an official during a particular phase of a construction project. The inspection may be done by a building inspector to ensure compliance with the building codes, by a representative of the lending institution to ensure that a particular level of work has been completed, or by another official with an interest in the building's construction progress.

Institute of Inspection, Cleaning and Restoration Certification (IICRC): An independent certification body helping to set and maintain standards of technical proficiency and ethical behavior for people and companies in the restoration industry.

insurance adjuster: The person representing the insurance company on an insurance-related loss. The adjuster inspects the loss, helps determine coverage, prepares and reviews estimates, oversees the many aspects of the loss claim and approves final payment to the home or building owner. Sometimes called an *insurance claims agent* or a *claims representative.*

insurance agent: A person who sells insurance. Agents may represent the policies of a single company, or they represent several different companies at the same time. Agents may also handle different types of policies, including homeowners, auto, life, health, and others.

insurance restoration: The repair of damage to structures and contents caused by an event that's covered by a standard homeowner's or business insurance policy.

intrusion: The sudden and accidental entry of water into a building. Water intrusion could be from broken or leaking plumbing, from storm damage, or from other sources.

ionization: The attraction of smoke particles with opposite charges to one another.

J

junction box: A metal or plastic box, in various sizes and configurations, used for splicing electrical wires in repair and remodeling situations.

K

kicker: A permanent or temporary structural support or brace, especially in an attic. Also see *knee kicker.*

kiln-dried (KD): Lumber that has been dried in an oven under controlled conditions to achieve a specific moisture content.

knee kicker: A carpet installation tool consisting of three basic parts: A horizontal head with a series of short, sharp, angled teeth on the bottom; a vertical plate with a pad; and an adjustable bar that connects the two. The teeth engage the carpeting, and knee pressure against the pad is used to move the carpeting into place. Sometimes called a *kicker.*

knob and tube: An old and no longer approved form of electrical wiring. It consists of individual wires for hot and neutral, with no ground. Wires were attached to framing members with a porcelain insulator called a knob. Where the wires passed through framing members, they were protected by a hollow porcelain tube.

knocked down (KD): An item that's premade and ready to put together, but is shipped unassembled with all the necessary hardware and instructions. Cabinets and furniture are often sent knocked down.

L

labor burden: The additional costs, such as insurance, taxes, and Social Security benefits, paid by the employer, on top of the employee's hourly wage.

laminate flooring: A synthetic flooring material, made up of several thin layers fused together using a lamination process. Laminate flooring simulates wood, with a photographic layer under a clear protective layer. The inner core is usually composed of melamine resin and fiberboard.

lateral: A load, such as the weight of snow, or a force, such as that from the wind, that pushes against the side of a building, or that pushes against the building from an angle.

Leadership in Energy and Environmental Design (LEED): A program developed by the U.S. Green Building Council (USGBC) that provides training in green building design and construction practices, as well as a Green Building Rating System.

LGR: See *low grain refrigerant.*

like kind and quality: A common insurance industry term, referring to contents or structural materials that are replaced with items of the same type and value as the original. For example, if a home had nylon carpet valued at $20 per square yard at the time of the loss, like kind and quality would be nylon carpet valued at $20 per square yard, regardless of color, style, manufacturer, etc.

lime: A product derived from calcined (burnt) limestone, which is a naturally-occurring rock extracted from quarries and underground mines all over the world. Lime is used in making cement, plaster, mortar, and many other building materials. It can also be used to treat certain odors. Also called *quicklime.*

limited liability company (LLC): A business structure similar to a sole proprietorship in the way that profits and losses pass through to the owner. It's also similar to a corporation in the way that it limits liabilities faced by the owner(s) should they be sued.

loss: An insurance term for any damage that occurs to a building or its contents. See also *covered loss.*

loss inventory: A detailed written list of nonsalvageable contents items.

loss mitigation: Steps taken, after a loss occurs, to prevent the loss damage from worsening. Examples of loss mitigation include drying a water-damaged building, or winterizing a building that's without heat.

low grain refrigerant (LGR): A type of refrigerant used in some dehumidifiers. Low grain refrigerants operate effectively at a lower relative humidity than standard or conventional refrigerants.

M

makeup air: Air that's brought into a building or room to replace air that's been displaced by combustion or exhaust appliances.

manufactured home: A home, constructed in a factory under controlled conditions, having a steel chassis with axles and wheels. It's intended to be towed to a site and set up either temporarily or permanently for use as living space. Sometimes referred to as a mobile home. However, by some definitions, a manufactured home is said to be one that was built after the 1976 HUD enactment of the National Manufactured Housing Construction and Safety Standards Act, and a mobile home would be one that was built before the Standards Act.

material allowance: The amount the insurance company has agreed to pay for certain items, such as carpet, vinyl, tile, or light fixtures. The allowance is based on the value of the materials that were in the building at the time of the loss.

Material Safety Data Sheet (MSDS): A document that gives handling, application, safety, and other important information for virtually any product. Retailers and wholesalers are required by law to provide an MSDS for any product upon request.

mattress bags: Large plastic bags designed specifically for storing mattresses after they've been cleaned.

MC: See *moisture content.*

medium density fiberboard (MDF): An engineered wood product made from wood fibers and resin that are formed under heat and pressure into a solid, dense material. Used for trim, as well as a variety of other uses.

micron: A unit of measurement, short for micrometer, equal to one millionth of a meter, or 1/1,000 of a millimeter. One micron is 0.000039 of an inch.

mil: A unit of measurement, equal to 1/1,000 (0.001) of an inch.

mini-boot: A type of short, narrow boot that can be clamped to the end of a 4-inch diameter hose, used in conjunction with an air mover and a splitter or snout adapter for drying.

mirror box: A wide, shallow cardboard carton used for storing and transporting mirrors and artwork.

mitigation: see *loss mitigation.*

mobile home: A home built in a factory on top of a steel chassis with axles and wheels, then towed to a site and set up for use as living space. Also called a manufactured home. Sometimes defined as being a manufactured home built prior to the 1976 HUD enactment of the National Manufactured Housing Construction and Safety Standards Act.

modular home: A home that's built in sections in a factory or other controlled conditions, then transported to a site on trucks. Once at the site, the sections are lifted into place with a crane and assembled into a finished home.

moisture content (MC): The amount of moisture that a material contains, expressed as a percentage. Moisture content is a common measurement used in lumber, hardwood flooring, trim, and other wood products.

moisture meter: Any of a variety of instruments used to detect the presence of moisture. See also *penetrating moisture meter* or *nonpenetrating moisture meter.*

mold remediation: The safe removal of visible mold and airborne mold spores, done in compliance with an industrial hygienist's written specifications and in compliance with currently accepted standards.

mover's dolly: A flat, square or rectangular dolly, usually made of hardwood or plastic, with a carpeted top and four swivel casters underneath. Used for moving furniture and other heavy objects around inside buildings or warehouses.

moving blanket: A large, padded blanket, used for covering and protecting furniture and other contents during transport and storage. Also called a *packing blanket.*

MSDS: See *Material Safety Data Sheet.*

N

NADCA: See *National Air Duct Cleaners Association.*

named insured: A person who's named on an insurance policy, and whose interests are protected by that policy.

named peril: A specifically-named threat or danger to a property covered in a homeowner's or renter's insurance policy.

National Air Duct Cleaners Association (NADCA): A nonprofit association formed in 1989 to help establish industry standards for the proper assessment, cleaning, and restoration of HVAC duct systems.

National Fire Protection Association (NFPA): An international nonprofit organization established in 1896 to advocate worldwide fire prevention and protection thorough codes and standards, research, training, and education. Their findings, or requirements, are published in the *National Electrical Code® (NEC®)*, published every three years.

National Institute for Occupational Safety and Health (NIOSH): A federal agency, and a division of the Centers for Disease Control, responsible for conducting research and making recommendations for the prevention of work-related injury and illness.

National Institute of Restoration (NIR): An association that assists restoration contractors and the insurance industry through education and information.

National Lightning Safety Institute (NLSI): An independent advocate of lightning safety, for both people and structures, promoting safety through education, forums, certification protocols, and other means.

National Manufactured Housing Construction and Safety Standards Act (NMHCSS): Legislation enacted in 1976 that allowed the U.S. Department of Housing and Urban Development (HUD) to begin overseeing and regulating the construction of manufactured homes.

natural smoke residue: See *smoke residue.*

natural ventilation: Allowing outside air to enter a building through open windows and doors, sometimes aided by ventilation fans.

NCR: See *No Carbon Required.*

NDA: See *nondisclosure agreement.*

NEC: See *National Fire Protection Association.*

negative pressure: A condition that occurs within a space when more air is exhausted from the space than is supplied to it. The result is a drop in the air pressure within the space to a pressure lower than the air in the space surrounding it.

NFPA: See *National Fire Protection Association.*

NIOSH: See *National Institute for Occupational Safety and Health.*

NIR: See *National Institute of Restoration.*

NLSI: See *National Lightning Safety Institute*

NMHCSS: See *National Manufactured Housing Construction and Safety Standards Act.*

No Carbon Required (NCR): A special type of paper used for making duplicate copies of documents without the use of carbon paper. It's often used for contracts, change orders, and other forms so that all parties can immediately have identical duplicate signed copies of the same form.

noncompete agreement: An agreement between the buyer and the seller of a business that states the seller of the business will not start up a new company that's directly in competition with the company he just sold; or an agreement between an employer and an employee that states the employee will not go to work for another employer directly in competition with the company he was working for. Noncompete agreements typically last for a specified length of time, and often define a specific geographical area.

nondisclosure agreement (NDA): An agreement between two or more parties stating that information that's been provided by one party will be kept secret by the other party. Nondisclosure agreements are commonly used when selling a business, when hiring for certain key management positions, and for many other situations. The agreement typically spells out the scope and the length of the agreement, and the penalties for breaching it. Also known as a *confidentiality agreement.*

nonpenetrating moisture meter: A type of moisture detection instrument that uses electrical-conducting pads or radio frequencies to detect the presence of moisture without creating holes in the surface. Also called a *noninvasive* or *pinless meter.*

O

O&P: See *overhead and profit.*

Occupational Safety & Health Administration (OSHA): An agency of the U.S. Department of Labor whose mission is to prevent work-related injuries, illnesses, and occupational fatalities by issuing and enforcing standards for workplace safety and health.

offgassing: See *outgassing.*

off-site: Work, storage, or other activities that take place at a location other than the location of the building where the loss occurred.

on-site: Work, storage, or other activities that take place at the location where the loss occurred.

on-site storage container: A portable, secure, weathertight storage container delivered by truck and placed at a loss site. Used for the temporary storage of contents, tools, building materials, or other items during reconstruction. Sometimes called a *portable storage container.*

open drying plan: Using outside air to increase the drying speed of a wet building. For an open drying plan to be effective, the outside humidity must be lower than the inside humidity, the outside temperature should be above 70 degrees, and there should be no weather or security concerns.

organic smoke residue: See *smoke residue, natural.*

organizational chart: A chart that shows, in a visual format, how authority and responsibility are set up within a company. Organizational charts help employees understand who their supervisors are, and who they should report to. Also called an *org chart.*

oriented strand board (OSB): A type of engineered wood product made from layering wood strands mixed with resin in alternating directions, and pressing them together under heat and pressure. Typically formed into sheets.

OSHA: See *Occupational Safety & Health Administration.*

outgassing: The slow release of gas trapped in a material, back into the surrounding air. Also called *offgassing.*

outside living expense: Payments made by an insurance company to the insured party to cover all or part of the insured's expenses for food and lodging outside their home while repairs are being completed.

overhead and profit (O&P): A figure that's added at the end of an insurance estimate to cover the contractor's operating expenses and profit. It's typically an additional 10 percent for overhead and 10 percent for profit of the subtotaled amount of the estimate.

ozone (O_3): A colorless, heavier-than-air gas made up of three atoms of oxygen. The third oxygen atom reacts with airborne particles that are near it, making it very effective at destroying odor molecules. Ozone can be created through the use of an ozone generator under carefully controlled conditions.

ozone generator: A device for creating ozone gas, typically through the use of corona discharge.

P

pack-in: The process of moving contents from a temporary storage facility back into the building that suffered a loss. The pack-in process typically includes unpacking the items and placing them back in their original locations.

pack-out: The process of packing and moving contents out of a building that suffered a loss and into a temporary facility for cleaning, restoration, and/or storage.

pack-out/pack-in inventory: An inventory list of content items packed out of a building for transport to another location for cleaning and/or storage. The same inventory list is typically used when packing the contents items back in.

packing blanket: See *moving blanket.*

pallet: A flat, portable platform made from wood, plastic, or other material, used for storing and moving materials.

pallet jack: A type of portable material-handling equipment, consisting of pallet forks on small wheels with a large handle in the rear for lifting. Used in warehouses to lift and move loaded pallets. Also called a *pallet truck.*

particle of incomplete combustion (PIC): A tiny particle of unburnt carbon suspended in a gas, which is a portion of what makes up smoke. The technical term for soot.

particulate matter (PM): Tiny particles of solid material suspended in a gas. The particles in smoke would be a common example.

passive ventilation: Natural air movement not requiring assistance from fans or other mechanical means.

pathogen: Any disease-producing agent, especially a virus, bacterium, or other microorganism, that can cause disease in humans, animals and plants.

penetrating moisture meter: A type of moisture-detection instrument that utilizes two small sharp pins to probe a material. If moisture is present, an electrical circuit is completed between the pins, registering on the meter.

per diem: A payment made to cover an employee's lodging, meals, and some other expenses that are unique to an out-of-town job. The Latin meaning is *per day* or *by the day.*

peril: A source of risk or danger. For insurance purposes, a peril is any cause for the damage or loss that might occur to a building.

perishables: Contents items that can spoil or die, such as food items, cosmetics, some types of toiletries, and house plants.

personal guarantee: The legal promise and obligation made by individuals to be personally responsible for the debts of their corporation.

personal protective equipment (PPE): Any of a wide variety of equipment used to protect workers from jobsite hazards. This includes respirators, gloves, goggles, boots, hardhats, and other types of safety equipment.

PIC: See *particle of incomplete combustion.*

pigmented shellac: Shellac to which clay and titanium dioxide have been added. This gives the shellac a solid white color.

pinless moisture meter: See *nonpenetrating moisture meter.*

pints per day: A rating of the capacity of a dehumidifier. It refers to the number of pints of moisture that the dehumidifier is capable of removing from the air within a 24-hour period, at an air temperature of 80 degrees F with 60 percent relative humidity.

PM: See *particulate matter.*

policies and procedures handbook: See *employee handbook.*

polycyclic aromatic hydrocarbons: One of the by-products of a fire known to cause cancer.

portable storage container: See *on-site storage container.*

post-construction cleaning: A complete cleaning of the interior and exterior of a building at the conclusion of a construction project, which is done to prepare the building for reoccupancy. Also called a *final cleaning.*

potable water: Water that's of sufficiently high quality that it can be used without risk of immediate or long term harm to health. Water that's safe for drinking without any other precautions or processing.

PPE: See *personal protective equipment.*

pre-existing condition: Damage or other conditions that were already present before an insurance-related loss occurred.

Preferred Contractor Program: A program, set up between an insurance company and a preselected group of restoration contractors, used to speed up the processing of claims. Contractors selected to participate in the program receive additional job referrals and assignments, subject to the policyholder's approval. In exchange, the contractors typically provide expedited service and often an extended warranty to the policy holder. Also called a *Preferred Vendor Program.*

prefilter: An inexpensive filter used to trap large particles of dust and other material, before they reach the finer filters, such as HEPA filters. They extend the life of the HEPA filter by preventing premature clogging.

process air: That part of the air stream entering a desiccant dehumidifier that passes through the silica gel rotor and is dried. Process air accounts for approximately 75 percent of the air entering the dehumidifier.

protein smoke residue: See *smoke residue.*

protocol: A written evaluation created by an industrial hygienist or other trained professional detailing the steps necessary for remediating mold and bacterial losses.

Psf: Pounds per square foot.

Psi: Pounds per square inch.

psychological smoke odor: Smoke odor that a person thinks he or she smells after a fire, but that isn't real.

psychrometry: The science and practice of air and water vapor mixtures in the atmosphere.

puff-back: A malfunction inside the combustion chamber of an oil furnace that releases a film of oil-based soot particles back through the duct system and into the house.

punch list: A list of final detail items remaining to be completed on a job.

pyrophoric carbon: Wood, which has been exposed to a heat source for a greatly extended period of time that has begun to carbonize. That carbonization lowers the ignition point of the wood.

Q

quartersawn: Wood that's been sawn at the mill at an approximately 90-degree angle in relation to the tree's growth rings. This type of sawing results in more waste, but often reveals a more attractive grain pattern.

quicklime: See *lime*.

R

R&I: See *remove and reinstall.*

R&R: See *remove and replace.*

R-value: Short for resistance value, the measured resistance of a material to allow heat to flow through it. The higher the R-value, the more the material resists the transfer of heat through it. Used as a rating of insulation efficiency.

reactivation air: That part of the air stream entering a desiccant dehumidifier that is used in the reactivation process. Reactivation air accounts for approximately 25 percent of the air entering the dehumidifier.

reactivation process: The process of heating the silica gel in a desiccant dehumidifier to remove the trapped moisture and exhaust it to the outside.

regional modifier: A plus or minus percentage adjustment made to a unit cost to reflect higher or lower labor rates or material costs within a particular part of a state.

relative humidity: The ratio of the amount of water vapor actually present in the air to the greatest amount possible at the same temperature. Shown as a percentage, if the air is currently holding half the amount of moisture it can actually hold at the given temperature, the relative humidity level is 50 percent.

remediation: The process of cleaning and restoring a building to use after its exposure to mold, sewage, or other contamination.

remove and reinstall (R&I): A restoration operation in which a structural item is taken out of a building, salvaged, and then later reinstalled for reuse.

remove and replace (R&R): A restoration operation in which a structural item is taken out of a building, discarded, and replaced with new materials of like kind and quality.

replacement cost policy: A type of homeowner's insurance policy that replaces a building and its contents at today's actual cost, without regard to depreciation.

reserve: A pool of money set aside by an insurance company when a large loss or set of losses occurs. The reserve is held by the company to cover anticipated expenses against that loss.

restoration contractor: A general contractor who specializes in the repair and restoration of structures and contents that are damaged by fire, water, smoke, storm damage, and other types of sudden and unforeseen occurrences.

restorative drying: An industry-recognized method of drying a building with minimal demolition and as little expenditure of time and expense as possible.

road barrier: A tough, flexible fabric used on the underside of a manufactured home to protect the insulation, plumbing, and other components from damage during transport to the set-up site. In older homes, materials other than fabric were used. Also known as a *bottom barrier*, *bottom board*, *rodent barrier*, *underbelly*, *transport barrier*, and various other names.

room break: A separation between two rooms, such as a doorway. Often used in determining where to separate two materials, such as carpeting.

rotomolding: A manufacturing process for creating a hollow item, typically from some form of plastic. Plastic is melted inside a heated mold, and the mold is then slowly rotated and cooled as the plastic flows around and sticks to the mold's walls.

S

S&P: See *seal and paint.*

S-Corporation (S-Corp): A type of business structure with shareholders instead of individual owners. Profits and losses are divided among the shareholders based on their percentages of ownership. S-Corporations also offer a higher level of liability protection for the shareholders. The full name is Subchapter S-Corporation.

salvage value: The value of any contents declared a total loss by the insurance company. Salvage value is typically set either by the insurance company or by an independent appraiser or other third party.

salvageable: Building materials or contents that can be successfully cleaned and repaired for reuse.

scope: A complete list of the items to be removed, repaired, and/or replaced on a loss. An abbreviation for the term *scope of repairs.*

scrubber: See *air scrubber.*

seal: Apply a sealer, encapsulant, or other material to a surface to prevent odor or particles from being released or to hide stains.

seal and paint (S&P): A common restoration procedure for surfaces such as walls and ceilings.

sealer: Any material used for sealing. There are a variety of sealers on the market formulated for both general purposes and specific uses.

secondary damage: Additional damage to contents or structures which is related to, but not directly caused by, the original loss. For example, if a piece of furniture gets wet due to water from a broken pipe, then begins to warp as it dries, the warping is known as secondary damage.

shellac: A type of natural resin, secreted by a tiny insect called the lac bug, found on trees in Asia and Thailand. The processed resin is sold as dry flakes, which are dissolved in denatured alcohol to make liquid shellac. Liquid shellac has several uses, including as a sealant and as a wood finish. Also see *pigmented shellac.*

sight line: An imaginary line used to determine whether one room or space can be seen from another room or space, even when they're physically connected to one another. Sight lines are often used in deciding where to break replacement materials such as carpet and paint.

single-wide: A manufactured or mobile home that's built as a single unit and then towed to a permanent site. A single-wide manufactured home is typically 18 feet or less in width and 90 feet or less in length. A double-wide is built as two separate units and towed to its permanent site, where it's assembled into one unit.

skirting: An enclosure of wood or metal siding panels that surround the underside of a manufactured home after it's been set up on site. The skirting provides weather and other physical protection for the underside of the home and the utility connections, as well as improving the home's appearance.

smoke: The visible gas and particles that result from the incomplete combustion of a material during a fire.

smoke deodorization: The removal of smoke particles and smoke odor from structures and contents.

smoke residue: The thin film of material that's left behind by different types of smoke after it cools. Smoke residues are generally broken down into three general classifications: natural, protein and synthetic.

smoke residue, natural: Produced by the burning of natural materials such as wood and paper. Sometimes called *organic residue.*

smoke residue, protein: Produced by the burning of meat, fish, or poultry.

smoke residue, synthetic: Produced by the burning of synthetic materials such as plastic.

smoke webs: Fine webs that often appear in building corners and other areas after a fire, resembling spider webs. They're the result of smoke particles with opposite charges that are attracted to one another and form chains. Also called *streamers.*

snout adaptor: See *splitter.*

sofa bags: Large plastic bags designed specifically for storing sofas, chairs, and other large pieces of furniture to protect them after they've been cleaned.

soft goods: Contents items such as clothing, shoes, bedding, rugs, linens, and similar items.

soot: A tiny particle of unburnt carbon material suspended in a gas, which is part of what makes up smoke. The technical term for soot is a *particle of incomplete combustion,* or *PIC.*

splitter: An accessory used with many types of air movers to adapt the snout of the air mover to receive one or more 4-inch-diameter hoses, which can then be directed to concealed areas for moisture removal. Also called a *snout adaptor.*

square foot: A two-dimensional unit of measurement equal to 1 foot in length by 1 foot in width. Used for measuring area.

square inch: A two-dimensional unit of measurement equal to 1 inch in length by 1 inch in width. Used for measuring area.

square yard: A two-dimensional unit of measurement equal to 3 feet in length by 3 feet in width. Used for measuring area.

stick built: A home or other building that's been constructed on site using traditional carpentry and construction methods, as opposed to a manufactured or modular home that's built in a factory using assembly-line techniques.

storage vault: An individual plywood storage container used for long-term storage of contents.

storm duty: An insurance industry term for insurance adjusters who've been assigned to a catastrophe team to work on multiple losses in the wake of a large natural disaster, such as a hurricane or an ice storm.

streamers: See *smoke webs.*

Subchapter S-Corporation: See *S-Corporation.*

subcontractor: A specialty contractor, such as an electrician or a plumber, who's responsible for only one specific aspect of a job, as opposed to a general contractor, who oversees the job as a whole.

subrogation: A process in which an insurance company, after paying a loss, seeks reimbursement for all or part of that payment from another party or another insurance company.

substantial completion: The point in a construction project at which the project is essentially done and ready to be turned back over to the building owner for use or occupancy. Substantial completion usually occurs when all utilities are operational, all inspections have been passed, a Certificate of Occupancy, if needed, has been issued, and only final details are left to complete.

substrate: A solid material that's used underneath another material as a supportive base.

sudden and catastrophic: An insurance industry term, referring to damage that's caused by a single, isolated and unforeseen event, such as a pipe that freezes and breaks, as opposed to damage caused by an ongoing event, such as a pipe with a slow leak that's gone unnoticed and unrepaired.

supplemental: A separate estimate submitted to the insurance company in the event that additional repairs, which were not addressed in the original estimate, are necessary. Also called a *supplemental estimate.*

supplemental loss: Additional damage to a building that's discovered after the initial damage has already been inspected and estimated. This additional damage is considered part of the original loss claim, and is not subject to a new claim or a new deductible.

synthetic smoke residue: See *smoke residue, synthetic.*

synthetic stucco: See *Exterior Insulation and Finish Systems.*

T

tabbing: A repair process for composition shingle roofing in which each individual shingle, or tab, is lifted up, a small dab of roofing cement applied to the shingle below the one that's been lifted, then the lifted shingle pressed back down and cemented in place.

temp a roof: The act of using tarps, plastic sheeting, and other materials to temporarily cover a roof for weather protection.

temporary covering: A cover of plastic sheeting, tarps, or other material used to cover a hole in a roof, exterior wall, or other area in order to provide weather protection after a loss. Often known simply as a *temp.*

temporary power: A nonpermanent source of electrical power for a building, such as a generator or a temporary electrical panel on a pole. The temporary power is used during construction and repairs until the permanent electrical power source is installed, inspected, and connected.

TES: See Thermal Energy System.

TEX: See Thermal Exchanger.

Thermal Energy System® (TES): Proprietary drying equipment that utilizes a combination of heat and air flow to speed the process of structural drying. Both propane-fired and electric-powered versions are currently available.

Thermal Exchanger (TEX): A proprietary piece of equipment, part of the Thermal Energy System, which works with a standard air mover to distribute heated air during the structural drying process.

thermal fog: A process for the deodorization of smoke and other organic odors. A highly-specialized machine is used to heat and atomize a liquid deodorant into an extremely fine mist that resembles fog, which is then allowed to completely fill the interior of a structure.

thermo-hygrometer: An instrument used to measure both temperature and humidity.

toekick cover: A strip of wood, usually 1/8-inch thick, which is used to cover drying holes in a cabinet toekick.

The wood is selected to be the same species as the cabinets, and is stained and finished to match.

transducer: A device for generating sound waves in an ultrasonic cleaning tank. The sound waves are typically in the range of 20,000 to 100,000 cycles per second.

transport barrier: See *road barrier.*

trash pump: A heavy-duty, gas-powered, portable water pump with a screened pickup hose used for removing water that is fouled with floating debris.

trauma scene cleanup: The cleaning, repair and restoration of structures that have been damaged or contaminated due to death or serious injury.

triple-wide: A manufactured home that's built and then towed to a site as three separate units. The three units are assembled at the site into one home.

truck-mount equipment: Carpet cleaning equipment that's permanently mounted in a vehicle, typically a van or small truck. The equipment is usually powered by the same motor that runs the vehicle.

U

ultrasonic: The use of high-frequency sound waves, passed through water or other cleaning solutions in a stainless steel tank, to clean contents. Ultrasonic cleaning is a highly-effective method for the cleaning and restoration of many types of contents that have been damaged by smoke and other materials.

underbelly: See *road barrier.*

unit: A standardized method or quantity of measurement for a construction material. For example, a unit of drywall is a square foot, and a unit of molding is a linear foot.

unit cost: The cost associated with the removal, installation, or purchasing of a unit of material. The unit cost includes everything associated with performing an operation using that unit of material, whatever the operation may be.

unit cost add-ons: See *add-ons.*

unit costing: A common estimating method used for insurance restoration, in which a repair is estimated based on a unit of measurement, such as a linear foot or a square foot. For example, the unit-cost for painting is estimated by the square foot, which includes the labor and materials to paint one square foot of surface area.

unoccupied: A home or other building that's furnished and is still in use by the owner or a tenant, but does not currently have anyone in it. In the case of a residence, the owners or tenants could be on vacation, or the home could be a second home or a vacation rental that's not a primary residence. In the case of a commercial building, the building could be unoccupied outside of normal business hours.

U.S. Green Building Council (USGBC): A nonprofit organization that promotes energy-saving green building design and construction techniques. The USGBC also developed the LEED green-building rating and certification system.

V

VA: See *Veterans Administration.*

vacant: A home or other building that's empty, unoccupied, and not currently in use.

valley flashing: A sheet metal flashing used for weather protection in the angled valley between two adjacent roof slopes.

vapor barrier: Any material that will resist condensation or the movement of moisture through it. Vapor barriers include such materials as plastic sheeting applied over the soil in a crawl space, and the paper or foil facing applied over insulation batts.

Veterans Administration (VA): The U.S. Department of Veterans Affairs provides federal benefits to veterans and their dependents, including patient care, survivor benefits, government-backed home loans, and educational assistance through the GI Bill.

W

wall boot: See *boot.*

wand: The handle and head combination used at the end of a water extraction vacuum or carpet cleaner.

wardrobe box: A large cardboard carton with a metal rod across the inside, used for temporarily hanging, storing, and transporting clothing.

water extractor: Equipment used to suck up standing water and then discharge it to another location. Also called a *water extraction vacuum.*

water intrusion: See *intrusion.*

water lift: The suction capacity of a vacuum measured in inches of water lift, which is the number of inches that the vacuum can pull water up into a tube under controlled conditions. The higher the number of inches of water lift, the more powerful the vacuum.

waterborne pathogens: Microorganisms such as bacteria, viruses, and protozoa that are present in water and that can cause disease in humans. Waterborne pathogens are a particular hazard when dealing with any type of sewage or contaminated water loss.

winterizing: Protecting a building and its contents against damage from freezing temperatures, especially the building's plumbing system. This is typically done when a building has lost its power and/or its heating system.

work authorization: A written, one-page document granting permission for the contractor to perform emergency services on a building. The work authorization typically explains hourly rates, payment terms, and other legal aspects of the agreement between the owner and the contractor.

X-Y-Z

Xactimate®: A specialized computer estimating and CAD software package, widely used within the insurance restoration industry.

X-O: The designations used for the sashes in sliding windows. X refers to the sash that slides, while O represents the fixed sash. The designation is always read from left to right, looking at the window from the exterior of the building.

yard: An abbreviation for one square yard of a material. Also, an abbreviation for one cubic yard of a material.

zero-clearance: A heat-producing appliance, such as a fireplace, that's designed and constructed to be placed directly on or against combustible materials, such as wood, with little or no clearance.

zoning: The state and/or local regulations, laws and ordinances that govern how and where a building may be located, constructed and used.

Index

A

B

F

G

H

I

J

K

L

M

N

Q

R

S

Practical References for Builders

Construction Forms for Contractors

This practical guide contains 78 practical forms, letters and checklists, guaranteed to help you streamline your office, organize your jobsites, gather and organize records and documents, keep a handle on your subs, reduce estimating errors, administer change orders and lien issues, monitor crew productivity, track your equipment use, and more. Includes accounting forms, change order forms, forms for customers, estimating forms, field work forms, HR forms, lien forms, office forms, bids and proposals, subcontracts, and more. All are also on the CD-ROM included, in *Excel* spreadsheets, as formatted Rich Text that you can fill out on your computer, and as PDFs. **360 pages, 8½ x 11, $48.50**

CD Estimator

CD Estimator puts at your fingertips over 150,000 construction costs for new construction, remodeling, renovation & insurance repair, home improvement, framing & finish carpentry, electrical, concrete & masonry, painting, earthwork & heavy equipment and plumbing & HVAC. Quarterly cost updates are available at no charge on the Internet. You'll also have the *National Estimator* program — a stand-alone estimating program for *Windows*™ that *Remodeling* magazine called a "computer wiz," and *Job Cost Wizard*, a program that lets you export your estimates to QuickBooks Pro for actual job costing. A 60-minute interactive video teaches you how to use this CD-ROM to estimate construction costs. And to top it off, to help you create professional-looking estimates, the disk includes over 40 construction estimating and bidding forms in a format that's perfect for nearly any *Windows*™ word processing or spreadsheet program.
CD Estimator is $108.50

Contractor's Guide to *QuickBooks Pro* 2010

This user-friendly manual walks you through *QuickBooks Pro's* detailed setup procedure and explains step-by-step how to create a first-rate accounting system. You'll learn in days, rather than weeks, how to use *QuickBooks Pro* to get your contracting business organized, with simple, fast accounting procedures. On the CD included with the book you'll find a *QuickBooks Pro* file for a construction company. Open it, enter your own company's data, and add info on your suppliers and subs. You also get a complete estimating program, including a database, and a job costingprogram that lets you export your estimates to *QuickBooks Pro*. It even includes many useful construction forms to use in your business.
344 pages, 8½ x 11, $57.00

See checklist for other available editions.

National Repair & Remodeling Estimator

The complete pricing guide for dwelling reconstruction costs. Reliable, specific data you can apply on every repair and remodeling job. Up-to-date material costs and labor figures based on thousands of jobs across the country. Provides recommended crew sizes; average production rates; exact material, equipment, and labor costs; a total unit cost and a total price including overhead and profit. Separate listings for high- and low-volume builders, so prices shown are specific for any size business. Estimating tips specific to repair and remodeling work to make your bids complete, realistic, and profitable. Includes a CD-ROM with an electronic version of the book with *National Estimator,* a stand-alone *Windows*™ estimating program, plus an interactive multimedia video that shows how to use the disk to compile construction cost estimates. **496 pages, 8½ x 11, $63.50. Revised annually**

National Home Improvement Estimator

Current labor and material prices for home improvement projects. Provides manhours for each job, recommended crew size, and the labor cost for the removal and installation work. Material prices are current, with location adjustment factors and free monthly updates on the Web. Gives step-by-step instructions for the work, with helpful diagrams, and home improvement shortcuts and tips from an expert. Includes a CD-ROM with an electronic version of the book, and *National Estimator*, a stand-alone *Windows*™ estimating program, plus an interactive multimedia tutorial that shows how to use the disk to compile home improvement cost estimates. **520 pages, 8½ x 11, $63.75. Revised annually**

National Construction Estimator

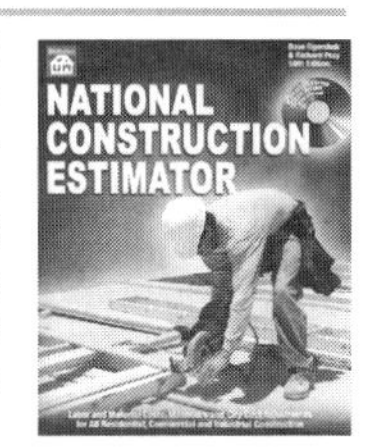

Current building costs for residential, commercial, and industrial construction. Estimated prices for every common building material. Provides manhours, recommended crew, and gives the labor cost for installation. Includes a CD-ROM with an electronic version of the book with *National Estimator*, a stand-alone *Windows*™ estimating program, plus an interactive multimedia video that shows how to use the disk to compile construction cost estimates.
672 pages, 8½ x 11, $62.50. Revised annually

National Renovation & Insurance Repair Estimator

Current prices in dollars and cents for hard-to-find items needed on most insurance, repair, remodeling, and renovation jobs. All price items include labor, material, and equipment breakouts, plus special charts that tell you exactly how these costs are calculated. Includes a CD-ROM with an electronic version of the book with *National Estimator*, a stand-alone *Windows*™ estimating program, plus an interactive multimedia video that shows how to use the disk to compile construction cost estimates.
488 pages, 8½ x 11, $64.50. Revised annually

National Painting Cost Estimator

A complete guide to estimating painting costs for just about any type of residential, commercial, or industrial painting, whether by brush, spray, or roller. Shows typical costs and bid prices for fast, medium, and slow work, including material costs per gallon; square feet covered per gallon; square feet covered per manhour; labor, material, overhead, and taxes per 100 square feet; and how much to add for profit. Includes a CD-ROM with an electronic version of the book with *National Estimator*, a stand-alone *Windows*™ estimating program, plus an interactive multimedia video that shows how to use the disk to compile construction cost estimates. **448 pages, 8½ x 11, $63.00. Revised annually**

Construction Contract Writer

Relying on a "one-size-fits-all" boilerplate construction contract to fit your jobs can be dangerous — almost as dangerous as a handshake agreement. *Construction Contract Writer* lets you draft a contract in minutes that precisely fits your needs and the particular job, and meets both state and federal requirements. You just answer a series of questions — like an interview — to construct a legal contract for each project you take on. Anticipate where disputes could arise and settle them in thecontract before they happen. Include the warranty protection you intend,the payment schedule, and create subcontracts from the prime contractby just clicking a box. Includes a feedback button to an attorney on the Craftsman staff to help should you get stumped — *No extra charge.* **$99.95.** Download the *Construction Contract Writer* at: http://www.constructioncontractwriter.com

The Complete Book of Home Inspections

This comprehensive manual covers every aspect of home inspection, from the tools required through the inspection of roofs, walls, interior rooms, windows and doors; garages, attics, basements and crawl spaces; paved areas around the structure, landscaping, insect damage and rot; electrical systems, HVAC systems, plumbing systems, and swimming pools. Covers energy considerations and environmental concerns such as radon and mold. Includes hundreds of photos and illustrations to help you understand, check, and identify potential problems. **290 pages, 8 x 10, $19.95**

Craftsman's Construction Installation Encyclopedia

Step-by-step installation instructions for just about any residential construction, remodeling or repair task, arranged alphabetically, from Acoustic tile to Wood flooring. Includes hundreds of illustrations that show how to build, install, or remodel each part of the job, as well as manhour tables for each work item so you can estimate and bid with confidence. Also includes a CD-ROM with all the material in the book, handy look-up features, and the ability to capture and print out for your crew the instructions and diagrams for any job. **792 pages, 8½ x 11, $65.00**

Estimating Home Building Costs, Revised

Estimate every phase of residential construction from site costs to the profit margin you include in your bid. Shows how to keep track of manhours and make accurate labor cost estimates for site clearing and excavation, footings, foundations, framing and sheathing finishes, electrical, plumbing, and more. Provides and explains sample cost estimate worksheets with complete instructions for each job phase. This practical guide to estimating home construction costs has been updated with digital *Excel* estimating forms and worksheets that ensure accurate and complete estimates for your residential projects. Enter your project information on the worksheets and *Excel* automatically totals each material and labor cost from every stage of construction to a final cost estimate worksheet. Load the enclosed CD-ROM into your computer and create your own estimate as you follow along with the step-by-step techniques in this book.
336 pages, 8½ x 11, $38.00

Contractor's Guide to the Building Code

Explains in plain, simple English just what the 2006 *International Building Code* and *International Residential Code* require. Building codes are elaborate laws, designed for enforcement; they're not written to be helpful how-to instructions for builders. Here you'll find down-to-earth, easy-to-understand descriptions, helpful illustrations, and code tables that you can use to design and build residential and light commercial buildings that pass inspection the first time. Written by a former building inspector, it tells what works with the inspector to allow cost-saving methods, and warns what common building shortcuts are likely to get cited. Filled with the tables and illustrations from the IBC and IRC you're most likely to need, fully explained, with examples to guide you. Includes a CD-ROM with the entire book in PDF format, with an easy search feature. **408 pages, 8½ x 11, $66.75**

Home Inspection Handbook

Every area you need to check in a home inspection — especially in older homes. Twenty complete inspection checklists: building site, foundation and basement, structural, bathrooms, chimneys and flues, ceilings, interior & exterior finishes, electrical, plumbing, HVAC, insects, vermin and decay, and more. Also includes information on starting and running your own home inspection business. **324 pages, 5½ x 8½, $24.95**

Building Code Compliance for Contractors & Inspectors

An answer book for both contractors and building inspectors, this manual explains what it takes to pass inspections under the 2009 *International Residential Code*. It includes a code checklist for every trade, covering some of the most common reasons why inspectors reject residential work — footings, foundations, slabs, framing, sheathing, plumbing, electrical, HVAC, energy conservation and final inspection. The requirement for each item on the checklist is explained, and the code section cited so you can look it up or show it to the inspector. Knowing in advance what the inspector wants to see gives you an (almost unfair) advantage. To pass inspection, do your own pre-inspection before the inspector arrives. If your work requires getting permits and passing inspections, put this manual to work on your next job. If you're considering a career in code enforcement, this can be your guidebook.
8½ x 11, 232 pages, $32.50

Residential Wiring to the 2008 *NEC*

This completely revised manual explains in simple terms how to install rough and finish wiring in new construction, alterations, and additions. It takes you from basic electrical theory to current wiring methods that comply with the 2008 *National Electrical Code*. You'll find complete instructions on troubleshooting and repairs of existing wiring, and how to extend service into additions and remodels. Hundreds of drawings and photos show you the tools and gauges you need, and how to plan and install the wiring. Includes demand factors, circuit loads, the formulas you need, and over 20 pages of *NEC* tables. Includes a CD-ROM with an Interactive Study Center that makes studying for the electrician's exam easy and fun. Also on the CD is the entire book in PDF format, with easy search features so you can quickly find answers to your residential wiring questions. **304 pages, 8½ x 11, $42.00**

Visual Handbook of Building and Remodeling

If you've ever had a question about different types of material, or the dimensions required by Code in most installations – you'll find the answers quickly and easily in this 632 page illustrated encyclopedia. This expanded third edition of the classic reference includes the latest Code information, new full-color drawings, and a new section on making homes green. 1,600 full-color drawings provide a clear look at every aspect of home construction and systems, to visualize exactly how to tackle any building project or problem. There are charts, dimensions, and illustrations on design, site and climate, masonry, foundations, wood, framing, sheathing, siding, roofing, windows and doors, plumbing, wiring, the thermal envelope, floors, walls and ceilings, storage, heating, cooling, passive solar, lighting, sound, and more. **632 pages, 8½ x 11, $29.95**

Handbook of Construction Contracting, Volume 1

Everything you need to know to start and run your construction business: the pros and cons of each type of contracting, the records you'll need to keep, and how to read and understand house plans and specs so you find any problems before the actual work begins. All aspects of construction are covered in detail, including all-weather wood foundations, practical math for the job site, and elementary surveying.
416 pages, 8½ x 11, $32.75

Concrete Construction

Just when you think you know all there is about concrete, many new innovations create faster, more efficient ways to do the work. This comprehensive concrete manual has both the tried-and-tested methods and materials, and more recent innovations. It covers everything you need to know about concrete, along with Styrofoam forming systems, fiber reinforcing adjuncts, and some architectural innovations, like architectural foam elements, that can help you offer more in the jobs you bid on. Every chapter provides detailed, step-by-step instructions for each task, with hundreds of photographs and drawings that show exactly how the work is done. To keep your jobs organized, there are checklists for each stage of the concrete work, from planning, to finishing and protecting your pours. Whether you're doing residential or commercial work, this manual has the instructions, illustrations, charts, estimating data, rules of thumb and examples every contractor can apply on their concrete jobs. **288 pages, 8½ x 11, $28.75**

Profits in Buying & Renovating Homes

Step-by-step instructions for selecting, repairing, improving, and selling highly profitable "fixer-uppers." Shows which price ranges offer the highest profit-to-investment ratios, which neighborhoods offer the best return, practical directions for repairs, and tips on dealing with buyers, sellers, and real estate agents. Shows you how to determine your profit before you buy, what "bargains" to avoid, and how to make simple, profitable, inexpensive upgrades. **304 pages, 8½ x 11, $24.75**

Working Alone

This unique book shows you how to become a dynamic one-man team as you handle nearly every aspect of house construction, including foundation layout, setting up scaffolding, framing floors, building and erecting walls, squaring up walls, installing sheathing, laying out rafters, raising the ridge, getting the roof square, installing rafters, subfascia, sheathing, finishing eaves, installing windows, hanging drywall, measuring trim, installing cabinets, and building decks. **152 pages, 5½ x 8½, $17.95**

DeWalt Residential Remodeling & Repair Professional Reference

Remodeling homes uses an assortment of skills from many parts of the building trades. Unless you're a jack-of-all-trades, you're likely to not know everything. This little book can save the day – it's packed to the brim with building requirements you're likely to need on a typical remodeling job. Here you'll find what you need to know on door framing and finishing, wiring diagrams, plumbing diagrams, paint and caulking applications, concrete and masonry, walls and finishes, roofing and siding, telephone and cable TV wiring, fastener weights and strengths, and more.
384 Pages, 4 x 6, $19.95

How to Succeed With Your Own Construction Business

Everything you need to start your own construction business: setting up the paperwork, finding the jobs, advertising, using contracts, dealing with lenders, estimating, scheduling, finding and keeping good employees, keeping the books, and coping with success. If you're considering starting your own construction business, all the knowledge, tips, and blank forms you need are here. **336 pages, 8½ x 11, $28.50**

Contractor's Plain-English Legal Guide

For today's contractors, legal problems are like snakes in the swamp — you might not see them, but you know they're there. This book tells you where the snakes are hiding and directs you to the safe path. With the directions in this easy-to-read handbook you're less likely to need a $200-an-hour lawyer. Includes simple directions for starting your business, writing contracts that cover just about any eventuality, collecting what's owed you, filing liens, protecting yourself from unethical subcontractors, and more. For about the price of 15 minutes in a lawyer's office, you'll have a guide that will make many of those visits unnecessary. Includes a CD-ROM with blank copies of all the forms and contracts in the book.
272 pages, 8½ x 11, $49.50

Estimating Electrical Construction, Revised

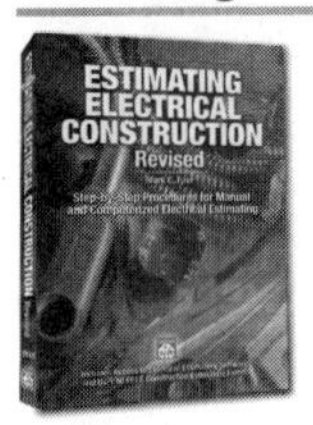

Estimating the cost of electrical work can be a very detailed and exacting discipline. It takes specialized skills and knowledge to create reliable estimates for electrical work. See how an expert estimates materials and labor for residential and commercial electrical construction. Learn how to use labor units, the plan take-off, and the bid summary to make an accurate estimate, how to deal with suppliers, use pricing sheets, and modify labor units. This book provides extensive labor unit tables and blank forms on a CD for estimating your next electrical job. **280 pages, 8½ x 11, $59.00**

Rough Framing Carpentry

If you'd like to make good money working outdoors as a framer, this is the book for you. Here you'll find shortcuts to laying out studs; speed cutting blocks, trimmers and plates by eye; quickly building and blocking rake walls; installing ceiling backing, ceiling joists, and truss joists; cutting and assembling hip trusses and California fills; arches and drop ceilings — all with production line procedures that save you time and help you make more money. Over 100 on-the-job photos of how to do it right and what can go wrong.
304 pages, 8½ x 11, $26.50

Finish Carpentry: Efficient Techniques for Custom Interiors

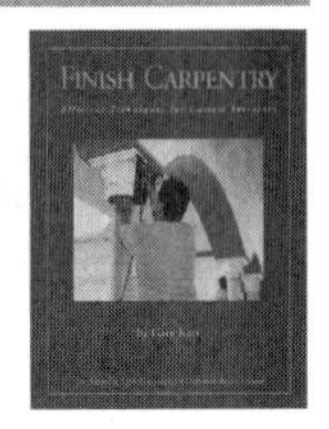

Professional finish carpentry demands expert skills, precise tools, and a solid understanding of how to do the work. This book explains how to install moldings, paneled walls and ceilings, and just about every aspect of interior trim – including doors and windows. Covers built-in bookshelves, coffered ceilings, and skylight wells and soffits, including paneled ceilings with decorative beams. **276 pages, 8½ x 11, $34.95**

Renovating & Restyling Older Homes

Any builder can turn a run-down old house into a showcase of perfection — if the customer has unlimited funds to spend. Unfortunately, most customers are on a tight budget. They usually want more improvements than they can afford — and they expect you to deliver. This book shows how to add economical improvements that can increase the property value by two, five or even ten times the cost of the remodel. Sound impossible? Here you'll find the secrets of a builder who has been putting these techniques to work on Victorian and Craftsman-style houses for twenty years. You'll see what to repair, what to replace and what to leave, so you can remodel or restyle older homes for the least amount of money and the greatest increase in value. **416 pages, 8½ x 11, $33.50**

Craftsman Book Company
6058 Corte del Cedro
P.O. Box 6500
Carlsbad, CA 92018

☎ 24 hour order line
1-800-829-8123
Fax (760) 438-0398

In A Hurry?

We accept phone orders charged to your
❍ Visa, ❍ MasterCard, ❍ Discover or ❍ American Express

Name ____________________

e-mail address (for order tracking and special offers) ____________________

Company ____________________

Address ____________________

City/State/Zip ❍ This is a residence

Total enclosed____________________(In California add 7.25% tax)

We pay shipping when your check covers your order in full.

Card# ____________________

Exp. date__________ Initials __________

Tax Deductible: Treasury regulations make these references tax deductible when used in your work. Save the canceled check or charge card statement as your receipt.

Order online www.craftsman-book.com
Free on the Internet! Download any of Craftsman's estimating costbooks for a 30-day free trial! www.craftsman-book/downloads

Prices subject to change without notice

❍ 32.50 Building Code Compliance for Contractors & Inspectors
❍ 108.50 CD Estimator
❍ 19.95 Complete Book of Home Inspections
❍ 28.75 Concrete Construction
❍ 48.50 Construction Forms for Contractors
❍ 57.00 Contractor's Guide to *QuickBooks Pro* 2010
❍ 56.50 Contractor's Guide to *QuickBooks Pro* 2009
❍ 54.75 Contractor's Guide to *QuickBooks Pro* 2008
❍ 53.00 Contractor's Guide to *QuickBooks Pro* 2007
❍ 49.75 Contractor's Guide to *QuickBooks Pro* 2005
❍ 48.50 Contractor's Guide to *QuickBooks Pro* 2004
❍ 47.75 Contractor's Guide to *QuickBooks Pro* 2003
❍ 46.50 Contractor's Guide to *QuickBooks Pro* 2002
❍ 45.25 Contractor's Guide to *QuickBooks Pro* 2001
❍ 66.75 Contractor's Guide to the Building Code
❍ 49.50 Contractor's Plain-English Legal Guide
❍ 65.00 Craftsman's Construction Installation Encyclopedia
❍ 19.95 DeWalt Residential Remodeling & Repair Professional Reference
❍ 59.00 Estimating Electrical Construction, Revised
❍ 38.00 Estimating Home Building Costs, Revised
❍ 34.95 Finish Carpentry: Efficient Techniques for Custom Interiors

10-Day Money Back Guarantee

❍ 32.75 Handbook of Construction Contracting, Volume 1
❍ 24.95 Home Inspection Handbook
❍ 28.50 How to Succeed With Your Own Construction Business
❍ 62.50 National Construction Estimator with FREE *National Estimator* on a CD-ROM
❍ 63.75 National Home Improvement Estimator with FREE *National Estimator on* a CD-ROM
❍ 63.00 National Painting Cost Estimator with FREE *National Estimator on* a CD-ROM
❍ 63.50 National Repair & Remodeling Estimator with FREE *National Estimator* on a CD-ROM
❍ 64.50 National Renovation & Insurance Repair Estimator with FREE *National Estimator* on a CD-ROM
❍ 24.75 Profits in Buying & Renovating Homes
❍ 42.00 Residential Wiring to the 2008 *NEC*
❍ 33.50 Renovating & Restyling Older Homes
❍ 26.50 Rough Framing Carpentry
❍ 29.95 Visual Handbook of Building and Remodeling
❍ 17.95 Working Alone
❍ 69.00 Insurance Restoration Contracting: Startup to Success
❍ FREE Full Color Catalog

Download all of Craftsman's most popular costbooks for one low price with the Craftsman Site License. www.craftsmansitelicense.com

Craftsman Book Company
6058 Corte del Cedro
P.O. Box 6500
Carlsbad, CA 92018

☎ 24 hour order line
1-800-829-8123
Fax (760) 438-0398

Name

e-mail address (for order tracking and special offers)

Company

Address

City/State/Zip ❍ This is a residence

Total enclosed________________________(In California add 7.25% tax)

We pay shipping when your check covers your order in full.

In A Hurry?

We accept phone orders charged to your

❍ Visa, ❍ MasterCard, ❍ Discover or ❍ American Express

Card#________________________

Exp. date____________Initials____________

Tax Deductible: Treasury regulations make these references tax deductible when used in your work. Save the canceled check or charge card statement as your receipt.

Order online www.craftsman-book.com
Free on the Internet! Download any of Craftsman's estimating costbooks for a 30-day free trial! www.craftsman-book.com/downloads

10-Day Money Back Guarantee

- ❍ 32.50 Building Code Compliance for Contractors & Inspectors
- ❍ 108.50 CD Estimator
- ❍ 19.95 Complete Book of Home Inspections
- ❍ 28.75 Concrete Construction
- ❍ 48.50 Construction Forms for Contractors
- ❍ 57.00 Contractor's Guide to *QuickBooks Pro* 2010
- ❍ 56.50 Contractor's Guide to *QuickBooks Pro* 2009
- ❍ 54.75 Contractor's Guide to *QuickBooks Pro* 2008
- ❍ 53.00 Contractor's Guide to *QuickBooks Pro* 2007
- ❍ 49.75 Contractor's Guide to *QuickBooks Pro* 2005
- ❍ 48.50 Contractor's Guide to *QuickBooks Pro* 2004
- ❍ 47.75 Contractor's Guide to *QuickBooks Pro* 2003
- ❍ 46.50 Contractor's Guide to *QuickBooks Pro* 2002
- ❍ 45.25 Contractor's Guide to *QuickBooks Pro* 2001
- ❍ 66.75 Contractor's Guide to the Building Code
- ❍ 49.50 Contractor's Plain-English Legal Guide
- ❍ 65.00 Craftsman's Construction Installation Encyclopedia
- ❍ 19.95 DeWalt Residential Remodeling & Repair Professional Reference
- ❍ 59.00 Estimating Electrical Construction, Revised
- ❍ 38.00 Estimating Home Building Costs, Revised
- ❍ 34.95 Finish Carpentry: Efficient Techniques for Custom Interiors
- ❍ 32.75 Handbook of Construction Contracting, Volume 1
- ❍ 24.95 Home Inspection Handbook
- ❍ 28.50 How to Succeed With Your Own Construction Business
- ❍ 62.50 National Construction Estimator with FREE *National Estimator* on a CD-ROM
- ❍ 63.75 National Home Improvement Estimator with FREE *National Estimator on a* CD-ROM
- ❍ 63.00 National Painting Cost Estimator with FREE *National Estimator on a* CD-ROM
- ❍ 63.50 National Repair & Remodeling Estimator with FREE *National Estimator* on a CD-ROM
- ❍ 64.50 National Renovation & Insurance Repair Estimator with FREE *National Estimator* on a CD-ROM
- ❍ 24.75 Profits in Buying & Renovating Homes
- ❍ 42.00 Residential Wiring to the 2008 *NEC*
- ❍ 33.50 Renovating & Restyling Older Homes
- ❍ 26.50 Rough Framing Carpentry
- ❍ 29.95 Visual Handbook of Building and Remodeling
- ❍ 17.95 Working Alone
- ❍ 69.00 Insurance Restoration Contracting: Startup to Success
- ❍ FREE Full Color Catalog

Prices subject to change without notice

Craftsman Book Company
6058 Corte del Cedro
P.O. Box 6500
Carlsbad, CA 92018

☎ 24 hour order line
1-800-829-8123
Fax (760) 438-0398

Name

e-mail address (for order tracking and special offers)

Company

Address

City/State/Zip ❍ This is a residence

Total enclosed________________________(In California add 7.25% tax)

We pay shipping when your check covers your order in full.

In A Hurry?

We accept phone orders charged to your

❍ Visa, ❍ MasterCard, ❍ Discover or ❍ American Express

Card#________________________

Exp. date____________Initials____________

Tax Deductible: Treasury regulations make these references tax deductible when used in your work. Save the canceled check or charge card statement as your receipt.

Order online www.craftsman-book.com
Free on the Internet! Download any of Craftsman's estimating costbooks for a 30-day free trial! www.craftsman-book.com/downloads

10-Day Money Back Guarantee

- ❍ 32.50 Building Code Compliance for Contractors & Inspectors
- ❍ 108.50 CD Estimator
- ❍ 19.95 Complete Book of Home Inspections
- ❍ 28.75 Concrete Construction
- ❍ 48.50 Construction Forms for Contractors
- ❍ 57.00 Contractor's Guide to *QuickBooks Pro* 2010
- ❍ 56.50 Contractor's Guide to *QuickBooks Pro* 2009
- ❍ 54.75 Contractor's Guide to *QuickBooks Pro* 2008
- ❍ 53.00 Contractor's Guide to *QuickBooks Pro* 2007
- ❍ 49.75 Contractor's Guide to *QuickBooks Pro* 2005
- ❍ 48.50 Contractor's Guide to *QuickBooks Pro* 2004
- ❍ 47.75 Contractor's Guide to *QuickBooks Pro* 2003
- ❍ 46.50 Contractor's Guide to *QuickBooks Pro* 2002
- ❍ 45.25 Contractor's Guide to *QuickBooks Pro* 2001
- ❍ 66.75 Contractor's Guide to the Building Code
- ❍ 49.50 Contractor's Plain-English Legal Guide
- ❍ 65.00 Craftsman's Construction Installation Encyclopedia
- ❍ 19.95 DeWalt Residential Remodeling & Repair Professional Reference
- ❍ 59.00 Estimating Electrical Construction, Revised
- ❍ 38.00 Estimating Home Building Costs, Revised
- ❍ 34.95 Finish Carpentry: Efficient Techniques for Custom Interiors
- ❍ 32.75 Handbook of Construction Contracting, Volume 1
- ❍ 24.95 Home Inspection Handbook
- ❍ 28.50 How to Succeed With Your Own Construction Business
- ❍ 62.50 National Construction Estimator with FREE *National Estimator* on a CD-ROM
- ❍ 63.75 National Home Improvement Estimator with FREE *National Estimator on a* CD-ROM
- ❍ 63.00 National Painting Cost Estimator with FREE *National Estimator on a* CD-ROM
- ❍ 63.50 National Repair & Remodeling Estimator with FREE *National Estimator* on a CD-ROM
- ❍ 64.50 National Renovation & Insurance Repair Estimator with FREE *National Estimator* on a CD-ROM
- ❍ 24.75 Profits in Buying & Renovating Homes
- ❍ 42.00 Residential Wiring to the 2008 *NEC*
- ❍ 33.50 Renovating & Restyling Older Homes
- ❍ 26.50 Rough Framing Carpentry
- ❍ 29.95 Visual Handbook of Building and Remodeling
- ❍ 17.95 Working Alone
- ❍ 69.00 Insurance Restoration Contracting: Startup to Success
- ❍ FREE Full Color Catalog

Prices subject to change without notice

Mail This Card Today For a Free Full Color Catalog

Over 100 books, annual cost guides and estimating software packages at your fingertips with information that can save you time and money. Here you'll find information on carpentry, contracting, estimating, remodeling, electrical work, and plumbing.

All items come with an unconditional 10-day money-back guarantee. If they don't save you money, mail them back for a full refund.

Name

e-mail address (for special offers)

Company

Address

City/State/Zip

Craftsman Book Company / 6058 Corte del Cedro / P.O. Box 6500 / Carlsbad, CA 92018

NO POSTAGE
NECESSARY
IF MAILED
IN THE
UNITED STATES

BUSINESS REPLY MAIL

FIRST CLASS MAIL PERMIT NO. 271 CARLSBAD, CA

POSTAGE WILL BE PAID BY ADDRESSEE

Craftsman Book Company
6058 Corte del Cedro
P.O. Box 6500
Carlsbad, CA 92018-9974

NO POSTAGE
NECESSARY
IF MAILED
IN THE
UNITED STATES

BUSINESS REPLY MAIL

FIRST CLASS MAIL PERMIT NO. 271 CARLSBAD, CA

POSTAGE WILL BE PAID BY ADDRESSEE

Craftsman Book Company
6058 Corte del Cedro
P.O. Box 6500
Carlsbad, CA 92018-9974

NO POSTAGE
NECESSARY
IF MAILED
IN THE
UNITED STATES

BUSINESS REPLY MAIL

FIRST CLASS MAIL PERMIT NO. 271 CARLSBAD, CA

POSTAGE WILL BE PAID BY ADDRESSEE

Craftsman Book Company
6058 Corte del Cedro
P.O. Box 6500
Carlsbad, CA 92018-9974